# Friction Stir Welding and Processing

Editors
Rajiv S. Mishra
Murray W. Mahoney

ASM International®
Materials Park, Ohio 44073-0002
www.asminternational.org

Micrograph image on cover: Friction stir welded 2519 wrought aluminum copper. Source: George F. Vander Voort, FASM, *Metallography and Microstructures*, Vol 9, *ASM Handbook*, 2004, p 510

First printing, March 2007
Second printing, September 2008
Third printing, January 2026



Comments, criticisms, and suggestions are invited, and should be forwarded to ASM International.

*Prepared under the direction of the ASM International Technical Book Committee (2006–2007), James C. Foley, Chair.*

*ASM International staff who worked on this project include Scott Henry, Senior Manager of Product and Service Development; Steven R. Lampman, Technical Editor; Diane Grubbs, Editorial Assistant; Bonnie Sanders, Manager of Production; Madrid Tramble, Senior Production Coordinator; and Kathryn Muldoon, Production Assistant*

**Library of Congress Cataloging-in-Publication Data**

Friction stir welding and processing / editors Rajiv S. Mishra, Murray W. Mahoney.
p. cm.
Includes bibliographical references and index.
ISBN-13: 978-0-87170-848-9
ISBN-10: 0-87170-848-5
1. Friction welding. I. Mishra, Rajiv S. II. Mahoney, Murray W., 1945–

TS228.9F739 2007
671.5′2—dc233

2007060365

SAN: 204-7586

ASM International®
Materials Park, OH 44073-0002
www.asminternational.org

Printed in the United States of America

# Contents

# Preface

This volume is the first comprehensive compilation of friction stir welding (FSW) and friction stir processing (FSP) data. It should be a valuable handbook to students studying joining and metal working practices, to welding engineers challenged to improving properties at reduced costs, to metallurgists needing new tools to locally improve properties, and to all engineers interested in sustainability; i.e., the ability to build structures while minimizing the negative impact to our environment. In the 15 years since the invention of FSW at TWI in 1991, research throughout the world has been extensive, and considerable information has been published in conference proceedings and technical journals. Although FSP arrived later, publications describing benefits attributable to FSP are increasing rapidly. Unfortunately, access to the data is not always convenient and thus potential users, students, educators, and others, cannot and are not fully aware of the benefits offered by FSW/P. This limits understanding and thus the ability of FSW/P to expand beyond the laboratory and into structural applications. Notable and successful exceptions to these are illustrated in the chapter on applications. However, the expansion into production applications can accelerate if an understanding of FSW/P and the associated benefits were more readily accessible. Within this volume, we have tried to compile the current understanding of the fundamentals of FSW/P as well as to provide a moderately extensive database. This volume includes an accessible state-of-the-art overview; essentially the starting point from which a more detailed search of the literature can be initiated when more information is required. It is not intended to provide all the data presented in the literature but does guide the reader to detailed references. Since both FSW and FSP are young technologies, this is just the beginning. Hopefully, this volume will be followed with an updated information base as more data become available and as we gain a more complete understanding of the process fundamentals.

FSW/P was first applied to aluminum alloys; therefore, the data presented herein for the most common aluminum alloys are quite extensive. These data illustrate mechanical, fatigue, and corrosion properties that can be realized for each alloy category when using sound FSW/P procedures. However, the reader is cautioned not to extrapolate even within the same alloy class because of subtle processing differences, which can result in considerable differences in resultant properties. A data base for the higher temperature alloys is growing. With the recent development and improvement in tooling capable of higher temperature operation, FSW/P is starting to be applied to the copper, iron, and nickel base alloys. The data for these alloys is not extensive, but what is available is presented herein. Due to the considerable benefits and ease of application, we anticipate friction stir welding and friction stir processing will be technologies of rapid growth following the normal progression of fundamental research, data development and eventually structural applications.

This initial reference volume is the collective effort of leading researchers who are active in the FSW/P community and whose by-lines appear throughout this volume. We gratefully thank and commend each of them for their contributions and considerable efforts. Each contributor to this volume has drawn on the research of colleagues and associates whose work is dutifully referenced. We thank these many researchers and their support staffs. Only through our composite efforts can we continue to advance science and engineering and contribute to society as a whole.

Rajiv S. Mishra
University of Missouri-Rollo
Murray W. Mahoney
Rockwell Scientific Company

# CHAPTER 1

# Introduction

Rajiv S. Mishra, Center for Friction Stir Processing, University of Missouri-Rolla
Murray W. Mahoney, Rockwell Scientific Company

FRICTION STIR WELDING (FSW) was invented at The Welding Institute (TWI) of the United Kingdom in 1991 as a solid-state joining technique and was initially applied to aluminum alloys (Ref 1, 2). The basic concept of FSW is remarkably simple. A nonconsumable rotating tool with a specially designed pin and shoulder is inserted into the abutting edges of sheets or plates to be joined and subsequently traversed along the joint line (Fig. 1.1). Figure 1.1 illustrates process definitions for the tool and workpiece. Most definitions are self-explanatory, but advancing and retreating side definitions require a brief explanation. Advancing and retreating side orientations require knowledge of the tool rotation and travel directions. In Fig. 1.1, the FSW tool rotates in the counterclockwise direction and travels into the page (or left to right). In Fig. 1.1 the advancing side is on the right, where the tool rotation direction is the same as the tool travel direction (opposite the direction of metal flow), and the retreating side is on the left, where the tool rotation is opposite the tool travel direction (parallel to the direction of metal flow).

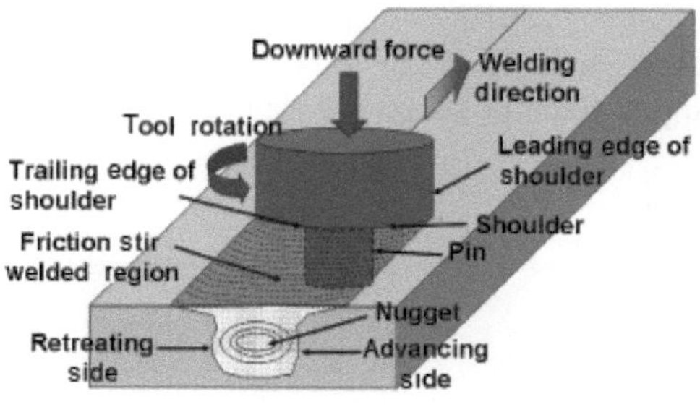


**Fig. 1.1** Schematic drawing of friction stir welding

The tool serves three primary functions, that is, heating of the workpiece, movement of material to produce the joint, and containment of the hot metal beneath the tool shoulder. Heating is created within the workpiece both by friction between the rotating tool pin and shoulder and by severe plastic deformation of the workpiece. The localized heating softens material around the pin and, combined with the tool rotation and translation, leads to movement of material from the front to the back of the pin, thus filling the hole in the tool wake as the tool moves forward. The tool shoulder restricts metal flow to a level equivalent to the shoulder position, that is, approximately to the initial workpiece top surface.

As a result of the tool action and influence on the workpiece, when performed properly, a solid-state joint is produced, that is, no melting. Because of various geometrical features on the tool, material movement around the pin can be complex, with gradients in strain, temperature, and strain rate (Ref 3). Accordingly, the resulting nugget zone microstructure reflects these different thermomechanical histories and is not homogeneous. In spite of the local microstructural inhomogeneity, one of the significant benefits of this solid-state welding technique is the fully recrystallized, equiaxed, fine grain microstructure created in the nugget by the intense plastic deformation at elevated temperature (Ref 4–7). As is seen within these chapters, the fine grain microstructure produces excellent mechanical properties, fatigue properties, enhanced formability, and exceptional superplasticity.

Like many new technologies, a new nomenclature is required to accurately describe observations. In FSW, new terms are necessary to adequately describe the postweld microstructures. The first attempt at classifying friction stir welded microstructures was made by Threadgill (Ref 8). Figure 1.2 identifies the different microstructural zones existing after FSW, and a brief description of the different zones is presented. Because the preponderance of work to date uses these early definitions (with minor modifications), this reference volume continues to do so. The system divides the weld zone into distinct regions, as follows:

- *Unaffected material or parent metal:* This is material remote from the weld that has not been deformed and that, although it may have experienced a thermal cycle from the weld, is not affected by the heat in terms of microstructure or mechanical properties.
- *Heat-affected zone:* In this region, which lies closer to the weld-center, the material has experienced a thermal cycle that has modified the microstructure and/or the mechanical properties. However, there is no plastic deformation occurring in this area.
- *Thermomechanically affected zone (TMAZ):* In this region, the FSW tool has plastically deformed the material, and the heat from the process will also have exerted some influence on the material. In the case of aluminum, it is possible to obtain significant plastic strain without recrystallization in this region, and there is generally a distinct boundary between the recrystallized zone (weld nugget) and the deformed zones of the TMAZ.
- *Weld nugget:* The fully recrystallized area, sometimes called the stir zone, refers to the zone previously occupied by the tool pin. The term *stir zone* is commonly used in friction stir processing, where large volumes of material are processed.

Friction stir welding is considered to be the most significant development in metal joining in decades and, in addition, is a "green" technology due to its energy efficiency, environmental friendliness, and versatility. As compared to the conventional welding methods, FSW consumes considerably less energy, no consumables such as a cover gas or flux are used, and no harmful emissions are created during welding, thereby making the process environmentally friendly. Further, because FSW does not involve the use of filler metal and because there is no melting, any aluminum alloy can be joined without concern for compatibility of composition or solidification cracking—issues associated with fusion welding. Also, dissimilar aluminum alloys and composites can be joined with equal ease (Ref 9–11).

In contrast to traditional friction welding, which is a welding process limited to small axisymmetric parts that can be rotated and pushed against each other to form a joint (Ref 12), FSW can be applied to most geometric structural shapes and to various types of joints, such as butt, lap, T-butt, and fillet shapes (Ref 13). The most convenient joint configurations for FSW are butt and lap joints. A simple square butt joint is shown in Fig. 1.3(a). Two plates or sheets with the same thickness are placed on a backing plate and clamped firmly to prevent the abutting joint faces from being forced apart. The backing plate is required to resist the normal forces associated with FSW and the workpiece. During the initial tool plunge, the lateral forces are also fairly large, and extra care is required to ensure that plates in the butt configuration do not separate. To accomplish the weld, the rotating tool is plunged into the joint line and traversed along this line, while the shoulder of the tool is maintained in intimate contact with the plate surface. Tool position and penetration depth are maintained by either position control or control of the applied normal force. On the other hand, for a lap joint configuration, two lapped plates or sheets are clamped, and a back-

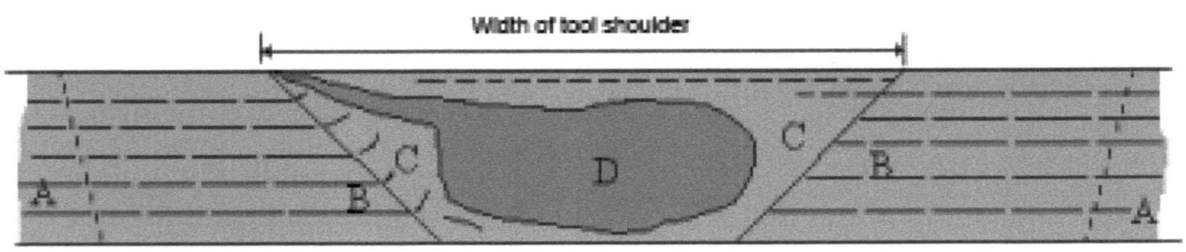


**Fig. 1.2** Various microstructural regions in the transverse cross section of a friction stir welded material. A, unaffected material or parent metal; B, heat-affected zone; C, thermomechanically affected zone; D, weld nugget

ing plate may or may not be needed, depending on the lower plate thickness. A rotating tool is vertically plunged through the upper plate and partially into the lower plate and traversed along the desired direction, joining the two plates (Fig. 1.3d). However, the tool design used for a butt joint, where the faying surfaces are aligned parallel to the tool rotation axis, would not be optimal for a lap joint, where the faying surfaces are normal to the tool rotation axis. The orientation of the faying surfaces with respect to the tool features is very important and is discussed in detail in Chapter 2. Configurations of other types of joint designs applicable to FSW are also illustrated in Fig. 1.3. Additional key benefits of FSW compared to fusion welding are summarized in Table 1.1.

This volume is the first comprehensive compilation of friction stir welding and friction stir processing data. This handbook should be valuable to students studying joining and metalworking practices, to welding engineers challenged to improve properties at reduced cost, to metallurgists needing new tools to locally improve properties, and to all engineers interested in sustainability, that is, the ability to build structures while minimizing the negative impact to our environment. The dual objectives of this first volume are to provide a ready reference to identify work completed to date and to provide an educational tool to understand FSW and how to both use and apply FSW. Not all process details can be presented within these pages, and readers are encouraged to obtain the original references for more details, especially weld parameters and appropriate boundary conditions.

To meet these objectives, the book is organized to first include a full description of tool materials and tool designs for both low- and high-temperature metals (Chapter 2). Understanding tools is a natural starting point to successfully use FSW. Chapter 3 provides an introduction to the fundamentals of FSW, including heat generation and metal flow. Although somewhat controversial at this time, Chapter 3 helps one visualize fundamental FSW characteristics and current

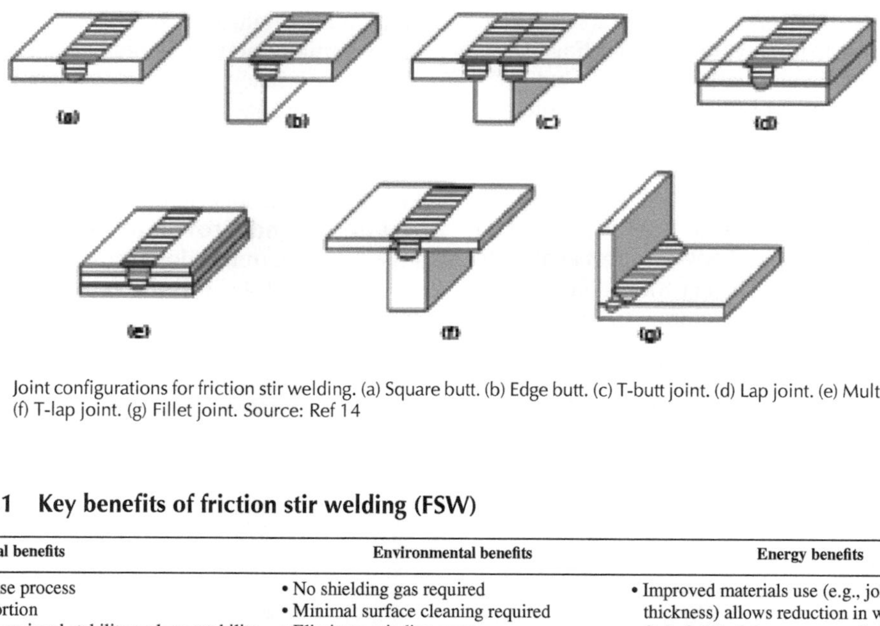


**Fig. 1.3** Joint configurations for friction stir welding. (a) Square butt. (b) Edge butt. (c) T-butt joint. (d) Lap joint. (e) Multiple lap joint. (f) T-lap joint. (g) Fillet joint. Source: Ref 14

**Table 1.1 Key benefits of friction stir welding (FSW)**

| Metallurgical benefits | Environmental benefits | Energy benefits |
|---|---|---|
| • Solid-phase process<br>• Low distortion<br>• Good dimensional stability and repeatability<br>• No loss of alloying elements<br>• Excellent mechanical properties in the joint area<br>• Fine recrystallized microstructure<br>• Absence of solidification cracking<br>• Replace multiple parts joined by fasteners<br>• Weld all aluminum alloys<br>• Post-FSW formability | • No shielding gas required<br>• Minimal surface cleaning required<br>• Eliminate grinding wastes<br>• Eliminate solvents required for degreasing<br>• Consumable materials saving, such as rugs, wire, or any other gases<br>• No harmful emissions | • Improved materials use (e.g., joining different thickness) allows reduction in weight<br>• Only 2.5% of the energy needed for a laser weld<br>• Decreased fuel consumption in lightweight aircraft, automotive, and ship applications |

Source: Ref 14

metal flow concepts. Because the preponderance of work has been performed on aluminum alloys, Chapter 4 presents microstructural evolution following FSW as an individual chapter. The ability to weld all aluminum alloys, including the 7*xxx* and metal-matrix composites, introduces new issues and benefits. In concert, Chapter 5 presents material properties for the common aluminum alloys, including the 2*xxx*, 3*xxx*, 5*xxx*, 6*xxx*, 7*xxx*, AlLi, and metal-matrix composites. Considerable data are available for hardness, mechanical properties, fatigue response, and, in some cases, fracture toughness and fatigue crack propagation. Chapter 5 provides a ready reference to identify what properties can be expected following FSW. Although the database is not as extensive, Chapter 6 presents microstructure and properties of ferrous and nickel-base alloys. With the development of high-temperature tooling, that is, polycrystalline cubic boron nitride tools, FSW is rapidly expanding into the welding of high-temperature alloys, and considerable growth is anticipated in this area. Chapter 7 continues the theme of high-temperature FSW but for titanium alloys. Titanium alloys offer unique difficulties, and although the available data are limited at this time, there is considerable interest. The challenge to identify long-life tooling to friction stir weld titanium alloys remains, but early results illustrate the metallurgical potential to apply FSW. Copper alloys (~1000 °C, or 1830 °F) are intermediate in FSW temperature between aluminum alloys (~500 °C, or 930 °F) and ferrous alloys (~1100 to 1200 °C, or 2010 to 2190 °F). Considerable FSW success has already been demonstrated (Chapter 8), and because of the intermediate temperature, different high-temperature flow, and different physical properties such as thermal conductivity, different lessons can be learned. Chapter 9 presents post-FSW corrosion properties of aluminum alloys. Compared to fusion welds, corrosion sensitivity following FSW is always equivalent or less. However, FSW does introduce local heat, creating heat-affected zones and potential segregation of second-phase particles at grain boundaries. Corrosion sensitivity following FSW should always be considered, as one would for any welding practice. Chapter 10 presents results from computational modeling of FSW. Modeling helps visualize fundamental behavior and allows for comparison of flow and temperature response for different weld parameters and boundary conditions without performing costly experiments and subsequent evaluation. The advancement of FSW out of the laboratory and into commercial practice is highlighted in Chapters 11 and 13. Chapter 11 illustrates the portability and versatility of FSW whereby it can be applied with robots. Further, Chapter 11 discusses current FSW machine capabilities. Chapter 12 presents an overview of friction stir spot welding (FSSW). The total cycle in FSSW is relatively short, and the dynamics of the process are close to the plunge part of FSW. The potential to produce solid-state spot welds is generating considerable interest in the automotive industry. Chapter 13 summarizes current FSW applications. It is anticipated that the number of applications will grow rapidly as fabricators learn the ease of application and property benefits attributable to FSW. Chapter 14 presents an outgrowth of FSW, that is, friction stir processing (FSP). Because of the creation of a fine grain microstructure and the ability to eliminate casting defects, FSP offers the ability to locally tailor properties within a structure such that the structure can survive better in its environment. For example, by applying FSP, local properties can be improved, such as abrasion resistance, strength, ductility, fatigue life, formability, and superplasticity. Friction stir processing is a growth technology that may become as important as FSW. Lastly, FSW and FSP are essentially new technologies not much beyond their infancy. The growth potential for the future can be considerable. Chapter 15 offers the authors' thoughts on technology gaps to be overcome to accelerate growth as well as some speculation on future opportunities and applications.

**Interest and Growth in FSW.** The field of FSW has seen tremendous growth in the last ten years. Figure 1.4 shows the increase in publica-

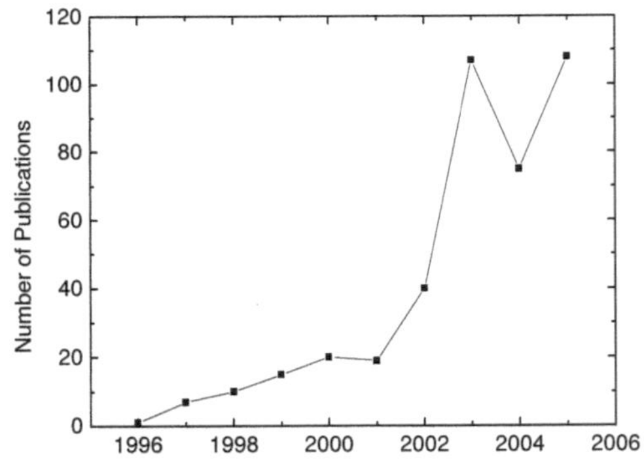


**Fig. 1.4** Significant increase in publications on friction stir welding/friction stir processing. This figure is based on the Institute for Scientific Information Web of Science database and does not include proceedings papers published in The Welding Institute international symposiums and TMS annual meeting symposiums.

tions in this field. This is a summary from the Institute for Scientific Information Web of Science database and does not include proceedings. The first international symposium was held at Rockwell Science Center and was organized by TWI in 1999. From that time, many symposiums have been organized, including three in TMS annual meetings, which have accompanying proceedings.

## REFERENCES


1. W.M. Thomas, E.D. Nicholas, J.C. Needham, M.G. Murch, P. Templesmith, and C.J. Dawes, G.B. Patent 9125978.8, Dec 1991
2. C. Dawes and W. Thomas, *TWI Bull.,* Vol 6, Nov/Dec 1995, p 124
3. B. London, M. Mahoney, B. Bingel, M. Calabrese, and D. Waldron, in *Proceedings of the Third Int. Symposium on Friction Stir Welding,* Sept 27–28, 2001 (Kobe, Japan)
4. C.G. Rhodes, M.W. Mahoney, W.H. Bingel, R.A. Spurling, and C.C. Bampton, *Scr. Mater.,* Vol 36, 1997, p 69
5. G. Liu, L.E. Murr, C.S. Niou, J.C. McClure, and F.R. Vega, *Scr. Mater.,* Vol 37, 1997, p 355
6. K.V. Jata and S.L. Semiatin, *Scr. Mater.,* Vol 43, 2000, p 743
7. S. Benavides, Y. Li, L.E. Murr, D. Brown, and J.C. McClure, *Scr. Mater.,* Vol 41, 1999, p 809
8. P.L. Threadgill, *TWI Bull.,* March 1997
9. L.E. Murr, Y. Li, R.D. Flores, and E.A. Trillo, *Mater. Res. Innov.,* Vol 2, 1998, p 150
10. Y. Li, E.A. Trillo, and L.E. Murr, *J. Mater. Sci. Lett.,* Vol 19, 2000, p 1047
11. Y. Li, L.E. Murr, and J.C. McClure, *Mater. Sci. Eng. A,* Vol 271, 1999, p 213
12. H.B. Cary, *Modern Welding Technology,* Prentice Hall
13. C.J. Dawes and W.M. Thomas, *Weld. J.,* Vol 75, 1996, p 41
14. R.S. Mishra and Z.Y. Ma, *Mater. Sci. Eng. R,* Vol 50, 2005, p 1

# CHAPTER 2

# Friction Stir Tooling: Tool Materials and Designs

Christian B. Fuller, Rockwell Scientific Company

FRICTION STIR WELDING AND PROCESSING (collectively referred to as friction stirring) is not possible without the nonconsumable tool. The tool produces the thermomechanical deformation and workpiece frictional heating necessary for friction stirring. A friction stir welding (FSW) butt joint is schematically illustrated in Figure 1 in Chapter 1, "Introduction," and the same steps are necessary for friction stir processing (Ref 1). During the tool plunge, the rotating FSW tool is forced into the workpiece. The friction stirring tool consists of a pin, or probe, and shoulder. Contact of the pin with the workpiece creates frictional and deformational heating and softens the workpiece material; contacting the shoulder to the workpiece increases the workpiece heating, expands the zone of softened material, and constrains the deformed material. Typically, the tool dwells (or undergoes only rotational motion) in one place to further increase the volume of deformed material. After the dwell period has passed, the tool begins the forward traverse along a predetermined path, creating a fine-grained recrystallized microstructure behind the tool. Forward motion of the tool produces loads parallel to the direction of travel, known as transverse load; normal load is the load required for the tool shoulder to remain in contact with the workpiece.

The initial aluminum FSW studies conducted at The Welding Institute (TWI) used a cylindrical threaded pin and concave shoulder tool machined from tool steel (Ref 2). Since that time, tools have advanced to complex asymmetric geometries and exotic tool materials to friction stir higher-temperature materials. This chapter uses two sections to examine the evolution of tool material and design since 1991. The first section describes tool materials, including the material characteristics needed for a tool material and a listing of published friction stir tool materials. The second section presents a history of friction stir welding and processing tool design, general tool design philosophy, and associated tool topics.

## 2.1 Tool Materials

Friction stirring is a thermomechanical deformation process where the tool temperature approaches the workpiece solidus temperature. Production of a quality friction stir weld requires the proper tool material selection for the desired application. All friction stir tools contain features designed for a specific function. Thus, it is undesirable to have a tool that loses dimensional stability, the designed features, or worse, fractures.

### *2.1.1 Tool Material Characteristics*

Selecting the correct tool material requires knowing which material characteristics are important for each friction stir application. Many different material characteristics could be considered important to friction stir, but ranking the material characteristics (from most to least important) will depend on the workpiece material, expected life of the tool, and the user's own experiences and preferences. In addition to the physical properties of a material, some practical considerations are included that may dictate the tool material selection.

**Ambient- and Elevated-Temperature Strength.** The candidate tool material must be able to withstand the compressive loads when the tool first makes contact with the workpiece and have sufficient compressive and shear strength at elevated temperature to prevent tool fracture or distortion for the duration of the friction stir weld. Currently, predicting the required tool strength requires complex computational simulations, so typically, the strength requirements are based on experience. At a minimum, the candidate tool material should exhibit an elevated- (workpiece solidus temperature) temperature compressive yield strength higher than the expected normal forces of the tool.

**Elevated-Temperature Stability.** In addition to sufficient strength at elevated temperature, the tool must maintain strength and dimensional stability during the time of use. Creep (and creep fatigue) is a consideration for long weld lengths, where poor creep resistance would change the tool dimensions during welding. Tool materials that derive their strength from precipitates, work hardening, or transformation hardening have defined maximum-use temperatures. Tools used above the maximum-use temperatures will, in time, exhibit a decrease in mechanical properties. The change in mechanical properties is due to overaging, annealing and recovery of dislocation substructures, or reversion to a weaker phase. In friction stirring, these microstructural changes will weaken the tool and either change the tool shape or fracture the tool. Thermal fatigue strength should be considered when the friction stirring tools are subjected to many heating and cooling cycles (e.g., friction stir spot welding or short production welds). However, in most cases, other tool material characteristics will cause failure before thermal fatigue.

**Wear Resistance.** Excessive tool wear changes the tool shape (normally by removing tool features), thus changing the weld quality and increasing the probability of defects. In friction stirring, tool wear can occur by adhesive, abrasive, or chemical wear (which is addressed subsequently as reactivity) mechanisms. The exact wear mechanism depends on the interaction between the workpiece and tool materials and the selected tool parameters. For example, in the case of polycrystalline cubic boron nitride (PCBN) tools, wear at low tool rotation speed is caused by adhesive wear (also known as scoring, galling, or seizing), while wear at high tool rotation speed is caused by abrasive wear (Ref 3).

**Tool Reactivity.** Tool materials must not react with the workpiece or the environment, which would change (generally in a negative way) the surface properties of the tool. Titanium is well known to be reactive at elevated temperatures; thus, any reaction of titanium with the tool material will change the tool properties and alter the joint quality. Environmental reactions of the tool (e.g., oxidation) could change the tool wear resistance or even produce toxic substances (i.e., formation of $MoO_3$). These environmental reactions can be mitigated with cover gases, but these can add complexity to the welding system. The workpiece can also exhibit environmental reactions; in the case of titanium alloys, a cover gas is needed to prevent workpiece oxidation.

**Fracture Toughness.** Tool fracture toughness plays a significant role during the tool plunge and dwell. The local stresses and strains produced when the tool first touches the workpiece are sufficient to break a tool, even when mitigation methods are used (pilot hole, slow plunge speed, and preheating of the workpiece). It is generally accepted that the tool plunge and dwell periods produce the most damage to a tool (Ref 4). The friction stir machine spindle runout (lateral movement during spindle rotation) should also be considered when selecting a tool material. Low-fracture-toughness tools, for example, ceramics, should only be used in friction stir machines that contain low spindle runout (less than 0.0051 mm, or 0.0002 in.) to avoid premature tool fracture.

**Coefficient of Thermal Expansion (Bimetal Tools).** Thermal expansion is a consideration in multimaterial tools. Large differences in the coefficient of thermal expansion (CTE) between the pin and shoulder materials lead to either expansion of the shoulder relative to the pin or expansion of the pin relative to the shoulder. Both of these situations increase the stresses between the pin and shoulder, thus leading to tool failure.

Additional consideration should be made when the pin and shoulder are made of one material, while the tool shank (portion of tool within the spindle) is a different material. One way to mitigate this situation is with a thermal barrier designed to prevent heat removal from the tool into the shank. An example of this is used with PCBN tools where a thermal barrier prevents heat from moving into the tungsten carbide shank (Ref 5). The CTE differences between the tool and workpiece are not found to have a significant influence on friction stirring.

**Machinability.** Many friction stir tools are designed with features that must be machined, ground, or electrodischarged machined into the tool. Any material that cannot be processed to the required tool design should not be considered.

**Uniformity in Microstructure and Density.** Tool materials are not useful if there are local variations in microstructure or density. These slight variations produce a weak region within the tool where premature fracture occurs. Powder metallurgical alloys are manufactured with different densities, so friction stirring tools should only be manufactured from a fully dense grade.

**Availability of Materials.** A tool material is not useful if a steady supply of tool material is not available. This is especially true in a production environment, where production specifications dictate the use of a specific material.

### 2.1.2 Published Tool Materials

This section considers all of the published tool materials listed for friction stir welding and processing. The listed tool materials should not be viewed as an exhaustive list, because many papers do not specify the tool material or claim the tool materials are proprietary. In instances where specific alloys are not cited, effort was made to include the class of tool materials used. The exception is tool steels, where many papers cite tool steels but not the specific alloy. Table 2.1 is a summary of the current tool materials used to friction stir the indicated materials and thicknesses. These data are assembled from the indicated literature sources.

**Tool Steels.** Tool steel is the most common tool material used in friction stirring (Ref 6–26). This is because a majority of the published FSW literature is on aluminum alloys, which are easily friction stirred with tool steels. The advantages to using tool steel as friction stir tooling material include easy availability and machinability, low cost, and established material characteristics. References cite AISI H13 (Ref 7, 8, 10–12, 14–16, 18–20, 24–27) more than any other steels. AISI H13 is a chromium-molybdenum hot-worked air-hardening steel and is known for good elevated-temperature strength, thermal fatigue resistance, and wear resistance. In addition to friction stir welding aluminum alloys, H13 tools have been used to friction stir weld both oxygen-free copper (Cu-OF) and phosphorus-deoxidized copper with high residual phosphorus (Cu-DHP) (Ref 25). However, the limited travel speed in Cu-DHP would limit the production use of H13. Another study found that tool steel FSW tools could weld 3 mm (0.12 in.) thick copper, but 10 mm (0.4 in.) thick copper filled the tool features and softened the tool steel, distorting the pin profile (Ref 28). Other tool steels used for FSW tools include oil-hardened 0-1 (Ref 13, 17, 29), D2 (Ref 30), SKD61 (Ref 23), Orvar Supreme (Ref 31), and Divar (Ref 32). The maximum-use temperature of tool steels depends on the type of tool steel: oil- and water-hardened tool steels can be used up to 500 °C (930 °F); secondary-hardened tool steels can be used up to 600 °C (1110 °F).

**Nickel- and Cobalt-Base Alloys.** High-temperature nickel- and cobalt-base alloys were developed to have high strength, ductility, creep resistance, and corrosion resistance. These alloys derive their strength from precipitates, so the use temperature must be kept below the precipitation temperature (typically 600 to 800 °C, or 1110 to 1470 °F) to prevent precipitate overaging and dissolution. Nickel- and cobalt-base alloys were initially designed for aircraft engine components, so much is known about the alloys, and a reasonable supply exists. It is reasonable to assume that new alloys will improve the quality and use temperature of nickel- and cobalt-base alloys, thus providing additional alloys for friction stirring. Nickel- and cobalt-base alloys can be difficult to machine, especially for the highly alloyed alloys. Several different nickel-base alloys have been used to friction stir weld copper alloys, including IN738LC, IN939 (Ref 26), MAR-M-002, Stellite 12, IN-100, PM 3030, Nimonic 90, Inconel 718, Waspalloy (Ref 33), and Nimonic 105 (Ref 33, 34). Aluminum alloys have been friction stirred with tools made

**Table 2.1 Summary of current friction stir welding tool materials**

| Alloy | Thickness, mm | Thickness, in. | Tool material |
|---|---|---|---|
| Aluminum alloys | <12 | <0.5 | Tool steel, WC-Co |
| | <26 | <1.02 | MP159 |
| Magnesium alloys | <6 | <0.24 | Tool steel, WC |
| Copper and copper alloys | <50 | <2.0 | Nickel alloys, PCBN(a), tungsten alloys |
| | <11 | <0.4 | Tool steel |
| Titanium alloys | <6 | <0.24 | Tungsten alloys |
| Stainless steels | <6 | <0.24 | PCBN, tungsten alloys |
| Low-alloy steel | <10 | <0.4 | WC, PCBN |
| Nickel alloys | <6 | <0.24 | PCBN |

(a) PCBN, polycrystalline cubic boron nitride

from the cobalt-nickel-base alloy MP 159 (Ref 14, 32, 35), which is readily machined. Figure 2.1 shows the ultimate tensile strength as a function of test temperature for selected nickel- and cobalt-base alloy bars (Ref 36, 37).

**Refractory Metals.** The refractory metals (tungsten, molybdenum, niobium, and tantalum) are used for their high-temperature capabilities (e.g., light bulb filaments) and high densities (ballistic projectiles). Many of these alloys are produced as a single phase, so strength is maintained to nearly the melting-point temperature. Therefore, refractory metals are among the strongest alloys between 1000 and 1500 °C (1830 and 2730 °F). However, tantalum and niobium have high solubility of oxygen at elevated temperatures, which quickly degrades the ductility. The drawbacks to using refractory metals include limited material availability, long lead times, cost, and difficult machining (typically involving grinding processes). Powder processing is the primary production method for refractory alloys. Occasionally, partially dense powder-processed material is manufactured, which produces a friction stir tool that easily fractures. Thus, care must be taken to ensure that the raw material is fully dense before machining.

Tungsten-base alloys have been used in the friction stirring of copper alloys, nickel-aluminum bronze, titanium alloys, and steels (Ref 4, 15, 26, 28, 33, 34, 38–48). The FSW of 1018 steel (Ref 4) and ultrahard 0.29C-Mn-Si-Mo-B 500 Brinell steel (Ref 40) caused tool wear on tungsten alloy FSW tools. Four tungsten-base materials have been specifically cited for friction stirring tools: W (Ref 5), W-25%Re (Ref 33, 39), Densimet (Ref 28, 33, 34, 41, 44, 47), and W-1%$LaO_2$ (Ref 48). Tungsten-rhenium has a high operational temperature, but machining features require grinding (more difficult than conventional machining), and tungsten-rhenium has a high cost. Densimet consists of small spheres of tungsten bound in a matrix containing either nickel-iron or nickel-copper combinations (Ref 49). Figure 2.2 demonstrates that the matrix of Densimet lowers the operational temperatures (relative to other tungsten-base alloys). However, in contrast to other tungsten-base alloys (i.e., tungsten-rhenium), Densimet is readily machined by conventional methods and has a lower raw material cost. The high thermal conductivity of Densimet has been cited as a reason to use this material for the shoulder of FSW tools (Ref 33, 34) used to weld 50 mm (2 in.) thick copper. Another tungsten-base alloy is W-1%$LaO_2$ (Ref 48), which has the cost and machinability of Densimet but the temperature range of tungsten-rhenium tools. The ultimate tensile strength temperature dependence of tungsten (Ref 50), W-27%Re (Ref 51), Densimet (Ref 49), and W-1%$LaO_2$ (Ref 52) is shown in Fig. 2.2.

Friction stir tools were also made from molybdenum-base alloys (Ref 4, 33). Cederqvist examined four molybdenum-base alloys to friction stir weld up to 50 mm thick copper plates (Ref 33). However, none of the alloys survived the plunge sequence and remained dimensionally unchanged after a 1 m (3 ft) long weld.

**Carbides and Metal-Matrix Composites.** Carbides (or cermets) are commonly used as

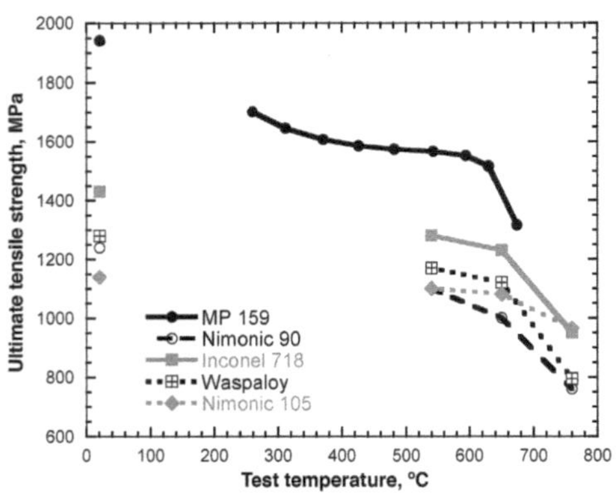


**Fig. 2.1** Elevated-temperature tensile properties for select nickel- and cobalt-base alloys. Source: Ref 36, 37

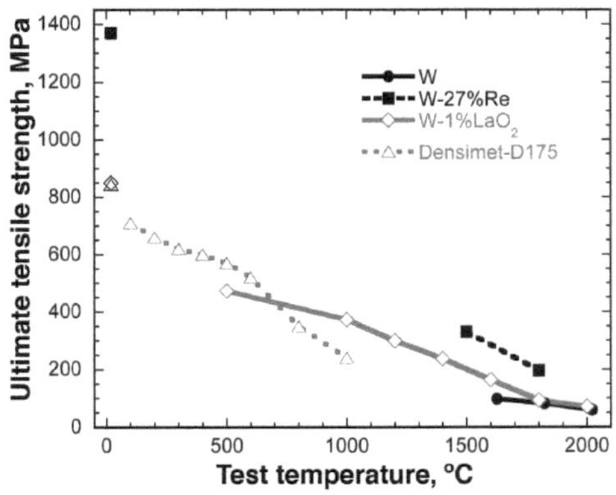


**Fig. 2.2** Elevated-temperature tensile properties for W, W-27%Re, Densimet D175, and W-1%$LaO_2$. Source: Ref 49–52

machining tools due to superior wear resistance and reasonable fracture toughness at ambient temperatures (especially when compared to other ceramics). Because they are made for machining tools, carbides perform well at elevated temperatures. Friction stirring tools made from tungsten carbide are reported to have smooth and uniform thread surfaces for the FSW of 6061 Al (Ref 10). The superior wear resistance of WC-Co allows threadless pins to friction stir weld 5 mm (0.2 in.) thick AC4A (aluminum-silicon alloy) + 30 vol% SiC with little wear (Ref 53). However, severe wear is observed when the tools contain threads. The high-temperature strength of WC and WC-Co tools was used to weld interstitial-free steel (Ref 23) and carbon S45C steel to 6064 Al (Ref 54, 55).

Metal-matrix composites using TiC as the reinforcing phase have also been used as tool materials for copper alloys (Ref 26). Both sintered TiC:Ni:W and hipped TiC:Ni:Mo alloys were used to friction stir copper alloys. However, both TiC-containing alloys produced brittle tools that fractured during the tool plunge.

**Cubic Boron Nitride.** Polycrystalline cubic boron nitride was originally developed for the turning and machining of tool steels, cast irons, and superalloys. Recently, PCBN has gained acceptance as a friction stir tool material, especially for high-temperature alloys (Ref 3, 5, 26, 33, 40, 44, 56–69). The PCBN was chosen as a friction stir tool based on its prior success in extreme machining applications. The manufacturing of PCBN occurs via an ultrahigh-temperature/high-pressure process, where the extreme temperatures and pressures limit the size of PCBN that can be produced. Only the shoulder and pin of the tool are produced from PCBN; the shank is made from tungsten carbide, and both are held together by a superalloy locking collar (Ref 3, 58). The high tool costs (due to the extreme manufacturing methods) and the low fracture toughness mean that care should be used with PCBN tools. The PCBN tools require a low eccentricity spindle to minimize tool fracture. Successful PCBN friction stir welds have been made with ferritic steels (Ref 5, 40, 54), dual-phase steels (Ref 5, 65), austenitic stainless steels (Ref 5, 56, 59, 60, 63, 64, 67), type 430 stainless steel (Ref 5), 2507 super duplex stainless steel (Ref 5), class 40 gray cast iron (Ref 68), nickel-base alloys (Ref 5), Narloy Z (Ref 5), Ni-Al bronze (Ref 5), Invar (Ref 5), copper (Ref 26, 33), sonoston (Ref 61), ultrafine-grained steels (Ref 62), and nitinol (Ref 44).

**Direct Comparison of Tool Materials.** Only a handful of published studies have examined the effect of different tool materials on FSW. Midling and Rorvik (Ref 31) examined how weld heat input changed with different tool shoulder materials using 6 mm (0.25 in.) thick 7109.50-T79 Al friction stir welds. To perform this task, they constructed a tool shank made of titanium, into which hardened tool steel (Orvar Supreme) pin and tool shoulder inserts were placed. Shoulder inserts consisted of Inconel 718, Nimonic 105, a zirconia engineering ceramic, 94%WC + 6%Co, and a $Ni_3(Si,Ti,Cr)$ intermetallic. All the metallic tool materials behaved similarly to the reference tool steel except at the slowest welding speed (5 mm · $s^{-1}$, or 0.2 in. · $s^{-1}$), where all the tool materials exhibited better heat generation than the reference tool steel. However, the zirconia ceramic insert produced 30 to 70% more heat than the reference Orvar Supreme tool steel. The higher heat input allowed the tool travel speed to increase from 12 to 18 to 30 mm · $s^{-1}$ (0.5 to 0.7 to 1.2 in. · $s^{-1}$), just by changing the tool shoulder material.

Savolanen et al. (Ref 25) examined how different tool materials were able to friction stir weld four different 10 to 11 mm (0.40 to 0.43 in.) thick copper alloys: Cu-OF, Cu-DHP, aluminum bronze, and Cu-25%Ni. The evaluated tool materials included H13 tool steel, IN738LC, IN939, IN738LCmod, sintered TiC:Ni:W (2:1:1), hipped TiC:Ni:Mo (3:2:1), pure tungsten, and PCBN. Tool steel (H13) and nickel-base alloy tools were only suitable for Cu-OF and CU-DHP, but the welding speeds with H13 tools were quite low. Both of the TiC-base alloys were too brittle, and the tungsten tools worked for only Cu-OF and Cu-DHP (a tungsten-base alloy was postulated to produce better results, Ref 25). The PCBN was the only tool material to produce quality friction stir welds in all four copper alloys.

Cederqvist studied 17 different tool materials to friction stir weld 50 mm thick copper (Ref 33), and the first material evaluations were for use as the tool pin. Tungsten carbide-cobalt pins provided the initial welding parameter development, but tool life issues (due to large spindle eccentricities) made this tool material impractical for production. Likewise, eccentricity issues caused PCBN, alumino-silicate, and yttria-stabilized zirconium oxide pins to fail within the plunge or dwell sequence of the friction stir welds. A majority of the pins manufactured from refractory metals (four molybdenum-base and three tungsten-base) did not have dimensional stability after the plunge sequence and 1 m of welding. The exception was the tungsten-rhenium alloy, which had the best performance of the refractory metals, but the cost of tungsten-rhenium was too high for the selected application. Pins made of cast (MAR-M-002 and Stellite 12) and powder-processed (PM 3030) superalloys produced 1.0 m long welds, and pins made from IN-100 fractured after 150 mm (6 in.) of weld length. Evaluations of these tool materials were stopped after metallurgical examination showed the presence of porosity (cast alloys) and carbide films (powder alloys). Pins made from Nimonic 90, Inconel 718, and Waspalloy produced welds 3.3 m (11 ft) long without fracture, but all of these tools had started to twist, causing a reduction in length. Nimonic 105 was able to produce 20 m (66 ft) long friction stir welds with no fracture or change in dimensions. Selection of Nimonic 105 was attributed to good creep rupture strength up to 950 °C (1740 °F) and consistent ductility up to 900 °C (1650 °F). Densimet was selected as the shoulder material based on higher thermal conductivity (130 W/m°C) than nickel-base (10 to 20 W/m°C) and cobalt-base alloys (70 W/m°C), where the author assumed that faster heating of the tool shoulder is preferred in FSW.

## 2.2 Friction Stir Tools

Each of the friction tool parts (pin and shoulder) has a different function. Therefore, the best tool design may consist of the shoulder and pin constructed with different materials. The workpiece and tool materials, joint configuration (butt or lap, plate or extrusion), tool parameters (tool rotation and travel speeds), and the user's own experiences and preferences are factors to consider when selecting the shoulder and pin designs. The tool designs shown in this chapter are a summary of those found in literature. New tool designs are in constant introduction, so the reader is encouraged to seek out recently published tool designs, especially for niche applications.

### *2.2.1 Friction Stirring Imperfections*

There are three common imperfections encountered in friction stirring: voids, joint line remnants, and root flaws (or incomplete root penetration). The presence of voids is easily detectable by current nondestructive testing methods, but joint line remnants and root flaws can be quite difficult to find (Ref 70). These defects must be considered when designing an FSW tool for a given application. While several different process variables (e.g., the tool design, tool rotation and travel speed, tool shoulder plunge depth, tool tilt angle, welding gap, and thickness mismatch) affect the quality of friction stir welds, this section focuses on how tool design and operation affect imperfections.

**Voids** are generally found on the advancing side of the weld, and they may or may not break through to the surface of the friction stir weld (Fig. 2.3). For a given tool design, void formation is due to insufficient forging pressure, too high of welding speed, and insufficient workpiece clamping (too large of joint gap) (Ref 71). Material deformed by the friction stir tool must be able to fill the void produced by a traversing pin. If the tool design is incorrect (i.e., pin diameter is too large for selected parameters) or the travel speed too fast, the deformed material will cool before the material can fully fill the region directly behind the tool. In addition, the shoulder is needed to apply sufficient heat generation to allow material flow around the tool; if insufficient heat is generated (through insufficient forging pressure or incorrect shoulder diameter), then material will not flow properly, and voids will form.

**Joint Line Remnant.** A joint line remnant defect (also known as a kissing bond, lazy S, or entrapped oxide defect) is due to a semicontinuous layer of oxide through the weld nugget (Fig. 2.4). The semicontinuous layer of oxide was initially a continuous layer of oxide on the faying surfaces of the plates to be joined. Joint line remnants form because of insufficient cleaning of workpieces prior to welding or insufficient deformation at the faying surface interface due

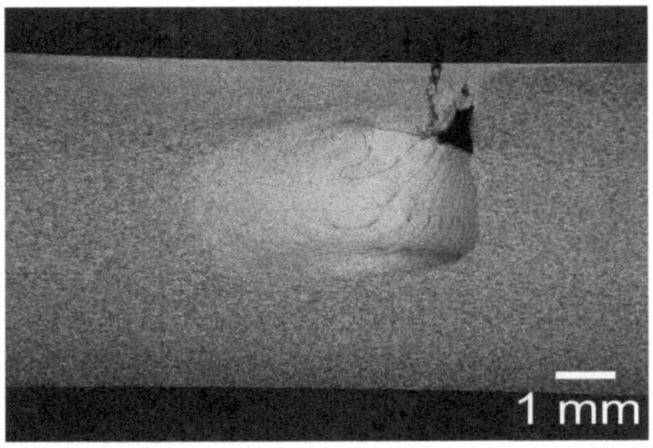


**Fig. 2.3** Macrograph showing void imperfection in a friction stir weld

to incorrect tool location relative to the joint line, too fast of welding speed, or too large of tool shoulder diameter (Ref 71).

**Incomplete Root Penetration.** There are several causes for incomplete root penetrations, including local variations in the plate thickness, poor alignment of tool relative to joint interface, and improper tool design. In the realm of tool design, incomplete root penetration occurs when the FSW pin is too distant from the support anvil. Thus, an undeformed region exists between the bottom of the tool and the bottom surface of the plate (Fig. 2.5). When subjected to a bending stress, the friction stir weld will fail along the lack of penetration line (Fig. 2.5b). The proper FSW of butt joints requires a sufficient depth of deformation (either through pin length or design) to eliminate the incomplete root penetration, while ensuring that the pin will not touch the backing anvil.

### 2.2.2 Design of Tool Shoulders

Tool shoulders are designed to produce heat (through friction and material deformation) to the surface and subsurface regions of the workpiece. The tool shoulder produces a majority of the deformational and frictional heating in thin sheet, while the pin produces a majority of the heating in thick workpieces. Also, the shoulder produces the downward forging action necessary for weld consolidation.

**Concave Shoulder.** The first shoulder design was the concave shoulder (Ref 2), com-

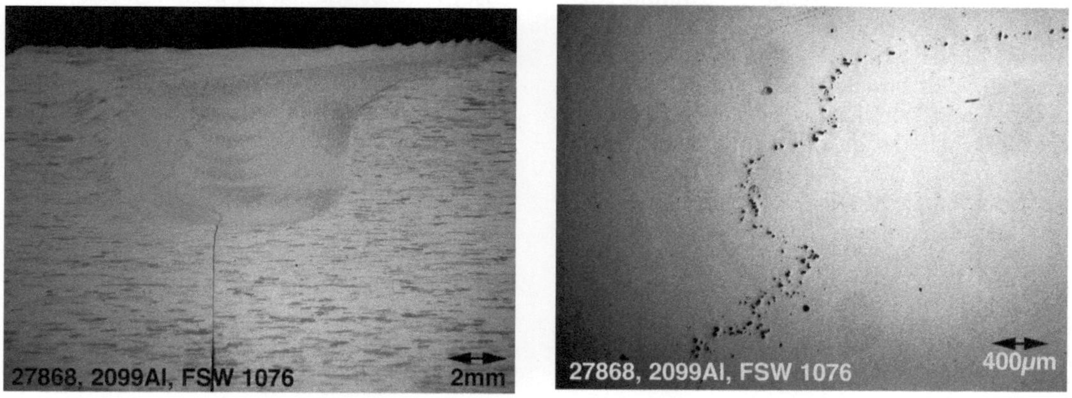


(a) (b)

**Fig. 2.4** Joint line remnant imperfection in a friction stir weld shown by (a) macrograph and (b) magnification of oxide debris that causes the joint line remnant

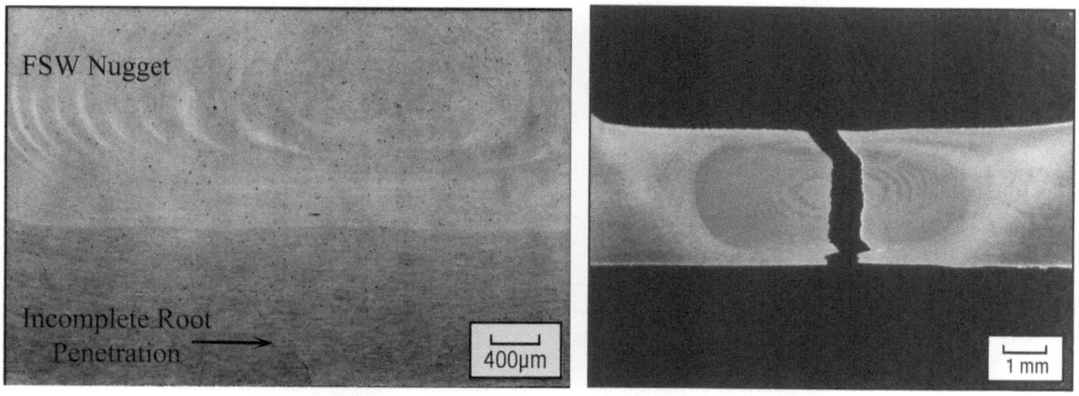


(a) (b)

**Fig. 2.5** Incomplete root penetration imperfection as demonstrated by (a) micrograph and (b) fracture path dictated by incomplete root penetration at the weld root. FSW, friction stir weld monly referred to as the standard-type shoulder, and is currently the most common shoulder design in friction stirring (Ref 3–13, 15–21, 23–26, 28, 33, 34, 38–48, 53–67, 72–78). Concave shoulders produce quality friction stir welds, and the simple design is easily machined. The shoulder concavity is produced by a small angle between the edge of the shoulder and the pin, between 6 and 10°. During the tool plunge, material displaced by the pin is fed into the cavity within the tool shoulder. This material serves as the start of a reservoir for the forging action of the shoulder. Forward movement of the tool forces new material into the cavity of the shoulder, pushing the existing material into the flow of the pin. Proper operation of this shoulder design requires tilting the tool 2 to 4° from the normal of the workpiece away from the direction of travel; this is necessary to maintain the material reservoir and to enable the trailing edge of the shoulder tool to produce a compressive forging force on the weld. A majority of the friction stir welds produced with a concave shoulder are linear; nonlinear welds are only possible if the machine design can maintain the tool tilt around corners (i.e., multiaxis FSW machine).

**Shoulder Features.** The FSW tool shoulders can also contain features to increase the amount of material deformation produced by the shoulder, resulting in increased workpiece mixing and higher-quality friction stir welds (Ref 79, 80). These features can consist of scrolls, ridges or knurling, grooves, and concentric circles (Fig. 2.6) and can be machined onto any tool shoulder profile (concave, flat, and convex). Currently, there are published examples of three types of shoulder features: scoops (Ref 80), concentric circles (Ref 9, 80), and scrolls (Ref 9, 14, 75, 76, 80, 81).

*Scroll Shoulder.* Scrolls are the most commonly observed shoulder feature. The typical scrolled shoulder tool consists of a flat surface with a spiral channel cut from the edge of the shoulder toward the center (Fig. 2.7). The channels direct deformed material from the edge of the shoulder to the pin, thus eliminating the need to tilt the tool. Removing the tool tilt simplified the friction stirring machine design and allowed for the production of complicated nonlinear weld patterns. Concave shoulder tools also have a tendency to lift away from the workpiece surface when the tool travel speed is increased. Replacing the concave shoulder with a scrolled shoulder reduces the tool lift and increases the welding speed. An additional advantage of the scrolled shoulder tool is elimination of the undercut produced by the concave tool and a corresponding reduction in flash. Also, because the tool is normal to the workpiece, the normal forces are lower than concave shoulder tools, which must apply load in both the normal and transverse directions to keep the shoulder in sufficient contact. In addition, the material within the channels is continually sheared from the plate surface, thereby increasing the deformation and frictional heating at the surface (Ref 80).

Scrolled shoulder tools are operated with only 0.1 to 0.25 mm (0.004 to 0.01 in.) of the tool in contact with the workpiece; any additional workpiece contact will produce significant amounts of flash. If the tool is too high (insufficient contact), the shoulder will ride on a cushion of material that will smear across the joint line and make a determination of weld quality difficult (Ref 80). Thus, use of the scrolled shoulder requires more positional care than the concave shoulder. The limitations of

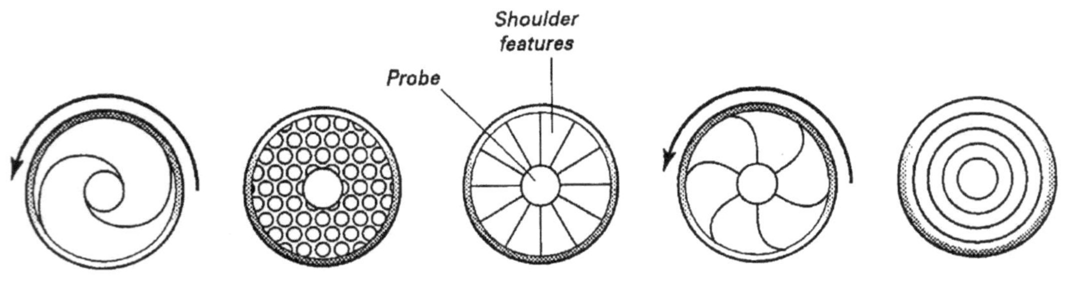


**Fig. 2.6** Different shoulder features used to improve material flow and shoulder efficiency. Source: Ref 79

scrolled shoulder tools include the inability to weld two plates with different thicknesses, an inability to accommodate for workpiece thickness variation in the length of the weld, and welding of complex curvatures (especially tight curvatures). Scrolled shoulder tools can weld two plates of different thicknesses, but some amount of material from the thicker plate is expelled in the form of flash.

**Convex Shoulders.** Friction stir tool shoulders can also have a convex profile (Ref 22, 79, 82–84). Early attempts at TWI to use a tool with a convex shoulder were unsuccessful, because the convex shape pushed material away from the pin. The only reported success with a smooth convex tool was with a 5 mm (0.2 in.) diameter shoulder tool that friction stir welded 0.4 mm (0.015 in.) sheet (Ref 22). Convex shoulder tools for thicker material were only realized with the addition of a scroll to the convex shape (Ref 82–84). Like the scrolls on the flat profile shoulders (see the section "Scroll Shoulder" in this chapter), the scrolls on the convex shoulders move material from the outside of the shoulder in toward the pin. The advantage of the convex shape is that the outer edge of the tool need not be engaged with the workpiece, so the shoulder can be engaged with the workpiece at any location along the convex surface. Thus, a sound weld is produced when any part of the scroll is engaged with the workpiece, moving material toward the pin. This shoulder design allows for a larger flexibility in the contact area between the shoulder and workpiece (amount of shoulder engagement can change without any loss of weld quality), improves the joint mismatch tolerance, increases the ease of joining different-thickness workpieces, and improves the ability to weld complex curvatures. The profile of the convex shoulder can be either tapered (Ref 82, 83) or curved (Ref 79, 84) (Fig. 2.8).

**Fig. 2.7** Photograph of a scrolled shoulder tool and a truncated cone pin containing three flats

### 2.2.3 Pin Designs

Friction stirring pins produce deformational and frictional heating to the joint surfaces. The pin is designed to disrupt the faying, or contacting, surfaces of the workpiece, shear material in front of the tool, and move material behind the tool. In addition, the depth of deformation and tool travel speed are governed by the pin design. The focus of this section is to illustrate the different pin designs found in the open literature, including their benefits and drawbacks. In addition to the pins presented in this section, many other viable pin designs are contained within patent or patent application documents that are not contained within the known literature (e.g., Ref 79). The reader is encouraged to search the patent literature for additional information about pins not contained within this chapter.

**Round-Bottom Cylindrical Pin.** The pin cited in the original FSW patent (Ref 2) consists of a cylindrical threaded pin with a round bottom (Fig. 2.9). This pin design was achieved

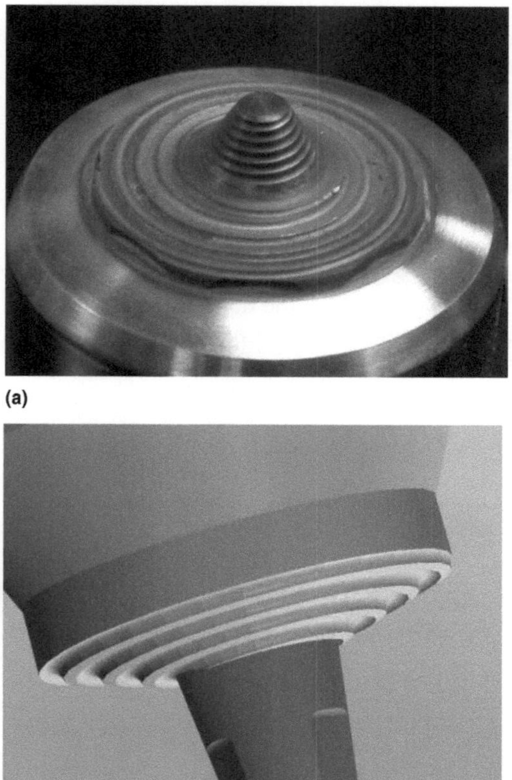

(a)

(b)

**Fig. 2.8** Depictions of the convex shoulder having either (a) curved or (b) tapered geometries during the TWI group-sponsored project number 5651 (Ref 85) and is commonly referred to as the 5651 tool in the friction stir community.

Threads are used to transport material from the shoulder down to the bottom of the pin; for example, a clockwise tool rotation requires left-handed threads. A round or domed end to the pin tool reduces the tool wear upon plunging and improves the quality of the weld root directly underneath the bottom of the pin. The best dome radius was specified as 75% of the pin diameter. It was claimed that as the dome radius decreased (up to a flat-bottom tool), a higher probability of poor-quality weld was encountered, especially directly below the pin (Ref 85). The versatility of the cylindrical pin design is that the pin length and diameter can readily be altered to suit the user's needs. Also, machining a radius at the bottom of the threads will increase tool life by eliminating stress concentrations at the root of the threads.

**Flat-Bottom Cylindrical Pin.** Contrary to the statements made in the previous section about the negative aspects of the flat-bottom cylindrical pin (Fig. 2.10), the flat-bottom pin design is currently the most commonly used pin design (Ref 8–10, 16, 17, 20, 53, 73, 74, 77, 78). Changing from a round-bottom to a flat-bottom pin is attributed to a geometrical argument (Ref 86). The surface velocity of a rotating cylinder increases from zero at the center of the cylinder to a maximum value at the edge of the cylinder. The local surface velocity coupled with the friction coefficient between the pin and the metal dictates the deformation during friction stirring. The lowest point of the flat-bottom pin tilted to a small angle to the normal axis is the edge of the pin, where the surface velocity is the highest (Fig. 2.11a). In contrast, the lowest point of a round-bottom pin is not far from the center of the pin exhibiting a slower surface velocity (Fig. 2.11b). The surface velocities at the lowest points of flat-bottom and round-bottom pins are compared in Table 2.2, assuming a 3° tool tilt, 5 mm (0.2 in.) diameter pin, and a 3.8 mm (0.15 in.) round-bottom pin radius. A larger round-bottom pin radius will reduce the velocity differential, while a smaller pin radius will increase the velocity differential. For this example, the flat-bottom pin has a surface velocity 27.9 times the round-bottom pin. The increased surface velocity at the bottom of the pin would increase the throwing power of the pin, or the ability of the pin to affect metal below the end of the pin. In addition, the flat-bottom pin is easier to machine, and the defects mentioned in the previous section can be eliminated with correct tool parameters and sufficient forging load.

**Truncated Cone Pins.** Cylindrical pins were found to be sufficient for aluminum plate up to 12 mm (0.5 in.) thick, but researchers wanted to friction stir weld thicker plates at faster travel speeds. A simple modification of a cylindrical pin is a truncated cone (Ref 14, 35, 81) (Fig. 2.12). Truncated cone pins have lower transverse loads (when compared to a cylindrical pin), and the largest moment load on a truncated cone is at the base of the cone, where it is the strongest.

A variation of the truncated cone pin is the stepped spiral pin (Fig. 2.13), a design developed for high-temperature materials (Ref 41, 47, 48, 66, 68, 87). During the friction stir processing (FSP) of Ni-Al bronze, a threaded profile distorted, and threadless tools did not produce sufficient material flow to obtain 6 mm (0.25 in.) deep deformation regions. Thus, the stepped spiral tool was designed with robust

**Fig. 2.9** Photograph of a concave shoulder with a round-bottom pin

**Fig. 2.10** Photograph of a flat-bottom pin

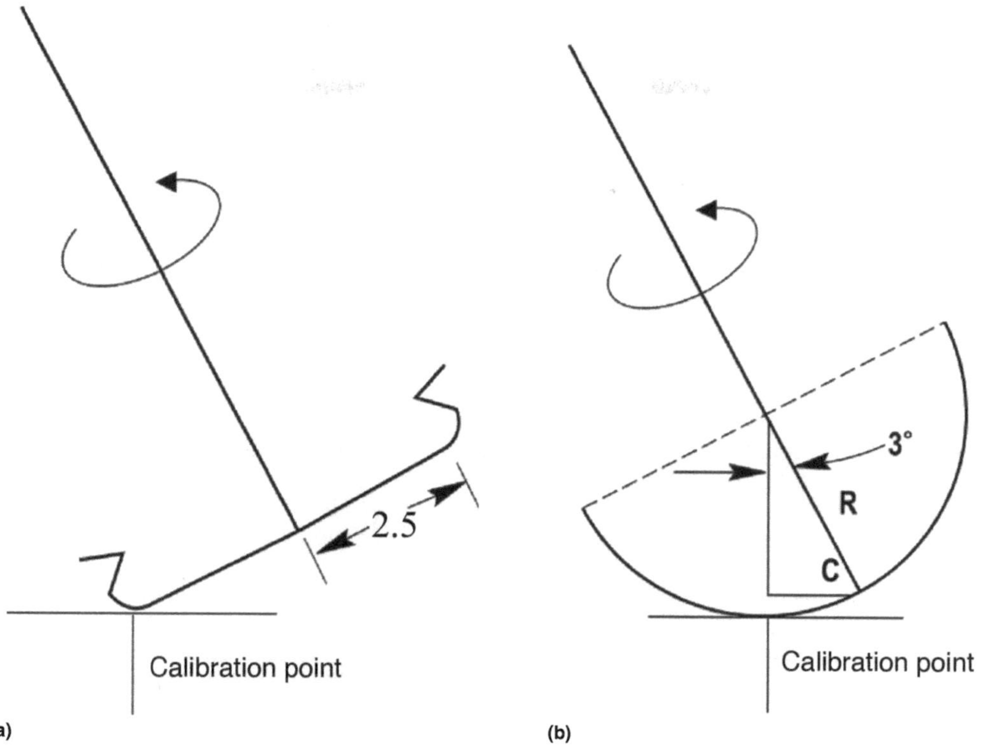


Fig. 2.11 Geometry used to compare surface velocities at calibration point for (a) flat-bottom and (b) round-bottom pins. Source: Ref 86

**Table 2.2 Calculated surface velocities of lowest pin locations**

| | Surface velocities, cm · min$^{-1}$ | |
|---|---|---|
| Tool rpm | Flat-bottom pin | Round-bottom pin |
| 200 | 314 | 11 |
| 400 | 628 | 22 |
| 600 | 942 | 34 |

Fig. 2.12 Truncated cone pin and convex shoulder friction stir welding tool

features that survived the 1000 °C (1830 °F) temperatures. The stepped spiral has a square edge and never forms a recess between a step and the following step. Also, the stepped spiral profile can be ground into ceramic tools, where threaded features are not possible. Thus, some PCBN tools contain a stepped spiral pin that increases the volume of material deformed by the pin (Ref 63, 68, 84).

**Addition of Machined Flats on Pins.** Thomas et al. (Ref 79) found that the addition of flat areas to a pin (as shown in Fig. 2.7) changes material movement around a pin. The effect of adding flat regions is to locally increase the deformation of the plasticized material by acting as "paddles" and producing local turbulent flow of the plasticized material. Colligan, Xu, and Pickens (Ref 14) used 25.4 mm (1 in.) thick 5083-H131 to demonstrate that a reduction in transverse forces and tool torque was directly proportional to the number of flats placed on a truncated cone (up to four flats). Recently, Zettler et al. (Ref 76) have examined the FSW of 4 mm (0.16 in.) thick 2024-T351 and 6056-T4 Al alloys as a function of FSW tool parameters for three different pin designs: a nonthreaded truncated cone pin, a threaded truncated cone pin, and a threaded truncated cone pin with flats. Welding trials quickly showed that the nonthreaded pin produced voids, while the two threaded pins (with and without flats) produced fully consolidated friction stir welds. Adding the flats was shown to increase the weld nugget area and the workpiece temperature measured at the plate midthickness 12.3 mm (0.5 in.) from the joint centerline when compared to the pin without flats.

**Whorl Pin.** The next evolution in pin design is the Whorl pin developed by TWI (Ref 88, 89). The Whorl pin reduces the displaced volume of a cylindrical pin of the same diameter by 60%. Reducing the displaced volume also decreases the traverse loads, which enables faster tool travel speeds. The key difference between the truncated cone pin and the Whorl pin is the design of the helical ridge on the pin surface. In the case of the Whorl pin, the helical ridge is more than an external thread, but the helical ridge acts as an auger, producing a clear downward movement. Variations of the Whorl pin include circular, oval, flattened, or re-entrant pin cross sections (Fig. 2.14) (Ref 89). The significant advantage of the Whorl pin is the ratio of the volume swept by the pin to the pin volume. Cylindrical pins have a ratio of 1.1 to 1, while the Whorl pin has a 1.8 to 1 ratio (when welding 25 mm, or 1 in., thick plate).

**Fig. 2.13** Photograph of a stepped spiral pin

**MX Triflute Pin.** The MX Triflute pin (TWI) is a further refinement of the Whorl pin (Fig. 2.15) (Ref 88, 89). In addition to the helical ridge, the MX Triflute pin contains three flutes cut into the helical ridge. The flutes reduce the displaced volume of a cylindrical pin by 70% and supply additional deformation at the weld line. Additionally, the MX Triflute pin has a pin volume swept to pin volume ratio of 2.6 to 1 (when welding 25 mm thick plate). Published examples using Triflute-type pins include FSW 5 mm (0.2 in.) thick 5251 Al (Ref 90) and up to 50 mm (2 in.) thick copper (Ref 33). Cederqvist (Ref 33) cited that changing to an MX Triflute increased the tool travel speed by 2.5 times over the previous tool design. In addition to welding thick-section copper, the MX Triflute has shown promise for thick-section aluminum alloys. Ma et al. (Ref 91) used the FSP of cast A356 Al to demonstrate that a modified Triflute pin (cylindrical pin with three flutes) is more effective in breaking up silicon particles and healing casting porosity than either cylindrical or truncated cone pins.

**Trivex Pin.** Two-dimensional (2-D) computational fluid dynamics simulations were used to examine material flow around a series of pin

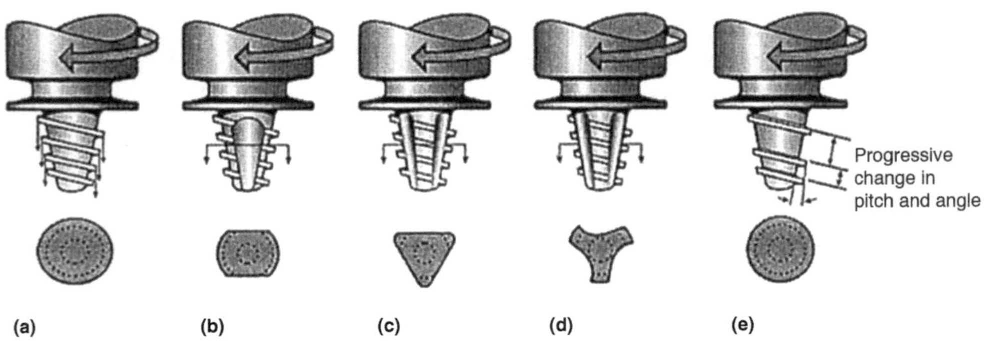


**Fig. 2.14** Schematics of the Whorl pin variations. (a) Oval-shaped probe. (b) Paddle-shaped probe. (c) Three-flat-sided probe. (d) Three-sided re-entrant probe. (e) Changing spiral form and flared probe. Source: Ref 89

designs (Ref 92, 93). The simulations used a novel slip model on the 2-D pin profiles to establish the profile that produced the minimum traverse force. The optimal 2-D pin profile was used to produce two versions: the featureless Trivex pin (TWI) and the threaded MX-Trivex pin (TWI) (Fig. 2.16). Friction stir welding experiments of 6.35 mm (0.25 in.) thick 7075-T7351 Al demonstrated that the Trivex and MX-Trivex pin produced an 18 to 25% reduction of traversing forces and a 12% reduction in forging (normal) forces in comparison to an MX Triflute pin of comparable dimensions (Ref 92, 93). In addition, both the Trivex and Triflute tools produced friction stir welds with comparable tensile strengths.

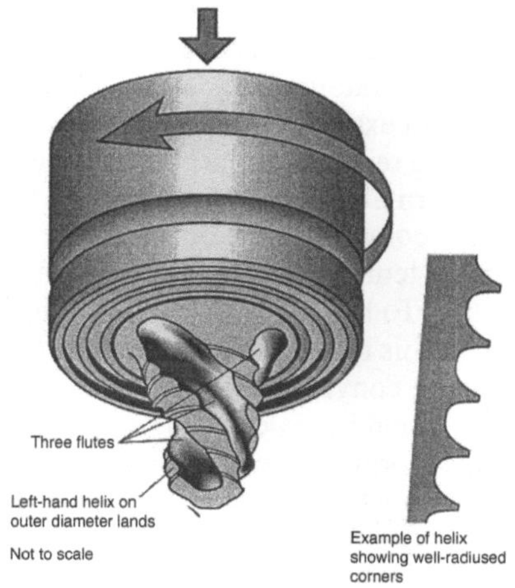


**Fig. 2.15** Schematic of MX Triflute pin. Source: Ref 89

**Threadless pins** are useful in specific FSW applications where thread features would not survive without fracture or severe wear. Tools operating under aggressive environments (high temperature or highly abrasive composite alloys) cannot retain threaded tool features without excessive pin wear; pins for these conditions typically consist of simple designs with robust features. For example, early PCBN pins designed to friction stir weld stainless steels consisted of a truncated cone with three flats at the tip (Fig. 2.17). Also, Loftus et al. used a featureless cylindrical pin to friction stir weld 1.2 mm (0.05 in.) thick beta 21S Ti (Ref 42). Tools used to friction stir weld thin sheet commonly have fine pins with little surface area for features. The addition of any threads would severely weaken the pin, causing premature pin failure. Thus, thin sheet, for example, 0.4, mm (0.015 in.) thick Mg AZ31 (Ref 22), is commonly friction stir welded with threadless tools. Threadless pins have also been used to purposely produce defective welds (Ref 9) and to study material flow (Ref 76).

**Retractable Pins.** The retractable pin tool (RPT) consisted of an actuated pin within a rotating shoulder (Ref 94, 95) to allow pin length adjustment during FSW (Fig. 2.18). The normal operational mode for these tools was to retract the pin at a prescribed rate as the tool traversed forward. This allowed the closure of the exit hole in circumferential friction stir welds.

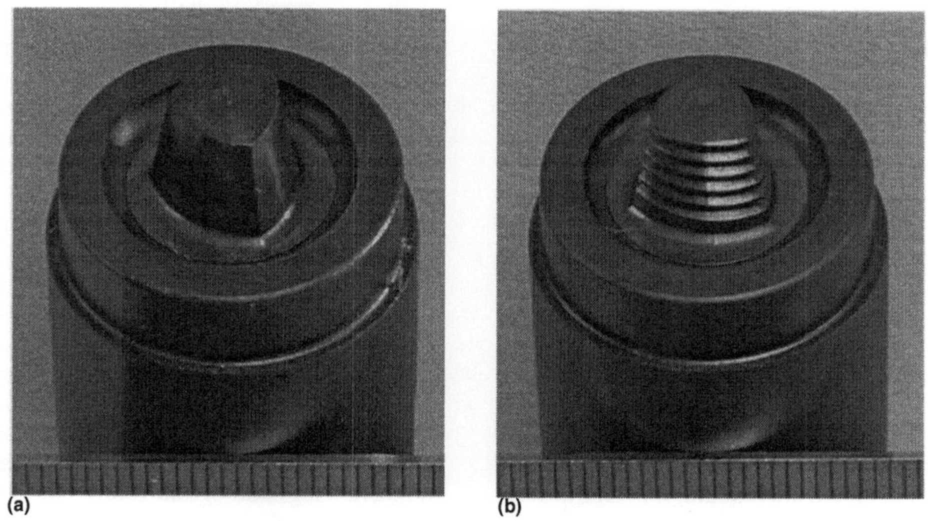


**Fig. 2.16** Photos showing details of (a) Trivex and (b) MX Trivex pins. Scale is in millimeters. Source: Ref 93

**Fig. 2.17** Example of a threadless pin tool. Polycrystalline cubic boron nitride pin tool with three flats at pin tip. Source: Ref 57

Also, pin lengths could be adjusted to ensure full penetration welds in workpieces with known thickness variations.

### 2.2.4 Bobbin Tools

Bobbin tools consist of two shoulders, one on the top surface and one on the bottom surface of the workpiece, connected by a pin fully contained within the workpiece (Fig. 2.19). The bobbin tool concept was included in the first FSW patent by TWI (Ref 2), but initial trials had problems with weld nugget containment due to improper shoulder design. The next iteration of bobbin tools used a fixed shoulder-to-shoulder distance and the scrolled shoulder tool (Ref 88), which eliminated the need to tilt the tool. However, subsequent FSW trials showed that the fixed shoulder distance bobbin tools had issues with pin fractures that were attributed to thermal expansion stresses between the tool and workpiece. The final bobbin tool iteration included the RPT (Ref 94), which allowed the relative movement between the shoulders to maintain a constant force between the shoulders. The bobbin tool works by placing the bottom or reacting scrolled shoulder onto the end of a retractable pin. This is typically done by first drilling a hole through the workpiece, inserting the threaded pin, and securing the second shoulder to the pin. During FSW, the bottom shoulder is drawn toward the top shoulder (using the RPT technology) until the desired force is reached. Because the two shoulders are reacting together to form the friction stir weld, the bobbin tool is also known as the self-reacting tool. The primary advantages of bobbin tools include ease of fixturing (no anvil is needed), the elimination of incomplete root penetration, and increased tool travel speeds due to heating from both shoulders (Ref 96). Fixed shoulder-to-shoulder distance bobbin tools are now possible with the convex scrolled shoulder (Ref 82, 83). This bobbin tool configuration does not require the bottom shoulder actuation (RPT) to produce a sound weld and simplifies the design of FSW machines.

Bobbin tools have successfully joined thick aluminum plates from 8 to 25.4 mm (0.3 to 1 in.) (Ref 96) and thin aluminum plate from 1.8 to 3 mm (0.07 to 0.12 in.) (Ref 97). However, several considerations must be made when dealing with the bobbin tools (Ref 96). Careful cleaning of the tools after each weld is necessary to maintain the needed load by actuating the pin and bottom shoulder. During welding, material can extrude between the pin and shoulder, making removal of the bottom shoulder difficult. Thermal comparisons between the bobbin and conventional tools show that the maximum temperature for the bobbin tools is 50 °C (90 °F) higher than the conventional tool (Ref 98). This behavior is attributed to the backing anvil in conventional FSW acting as a heat sink. As would be expected with higher temperatures, the forging forces were 4 to 8 times less for the bobbin tool than conventional FSW.

### 2.2.5 Lap Joint Tools

While many have demonstrated that two plates in the butt joint configuration can be readily friction stir welded, it is the lap joint that offers the most applications. Lap joints are frequently used in industry, and the replacement of fasteners (rivets or bolts) with FSW would be faster if significant modifications of current production parts were not necessary. The lap joint interface (and corresponding surface oxides) resides in a horizontal layer that is more difficult to break up than the vertical interface encountered in butt joints.

Cylindrical pin butt joint tools were used in the first friction stir lap welding (FSLW) attempts. These tools produced uplift adjacent to the friction stir zone and thinning of the upper sheet (Fig. 2.20). Interface uplift is produced by vertical flow adjacent to the pin, which sharply moves the joint interface upward (typically on the advancing side of the tool). The angle of uplift can reach 90° and greatly reduces the fatigue resistance of the joint. Thinning of the

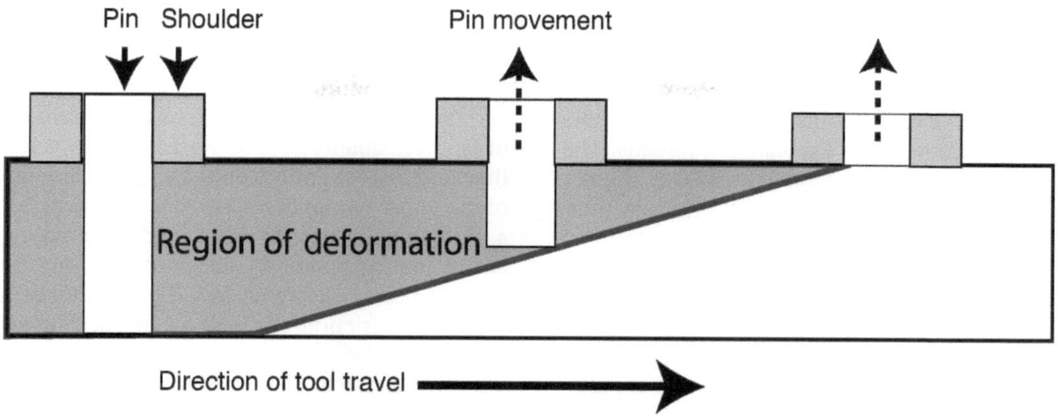


**Fig. 2.18** Example of the retractable pin tool technology, where the pin is fully withdrawn into the shoulder (from left to right), thereby eliminating the exit hole (as shown by the region of deformation)

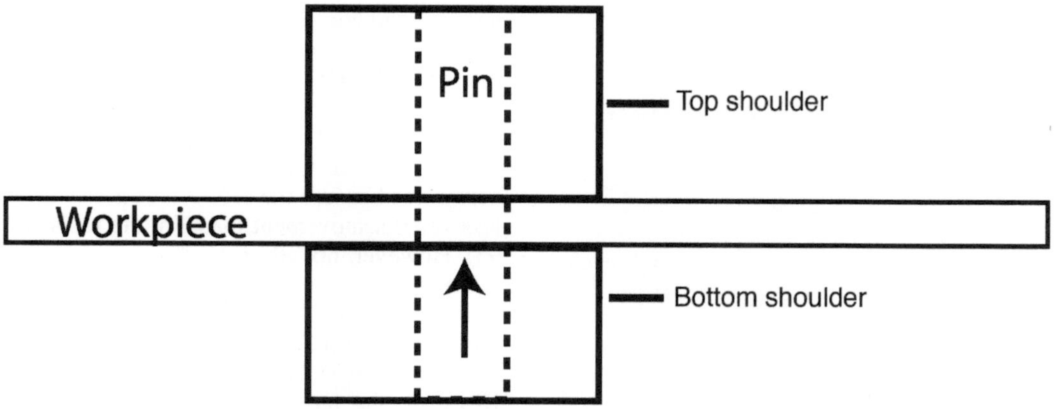


**Fig. 2.19** Schematic of a bobbin tool consisting of a top shoulder, pin, and bottom shoulder attached to the pin. The friction stir weld is produced when the pin is moved upward, forcing the bottom shoulder to react against the top shoulder.

upper workpiece occurs concomitantly with a continuous layer of oxide on the retreating side of the pin. The retreating side material flow produced by cylindrical threaded pins produces uplift and insufficient deformation on the retreating side of the pin. Combining the severe thinning and continuous oxide produces lap joints with low tensile and peel strengths. A small amount of uplift and top sheet thinning can be tolerated, depending on how the joint is loaded.

**Modification of Butt Joint Pins.** Derivatives of butt joint tools have shown some promise to produce quality lap joints. One such tool used a scrolled shoulder and a partially threaded pin. Threads were on the upper part of the pin but not the lower part. The lack of threads on the bottom of the pin changed the material flow in the bottom sheet. Removing the threads eliminated the horizontal material flow induced by the threads, which the authors claim is the primary cause of weld defects (Ref 99).

**Threadless Tools.** A simple lap welding tool consisting of a tool with two shoulders (Fig. 2.21) was developed by TWI and designated the MultiStage tool (Ref 100). The first shoulder rested on the top surface of the overlapping plates. The second shoulder was located at the interface between the two lapped plates and was designed to disrupt the oxides at the lap joint interface. A variation of the MultiStage tool was later used to friction stir weld 2.4 mm (0.09 in.)

thick 7075-T7351 Al (Ref 101). A series of threadless tools were used to friction stir lap weld 2.11 mm (0.0831 in.) thick aluminum-clad 2024-T3 to 2.16 mm (0.0850 in.) thick 7075-T6 (Ref 102). The short pin lengths used in this work necessitated the exclusion of threads. The shear strength was maximized with two slightly offset FSW passes, one pass in opposite rotational direction of the other, such that the two retreating sides were on the edges of the FSW nugget and the advancing sides were on the interior of the FSW nugget. Sheet thinning on the retreating side was minimized with shorter pin lengths; this was attributed to less vertical material flow near the bottom of the pin than the middle of the pin.

**MX Triflute and Flared-Triflute Pins.** Two studies have examined the use of Flared-Triflute and MX Triflute-based pins (TWI) for lap joints. In a Flared-Triflute pin, the bottom of the pin is flared outward, causing a whisk-type profile (Fig. 2.22) (Ref 89). This profile increases the swept and static volumes of the pin and changes the flow pattern around the bottom of the pin for improved FSLW quality. Mishina and Norlin (Ref 103) compared the difference of lap weld quality in 6082 Al using an MX Triflute and Flared-Triflute pins. Lap joint thinning of the upper workpiece was reduced with either a double-pass friction stir weld (alternating advancing and retreating side of the tool on subsequent passes) using an MX Triflute tool or a single-pass friction stir weld using the Flared-Triflute pin. Ericsson and Sandström (Ref 104) used two different MX Triflute pins to produce lap welds; one MX Triflute pin had a convex bottom, and the other MX Triflute pin had a concave bottom. The best lap joint fatigue results were observed with a larger shoulder (18 mm, or 0.7 in., diameter) and concave pin. The improved fatigue results were attributed to an increased contact area from the shoulder and improved flow path at the hollowed-out end of the pin (Ref 104).

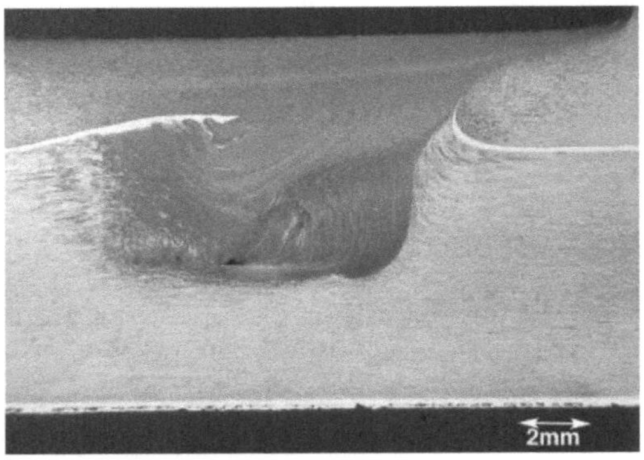


Fig. 2.20 Friction stir lap weld produced by cylindrical pin tool

**Trivex Pins.** Friction stir lap welding trials with the MX Trivex pins (see the section "Trivex Pin" in this chapter) (Ref 105) were performed because the MX Trivex pins showed less vertical movement than Triflute tools (Ref 93). However, not all of the downward interfacial movement could be eliminated with the MX Trivex pins. The MX Trivex pins produced as much as 1 mm (0.04 in.) of retreating side plate thinning (on 6.35 mm, or 0.25 in., thick plates). However, the fatigue results demonstrate that the MX Trivex pin is comparable to the A-Skew pin (see the section "Skew-Stir Tool" in this chapter). The best fatigue results were produced with the Re-Stir tool (see the section "Re-Stir

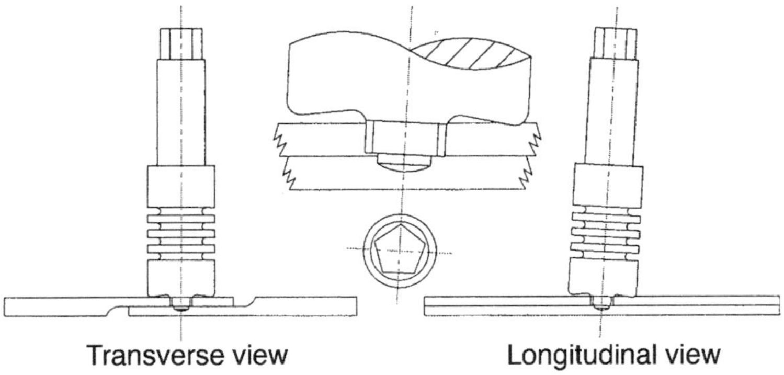


Fig. 2.21 MultiStage friction stir lap welding probe tool. Source: Ref 101

Tool" in this chapter), which requires complex machinery.

## 2.2.6 Complex Motion Tools

Wayne Thomas at TWI has recently focused on FSW tool designs that increase the tool travel speed, increase the volume of material swept by pin-to-pin volume ratio, and/or increase the weld symmetry (Ref 106, 107). Many of these tool designs have focused on tool motion and not specifically on the tool pin design, although each type of complex motion can have an optimal tool design. Most complex motion tools require specialized machinery or specially machined tools, making these tools unsuitable for basic applications.

**Skew-Stir Tool.** The Skew-Stir tools (TWI) increase the volume of material swept by pin-to-pin volume ratio by offsetting the axis of the pin from the axis of the spindle (Fig. 2.23), thus pro-

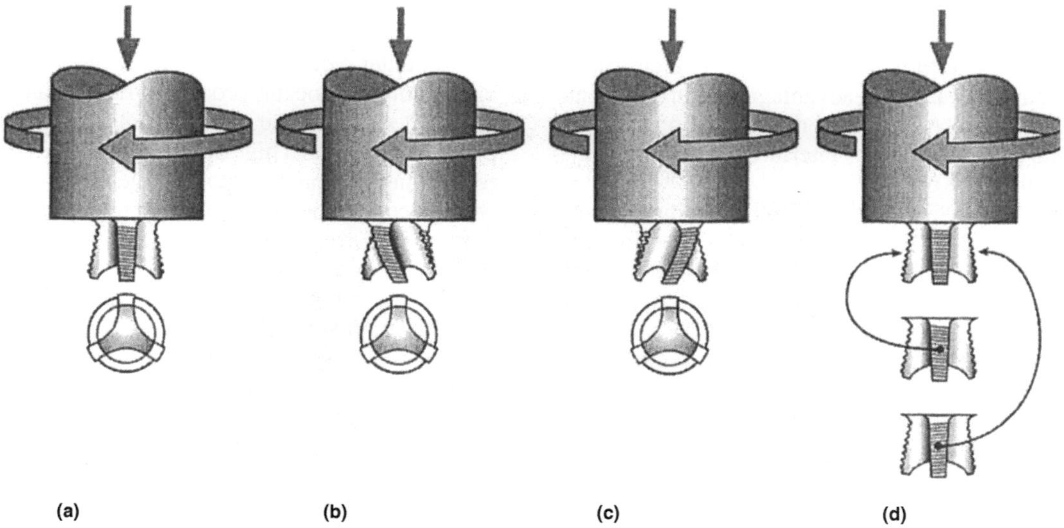


**Fig. 2.22** Schematics of four different Flared-Triflute pin tool variations containing (a) neutral flutes, (b) left-hand flutes, (c) right-hand flutes, and (d) ridge detail, showing that the ridge grooves (threads) can be neutral, left, or right handed. Source: Ref 89

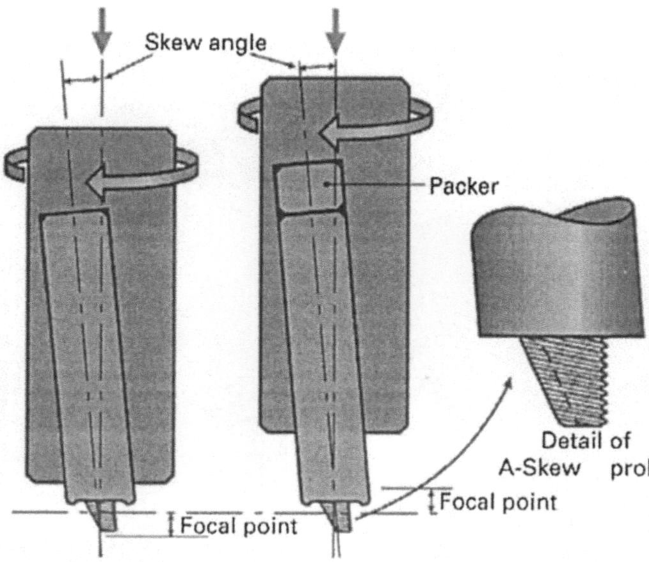


**Fig. 2.23** Schematic of Skew-Stir tool showing different focal points and detail of the A-Skew pin. Source: Ref 89

ducing an orbital motion (Ref 89, 106, 107). The focal length of the tool can be changed to alter the amplitude of tool motion, ranging from rotary to orbital motion. Due to the orbital tool motion, only a portion of the pin is in constant contact with the workpiece. Thus, the A-Skew pin (TWI) takes advantage of the partial pin contact by removing the inner portion of the tool and improves the material flow during friction stir (Ref 106, 107). The resulting weld nugget produced by the Skew-Stir tool is greater than the pin diameter. Also, the orbital motion creates more deformation at the bottom of the pin, decreasing the incidence of root defects. Skew-Stir tools are also advantageous for lap joints, where the A-Skew pin orbital motion produces no plate thinning or interfacial movement adjacent to the pin.

**Com-Stir tools (TWI)** combine rotary motion (tool shoulder) with orbital motion (tool pin) to maximize the volume of material swept by pin-to-pin volume ratio (Ref 108) (Fig. 2.24). Moving the pin in an orbital motion produces a wider weld and increases oxide fragmentation on the interfacial (also known as faying) surfaces. In addition, the motion of the Com-Stir tool produces lower torque than the typical rotary motion FSW tool, reducing the amount of fixturing necessary to secure the workpiece.

**Re-Stir Tool.** The Re-Stir tool (TWI) avoids the inherent asymmetry produced during friction stirring by alternating the tool rotation, either by angular reciprocation (direction reversal during one revolution) or rotary reversal (direction reversal every one or more revolutions) (Ref 109). Alternating the tool rotation produces alternating regions of advancing and retreating side material through the length of the weld, thus eliminating the asymmetry issues (e.g., lack of deformation on the retreating side) found in rotary friction stir welds. An example of the microstructure produced by a Re-Stir tool and a Flared-Triflute pin is shown in Fig. 2.25.

**Dual-Rotation Tool.** In dual-rotation tools, the pin and shoulder rotate separately at different speeds and/or in different directions (Ref 110). In conventional FSW, the pin and shoulder are rotated at the same speed, so the velocity at the edge of the shoulder is much higher than the velocity at the edge of the pin. When the shoulder velocity is too high, workpiece overheating can occur, producing defects along the weld surface. The dual rotation allows the pin to be rotated at a high speed without the corresponding increase in shoulder velocity. Peak workpiece temperature measurements show that the dual-rotation tool produces as much as 66 °C (119 °F) lower temperatures in 7050-T7451 Al, when compared to conventional rotary friction stir welds produced with similar pin design and process condition. The decrease in workpiece temperature produced an increase in microhardness after two months of natural aging and a reduction in corrosion susceptibility (Ref 110).

**Two or More FSW Tools.** The speed and efficiency of FSW can be improved with the use of two or more FSW tools (Ref 111). Thick plates can be welded with two counterrotating FSW tools on either side of the plate. Counter-rotating tools reduce the fixturing required to

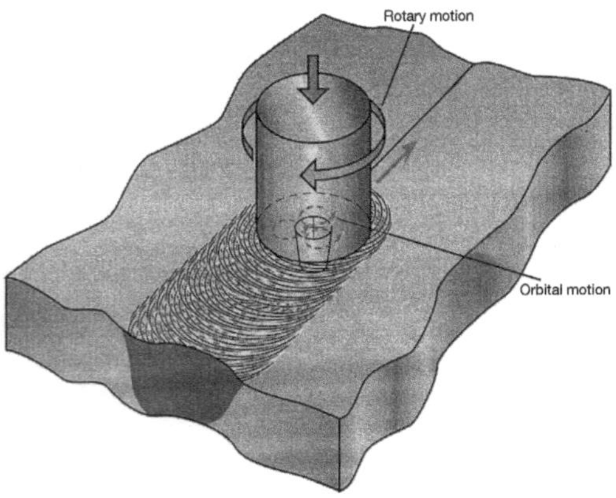


**Fig. 2.24** Principle of the Com-Stir tool. Source: Ref 109

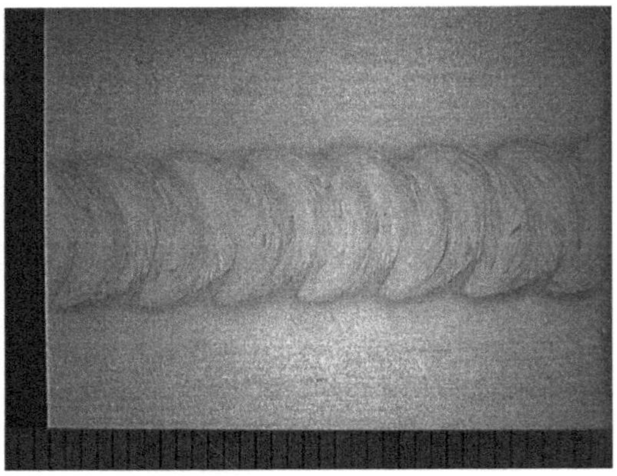

**Fig. 2.25** Plane view (top view) of microstructure produced by Re-Stir friction stir welding technique in 6 mm (0.24 in.) thick 5083-H111 Al plates at 10 reversals per interval, at a welding speed of 198 mm · min$^{-1}$ (7.8 in. · min$^{-1}$). Source: Ref 110

secure the workpiece due to a decrease in torque, as observed with the bobbin tool, but do not need the added complexity produced by the RPT. The use of two FSW tools in close proximity was initially suggested in 1999 (Ref 112) and later saw additional attention (Ref 113, 114). Currently, the two-FSW-tool concept is being developed at TWI in several variations and is referred to as Twin-Stir (Ref 111). Parallel Twin-Stir uses two counterrotating side-by-side tools to produce lap welds. The two tools locate the retreating side plate thinning defects between the two tools, where the thinning will not affect the mechanical properties of the joint. Tandem Twin-Stir uses two FSW tools (with or without counterrotation) positioned one in front of the other to reduce workpiece fixturing, improve the welding speed, and increase deformation and fragmentation of the faying surfaces oxide layer. The motion produced by the counterrotating tandem Twin-Stir is similar to the Re-Stir tool, but the Twin-Stir produces faster travel speeds. The third Twin-Stir variation is with two staggered tools (one tool positioned slightly in front of and to the side of the other tool) that together produce an extremely wide weld nugget. Lap welds will benefit from the increased oxide dispersion and wide nugget width produced by the staggered Twin-Stir tool. The wide nugget is also advantageous for FSP, where overlapping passes are commonly used to friction stir process a desired region.

## 2.2.7 *Tool Dimensions*

The pin length is determined by the workpiece thickness, the tool tilt, and the desired clearance between the end of the pin and the anvil. Pin diameters need to be large enough to not fracture due to the traverse loads but small enough to allow consolidation of the workpiece material behind the tool before the material cools. In the early TWI work (Ref 85), an optimal ratio of shoulder diameter to pin diameter was suggested to assist with tool design. However, the ratio (between 2.5 to 1 and 3 to 1) (Ref 85) was dependent on the aluminum alloy composition and only applied to 6 mm (0.24 in.) thick plate. As the workpiece thickness increases, the thermal input from the shoulder decreases, and the pin must supply more thermal energy. Thus, while the ratio of shoulder diameter to pin diameter determined for 6 mm plate may produce a void-free weld, this may not be the optimal ratio for plates thicker than 12 mm. Also, workpiece materials with lower thermal conductivity values than aluminum can be friction stirred with smaller shoulder diameters (reducing normal loads) than tools used in aluminum. An example of some tool dimensions for only flat-bottom pins is given in Table 2.3.

Several researches have examined the effect of tool dimensions on friction stir weld quality. Reynolds and Tang (Ref 11) used several different variations of cylindrical pins with a concave shoulder to show that defect-free friction stir welds in 8.1 mm (0.32 in.) thick 2195 aluminum alloys could be produced with pin diameter to shoulder diameter ratios ranging from 2 to 1 to 3.125 to 1. Peel et al. (Ref 115) evaluated cylindrical pins with either a standard metric M5 thread (5 mm wide with 0.8 mm pitch) or a wider pin (6 mm wide) with a coarser thread (1 mm pitch). At higher travel speeds (200 mm/min, or 8 in./min), the broader 6 mm tool with the coarser threads was more effective in disrupting the faying interface between the two joined workpieces. This change of pin design

**Table 2.3 Summary of friction stirring tool dimensions for a given workpiece material**

| Shoulder diameter | | Cylindrical pin diameter | | | | |
|---|---|---|---|---|---|---|
| mm | in. | mm | in. | Shoulder-to-pin ratio | Workpiece material and thickness, mm | Ref |
| 13 | 0.5 | 5 | 0.2 | 2.6:1 | 6061-T6 Al, 3 mm | 9 |
| 20–30 | 0.8–1.2 | 8–12 | 0.3–0.5 | 2.5:1, 1.6:1 | 7050, 2195, 5083, 2024, 7075 Al, 6.35 mm | 11 |
| 23 | 0.9 | 8.2 | 0.32 | 2.8:1 | 2024-T351 Al, 6.4 mm | 20 |
| 20,16 | 0.8, 0.6 | 6 | 0.24 | 3.3:1, 2.7:1 | 5083 and 6061 Al, 5.5 mm | 21 |
| 12 | 0.5 | 4 | 0.16 | 3:1 | 1050 Al and oxygen-free copper, 1.8 mm | 23 |
| 25.4 | 1.0 | 7.87 | 0.31 | 3.22:1 | 7075-T7351 Al, 9.53 mm | 24 |
| 23 | 0.9 | 8.4 | 0.33 | 2.7:1 | 2524-T351 Al, 6.4 mm | 26 |
| 20 | 0.79 | 4 | 0.16 | 5:1 | 6064 Al to carbon steel, 4.5 mm | 54 |
| 23 | 0.9 | 8.2 | 0.32 | 2.8:1 | 2024-T351, 7 mm | 72 |
| 10 | 0.4 | 3.8 | 0.15 | 2.6:1 | 2095 Al, 1.63 mm | 74 |
| 25 | 1.0 | 9 | 0.35 | 2.8:1 | 5251 Al, 5 mm | 101 |

produced a 16% increase in joint efficiency (tensile strength of weld divided by tensile strength of base material).

### 2.2.8 Friction Stir Spot Welding Tools

Friction stir spot welding (FSSW) uses the deformation produced by a rotating tool to locally join overlapping parts. Applications for FSSW include substitution for rivets or resistance spot welding and the elimination of fixturing by tacking parts prior to FSW. There are two different FSSW methods: one that retains the exit hole produced by the FSSW tool, sometimes referred to as "poke or plunge" FSSW, and one that fills in the exit hole, known as "filled" FSSW. The filled method of FSSW was developed by GKSS (Ref 116) and requires a tool with three parts: pin, shoulder, and outside retaining clamp (Fig. 2.26). First, the pin is plunged into the workpiece, and displaced material fills the gap between the shoulder and the workpiece. When the pin reaches the desired plunge depth, the pin is then retracted as the shoulder is pushed down to the workpiece, pushing the displaced material back into the workpiece. Once the shoulder has reached the workpiece surface, the pin and shoulder dwell to ensure proper mixing and the production of a defect-free FSSW. Finally, the retaining clamp, shoulder, and pin are retracted from the workpiece, leaving an FSSW. This FSSW method is quite effective but requires complex control to produce an optimal weld (Ref 116, 117).

Poke or plunge spot welds are produced by plunging and retracting the FSSW tool. Due to the simple control and implementation, this type of FSSW has seen more research (Ref 118–120) than the filled exit hole (Ref 116, 117). Any pin design can be used to produce friction stir spot welds. However, Addison and Robelou (Ref 120) demonstrated that an MX Triflute pin produces higher failure loads in 2 mm (0.08 in.) thick 6061-T4 Al, than either a Flared-Triflute or threaded cylindrical round-bottom pin.

### 2.2.9 Friction Stir Processing Tools

In certain applications, it is desirable to friction stir process a large surface area, which requires many overlapping passes. While many of the pin tools described previously can produce a fine-grained microstructure beyond 12 mm deep, some applications call for a layer of fine-grained material across a large surface area. One way to produce the thin, fine-grained layer is with pinless tools (Ref 121, 122). The advantage to the pinless tools includes less passes (due to large shoulder diameter) and lower transverse forces, allowing increased tool travel speeds over tools with pins. Shinoda and Kawai (Ref 121) friction stir processed a cast aluminum (AC2B, 6% Si, and 3.2% Cu) using a 20 mm (0.8 in.) diameter cylinder. The influence of the tool was observed as far as 4 mm (0.16 in.) into the plate and was directly proportional to the normal load. Later Fuller, Mahoney, and Bingel (Ref 122) used a 38 mm (1.5 in.) diameter scrolled shoulder tool to produce a 3 mm (0.12 in.) deep fine-grained region in 6061/5356 Al fusion welds.

## 2.3 Tool Coatings

Tool coatings are commonly used for machining tools to improve tool life by decreas-

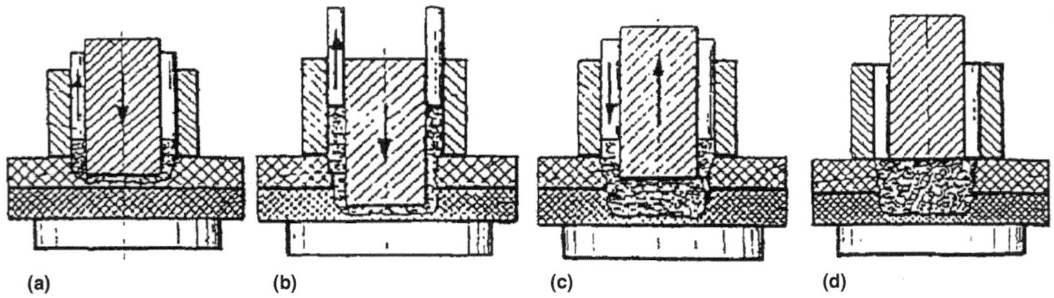


**Fig. 2.26** Schematic of filled exit hole friction stir spot welding showing (a) initial penetration of the pin, (b) continued penetration of the pin into the bottom sheet and withdrawal of the shoulder to allow material to flow around the pin and under the shoulder, (c) withdrawal of the pin with a concomitant plunging of the shoulder to push material back into the void left by the pin, and (d) a completed friction stir spot weld. Adapted from: Ref 117

ing tool wear and thermally protecting the tool. However, traditional coatings have difficulty surviving the aggressive thermal and stress conditions produced during friction stirring, which are extreme environments for most tool coatings, especially with composite and high-temperature materials. A coating of Ti:N was used on H13 tool steel tools used to friction stir weld 2195-T8, 5083-O, 6061-T6, 2219-T8, 2024-T3, 7075-T6, and 7050-T7 Al alloys, but no comment is made on the condition of the coating after FSW (Ref 11). A B4C coating was used on H13 tool steel to reduce wear during the FSW of 6092 Al 17.5% SiC composites, but the coating was worn away after only a few centimeters of welding (Ref 123). Proprietary General Electric chemical vapor deposition and physical vapor deposition tool coatings were used to friction stir weld Ti-17 and Ti-6-4 alloys (Ref 124), but neither coating reduced tool wear, because minor tool wear was noticed on the pin, and debris was detected in the stir zone of the weld. Currently, there is no published work that carefully examines the benefits and impacts of tool coatings.

## 2.4 Thermal Management

The thermal management system consists of the tool (and connection to spindle), workpiece, and backing anvil. Proper thermal management concentrates sufficient heat to the friction stir region to allow efficient thermomechanical deformation while dissipating heat from unwanted regions in the friction stir system (e.g., spindle and machine bearings). Depending on the type of workpiece material, the friction stirring tools and anvil can either be heated or cooled. Tools can be cooled by ambient air, forced air, or a circulating coolant, or tools can be electrically heated. The anvil can be cooled by ambient air, forced air, or a circulating coolant and heated with resistance heaters. In addition, thermal conductivity of the anvil and tool affects the heat input into the workpiece.

Aluminum and magnesium alloys are commonly friction stir welded with ambient air-cooled tools and anvils. Coolant cooling of the tool is not required with aluminum, and magnesium alloys, but the coolant does provide an equilibrium tool temperature for the entire tool usage, especially for long welds, and rapid tool changes are easily performed. Midling and Rorvik (Ref 31) demonstrated that a zirconia-coated steel anvil retained more heat, with the workpiece allowing the tool to travel three times faster to obtain the same heat-affected zone width produced with a steel anvil. A statistical analysis of nine input parameters determined that cooling the anvil had a minimal impact on the friction stir weld; in fact, tool rotation rate, travel speed, and tool depth were more important (Ref 24). Weld quality and performance is affected by differences in heat transfers observed when comparing the friction stir welding of flat plate versus extrusions (Ref 97). Extrusions typically have complicated cross sections, with features that quickly draw heat away from the friction stir weld. This dissipation of heat through the extrusion increases the tool heat input necessary to create a quality friction stir weld.

Steel, titanium, stainless steel, and other higher-temperature alloys are commonly friction stirred with coolant-cooled tools. The higher temperatures produce a dynamic FSW process that is subject to large temperature and load gradients. As opposed to the lower-temperature aluminum alloys where the workpiece governs the heat flow of the welding process, in FSW of the higher-temperature alloys, the tool governs the heat flow (Ref 58). Cooling of the friction stirring tools is necessary to produce a consistent heat flow at the tool and to prevent thermal energy from moving into the FSW system spindle and away from the workpiece. The FSW trials by Packer et al. (Ref 58) demonstrated that passive cooling (cooling of only the spindle bearings) or no liquid cooling of the tool produced excessive heating of the spindle, and a steady-state FSW condition was not achieved.

In contrast to cooling the tool during the weld, other published thermal management methods include the heating of the workpiece or tool (Ref 66, 112, 125–127). The heating is performed to minimize tool wear (especially in the plunge) and increase the tool travel speed. The key to heating the workpiece is to not input too much thermal energy to allow surface melting to occur and to localize the thermal input to the FSW region. Preheating of the 6 mm thick 1018 steel workpiece with induction heating reduced the thrust load by 30%, the side load by 110%, the normal load by 10%, and the tool torque by 20% (Ref 66). Workpiece surface heating during FSW for improved tool travel speed has been demonstrated with flame or arc/plasma

(Ref 112, 125) and lasers (Ref 126). Also, a finite element model has shown that a current passing between the tool and anvil can reduce the normal forces during tool plunge and at least double the tool travel speed, when compared to conventional FSW (Ref 127).

## REFERENCES


1. M.W. Mahoney, Science Friction, *Weld. Join.,* Jan/Feb 1997, p 18–20
2. W.M. Thomas et al., Friction Stir Butt Welding, International Patent Application PCT/GB92/02203 and G.B. Patent Application 9125978.8, Dec 1991
3. M. Collier, R. Steel, T. Nelson, C. Sorensen, and S. Packer, Grade Development of Polycrystalline Cubic Boron Nitride for Friction Stir Processing of Ferrous Alloys, *Proceedings of the Fourth International Conference on Friction Stir Welding,* May 14–16, 2003 (Park City, UT), TWI, paper on CD
4. T.J. Lienert, W.L. Stellwag, Jr., B.B. Grimmett, and R.W. Warke, Friction Stir Welding Studies on Mild Steel, *Weld. J.,* Jan 2003, p 1-s to 9-s
5. C.D. Sorenson, T.W. Nelson, S.M. Packer, and R.J. Steel, Innovative Technology Application in FSW of High Softening Temperature Materials, *Proceedings of the Fifth International Conference on Friction Stir Welding,* Sept 14–16, 2004 (Metz, France), TWI, paper on CD
6. A.P. Reynolds and W.D. Lockwood, Digital Image Correlation for Determination of Weld and Base Metal Constitutive Behavior, *Proceedings of the First International Conference on Friction Stir Welding,* June 14–16, 1999 (Thousand Oaks, CA), TWI, paper on CD
7. T.W. Nelson, B. Hunsaker, and D.P. Field, Local Texture Characterization of Friction Stir Welds in 1100 Aluminum, *Proceedings of the First International Conference on Friction Stir Welding,* June 14–16, 1999 (Thousand Oaks, CA), TWI, paper on CD
8. T.W. Nelson, H. Zhang, and T. Haynes, Friction Stir Welding of Aluminum MMC 6061-Boron Carbide, *Proceedings of the Second International Conference on Friction Stir Welding,* June 26–28, 2000 (Gothenburg, Sweden), TWI, paper on CD
9. S. Brinckmann, A. von Strombeck, C. Schilling, J.F. dos Santos, D. Lohwasser, and M. Koçak, Mechanical and Toughness Properties of Robotic-FSW Repair Welds in 6061-T6 Aluminum Alloys, *Proceedings of the Second International Conference on Friction Stir Welding,* June 26–28, 2000 (Gothenburg, Sweden), TWI, paper on CD
10. K. Colligan, Dynamic Material Deformation During Friction Stir Welding of Aluminum, *Proceedings of the Second International Conference on Friction Stir Welding,* June 26–28, 2000 (Gothenburg, Sweden), TWI, paper on CD
11. A.P. Reynolds and W. Tang, Alloy, Tool Geometry, and Process Parameter Effects on Friction Stir Welding Energies and Resultant FSW Joint Properties, *Friction Stir Welding and Processing,* K.V. Jata, M.W. Mahoney, R.S. Mishra, S.L. Semiatin, and D.P. Field, Ed., TMS, 2001, p 15–23
12. A. Askari, S. Silling, B. London, and M. Mahoney, Modeling and Analysis of Friction Stir Welding Processes, *Friction Stir Welding and Processing,* K.V. Jata, M.W. Mahoney, R.S. Mishra, S.L. Semiatin, and D.P. Field, Ed., TMS, 2001, p 43–54
13. R.A. Prado, L.E. Murr, D.J. Shindo, and J.C. McClure, Friction Stir Welding: A Study of Tool Wear Variation in Aluminum Alloy 6061-20%$Al_2O_3$, *Friction Stir Welding and Processing,* K.V. Jata, M.W. Mahoney, R.S. Mishra, S.L. Semiatin, and D.P. Field, Ed., TMS, 2001, p 105–116
14. K.J. Colligan, J. Xu, and J.R. Pickens, Welding Tool and Process Parameter Effects in Friction Stir Welding of Aluminum Alloys, *Friction Stir Welding and Processing II,* K.V. Jata, M.W. Mahoney, R.S. Mishra, S.L. Semiatin, and T. Lienert, Ed., TMS, 2003, p 181–190
15. P.R. Subramanian, N.V. Nirmalan, L.M. Young, P. Sudkamp, M. Larsen, P.L. Dupree, and A.K. Shukla, Effect of Microstructural Evolution in Mechanical and Corrosion Behavior of Friction Stir-Processed Aluminum Alloys, *Friction Stir Welding and Processing II,* K.V. Jata, M.W. Mahoney, R.S. Mishra, S.L. Semi-

atin, and T. Lienert, Ed., TMS, 2003, p 235–242

16. B. London, M. Mahoney, M. Bingel, M. Calabrese, R.H. Bossi, and D. Waldron, Material Flow in Friction Stir Welding Monitored with Al-SiC and Al-W Composite Markers, *Friction Stir Welding and Processing II,* K.V. Jata, M.W. Mahoney, R.S. Mishra, S.L. Semiatin, and T. Lienert, Ed., TMS, 2003, p 3–12
17. R.A. Prado, L.E. Murr, A.R. Rivera, D.J. Shindo, K.F. Soto, G.J. Fernandez, and J.C. McClure, Tool Wear and Shape Optimization in the Friction-Stir Welding of Aluminum Metal-Matrix Composite, *Friction Stir Welding and Processing II,* K.V. Jata, M.W. Mahoney, R.S. Mishra, S.L. Semiatin, and T. Lienert, Ed., TMS, 2003, p 23–32
18. T.J. Lienert, W.L. Stellwag, Jr., and L.R. Lehman, Heat Inputs, Peak Temperatures, and Process Efficiencies for FSW, *Proceedings of the Fourth International Conference on Friction Stir Welding,* May 14–16, 2003 (Park City, UT), TWI, paper on CD
19. G. Luan, C. Sun, H. Guo, and Y. Yu, Friction Stir Welding of Pure Aluminum and Pure Copper, *Proceedings of the Fourth International Conference on Friction Stir Welding,* May 14–16, 2003 (Park City, UT), TWI, paper on CD
20. M.A. Sutton, A.P. Reynolds, B. Yang, and R. Taylor, Mixed Mode I/II Fracture of 2024-T3 Friction Stir Welds, *Eng. Fract. Mech.,* Vol 70, 2003, p. 2215–2234
21. Z.W. Chen and R. Maginness, Formation of Weld Zones During Friction Stir Welding of Aluminum Alloys, *Proceedings of the Fifth International Conference on Friction Stir Welding,* Sept 14–16, 2004 (Metz, France), TWI, paper on CD
22. T. Hishihara and Y. Nagasaka, Development of Micro-FSW, *Proceedings of the Fifth International Conference on Friction Stir Welding,* Sept 14–16, 2004 (Metz, France), TWI, paper on CD
23. H. Fujii, Y. Takada, N. Tsuji, and K. Nogi, Friction Stir Welding of Ultrafine Grained Materials, *Proceedings of the Fifth International Conference on Friction Stir Welding,* Sept 14–16, 2004 (Metz, France), TWI, paper on CD
24. J.H. Record, J.L. Covington, T.W. Nelson, C.D. Sorensen, and B.W. Webb, Fundamental Characterization of Friction Stir Welding, *Proceedings of the Fifth International Conference on Friction Stir Welding,* Sept 14–16, 2004 (Metz, France), TWI, paper on CD
25. K. Savolanen, J. Mononen, T. Saukkonen, H. Hänninen, and J. Koivula, Friction Stir Weldability of Copper Alloys, *Proceedings of the Fifth International Conference on Friction Stir Welding,* Sept 14–16, 2004 (Metz, France), TWI, paper on CD
26. M.A. Sutton, A.P. Reynolds, J. Yan, B. Yang, and N. Yuan, Microstructure and Mixed Mode I/II Fracture of AA2524-T351 Base Material and Friction Stir Welds, *Eng. Fract. Mech.,* Vol 73 (No. 4), 2006, p 391–407
27. A.L. Stahl and C.D. Sorensen, Experimental Measurements of Load Distributions on Friction Stir Welding Pin Tools, *Friction Stir Welding and Processing III,* K.V. Jata, M.W. Mahoney, R.S. Mishra, and T.J. Lienert, Ed., TMS, 2005, p 179–190
28. C.G. Andersson and R.E. Andrews, Fabrication of Containment Canisters for Nuclear Waste by Friction Stir Welding, *Proceedings of the First International Conference on Friction Stir Welding,* June 14–16, 1999 (Thousand Oaks, CA), TWI, paper on CD
29. S.H. Kazi and L.E. Murr, Complex Flow Phenomena Associated with Friction-Stir Welding of Aluminum Alloys, *Friction Stir Welding and Processing,* K.V. Jata, M.W. Mahoney, R.S. Mishra, S.L. Semiatin, and D.P. Field, Ed., TMS, 2001, p 139–150
30. M. Guerra, J.C. McClure, L.E. Murr, and A.C. Nunes, Metal Flow During Friction Stir Welding, *Friction Stir Welding and Processing,* K.V. Jata, M.W. Mahoney, R.S. Mishra, S.L. Semiatin, and D.P. Field, Ed., TMS, 2001, p 25–34
31. O.T. Midling and G. Rorvik, Effect of Tool Shoulder Material on Heat Input During Friction Stir Welding, *Proceedings of the First International Conference on Friction Stir Welding,* June 14–16, 1999 (Thousand Oaks, CA), TWI, paper on CD

32. P.A. Colegrove and H.R. Shercliff, Experimental and Numerical Analysis of Aluminum Alloy 7075-T7351 Friction Stir Welds, *Sci. Technol. Weld. Join.*, Vol 8 (No. 5), 2003, p 360–368
33. L. Cederqvist, A Weld That Lasts for 100,000 Years: FSW of Copper Canisters, *Proceedings of the Fourth International Conference on Friction Stir Welding,* May 14–16, 2003 (Park City, UT), TWI, paper on CD
34. L. Cederqvist, FSW to Seal 50 mm Thick Copper Canisters—A Weld That Lasts for 100,000 Years, *Proceedings of the Fifth International Conference on Friction Stir Welding,* Sept 14–16, 2004 (Metz, France), TWI, paper on CD
35. S.P. Vaze, J. Xu, R.J. Ritter, K.J. Colligan, J.J. Fisher, Jr., and J.R. Pickens, Friction Stir Processing of Aluminum Alloy 5083 Plate for Cold Bending, *Mater. Sci. Forum,* Vol 426–432, 2003, p 2979–2986
36. F.R. Morral, Ed., Wrought Superalloys, *Properties and Selection: Stainless Steels, Tool Materials, and Special-Purpose Metals,* Vol 3, *Metals Handbook,* 9th Ed., American Society for Metals, 1980, p 207–237
37. "MP159 Data Sheet," trade literature, Timken Latrobe Steel, Latrobe, PA, 1986
38. A.P. Reynolds, W. Tang, T. Gnaupel-Herold, and H. Prask, Structure, Properties, and Residual Stress of 304L Stainless Steel Friction Stir Welds, *Scr. Metall.,* Vol 48, 2003, p 1289–1294
39. T. Lienert, W. Tang, J.A. Hogeboom, and L.G. Kvidahi, Friction Stir Welding of DN-36 Steel, *Proceedings of the Fourth International Conference on Friction Stir Welding,* May 14–16, 2003 (Park City, UT), TWI, paper on CD
40. P. Konkol, Characterization of Friction Stir Weldments in 500 Brinell Hardness Quenched and Tempered Steel, *Proceedings of the Fourth International Conference on Friction Stir Welding,* May 14–16, 2003 (Park City, UT), TWI, paper on CD
41. M.W. Mahoney, W.H. Bingel, S.R. Sharma, and R.S. Mishra, Microstructural Modification and Resultant Properties of Friction Stir Processed Cast NiAl Bronze, *Mater. Sci. Forum,* Vol 426–432, 2003, p 2843–2848
42. Z. Loftus, J. Takeshita, A. Reynolds, and W. Tang, An Overview of Friction Stir Welding Beta 21S Titanium, *Proceedings of the Fifth International Conference on Friction Stir Welding,* Sept 14–16, 2004 (Metz, France), TWI, paper on CD
43. M. Hirakawa, H. Yamamoto, T. Shinoda, and H. Takegami, Mechanical Properties of Friction Stir Welding Joint for Mild Steel, *Proceedings of the Fifth International Conference on Friction Stir Welding,* Sept 14–16, 2004 (Metz, France), TWI, paper on CD
44. B. London, J. Fino, A. Pelton, C. Fuller, and M. Mahoney, Friction Stir Processing of Nitinol, *Friction Stir Welding and Processing III,* K.V. Jata, M.W. Mahoney, R.S. Mishra, and T.J. Lienert, Ed., TMS, 2005, p 67–74
45. B.K. Jasthi, S.M. Howard, W.J. Arbegast, G.J. Grant, S. Koduri, and D.R. Herling, Friction Stir Welding of Ma957 Oxide Dispersion Strengthened Ferritic Steel, *Friction Stir Welding and Processing III,* K.V. Jata, M.W. Mahoney, R.S. Mishra, and T.J. Lienert, Ed., TMS, 2005, p 75–79
46. Y. Li, E. Hood, W. Tang, and A.P. Reynolds, Friction Stir Welding of TiMetal 21S, *Friction Stir Welding and Processing III,* K.V. Jata, M.W. Mahoney, R.S. Mishra, and T.J. Lienert, Ed., TMS, 2005, p 81–89
47. L. Christodoulou, W. Palko, and C. Fuller, Equipment and Processing Variables Affecting Friction Stir Processing of NiAl Bronze, *Friction Stir Welding and Processing III,* K.V. Jata, M.W. Mahoney, R.S. Mishra, and T.J. Lienert, Ed., TMS, 2005, p 57–66
48. M.W. Mahoney, C.B. Fuller, W.H. Bingel, and M. Calabrese, "Friction Stir Processing of Cast NiAl Bronze," THERMEC 2006, July 4–8, 2006 (Vancover, Canada), TMS
49. "Densimet Data Sheet," trade literature, Plansee GmbH, Germany
50. A. Luo, K.S. Shin, and D.L. Jacobson, Hafnium Carbide Strengthening in a Tungsten-Rhenium Matrix at Ultrahigh Temperatures, *Acta Metal. Mater.,* Vol 40 (No. 9), 1992, p 2225–2232
51. E.M. Savitskii, M.A. Tylkina, and A.M. Levin, Rhenium Alloys in Electronics, *Izd. Energ.,* 1980, p 216

52. "W-1%$La_2O_3$ Data Sheet," trade literature, Plansee GmbH, Germany
53. H.J. Liu, H. Fujii, and K. Nogi, Wear Behavior of Hard Alloy Tools in the Friction Stir Welding of AC4A + 30 vol.% SiCp Aluminum Matrix Composite, *Proceedings of the Fifth International Conference on Friction Stir Welding,* Sept 14–16, 2004 (Metz, France), TWI, paper on CD
54. T. Yasui, T. Ishii, Y. Shimoda, M. Tsubaki, M. Fukumoto, and T. Shinoda, Friction Stir Welding Between Aluminum and Steel with High Welding Speed, *Proceedings of the Fifth International Conference on Friction Stir Welding,* Sept 14–16, 2004 (Metz, France), TWI, paper on CD
55. M. Fukumoto, T. Yasui, Y. Shimoda, M. Tsubaki, and T. Shinoda, Butt Welding Between Dissimilar Metals by Friction Stirring, *Proceedings of the Fifth International Conference on Friction Stir Welding,* Sept 14–16, 2004 (Metz, France), TWI, paper on CD
56. C.D. Sorensen, T.W. Nelson, and S.M. Packer, Tool Material Testing for FSW of High-Temperature Alloys, *Proceedings of the Third International Conference on Friction Stir Welding,* Sept 27–28, 2001 (Kobe, Japan), TWI, paper on CD
57. C.D. Sterling, T.W. Nelson, C.D. Sorensen, R.J. Steel, and S.M. Packer, Friction Stir Welding of Quenched and Tempered C-Mn Steel, *Friction Stir Welding and Processing II,* K.V. Jata, M.W. Mahoney, R.S. Mishra, S.L. Semiatin, and T. Lienert, Ed., TMS, 2003, p 165–171
58. S. Packer, T. Nelson, C. Sorensen, R. Steel, and M. Matsunaga, Tool and Equipment Requirements for Friction Stir Welding Ferrous and Other High Melting Temperature Alloys, *Proceedings of the Fourth International Conference on Friction Stir Welding,* May 14–16, 2003 (Park City, UT), TWI, paper on CD
59. K. Okamoto, S. Hirano, M. Inagaki, S.C. Park, Y.S. Sato, H. Kokawa, T.W. Nelson, and C.D. Sorensen, Metallurgical and Mechanical Properties of Friction Stir Welded Stainless Steels, *Proceedings of the Fourth International Conference on Friction Stir Welding,* May 14–16, 2003 (Park City, UT), TWI, paper on CD
60. M. Collier, R. Steel, T. Nelson, C. Sorensen, and S. Packer, Grade Development of Polycrystalline Cubic Boron Nitride for Friction Stir Processing of Ferrous Alloys, *Mater. Sci. Forum,* Vol 426–432, 2003, p 3011–3016
61. S. Lynch, D.P. Edwards, A. Majumdar, S. Moutsos, and M.W. Mahoney, Friction-Stir Processing of a High-Damping Mn-Cu Alloy Used for Marine Propellers, *Mater. Sci. Forum,* Vol 426–432, 2003, p 2903–2908
62. S. Hirano, K. Okamoto, K. Aota, and M. Inagaki, Metallurgical and Mechanical Properties of Friction Stir Welded Ultra Fine Grained Steel, *Proceedings of the Fifth International Conference on Friction Stir Welding,* Sept 14–16, 2004 (Metz, France), TWI, paper on CD
63. S.H.C. Park, Y.S. Sato, H. Kokawa, K. Okamoto, S. Hirano, and M. Inagaki, Effect of Microstructure on Corrosion of Friction Stir Welded 304 Stainless Steel, *Proceedings of the Fifth International Conference on Friction Stir Welding,* Sept 14–16, 2004 (Metz, France), TWI, paper on CD
64. S.H.C. Park, Y.S. Sato, H. Kokawa, K. Okamoto, S. Hirano, and M. Inagaki, Corrosion Resistance of Friction Stir Welded 304 Stainless Steel, *Scr. Metall.,* Vol 51, 2004, p 101–105
65. M.P. Miles, J. Pew, T.W. Nelson, and M. Li, Formability of Friction Stir Welded Dual Phase Steel Sheets, *Friction Stir Welding and Processing III,* K.V. Jata, M.W. Mahoney, R.S. Mishra, and T.J. Lienert, Ed., TMS, 2005, p 91–96
66. B.M. Tweedy, W. Arbegast, and C. Allen, Friction Stir Welding of Ferrous Alloys Using Induction Preheating, *Friction Stir Welding and Processing III,* K.V. Jata, M.W. Mahoney, R.S. Mishra, and T.J. Lienert, Ed., TMS, 2005, p 97–104
67. Y.S. Sato, T.W. Nelson, and C.J. Sterling, Recrystallization in Type 304L Stainless Steel During Friction Stirring, *Acta Mater.,* Vol 53, 2005, p 637–645
68. U. Ramasubramanian, W.J. Arbegast, G.A. Stone, and G. Grant, Friction Stir Processing of Class 40 Grey Cast Iron, *Friction Stir Welding and Processing III,* K.V. Jata, M.W. Mahoney, R.S. Mishra, and T.J. Lienert, Ed., TMS, 2005, p 115–122

69. S.M. Howard, B.K. Jasthi, W.J. Arbegast, G.J. Grant, and D.R. Herling, Friction Surface Reaction Processing in Aluminum Substrates, *Friction Stir Welding and Processing III,* K.V. Jata, M.W. Mahoney, R.S. Mishra, and T.J. Lienert, Ed., TMS, 2005, p 139–146
70. C. Bird, Ultrasonic Phased Array Inspection Technology for the Evaluation of Friction Stir Welds, *Proceedings of the Fourth International Conference on Friction Stir Welding,* May 14–16, 2003 (Park City, UT), TWI, paper on CD
71. A.J. Leonard and S.A. Lockyer, Flaws in Friction Stir Welds, *Proceedings of the Fourth International Conference on Friction Stir Welding,* May 14–16, 2003 (Park City, UT), TWI, paper on CD
72. M.A. Sutton, A.P. Reynolds, B. Yang, and R. Taylor, Mode I Fracture and Microstructure for 2024-T3 Friction Stir Welds, *Mater. Sci. Eng. A,* Vol 354, 2003, p 6–16
73. P. Volovitch, J.E. Masse, T. Baudin, B. Da Costa, J.C. Goussain, W. Saikaly, and L. Barrallier, Microstructure and Mechanical Properties of Friction Stir Welded Mg Alloy AZ91, *Proceedings of the Fifth International Conference on Friction Stir Welding,* Sept 14–16, 2004 (Metz, France), TWI, paper on CD
74. M. Attallah and H.G. Salem, Effect of Friction Stir Welding Process Parameters on the Mechanical Properties of the As-Welded and Post-Weld Heat Treated AA2095, *Proceedings of the Fifth International Conference on Friction Stir Welding,* Sept 14–16, 2004 (Metz, France), TWI, paper on CD
75. J. Lumsden, G. Pollock, and M. Mahoney, Effect of Tool Design on Stress Corrosion Resistance of FSW AA7050-T7451, *Friction Stir Welding and Processing III,* K.V. Jata, M.W. Mahoney, R.S. Mishra, and T.J. Lienert, Ed., TMS, 2005, p 19–25
76. R. Zettler, S. Lomolino, J.F. dos Santos, T. Donath, F. Beckmann, T. Lipman, and D. Lohwasser, A Study of Material Flow in FSW of AA2024-T351 and AA 6056-T4 Alloys, *Proceedings of the Fifth International Conference on Friction Stir Welding,* Sept 14–16, 2004 (Metz, France), TWI, paper on CD
77. I. Charit and R.S. Mishra, Low Temperature Superplasticity in a Friction Stir Processed Ultrafine Grained Al-Zn-Mg-Sc Alloy, *Acta Mater.,* Vol 53, 2005, p 4211–4223
78. J. Yan, M.A. Sutton, and A.P. Reynolds, Process-Structure-Property Relationships for Nugget and Heat Affected Zone Regions of AA2524-T351 Friction Stir Welds, *Sci. Technol. Weld. Join.,* Vol 10 (No. 6), 2005, p 725–736
79. W.M. Thomas et al., Friction Stir Welding, U.K. Patent Application 2,306,366, Oct 17, 1996
80. C.J. Dawes and W.M. Thomas, Development of Improved Tool Designs for Friction Stir Welding of Aluminum, *Proceedings of the First International Conference on Friction Stir Welding,* June 14–16, 1999 (Thousand Oaks, CA), TWI, paper on CD
81. R.W. Fonda, J.F. Bingert, and K.J. Colligan, Texture and Grain Evolutions in a 2195 Friction Stir Weld, *Proceedings of the Fifth International Conference on Friction Stir Welding,* Sept 14–16, 2004 (Metz, France), TWI, paper on CD
82. K. Colligan, Tapered Friction Stir Welding Tool, U.S Patent 6,669,075, Dec 30, 2003
83. K.J. Colligan and J.R. Pickens, Friction Stir Welding of Aluminum Using a Tapered Shoulder Tool, *Friction Stir Welding and Processing III,* K.V. Jata, M.W. Mahoney, R.S. Mishra, and T.J. Leinert, Ed., TMS, 2005, p 161–170
84. S. Packer and R. Steel, "Megastir Technologies," unpublished research, 2006
85. C.J. Dawes, P.L. Threadgill, E.J.R. Spurgin, and D.G. Staines, "Development of the New Friction Stir Technique for Welding Aluminum—Phase II," TWI member report 5651/35/95, Nov 1995
86. M. Mahoney and W. Bingel, "Tool Development for Friction Stir Welding," Rockwell Scientific Co. internal report, Thousand Oaks, CA, 2002
87. C.B. Fuller, M.W. Mahoney, W.H. Bingel, M. Calabrese, and B. London, "Tensile and Fatigue Properties of Friction Stir Processed NiAl Bronze," THERMEC 2006, July 4–8, 2006 (Vancouver, Canada), TMS
88. W.M. Thomas, E.D. Nicholas, and S.D.

Smith, Friction Stir Welding-Tool Developments, *Aluminum 2001, Proceedings of the TMS 2001 Aluminum Automotive and Joining Sessions,* S.K. Das, J.G. Kaufman, and T.J. Lienert, Ed., TMS, 2001, p 213

89. W.M. Thomas, D.G. Staines, I.M. Norris, and R. de Frias, "Friction Stir Welding-Tools and Developments," FSW Seminar, Dec 3, 2002 (Porto, Portugal), Instituto de Soldadura e Qualidade
90. M. Strangwood, C. Davis, and M. Attallah, Microstructural Development and Thermal Fields Modelling in Friction Stir Welds of Strain-Hardenable Al-Alloys, *Proceedings of the Fifth International Conference on Friction Stir Welding,* Sept 14–16, 2004 (Metz, France), TWI, paper on CD
91. Z.Y. Ma, S.R. Sharma, R.S. Mishra, and M.W. Mahoney, Microstructural Modification of Cast Aluminum Alloys via Friction Stir Processing, *Mater. Sci. Forum,* Vol 426–432, 2003, p 2891–2896
92. P.A. Colegrove, H.R. Shercliff, and P.L. Threadgill, Modeling and Development of the Trivex Friction Stir Welding Tool, *Proceedings of the Fourth International Conference on Friction Stir Welding,* May 14–16, 2003 (Park City, UT), TWI, paper on CD
93. P.A. Colegrove and H.R. Shercliff, Development of Trivex Friction Stir Welding Tool, Part 1: Two-Dimensional Flow Modeling, *Sci. Technol. Weld. Join.,* Vol 9 (No. 4), 2004, p 345–351
94. R.J. Ding and P.A. Oelgoetz, Autoadjustable Pin Tool for Friction Stir Welding, U.S. Patent 5893507, April 13, 1999
95. R.J. Ding, Force Characterization on the Welding Pin of a Friction Stir Welding Retractable Pin-Tool Using Aluminum-Lithium 2195, *Proceedings of the Second International Conference on Friction Stir Welding,* June 26–28, 2000 (Gothenburg, Sweden), TWI, paper on CD
96. M. Skinner and R.L. Edwards, Improvements to the FSW Process Using the Self-Reacting Technology, *Mater. Sci. Forum,* Vol 426–432, 2003, p 2849–2854
97. G. Sylva, R. Edwards, and T. Sassa, A Feasibility Study for Self Reacting Pin Tool Welding of Thin Section Aluminum, *Proceedings of the Fifth International Conference on Friction Stir Welding,* Sept 14–16, 2004 (Metz, France), TWI, paper on CD
98. F. Marie, D. Allehaux, and B. Esmiller, Development of the Bobbin Tool Technique on Various Aluminum Alloys, *Proceedings of the Fifth International Conference on Friction Stir Welding,* Sept 14–16, 2004 (Metz, France), TWI, paper on CD
99. K. Erbsloh, C. Dalle Donne, and D. Lohwasser, Fiction Stir Welding of T-Joints, *Mater. Sci. Forum,* Vol 426–432, 2003, p 2965–2970
100. S. Kallee, D. Nicholas, and W. Thomas, Friction Stir in Aerospace—The Industrial Way, *TWI Bull.,* Vol 43 (No. 3), May/June 2002, p 13–16
101. M.J. Brooker, A.J.M. van Deudekom, S.W. Kallee, and P.D. Sketchley, Applying Friction Stir Welding to the Arian 5 Main Motor Thrust Frame, *Proceedings of the Second International Conference on Friction Stir Welding,* June 26–28, 2000 (Gothenburg, Sweden), TWI, paper on CD
102. L. Cederqvist and A.P. Reynolds, Properties of Friction Stir Welded Aluminum Lap Joints, *Proceedings of the Second International Conference on Friction Stir Welding,* June 26–28, 2000 (Gothenburg, Sweden), TWI, paper on CD
103. O.K. Mishina and A. Norlin, Lap Joints Produced by FSW on Flat Aluminum EN AW-6082 Profiles, *Proceedings of the Fourth International Conference on Friction Stir Welding,* May 14–16, 2003 (Park City, UT), TWI, paper on CD
104. M. Ericsson and R. Sandström, Fatigue of FSW Overlap Joints in Aluminum Welded with Different Tool Designs, *Proceedings of the Fifth International Conference on Friction Stir Welding,* Sept 14–16, 2004 (Metz, France), TWI, paper on CD
105. P. Colegrove, H.R. Shercliff, and T. Hyoe, Development of the Trivex Friction Stir Welding Tool for Making Lap Welds, *Proceedings of the Fifth International Conference on Friction Stir Welding,* Sept 14–16, 2004 (Metz, France), TWI, paper on CD
106. W.M. Thomas, A.B.M. Braithwaite, and R. John, Skew-Stir Technology, *Proceed-*

*ings of the Third International Conference on Friction Stir Welding,* Sept 27–28, 2001 (Kobe, Japan), TWI, paper on CD

107. G.M.D. Cantin, S.A. David, W.M. Thomas, E. Lara-Curzio, and S.S. Babu, Friction Skew-Stir Welding of Lap Joints in 5083-O Aluminum, *Sci. Technol. Weld. Join.,* Vol 10 (No. 3), 2005, p 268–280
108. W.M. Thomas, D.G. Staines, K.I. Johnson, and P. Evans, "Com-Stir-Compound Motion for Friction Stir Welding and Machining," TWI Ltd. report, Cambridge, U.K., March 6, 2003
109. W.M. Thomas, "Friction Stir Welding—Recent Developments," THERMEC 2003, July 7–11, 2003 (Madrid, Spain), TMS
110. W.M. Thomas, C.J. Goodfellow, D.G. Staines, and E.R. Watts, "Preliminary Trials Using Dual-Rotation Stir Welding," Romanian Welding Society Conference, Sept 28–30, 2005 (Galati, Romania)
111. W.M. Thomas, D.J. Staines, E.R. Watts, and I.M. Norris, "The Simultaneous Use of Two or More Friction Stir Welding Tools," TWI Ltd. report, Cambridge, U.K., Jan 13, 2005
112. W.M. Thomas, Friction Stir Welding of Ferrous Materials: A Feasibility Study, *Proceedings of the First International Conference on Friction Stir Welding,* June 14–16, 1999 (Thousand Oaks, CA), TWI, paper on CD
113. H. Mitsuo, Friction Agitation Joining Method and Friction Agitation Joining Device, Japan Patent Application 2000-094156, April 4, 2000
114. K. Atsuo, O. Yoshinori, and Y. Mutsumi, Friction Stir Welding Method, Japan Patent Application 2003-112271 and 2003-112272, April 15, 2003
115. M. Peel, A. Steuwer, M. Preuss, and P.J. Withers, Microstructure, Mechanical Properties and Residual Stresses as a Function of Welding Speed in Aluminum AA5083 Friction Stir Welds, *Acta Mater.,* Vol 51, 2003, p 4791–4801
116. C. Schilling and J. dos Santos, Method and Device for Joining at Least Two Adjoining Work Pieces by Friction Welding, U.S. Patent 6,722,556, April 20, 2004; DE Patent Application 199-55-737, 18, 1999
117. S.S.T. Kakarla, K.H. Muci-Küchler, W.A. Arbegast, and C.D. Allen, Three-Dimensional Finite Element Model of the Friction Stir Spot Welding Process, *Friction Stir Welding and Processing III,* K.V. Jata, M.W. Mahoney, R.S. Mishra, and T.J. Lienert, Ed., TMS, 2005, p 213–220
118. R. Sakano, K. Murakami, K. Yamashita, T. Hyoe, M. Fujimoto, M. Inuzuka, Y. Nagao, and H. Kashiki, Development of FSW Robot System for Automobile Body Members, *Proceedings of the Third International Conference on Friction Stir Welding,* Sept 27–28, 2001 (Kobe, Japan), TWI, paper on CD
119. J.F. Hinrichs, C.B. Smith, B.F. Orsini, R.J. DeGeorge, B.J. Smale, and P.C. Ruehl, Friction Stir Welding for the 21st Century Automotive Industry, *Proceedings of the Fifth International Conference on Friction Stir Welding,* Sept 14–16, 2004 (Metz, France), TWI, paper on CD
120. A.C. Addison and A.J. Robelou, Friction Stir Spot Welding: Principal Parameters and Their Effects, *Proceedings of the Fifth International Conference on Friction Stir Welding,* Sept 14–16, 2004 (Metz, France), TWI, paper on CD
121. T. Shinoda and M. Kawai, Proposals of Novel Surface Modification Technology Using Friction Stir Welding Phenomenon, *Mater. Sci. Forum,* Vol 426–432, 2003, p 2837–2842
122. C. Fuller, M. Mahoney, and W. Bingel, A Study of Friction Stir Processing Tool Designs for Microstructural Modifications as Demonstrated by Aluminum Fusion Welds, *Proceedings of the Fifth International Conference on Friction Stir Welding,* Sept 14–16, 2004 (Metz, France), TWI, paper on CD
123. B.N. Bhat, R.W. Carter, R.J. Ding, K.G. Lawless, A.C. Nunes, Jr., C.K. Russell, and S.R. Shah, Friction Stir Welding Development at NASA-Marshall Space Flight Center, *Friction Stir Welding and Processing,* K.V. Jata, M.W. Mahoney, R.S. Mishra, S.L. Semiatin, and D.P. Field, Ed., TMS, 2001, p 117–128
124. T. Trapp, E. Helder, and P.R. Subramanian, FSW of Titanium Alloys for Aircraft Engine Components, *Friction Stir Welding and Processing II,* K.V. Jata, M.W. Mahoney, R.S. Mishra, S.L. Semiatin, and T. Lienert, Ed., TMS, 2003, p 173–176
125. O. Midling, Modified Friction Stir Weld-

ing, International Patent Application PCT /NO99/000042

126. G. Kohn, W. Greeberg, I. Makover, and A. Munitz, Laser-Assisted Friction Stir Welding, *Weld. J.*, Vol 81, 2002, p 46–48

127. X. Long and S.K. Khanna, Modeling of Electrically Enhanced Friction Stir Welding Process Using Finite Element Method, *Sci. Technol. Weld. Join.*, Vol 10 (No. 4), 2005, p 482–487

## CHAPTER 3

# Temperature Distribution and Resulting Metal Flow

J.A. Schneider, Mechanical Engineering Department
Mississippi State University

A PHYSICAL UNDERSTANDING of the friction stir welding (FSW) process can be described by combining the complementary efforts of experimental examination and analytical modeling. Early experimental work on FSW was done primarily to refine, not understand, the FSW process. As the process was refined, attention turned to understanding the mechanisms of joint formation and how they were influenced by weld process parameters, tool design, and materials. Generalized assessments were made of the temperature field during welding and the path of material flow. The next phase of development has been to quantify the effects of process parameters, tool design, and materials on the temperature and flow path. Research still remains to determine the level of plasticity required for the FSW process to be effective and the role of flow mixing in obtaining a good weld.

The coupling of thermal and mechanical work in the FSW process produces an asymmetrical weld nugget. Heating softens the metal for the subsequent stirring and/or extrusion processes. The conventional FSW tool, discussed in Chapter 2, "Friction Stir Tooling: Tool Materials and Designs," incorporates a shoulder and pin. The weld tool may form an angle with the workpiece, especially if the shoulder is smooth. To form a butt weld, two metal plates are clamped to a backing anvil. The shoulder rides on the surface of the metal plates being joined, while the pin penetrates into the metal plate thickness. Unique to this process is the generation of heat produced by friction between the tool and workpiece and from plastic dissipation within the workpiece.

In spite of the promise of this joining technique, very little information exists on actual material behavior under FSW conditions. The vast differences in the pin tool geometry and materials used in the various experimental and modeling studies have made it difficult to correlate the processing parameters with the microstructure development. However, some important aspects of FSW formation mechanisms have been illuminated that provide an effective framework for more focused investigations into some of the fundamentals of the joining process.

## 3.1 Generation of Heat

In FSW, heat is generated by a combination of friction and plastic dissipation during deformation of the metal. The dominating heat-generation mechanism is influenced by the weld parameters, thermal conductivities of the workpiece, pin tool and backing anvil, and the weld tool geometry. General guidelines apply to selection of the weld parameters that empirically correlate hotter welds with high rpm and low travel speed, and colder welds with low rpm and high travel speeds. The temperature field around the pin tool is asymmetric, with slightly higher temperatures reported on the retreating side (RS) of the FSW in aluminum alloys (Ref 1). This correlates with tensile test failures that occur predominantly on the RS of the FSW in the heat-affected zone (HAZ) region (Ref 2). To avoid overheating in welds at higher rotational speeds (>15,000 rpm), successful welds have been made with a nonrotating shoulder (Ref 3).

Early experimental studies showed that the majority of the heat generation occurs at the shoulder/workpiece interface (Ref 4). The controlling mechanism of heating is due to either friction or plastic dissipation, depending on the contact conditions between the two surfaces. The weld tool geometric features of both the pin and the shoulder influence whether the two surfaces slide, stick, or alternate between the two modes. More recent analytical studies have indicated that the heat generated between the pin tool and the workpiece is not insignificant and should be included in defining the heat field.

Mechanisms of heat generation between the pin tool and the workpiece are also due to friction or plastic dissipation, depending on whether slide or stick conditions prevail at the interface. The amount of heat input from deformational heating around the pin tool has been estimated to range from 2% (Ref 5) to 20% (Ref 6).

Two experimental approaches have been reported toward understanding the temperature field generated during FSW. The first is interpretation of the microstructure by comparison with aging curves for the alloy investigated. Transmission electron microscopy studies (Ref 7–10) have attempted to correlate the precipitation sequence of the microstructure in 6061 (Ref 7, 8), 6063 (Ref 9), and 2195 (Ref 10) aluminum alloys with the weld temperature of the metal. Variations in the precipitation state are reported that range from complete dissolution of precipitates in the weld nugget to incomplete dissolution, with the possibility of precipitate overaging. The temperature field during the FSW process is transient, thus making it difficult to correlate an exact temperature with precipitation sequence data obtained from the steady-state experimental aging curves. However, these microstructural studies do give an approximation of the maximum temperature in the weld of these aluminum alloys to be in the range of 723 to 753 K. No evidence of localized melting was reported in the microstructural studies cited (Ref 7–10).

Detailed temperature measurements with embedded thermocouples (TCs) have been used to map out the temperature field (Ref 4, 11–14). Interpretation of these measurements is affected by the coupled thermal conductivity of the workpiece, the backing anvil, and the weld tool. Table 3.1 summarizes the variation in peak temperature recorded in experimental studies in three aluminum alloys (Ref 4, 11, 14). The welds were made in panels with different thicknesses and different weld tool geometries. Embedded type K thermocouples were placed at various locations and depths in the weld panels.

Depending on the TC location, embedded TCs near the pin tool are generally consumed in the weld process. The measured data suggest that the region near the pin tool is nearly isothermal, indicating that the maximum temperature may occur in the shearing at the boundary of a rotating plug of metal around the weld pin tool (Ref 4). In thicker materials, measured temperature gradients suggest a limit to the depth affected by the surface heating of the shoulder/workpiece interaction (Ref 11).

Figure 3.1 summarizes the peak temperatures measured around the stirred zone as a function of distance from the stirred zone and through the

**Table 3.1 Measured temperatures in conventional friction stir welding of aluminum alloys**

| Thickness | | Shoulder diameter | | Pin diameter | | | | | |
|---|---|---|---|---|---|---|---|---|---|
| mm | in. | mm | in. | mm | in. | rpm | Travel mm/s | Peak temperature, K | Temperature difference (through thickness), K |
| **AA2195-T8 (Ref 11)** | | | | | | | | | |
| 8.1 | 0.32 | 25.4 | 1.0 | 10 | 0.4 | 240 | 2.4 | 693 | 60 |
| 8.1 | 0.32 | 25.4 | 1.0 | 10 | 0.4 | 240 | 3.3 | 698 | 60 |
| **AA6061-T6 (Ref 4)** | | | | | | | | | |
| 6.35 | 0.25 | 19 | 0.75 | 6.4 | 0.25 | 300 | 2 | 698 | None |
| 6.35 | 0.25 | 19 | 0.75 | 6.4 | 0.25 | 400 | 2 | 723 | None |
| 6.35 | 0.25 | 19 | 0.75 | 6.4 | 0.25 | 1000 | 2 | 750 | None |
| **AA70575-T651 (Ref 14)** | | | | | | | | | |
| 6.35 | 0.25 | N/A | N/A | | N/A | N/A | 2.1 | 748 | 60 |

thickness of a 6.35 mm (0.25 in.) thick AA7075-T651 plate. The temperature mapping shows the highest temperatures adjacent to the stir zone and near the top surface. A through-thickness temperature decrease is observed from the shoulder to the bottom, in addition to temperature decreases as the distance from the shear zone increases.

Recently, London and Mahoney (Ref 15) have made significant efforts to measure temperature in Ni-Al bronze. Figure 3.2 shows an illustrative example from this study. The measurement was taken from three locations in the nugget: centerline, advancing side, and retreating side. The peak temperature approaches 1000 °C (1830 °F) in Ni-Al bronze.

Because of the difficulty in obtaining spatial resolution with embedded TCs, the experimental data are often interpreted through the use of analytical and mathematical models. A reverse engineering approach is used to select the boundary conditions, either stick or slide or a combination, in an effort to match the temperature field obtained from the embedded TCs. The various process modeling approaches are discussed in Chapter 10, "Process Modeling."

As the temperature of the weld metal rises, the metal softens, torque is reduced, and less heat is imparted to the metal by mechanical work (Ref 16). This constitutes a temperature-regulating mechanism that tends to stabilize the temperature and avoid melting of the weld metal. Control of the temperature may occur by alternating the conditions at the interface between stick and slide. As the metal cools below a critical temperature, where the deformational flow stress rises above the frictional slip stress, the interaction between the weld tool and workpiece may change from deformational to frictional. If slide occurs between the weld tool and the workpiece, the heat input could decrease and reduce the temperature of the material (Ref 17). Alternating boundary conditions at the interface may act to destabilize the temperature and may cause stick-slide oscillations. Figure 3.3 illustrates how the boundary condition at the weld tool shoulder is theorized to affect the material flow nugget.

## 3.2 Metal Flow

The sharp temperature gradient at or near the tool/workpiece interface constrains the thermally softened, plasticized zone within the region bounded by the tool shoulder, anvil, and parent material. Weld parameters, coupled with the pin tool design and materials, control the volume of metal heated, of which a portion is then swept by the mechanical working portion

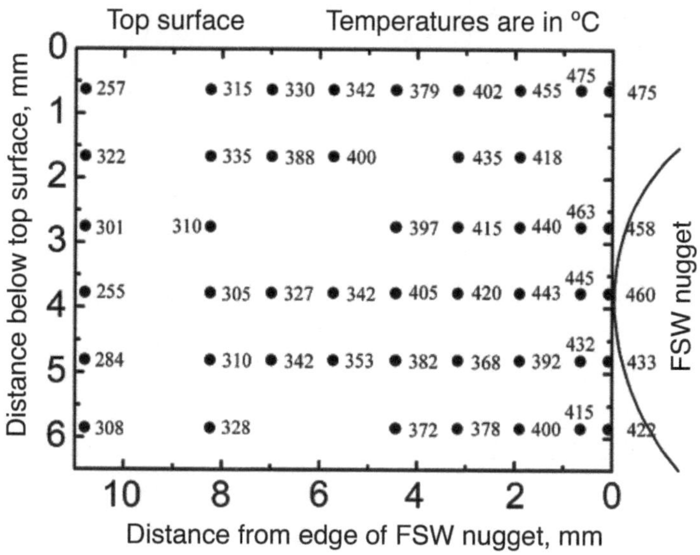


**Fig. 3.1** Peak temperature distribution adjacent to a friction stir weld (FSW) in 7075Al-T651. The line on the right side of the figure shows the nugget boundary. Source: Ref 14

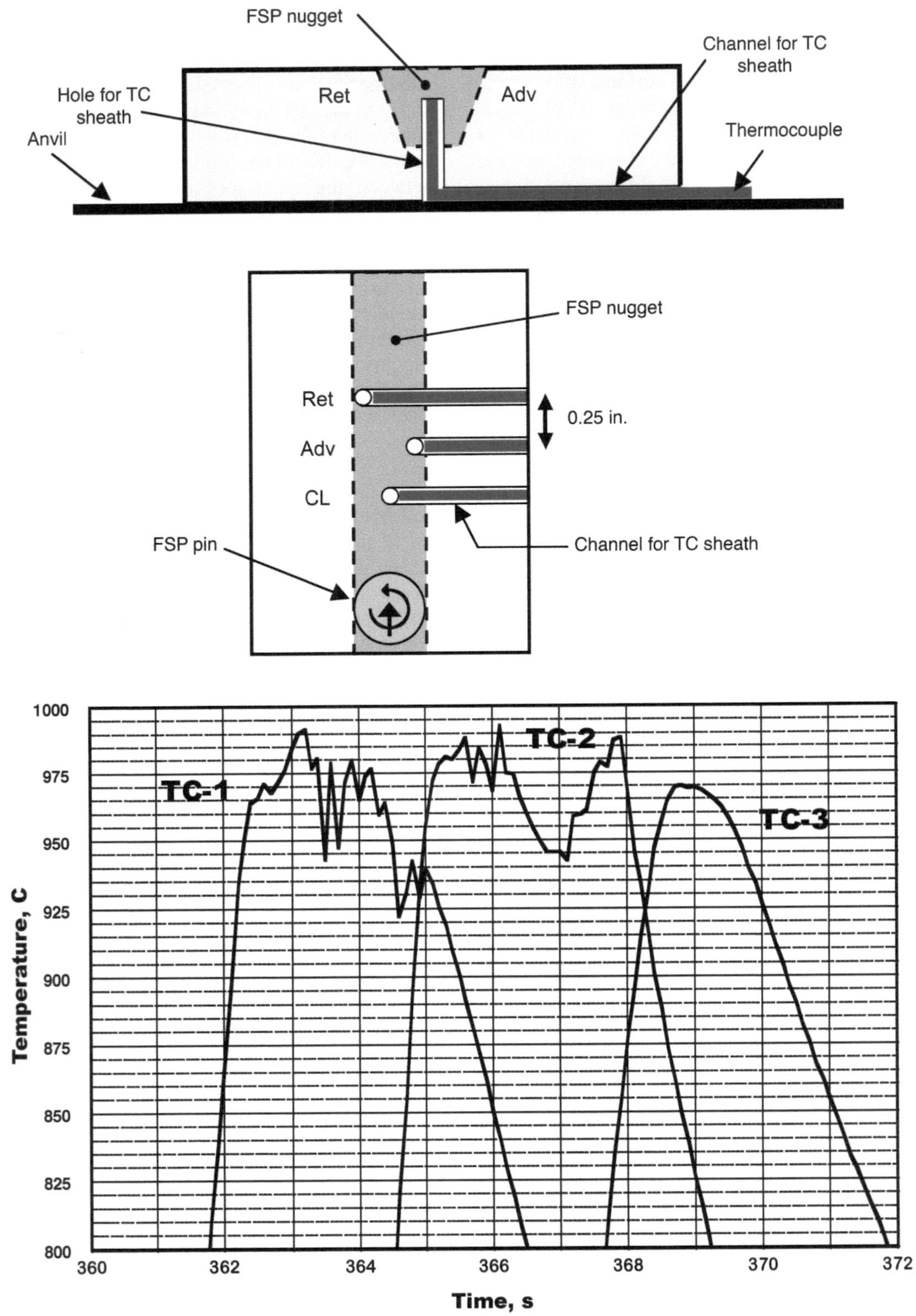


**Fig. 3.2** Location of thermocouples (TCs) and temperature plots showing maximum temperatures for friction stir process (FSP) 1429 (1000/6). TC-1: centerline; TC-2: advancing side; TC-3: retreating side. Source: Ref 15

of the process. The thermally softened material is transported around the tool in the direction of rotation and deposited in bands in the wake of the weld. Viewed in the plan section of an FSW, the spacing of the bands left in the wake of the FSW are equivalent to the longitudinal distance the weld tool travels during a single rotation, as illustrated in Fig. 3.4. Geometric and microstructural differences within the refined weld nugget reflect asymmetrical flow processes that occur around the weld centerline. This flow, or thermomechanical hot working of the metal in the weld zone, results in various microstructural evolutions that are discussed in Chapters 4, 6, 7, and 8, covering low- and high-melting-temperature alloys.

The microstructure of a transverse section of an FSW is presented in Fig. 3.5. The weld nugget is bounded by the HAZ and the thermomechanically affected zone. The generated heat controls the size of the swept volume, because hotter welds are reported to have a larger nugget than colder welds. The "onion ring" pattern (Ref 18) observed in the weld nugget in Fig. 3.5 is not always apparent in the weld macrostructure. Studies document visible patterns in colder welds, with no discernable ring pattern at hotter welds (Ref 19). The disappearance of the onion rings may result from slide conditions existing at the tool/workpiece interface at higher temperatures, when the FSW process becomes dominated by extrusion (Ref 18). Crystallographic orientation texture maps have shown the onion ring pattern corresponds to bands of shear-induced fiber texture in the weld nugget (Ref 20–22). Although the onion ring pattern is of benefit in interpreting the thermomechanical processing of the metal in the FSW process, there has been no reported correlation with the resulting quality of the weld nugget (Ref 4, 23).

Although the coupling between the metal flow, the heat-generation model, the weld tool material, and features of the shoulder and pin is complicated, some generalizations have been made regarding the mechanisms of the metal flow. Most of what is known about the deformation flow path is deduced from the asymmetric flow patterns inferred from tracer studies. Initial tracer studies used preferential etching to study the mixing of dissimilar alloys (Ref 23, 24). Definition of the flow paths in the FSW process was first obtained in a study by Colligan (Ref 25), in which the faying surface of the weld joint was embedded with 0.38 mm (0.015 in.) diameter steel balls placed at various linear positions through the weld thickness and to either side of the weld tool. Postweld positioning of the steel balls, as investigated by x-ray radiography, suggested an orderly flow of the metal around the pin tool. Based on the entrance into the weld zone, only some of the metal flow appeared to be forced downward by the threaded pin, while the rest appeared to be simply rotated from the front to the back of the pin tool (Ref 25).

Subsequent studies have looked at inserted copper foil, plated surfaces, and composite markers to further investigate these observations (Ref 26–28). All studies indicated that the flow was orderly, with the weld metal appearing to flow along defined paths or streamlines. Variations were observed in individual streamlines at

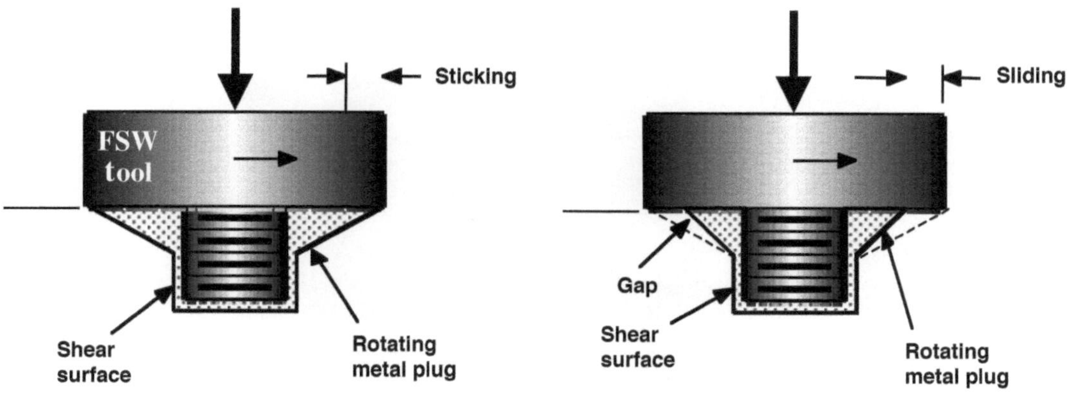


**Fig. 3.3** Alternating boundary conditions at the interface of the weld tool shoulder and the workpiece affect the boundary conditions for heat generation.

some weld parameters, with differences observed in the deposition dependent on advancing side (AS) versus RS insertion into the weld zone. These variations were attributed to metal either being stirred or extruded around the pin.

The marker approach has been further refined by placing 25 μm (1.0 mil) diameter wires at various positions within the weld panel (Ref 29). Figure 3.6 shows inverted x-ray radiographs of the postweld FSW segments, with the position of the wire marker enhanced. The white area in Fig. 3.6 denotes where the weld tool was removed at the E-stop termination of the weld. These FSWs were made using weld parameters of 36 kN (8100 lbf), 200 rpm, and 2 mm/s (0.08 in./s). The wire segments can be observed to follow streamlines that are either stirred, when introduced at the weld center or toward the AS, or extruded, when introduced toward the RS (Ref 29). The wire marker was introduced mid-material thickness into the weld nugget, with the pin tool located at the center of the panels in weld C08, whereas in weld C05, the pin tool was offset so the wire was introduced into the weld nugget at 3 mm (0.12 in.) RS. Figure 3.7 presents regular x-ray radiographs of the weld termination area. In C05, the side view of the weld shows minimal movement in the through-thickness position of the wire, whereas in C08, the wire can be observed to be pulled upward and deposited near the shoulder surface.

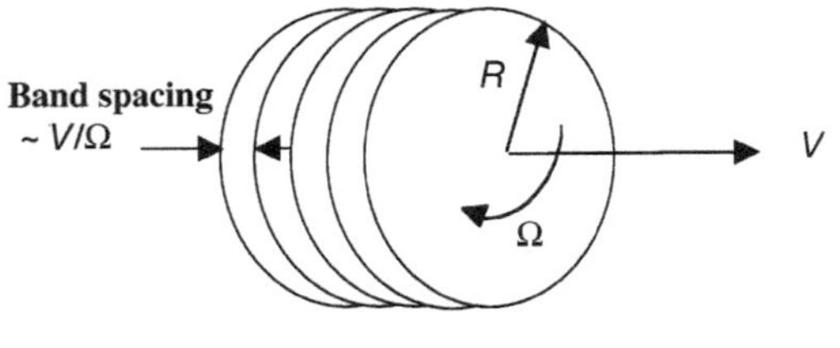


Fig. 3.4 The spacing of the bands, formed by weld material swept around the pin tool and deposited in the wake, is approximately equal to the longitudinal velocity (V) divided by the rotational speech (|).

Figure 3.8 presents the x-ray radiographs of a weld panel where the tracer wire was introduced 1.27 mm (0.05 in.) below the tool shoulder at the weld centerline. Unlike Fig. 3.6, the inverted x-ray radiographs shown in Fig. 3.8(a) and (b) show an unorganized scattering of tracer wire segments that range from the AS to the RS. The normal x-ray radiograph of the side view of the exit hole, Fig. 3.8(c), shows the tracer wire being drawn up toward the shoulder and then pushed downward, exiting in the wake at the AS of the weld panel.

Each band in the FSW zone of the welds in Fig. 3.6(b) contains one marker wire segment. In Fig. 3.8(c), where the wire markers show evidence of the metal being pushed downward from the shoulder surface to the anvil surface close to the pin tool, the scattered wire markers no longer correlate with the banded structure. A metallograph shown in Fig. 3.9. (Ref 30) shows the postweld positioning of a composite marker in which tracers of the marker can be observed upstream of the weld zone. These data suggest that not all the metal in an FSW zone is simply rotated around the pin tool, exiting in the wake of the FSW. Figure 3.8(c) shows that some of the metal is pushed downward in the material thickness direction. Figure 3.9 shows evidence that some metal may rotate multiple times around the weld pin tool before exiting in the wake.

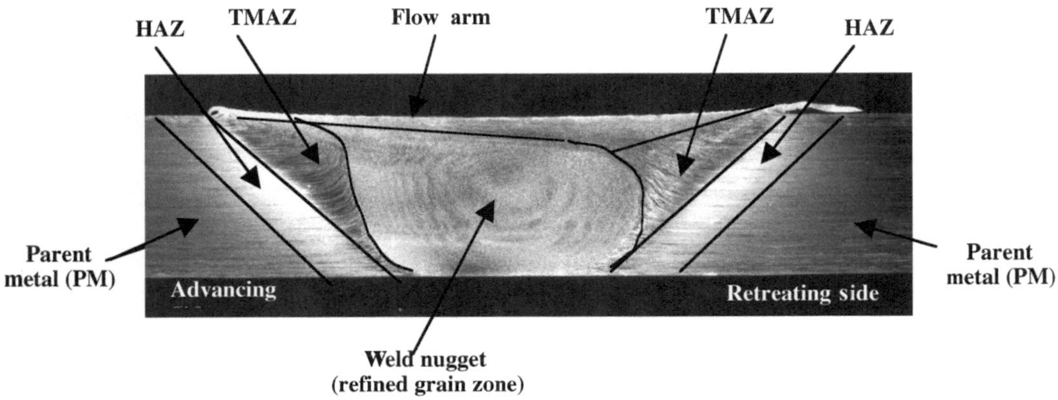


Fig. 3.5 Transverse section of a friction stir weld showing different regions of the weld. HAZ, heat-affected zone; TMAZ, thermo-mechanically affected zone

Computed tomography was used in a recent study to record the postweld position of a lead tracer wire (Ref 31). At the expected welding temperature of aluminum alloys, the lead wire would be molten. The constraint of the molten metal between the rotating plug of metal and the parent material allowed the lead to trace out a continuous flow path. The FSW was made with a 250 μm (10 mils) lead wire placed in the faying surfaces 1.3 mm (0.05 in.) from the shoulder surface and 6.4 mm (0.25 in.) offset to the AS in a 0.82 cm (0.32 in.) thick panel of 2195-T81 alloy. An inverted x-ray radiograph of the postweld position of a lead marker wire is presented in Fig. 3.10. Traces of the lead wire can be seen in the plan view that, similar to Fig. 3.8, do not correspond with the normal banded structure of the weld metal in the wake. Evidence of the movement of the lead wire through the material thickness can be observed in a side view of the weld panel, shown in Fig. 3.10(b). The initial placement of the lead wire can be observed on the AS of the exit hole of the weld tool. The grouping of lead traces through the weld metal

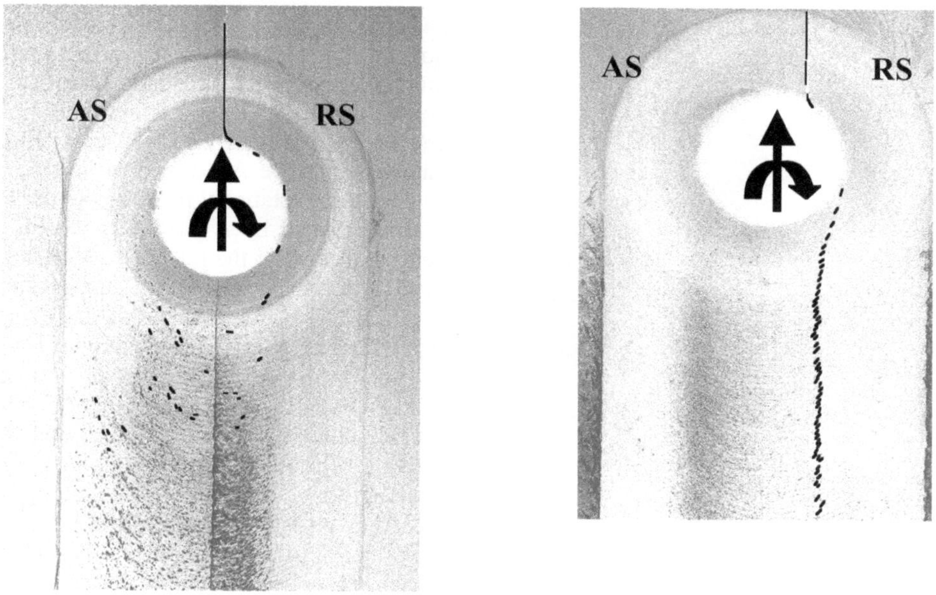


(a) (b)

**Fig. 3.6** Inverted x-ray radiograph of the plan view of the friction stir welded segments showing the variation in weld marker placement with respect to the entrance into the weld zone. The white circle is the hole left after E-stop removal of the weld tool. (a) C08 plan view. (b) C05 plan view. AS, advancing side; RS, retreating side

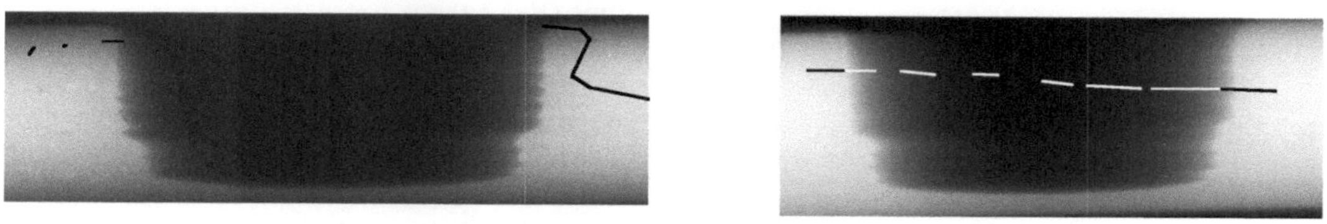

(a) (b)

**Fig. 3.7** Normal x-ray radiographs of the side view of the exit hole of the friction stir weld where the weld tool was removed following an E-stop. The initial wire placement is observed on the right side of the images. (a) C08 side view. (b) C05 side view may also be an indication that stick-slide conditions are operating.

Based on the experimental studies using tracers, two models have been published that describe the metal flow as influenced by the processing parameters and weld tool geometry (Ref 22, 32, 33).

**Nunes Kinematic Model.** Nunes (Ref 22, 32) has based his physical model of the metal flow in the friction stir process in terms of kinematics describing the metal motion. Figure 3.11 illustrates the deconvolution of the FSW process into three incompressible flow fields that combine to form two distinct currents. In

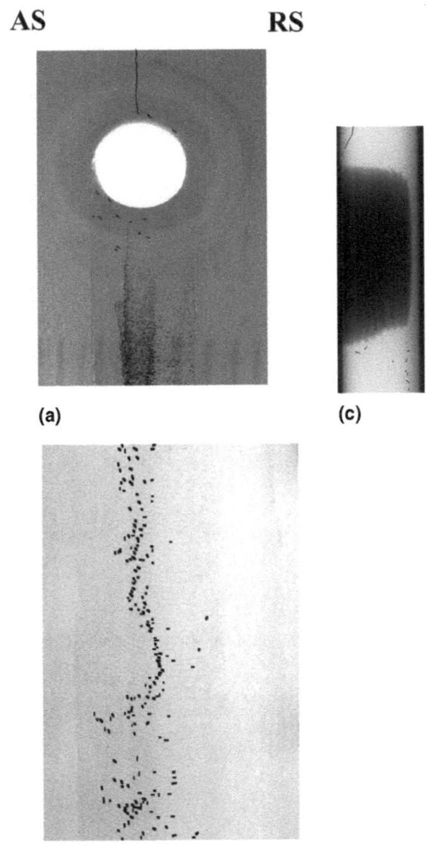


**Fig. 3.8** Weld panel C22 with weld parameters of 31 kN (7000 lbf), 114 mm/m (1.37 in./ft), and varying tool rotation. Tracer wire is introduced to the weld nugget at the panel centerline and at a depth of 1.27 mm (0.05 in.). (a) Inverted x-ray radiograph of the weld termination showing pin tool exit hole with tool rotation of 300 rpm. AS, advancing side; RS, retreating side. (b) Inverted x-ray radiograph of the section of plan view at 150 rpm tool rotation. (c) Normal x-ray radiograph of the side view of the pin tool exit hole at 300 rpm

this model, a rigid body rotation field imposed by the axial rotation of the pin tool is modified by a superimposed ring vortex field encircling the pin imposed by the pitch of the weld pin threads. These two flow fields, bound by a shear zone, are uniformly translated down the length of the weld panel. Metal not entrained in the ring vortex flow simply passes around the pin tool in a straight-through current, while metal entrained in the ring vortex flow experiences a high degree of thermomechanical processing, because it may pass around the pin tool more than once. Variations in features on the pin tool are reflected in the upward or downward motion of the metal as described by the vortex flow.

In the Nunes kinematic model, the metal on the RS of the weld is picked up at the front of the tool and deposited directly behind the tool, with minimal residence time in the rotational field around the tool. This is referred to as the straight-through current flow of metal. The weld material from the AS of the pin resides long enough in the rotational flow around the tool to become trapped by a gradual radial influx of metal at the top of the pin. The radial influx of metal is part of a ring vortex circulation induced by threads on the pin. The circulation drives the trapped metal down the pin. Superposition of the rotation around the pin with this downward flow results in a whirlpool pattern or maelstrom current, where the flow of weld metal emerges further down the pin as the circulation begins to move outward. Reversal of the direction of the threads reverses the direction of the flow of the weld metal in the maelstrom current from downward to upward along the pin. The two currents proposed by the Nunes kinematic model would impose a variation in the amount of thermomechanical processing experienced by each metal flow current. Variations in the hot working history have been used to explain the resulting textures, or onion rings, observed in the FSW nugget (Ref 32). This model has also been used to explain the occurrence of a weld defect reported to be based on entrained oxide films (Ref 34).

The interleaving of the two flow paths proposed by Nunes is illustrated in the plan view shown in Fig. 3.12. Similar macrostructures have been observed in welds of dissimilar metals (Ref 23). The occurrence of stick-slide modes due to the interface between the weld tool and the workpiece would explain the origin of interleaving. The residue of the straight-through current metal flow predominates on the RS and the upper regions of the tool; the maelstrom residue of metal flow predominates on the AS and lower regions of the tool. This variation on the AS of the weld is in agreement with marker studies showing a chaotic streamline when the marker is introduced on the weld AS (Ref 25, 29).

**Arbegast Metalworking Model.** The Arbegast model (Ref 33) treats the FSW as a metal-

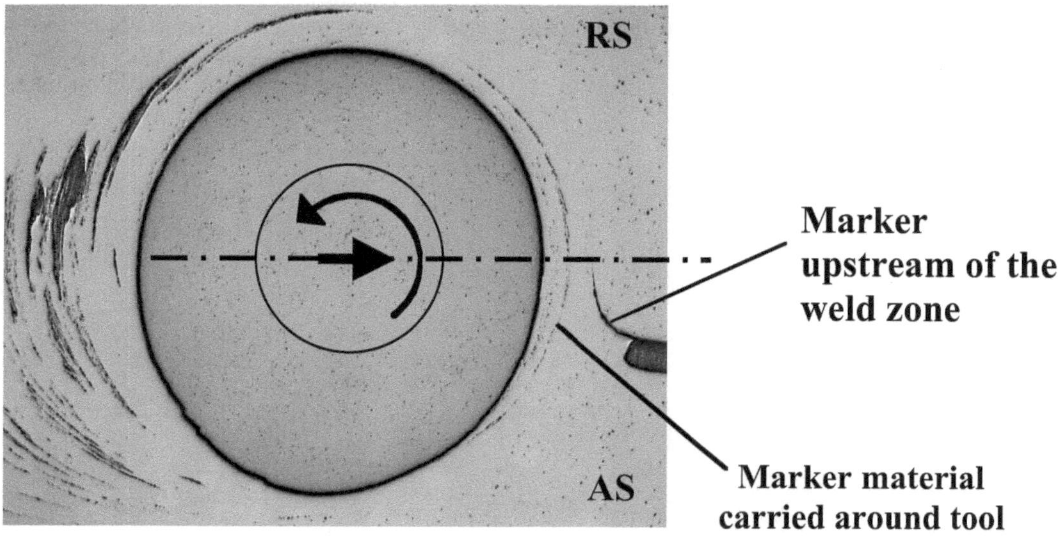


**Fig. 3.9** Continuous marker study (introduced 0.9 mm, or 0.035 in., to the advancing side, or AS, plate midplane) shows evidence of marker material being transported multiple times around the weld pin tool. RS, retreating side. Source: Ref 30

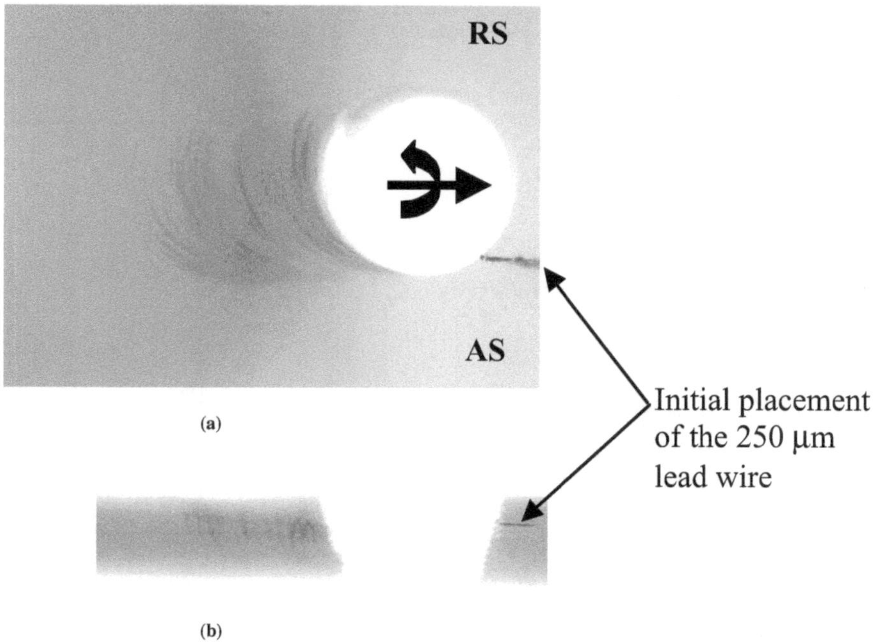


**Fig. 3.10** Inverted x-ray radiograph of the postweld position of a lead marker wire in the (a) plan view and (b) side view. Note the through-material thickness traces of the lead wire in the side view. The initial placement of the lead wire can be observed on the right side of the exit hole of the weld tool. RS, retreating side; AS, advancing side

working process that involves five zones: preheat, initial deformation, extrusion, forging, and postweld cooldown. These zones are illustrated in Fig. 3.13. The heat generated by the rotating weld tool preheats the metal in advance of the weld tool travel. The rotating motion of the weld tool forms the initial deformation zone in the softened metal. In this zone, the metal is forced upward into the shoulder and then downward into the extrusion zone. In the extrusion zone, the metal in front is moved around the pin tool to the exiting wake of the weld in the cavity being vacated by the pin as it moves forward. This model provides for an interleaving effect between the upper and lower extrusion zones. The back or heel of the shoulder passes over the metal exiting the extrusion zone and forges it, ensuring consolidation. As the weld tool leaves the area, the metal is cooled by either passive or forced means, analogous to quenching during

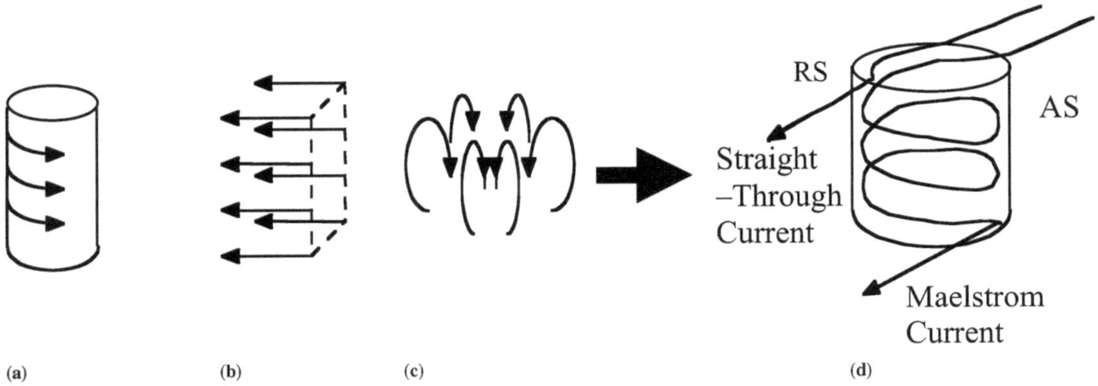


**Fig. 3.11** Three incompressible flow fields of the friction stir weld. (a) Rigid body rotation, (b) uniform translation, and (c) ring vortex combine to form (d) two flow currents. RS, retreating side; AS, advancing side

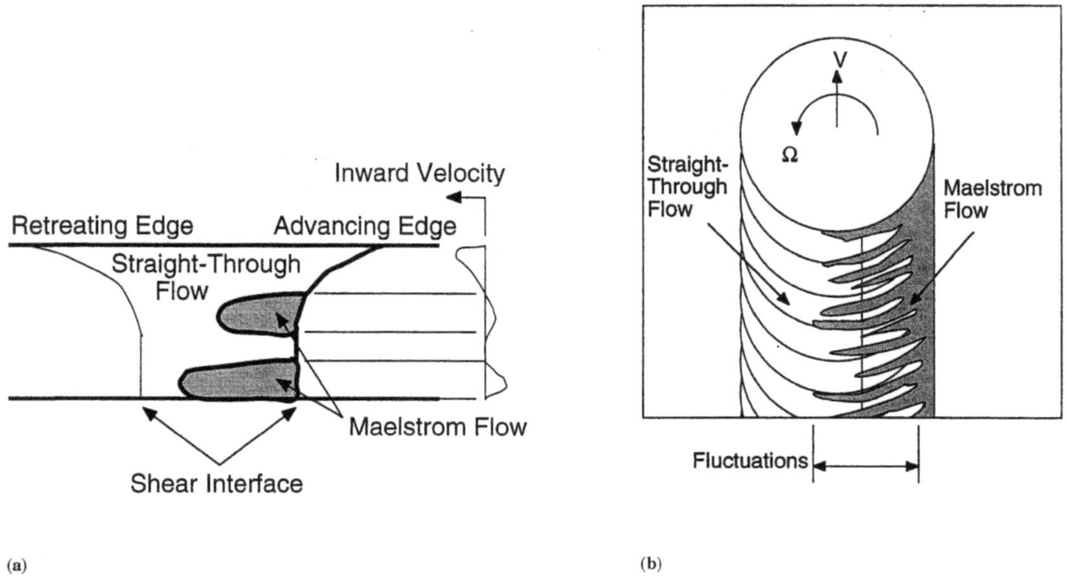


**Fig. 3.12** (a) Side view of two flow streams. (b) Plan view showing interleaving

heat treating operations. Marker studies by Reynolds (Ref 26, 35) also describe the process as one of extrusion followed by forging.

The Arbegast model can be used to explain two of the more common weld defects in terms of the processing parameters. The first is a wormhole or tunnel defect that runs the length of the weld and is attributed to insufficient forming pressure under the tool shoulder, which prevents the material from consolidating. The second defect is a lack of penetration on the root surface next to the anvil. This can result when the weld tool does not sufficiently penetrate into the metal plates, most likely from a too short pin tool.

## 3.3 Thermomechanical Working—The Coupled Process

The conceptual understanding of the FSW process is dominated by the mechanical deformation of the hot weld metal. To complete the physical understanding of this process, the hot metal deformation must be coupled with auxiliary issues, including grain size refinement, dislocation theory, thermophysical properties, and mixing. The next area of physical understanding is the influence of the tool pin geometry and material on the temperature field, microstructural refinement, resulting material flow, and the influence of flow variations on the subsequent mechanical properties of FSW butt joints in various materials.

The hot metal deformation portion of the FSW is complicated, and methods to decouple physical interactions are required to verify and validate the physical models. The resulting flow process may change from being mixing-dominated to extrusion-dominated or a mixture of the two flow paths as the processing parameters, weld tool geometry, and workpiece metal are changed. The challenge remains to understand the level of plasticity required for an effective FSW process and the role of mixing in obtaining a good weld.

## REFERENCES


1. J.E. Gould and Z. Feng, Heat Flow Model for Friction Stir Welding of Aluminum Alloys, *J. Mater. Process. Manuf. Sci.*, Vol 7, Oct 1998, p 185–194
2. A.P. Reynolds, W.D. Lockwood, and T.U. Seidel, Processing-Property Correlation in Friction Stir Welds, *Mater. Sci. Forum,* Vol 331–337, 2000, p 1719–1724
3. R. Rao, H. Raikoty, and G. Talia, High Speed Friction Stir Welding Using Rotating and Non-Rotating Shoulder Tool, *Proc. 46th AIAA/ASME/ASCE/AHS/ ASC Structures, Structural Dynamics and Materials Conf.*, April 2005
4. W. Tang, X. Guo, J.C. McClure, L.E. Murr, and A. Nunes, Heat Input and Temperature Distribution in Friction Stir Welding, *J. Mater. Process. Manuf. Sci.*, Vol 7, Oct 1998, p 163–172
5. M.J. Russell and H.R. Shercliff, Analytical Modeling of Microstructure Develop-


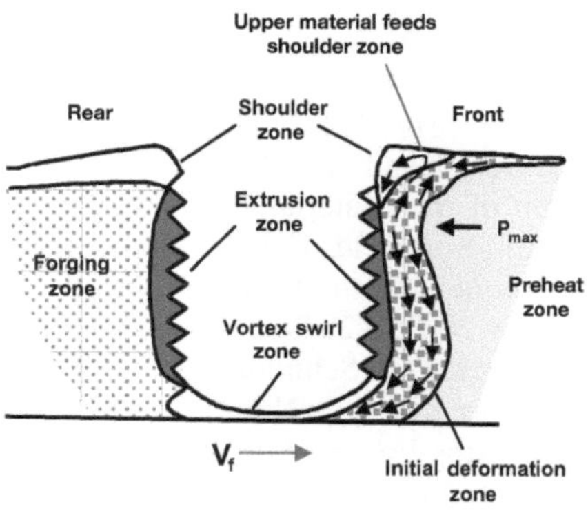


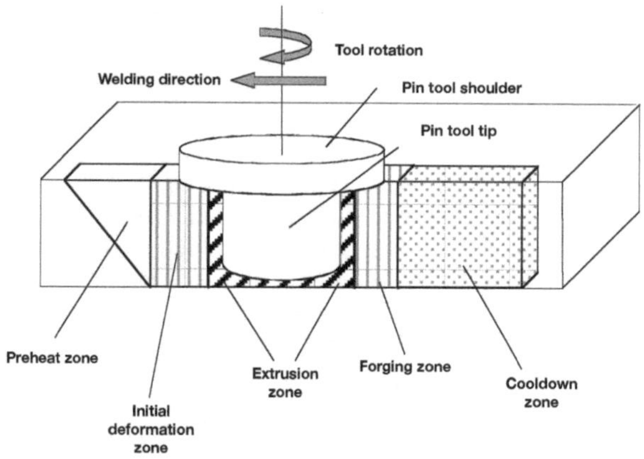


**Fig. 3.13** Metallurgical processing zones developed during friction stir joining. Adapted from: Ref 33

ment in Friction Stir Welding, *Proc. First Int. Symp. on Friction Stir Welding,* June 1999 (Thousand Oaks, CA)
6. P. Colegrove, M. Painter, D. Graham, and T. Miller, 3-Dimensional Flow and Thermal Modelling of the Friction Stir Welding Process, *Second Int. Symp. on Friction Stir Welding,* June 2000 (Gothenberg, Sweden)
7. G. Liu, L.E. Murr, C.-S. Niou, J.C. McClure, and F.R. Vega, Microstructural Aspects of the Friction-Stir Welding of 6061-T6 Aluminum, *Scr. Metall.,* Vol 37 (No. 3), 1997, p 355–361
8. L.E. Murr, G. Liu, and J.C. McClure, A TEM Study of Precipitation and Related Microstructures in Friction-Stir Welded 6061 Aluminum, *J. Mater. Sci.,* Vol 33, 1998, p 1243–1251
9. Y.S. Sato, H. Kokawa, M. Enomoto, S. Jogan, and T. Hashimoto, Precipitation Sequence in Friction Stir Weld of 6063 Aluminum During Aging, *Metall. Mater. Trans. A,* Vol 30, Dec 1991, p 3125–3130
10. J.A. Schneider, A.C. Nunes, Jr., P.S. Chen, and G. Steele, TEM Study of the FSW Nugget in AA2195-T81, *J. Mater. Sci.,* Vol. 40, 2005, p 1–5
11. Y.J. Chao, X. Qi, and W. Tang, Heat Transfer in Friction Stir Welding—Experimental and Numerical Studies, *Trans. ASME,* Vol 125, Feb 2003, p 138–145
12. M. Song and R. Kovacevic, Numerical and Experimental Study of the Heat Transfer Process in Friction Stir Welding, *Proc. Ins. Mech. Eng., B, J. Eng. Manuf., IMECHE 2003,* Vol 217, p 73–85
13. W.J. Arbegast and P.J. Hartley, Friction Stir Weld Technology Development at Lockheed Martin Michoud Space System—An Overview, *Proc. Fifth Int. Conf. on Trends in Welding Research,* June 1998 (Pine Mountain, GA), p 541–546
14. M.W. Mahoney, C.G. Rhodes, J.G. Flintoff, R.A. Spurling, and W.H. Bingel, Properties of Friction-Stir-Welded 7075 T651 Aluminum, *Metall. Mater. Trans. A,* Vol 29, 1998, p 1955–1964
15. B. London and M.W. Mahoney, personal communications, 2006
16. T.U. Seidel and A.P. Reynolds, A 2-D Friction Stir Welding Process Model Based on Fluid Mechanics, *Sci. Technol. Weld. Join.,* Vol 8, (No. 3), 2003, p 175–183
17. P.A. Colegrove and H.R. Shercliff, Experimental and Numerical Analysis of Aluminum Alloy 7075-T7351 Friction Stir Welds, *Sci. Technol. Weld. Join.,* Vol 8 (No. 5), 2003, p 360–368
18. K.N. Krishnan, On the Formation of Onion Rings in Friction Stir Welds, *Mater. Sci. Eng. A,* Vol 327, 2002, p 246–251
19. G. Biallas, R. Braun, C. Dalle Donne, G. Staniek, and W.A. Kaysser, Mechanical Properties and Corrosion Behavior of Friction Stir Welded 2024-T3, *First Int. Conf. on FSW,* June 1999 (Thousand Oaks, CA)
20. Y.S. Sato, H. Kokawa, K. Ikeda, M. Enomoto, S. Jogon, and T. Hashimoto, Microtexture in the Friction-Stir Weld of an Aluminum Alloy, *Metall. Mater. Trans. A,* Vol 32, 2001, p 941–948
21. D.P. Field, T.W. Nelson, Y. Hovanski, and K.V. Jata, Heterogeneity of Crystallographic Texture in Friction Stir Welds of Aluminum, *Metall. Mater. Trans. A,* Vol 32, 2001, p 2869–2877
22. J.A. Schneider and A.C. Nunes, Jr., Characterization of Plastic Flow and Resulting Microtextures in a Friction Stir Weld, *Metall. Mater. Trans. B,* Vol 35, 2004, p 777–783
23. Y. Li, L.E. Murr, and J.C. McClure, Solid-State Flow Visualization in the Friction Stir Welding of 2024 Al to 6061 Al, *Scr. Mater.,* Vol 40 (No. 9), 1999, p 1041–1046
24. L.E. Murr, Y. Li, R.D. Flores, E.A. Trillo, and J.C. McClure, Intercalation Vortices and Related Microstructural Features in the Friction-Stir Welding of Dissimilar Metals, *Mater. Res. Innovat.,* Vol 2, 1998, p 150–163
25. K. Colligan, Material Flow Behavior during Friction Stir Welding of Aluminum, *Weld. Res. Suppl.,* July 1999, p 229s–237s
26. T.U. Seidel and A.P. Reynolds, Visualization of the Material Flow in AA2195 Friction-Stir Welds Using a Marker Insert Technique, *Metall. Mater. Trans. A,* Vol 32, Nov 2001, p 2879–2884
27. M. Guerra, C. Schmidt, J.C. McClure, L.E. Murr, and A.C. Nunes, Jr., Flow Patterns during Friction Stir Welding, *Mater. Charact.,* Vol 49, 2003, p 95–101
28. B. London, M. Mahoney, W. Bingel, M. Calabrese, R.H. Bossi, and D. Waldron, Material Flow in Friction Stir Welding

Monitored with Al-SiC and Al-W Composite Markers, *Proc. Symp. on FSW and Processing II,* K.W. Jata, M.W. Mahoney, R.S. Mishra, S.L. Semiatin, and T. Lienert, Ed., TMS, 2003, p 3–12
29. J.R. Sanders, J.A. Schneider, and A.C. Nunes, Jr., Tracing Material Flow Paths in Friction Stir Welds, *Mater. Sci. Technol.,* (MS&T)/TMS, Sept 2005 (Pittsburgh, PA)
30. M.W. Mahoney, Senior Scientist, Rockwell Scientific, unpublished data, 2005
31. J.A. Schneider, R. Beshears, and A.C. Nunes, Jr., Interfacial Sticking and Slipping in the Friction Stir Welding Process, *Mater. Sci. Eng. A,* Vol 435–436, 2006, p 297–304
32. A.C. Nunes, Jr., Wiping Metal Transfer in Friction Stir Welding, *Aluminum 2001: Proc. 2001 TMS Annual Meeting Automotive Alloys and Joining Aluminum Symp.,* G. Kaufman, J. Green, and S. Das, Ed., TMS, p 235–248
33. W.J. Arbegast, Modeling Friction Stir Joining as a Metal Working Process, *Hot Deformation of Aluminum Alloys,* Z. Jin, Ed., TMS, 2003
34. Y.S. Sato, H. Takauchi, S.H.C. Park, and H. Kokawa, Characteristics of the Kissing-Bond in Friction Stir Welded Al Alloy 1050, *Mater. Sci. Eng. A,* Vol 405, 2005, p 333–338
35. A.P. Reynolds, Visualization of Material Flow in Autogenous Friction Stir Welds, *Sci. Technol. Weld. Join.,* Vol 5 (No. 2), 2000, p. 120–124

# CHAPTER 4

# Microstructure Development in Aluminum Alloy Friction Stir Welds

A.P. Reynolds, Department of Mechanical Engineering
University of South Carolina

THE MICROSTRUCTURE and consequent property distributions produced during friction stir welding (FSW) of aluminum alloys are dependent on several factors. The contributing factors include alloy composition, alloy temper, welding parameters, gage of the welded plate, and other geometric factors. Alloy composition determines the available strengthening mechanisms and how the material will be affected by the temperature and strain history associated with FSW. The alloy temper dictates the starting microstructure, which can have an important effect on the alloy response to FSW, particularly in the heat-affected zone (HAZ). Welding parameters (e.g., tool rotation rate and welding speed) dictate, for given tool geometry and thermal boundary conditions, the temperature and strain history of the material being welded. Plate gage and other geometric factors (e.g., shoulder size, heat sinks associated with clamping, etc.) may affect the temperature distribution within the weld zone and, in particular, through the thickness of the welded plates.

In this chapter, the FSW process parameters that can affect microstructure/property distributions in aluminum alloy friction stir welds are described. The chapter includes a brief description of the main classes of aluminum alloys, the processing routes (thermomechanical treatments) typically associated with each class, and how FSW parameters can be manipulated, in a general way, to modify the microstructure and property distribution in friction stir welds of each class of alloy.

## 4.1 Aluminum Alloy Metallurgy

For the purposes of discussion, it is convenient to first classify aluminum alloys by their available strengthening mechanisms (Ref 1).

**Non-Heat-Treatable Alloys.** Non-heat-treatable aluminum alloys are defined primarily by what they are not. They are not strengthened by second-phase particles and may be better described as non-precipitation-hardening alloys. The non-heat-treatable alloy classes are the 1*xxx*, 3*xxx*, and 5*xxx* alloys.

The simplest aluminum alloys are the 1*xxx* series. These are essentially commercially pure aluminum and are strengthened either by strain hardening (cold work) or by microstructural refinement (i.e., reduction of grain size or substructure formation). The 3*xxx*-series alloys are very similar to the 1*xxx*, but due to the addition of a small amount of manganese, some dispersoid is formed that affects the grain size, crystallographic texture, and grain morphology. The 1*xxx* and 3*xxx* alloys are relatively low strength. The 5*xxx* alloys contain a substantial amount of magnesium, which is a potent solid-solution strengthener. As such, the 5*xxx* alloys are strong relative to the other non-heat-treatable alloys. In addition to the solid-solution strengthening due to the magnesium content, the strength of 5*xxx* alloys can be improved by strain hardening. Under some conditions, second-phase $Mg_2Al_3$ precipitates may form in 5*xxx* alloys; however, these precipitates do not provide a strengthening increment, and their formation is generally undesirable.

The only heat treatment applicable to non-heat-treatable alloys is an annealing heat treatment. The processes that may occur during annealing of non-heat-treatable alloys include recovery, recrystallization, and grain growth. Generally, an annealing heat treatment is applied to a cold-worked material in order to reduce hardness and increase the capacity for further deformation. A fully annealed alloy is designated as "O temper," while alloys with some level of strengthening by cold deformation are designated as "H*xxx*," where the *xxx* are numbers indicating the amount of cold work.

**Heat Treatable (Precipitation-Hardening) Alloys.** The heat treatable alloys are drawn from the 2*xxx*, 6*xxx*, and 7*xxx* alloy families. The primary alloying elements for the three alloy series are, respectively, copper (2*xxx*), magnesium and silicon (6*xxx*), and magnesium and zinc (7*xxx*). Typically, a particular alloy will be strengthened mainly by a single precipitate phase; however, there may be multiple precipitate phases present. The situation is further complicated by the fact that, in general, the strengthening phase is not an equilibrium phase. Although the details of the phase distributions for the three alloy families may be quite complex, simply put, for optimal strengthening, it is critical to obtain a homogeneous distribution of very fine second-phase particles (precipitates).

The general form of the heat treatments required for obtaining the desired structure is the same for all three classes of alloys (although it may differ greatly in detail from alloy to alloy). The first step in forming a precipitation-hardened structure is the solution heat treatment (SHT). The SHT is a high-temperature step that is meant to put the alloy into a single-phase solid-solution condition. For many technologically important alloys, a single phase cannot be obtained; regardless, as much of the alloying content should be put into solution as possible without inducing local melting. Subsequently, the solution heat treated alloy is quenched (normally to room temperature, or $T$), producing a supersaturated solid solution. After quenching, the supersaturated solid solution is allowed to decompose into a two-phase mixture of the matrix solid solution and a strengthening phase (the precipitate). The decomposition may take place either at room temperature (natural aging) or at a somewhat elevated temperature (artificial aging). The aging time and temperature are chosen in order to develop particular desired combinations of properties. For some alloys (notably, many 2*xxx*), the precipitation process is enhanced by the application of limited cold work (normally 1.5 to 3%) prior to the aging treatment; the cold work increases the dislocation density, and the dislocations provide sites for heterogeneous nucleation of precipitate particles. Artificial aging for a period of time less than necessary to obtain the peak strength results in an underaged microstructure; aging for a time greater than that required for peak strength is overaging. Excessive aging times or temperatures can result in greatly degraded properties relative to the peak strength. This occurs as a result of coarsening of the precipitate distribution and/or excessive precipitation on grain boundaries.

Common temper designations in the heat treatable alloys are as follows:

- T3: SHT + cold working + natural aging
- T4: SHT + natural aging
- T6: SHT + artificial aging to the peak strength
- T8: SHT + cold working + artificial aging to peak strength
- T7: SHT + artificial aging beyond the peak aging time

The 6*xxx* alloys are normally used in the T6 condition. The 7*xxx* alloys are used in the T6 or T7 conditions. Some alloys of the 2*xxx* series are used in the T3 condition, while others are used in the T6 or T8 conditions. The T4 condition is typically only an intermediate stage; parts formed or assembled in the T4 condition are normally subsequently heat treated to the T6. The 7*xxx* alloys may also be provided in a "W" temper, which is an unstable, naturally aging temper. The properties of alloys in the W temper may continue to evolve over the course of years of natural aging. The W temper is not used in service.

**Aluminum Alloy Texture and Grain Structure.** Aluminum alloys exhibit a variety of grain sizes, grain morphologies, and crystallographic textures that depend not only on the composition but also on the product form and temper. While the texture and grain structure of an aluminum alloy may have a significant effect on its properties, the texture and grain structure in the highly deformed region of a friction stir weld are not profoundly affected by the starting condition.

**Recrystallization in Aluminum Alloys.** The subject of recrystallization mechanisms in aluminum alloys is somewhat contentious. The prevailing understanding is that under "normal" circumstances, meaning conditions encountered during conventional thermomechanical processing, aluminum alloys do not dynamically recrystallize in the traditional sense (Ref 2, 3). This is believed to be due to the very high stacking fault energy in aluminum, which facilitates cross slip of screw dislocations, easing recovery at the expense of recrystallization. On the other hand, the process of continuous dynamic recrystallization (CDRX) (sometimes called extended recovery) has been suggested to explain the production of small, relatively equiaxed grains separated by high-angle boundaries (Ref 4). The grains are believed to develop from a cellular deformation structure by a gradual process of deformation-induced grain rotation. Static recrystallization (SRX) is the formation of new grains after the cessation of deformation. The SRX may occur upon heating after cold deformation or, potentially, after high-rate deformation at elevated temperature (when the deformation rate is high enough so that at the end of deformation, there is still a substantial dislocation density). Generally, in aluminum alloys, it can be difficult to unambiguously distinguish between SRX, DRX, CDRX, and subsequent grain growth processes.

## 4.2 Thermomechanical Processes Associated with FSW

A fundamental difference between FSW and conventional fusion welding techniques is that in a fusion weld, the highest temperature experienced by solid metal is the melting temperature. Hence, at the weld pool boundary, it can be unambiguously determined that the temperature in the solid was the melting temperature of the alloy. In a friction stir weld, the highest temperature experienced by the material comprising the weld may be significantly lower than the bulk alloy melting temperature (Ref 5, 6). The potential for variation of the peak temperature in FSW can enable the production of a wide range of microstructure and properties that cannot be achieved in fusion welds. The processes that occur in some of the FSW regions will be dependent on the peak temperature achieved in the weld. In this section, a general overview of the possible thermomechanical processes and resulting microstructures is broken down by weld region. In a subsequent section, the various possibilities are illustrated by examples from the various alloy classes; these examples include details regarding the effects of welding parameter variations on the microstructures and properties.

**Thermomechanical Processes Occurring in the Weld Nugget.** The weld nugget is typically described as the region of the thermomechanically affected zone that has experienced sufficient deformation at elevated temperature to undergo recrystallization (by whatever mechanism). The region will be narrower in recrystallization-resistant alloys than in those alloys that are readily recrystallized (e.g., 2195 versus 6061). The two key variables that determine the properties of the material in the weld nugget are the peak temperature and the quenching rate from that temperature.

According to Sato et al. (Ref 5), the statically recrystallized grain size in the nugget region is determined predominantly by the peak temperature in the weld; the higher the peak temperature, the larger the grain size. Some effect of welding speed may also be involved, but because the grain size (for static grain growth) is exponential with temperature and linear with time, the peak temperature will exert the dominant influence. Similar functional relationships between time, temperature, and grain size are also expected for CDRX (Ref 7); however, estimates of the strain and temperature history for materials in the weld nugget are not well established (Ref 8). While it is conceivable that FSW could be performed without producing a recrystallized structure in the weld nugget (by, for example, welding with a very low tool rotation rate), to this author's knowledge, this has never been successfully performed. It has, however, been shown that a very wide range of nugget grain sizes can be achieved by manipulation of welding process parameters. Grain sizes on the order of 10s of micrometers and less than 1 μm have been reported (Ref 9, 10).

The important processes occurring in the weld nugget (other than recrystallization) will differ somewhat, depending on the type of alloy considered. For non-heat-treatable alloys, the only heat treatment that can occur in the nugget is an annealing cycle. If the starting temper of the alloy is O, then the properties in the weld region will be similar to those in the base metal. Depending on the nugget grain size, there may be some increment of strengthening due to microstructural refinement. If the base metal is in a strain-hardened condition, then the recrystallized nugget region will normally exhibit a substantial reduction in hardness relative to the base metal.

In heat treatable alloys, the processes occurring in the nugget may be more complex. Depending on the particular combination of alloy and welding parameters, the nugget may be left in an overaged condition, a partially solution heat treated condition, or a single-phase solid solution (Ref 6, 9, 11). The weld nugget microstructural condition may be assessed directly (e.g., by transmission electron microscopy) or inferred by its response to a postweld aging treatment. If the weld nugget is overaged, then one expects that an aging treatment will have either no effect or a negative effect on the nugget hardness. If the nugget is partially solution treated, then some hardening should result from the postweld aging. If the nugget has been left in a solid-solution condition, then postweld aging should enable recovery of properties similar to that of the base metal.

While FSW is a nominally solid-state process, in a heterogeneous material (essentially all technologically important alloys are heterogeneous on some scale), there may be low-melting regions distributed within a higher-melting bulk. Deformation heating in the bulk may, in some cases, result in the temperature exceeding the melting temperature of some low-melting phases. This may, in turn, result in grain-boundary liquation and the formation of brittle structures within the weld region. This local melting phenomenon may be described as overheating (Ref 12, 13).

**Thermal Processes in the HAZ.** The HAZ is, by definition, not mechanically deformed, so processes occurring in the HAZ are the result only of a temperature transient. The transient is, of course, more severe close to the weld centerline and lessens in severity farther from the weld. At some distance from the weld, depending again on the peak temperature and the temporal length of the transient, the effect of the transient will be negligible, and the HAZ will have transitioned to base metal. As for the nugget, the processes that occur in the HAZ will depend on the type of alloy being considered.

For non-heat-treatable alloys in the O temper, there will normally be no effect of the thermal transient. The material is already as soft as it can be, and further heating does not lower its hardness; however, it is possible that the temperature transient could lead to grain growth. If the alloy is in a strain-hardened temper, then there will generally be a range of microstructural transformation occurring, with a dependence based on the distance from the weld centerline. Close to the nugget, the strain-hardened material will likely be completely recrystallized. The fraction of recrystallized material will fall to zero as the distance from the weld increases, at which point there will normally be a recovered zone that will transition to the base metal.

In heat treatable alloys, the processes will depend on the starting temper also. For alloys in a peak or overaged condition (T6, T7, or T8), there will normally be a region of reduced hardness (relative to the base metal) in the HAZ. In this region, the thermal transient was such that the precipitate distribution has been significantly coarsened; overaging of the alloy has occurred. Depending on the welding parameters, the minimum hardness region will be found at various distances from the weld nugget and will have varying depths (minimum hardness values). The HAZ hardness minimum may have the same hardness as the nugget (and be adjacent to it), or it may be substantially softer, depending on the thermomechanical processing experienced by the nugget (Ref 6). If the alloy was welded in a naturally aged condition, the situation is more complex. For naturally aged materials, there may be two local hardness minima surrounding a local maximum (Ref 14). In the inner minimum (the one closest to the weld), overaging is the operating process (as in T6, T7, and T8 alloys). The local maximum occurs as a result of precipitation of a strengthening phase by a process of artificial aging (with a very short aging time). The mechanism by which the outer minimum hardness region is produced is not clear but may be due to re-solution of Guinier-Preston zones or recovery of cold work (in T3 materials).

## 4.3 Illustrative Examples

**Non-Heat-Treatable Alloys: AA5454.** The aluminum alloy AA5454 is a typical non-heat-treatable alloy that is solid-solution strengthened by additions of magnesium and may be obtained in strain-hardened or annealed tempers. The nominal composition of the alloy in weight percent is 2.7% Mg, 0.8% Mn, 0.12% Cr, and balance Al (Ref 1). In this section, data are presented for 3.8 mm (0.15 in.) thick 5454 sheet, friction stir welded in both the annealed temper (O) and a strain-hardened (H32) temper (Ref 15). Yield and tensile strengths for the O-temper base metal are, respectively, 115 and 220 MPa (16.7 and 32 ksi). For the H32 temper, the yield and tensile strengths are 230 and 300 MPa (33 and 43.5 ksi), respectively. Welds were made in the H32 at several welding speed and rpm combinations. The O temper was studied less extensively. Figure 4.1 shows etched and scanned cross sections of the O-temper and H32 welds. The O-temper base metal is comprised of equiaxed recrystallized grains. The weld nugget exhibits a finer grain structure than does the base metal. The H32 base metal is comprised of unrecrystallized, pancake-shaped grains that result from the cold rolling process. The H32 weld nugget is similar to the O-temper nugget; however, the H32 weld exhibits a gradual transition from the nugget to the base-metal grain structure. From the edge of the heavily deformed nugget to the base metal, there is a continuously declining area fraction of recrystallized material. The recrystallized material that is outside of the weld nugget is, presumably, material that is produced by the thermal transient associated with the welding process. Recrystallization in this region is driven by the cold work that is already present in the base metal and not by the deformation associated with the FSW process.

Figure 4.2 shows hardness traverses for welds made in the H32 and O-temper material (both welds made at 4.2 mm/s, or 0.17 in./s). The hardness distributions are quite typical for the two starting temper conditions. In the figure, it can be seen that the O-temper weld exhibits a very slight hardness maximum in the grain-refined region of the weld (the nugget). Outside of this region, the hardness decreases to the base-metal value. The H32 nugget has a similar hardness to that of the O-temper nugget (indicating similar grain size). Outside the H32 nugget, the hardness transitions smoothly to that of the strain-hardened base metal. In some cases, the H32 weld nugget may be placed in a mild local hardness maximum due to the presence of the undeformed but recrystallized material in the HAZ, as described in the preceding paragraph.

Figure 4.3 shows the tensile and yield strengths of a series of H32 welds made using a

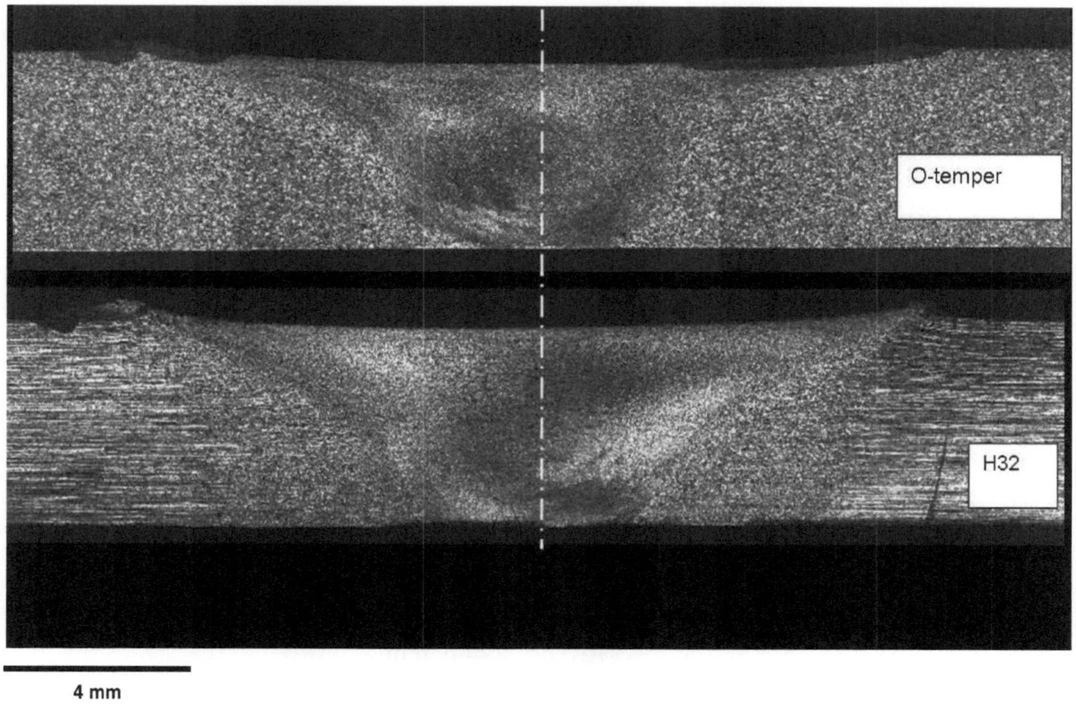


**Fig. 4.1** Etched and scanned cross sections of 5454-O (top) and 5454-H32 friction stir welds

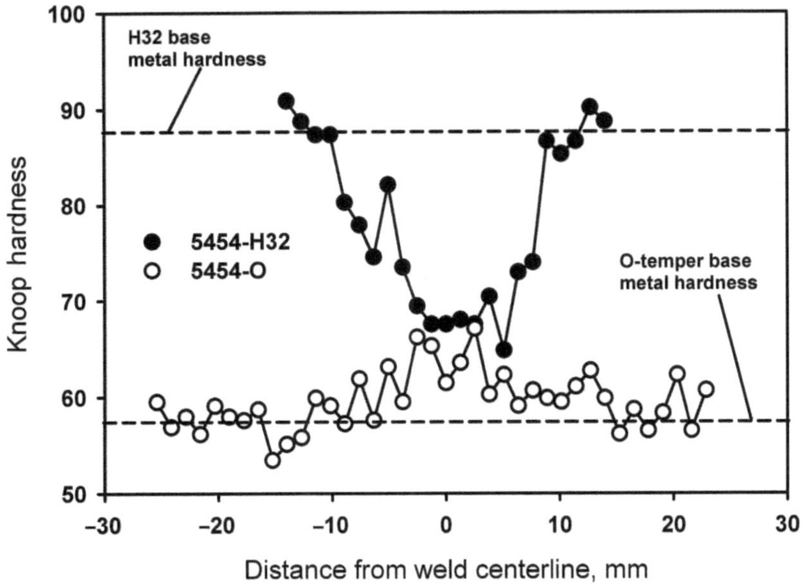


**Fig. 4.2** Hardness distributions on transverse cross sections from friction stir welds in 5454-O (open symbols) and 5454-H32 (closed symbols)

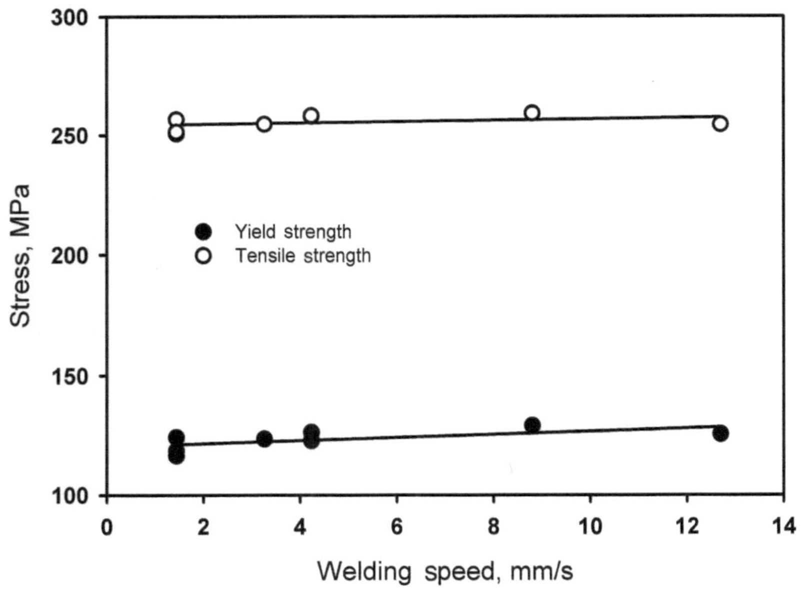


**Fig. 4.3** Transverse yield and transverse tensile strengths of 5454-H32 friction stir welds produced at a range of welding speeds

range of welding speeds between 1.4 and 12.7 mm/s (0.06 and 0.5 in./s). These properties are very insensitive to the welding speed and reflect the properties of the weld nuggets, which are uniformly low and similar to the O-temper base-metal properties. In order to gain a fuller understanding of the properties of the nugget and HAZ regions, digital image correlation (DIC) was used to measure the full-field surface strain on transversely loaded weld specimens (Ref 16). As described in several publications, the local strain derived from DIC can be mapped to the global stress to provide a reasonable approximation of the local constitutive behavior of the weld regions. Figure 4.4 shows the following information: (1) an O-temper base-metal tensile stress-strain curve (solid curve), (2) an H32 base-metal curve (solid curve), (3) the global tensile response of the transversely loaded weld (solid curve), (4) a DIC-derived local stress-strain curve from the nugget region (closed circles, labeled "DRZ" in the figure), (5) a DIC-derived local stress-strain curve from the partially recrystallized HAZ (closed squares), and (6) a DIC-derived local stress-strain curve from the recovered but not recrystallized HAZ (closed triangles). Important points include the local curve for the nugget region is nearly identical to the O-temper base-metal curve with respect to both the strength levels and the fracture strains, while the partially recrystallized HAZ and the recovered HAZ have properties intermediate to the H32 and O-temper base metals. The strain levels observed in the partially recrystallized and recovered HAZ regions are limited by the strength of the nugget region. That is, in a transversely loaded weld, no stress greater than the tensile strength of the weakest region can be transmitted to any other region.

In summary, non-heat-treatable alloys are relatively insensitive to the welding parameters (so long as no weld defects are generated). The strength of a transversely loaded weld in O-temper material will be similar to the base-metal strength, and the failure location could be anywhere. Conversely, the strength of a transversely loaded weld in strain-hardened material (e.g., H32 temper) will be similar to that of O-temper base-metal material, but the strain will be localized in the weld and HAZ, as will the fracture location.

**Peak or Overaged Heat Treatable Alloys: 7050-T7651.** Alloy 7050 is a high-strength, heat treatable alloy with a nominal composition of Al-6.2%Zn-2.3%Cu-2.2%Mg and 0.12% Zr. It is normally used in a slightly overaged temper

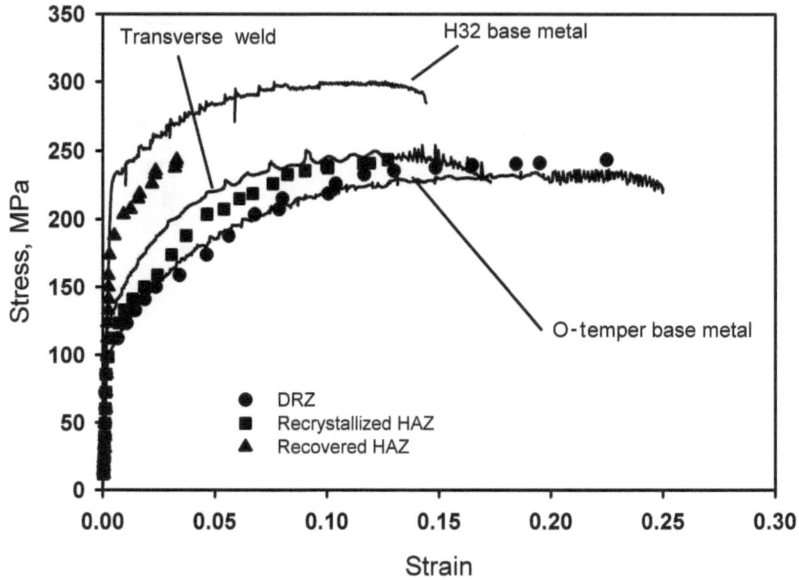


**Fig. 4.4** Standard and digital image correlation-derived tensile stress-strain curves for base metal (O temper and H32), overall transverse H32 weld, and local regions of the H32 weld. DRZ, dynamic recrystallization zone; HAZ, heat-affected zone

(T7*xx*) designed to provide a good combination of strength, fracture toughness, and stress-corrosion cracking resistance (Ref 1). The work described in this section was performed on 6.4 mm (0.25 in.) thick 7050-T7451 plate.

In an attempt to elucidate relationships between FSW parameters and weld nugget and HAZ hardness values in 7050, a series of welds was made in 6.4 mm thick 7050-T7451 plate (Ref 6). Welds were made at speeds between 0.86 and 5.1 mm/s (0.034 and 0.20 in./s), using three different ratios of welding speed to tool rotation rate (advance per revolution, or APR): 0.56, 0.42, and 0.28 mm/rev (0.022, 0.017, and 0.011 in./rev). All welds were performed under *z*-axis force control; the *z*-axis force was adjusted for the different welding speeds and tool rotation rates so as to produce good-quality welds. An FSW tool having a threaded cylindrical pin and a dished shoulder was used for all welding. The shoulder diameter was 20.3 mm (0.80 in.), the pin diameter was 7.1 mm (0.28 in.), the pin length was 6.1 mm (0.24 in.), and the thread pitch on the pin was 0.85 mm/thread (0.033 in./thread). A lead angle of 2.5° was used for all welds. During the welding process, the torque supplied to the spindle motor was monitored continuously; the spindle torque, after subtraction of the free running torque, may be used to calculate the weld power. For each weld, the Vickers hardness distribution on a transverse section with and without postweld heat treatment (PWHT) was determined. The PWHT was 24 h at 121 °C (250 °F). A finite element modeling (FEM) simulation was used to calculate the time/temperature history for a subset of the welds.

Figure 4.5 shows two typical hardness distributions (prior to PWHT) from the 7050 welds. The two welds shown were made at welding speeds of 0.85 and 3.8 mm/s (0.033 and 0.15 in./s) at the same APR, 0.42 mm/rev. Hence, the spindle rotation rate for the slower weld was 120 rpm and for the faster weld, 540 rpm. The distinctive "W"-shaped hardness distribution is typical of many FSWs in precipitation-hardening alloys (Ref 5, 6, 9, 11–13). In the following, important features of the hardness distribution include the average hardness in the central local maximum (located in the weld nugget) and the minimum hardness (located in the HAZ). For these two welds, it is quite clear that the different weld parameters have a substantial effect on the hardness distributions. The faster weld exhibits higher nugget and HAZ minimum hardness than does the slower weld. Also, the HAZ minimum hardness is located farther from the weld centerline in the fast weld than in the slow weld. All other things being equal, the higher hardness in

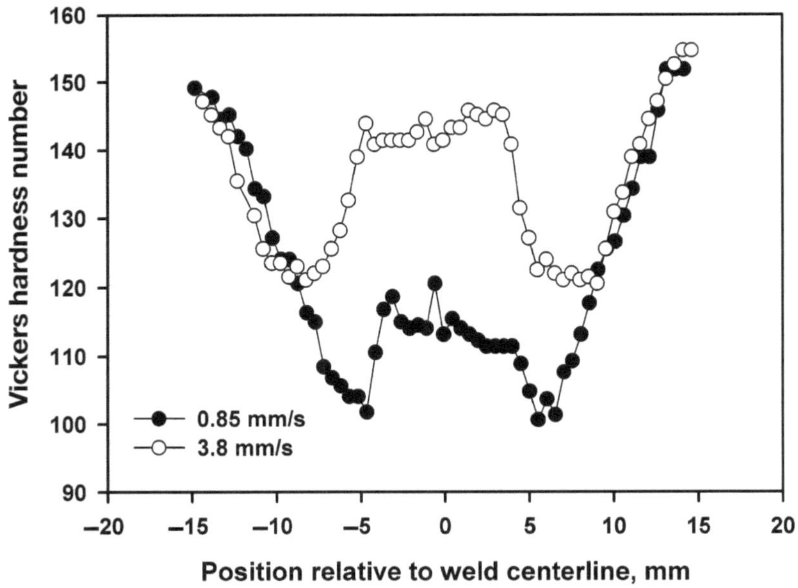


Fig. 4.5 Typical hardness distributions from transverse sections of 7050-T7451 friction stir welds made at two different welding speeds (no postweld heat treatment)

the faster weld will result in greater transverse tensile strength than in the slower weld. The critical questions to be answered are: (1) How do the welding parameters affect the weld hardness distributions? and (2) Why?

In order to answer question 2 posed in the preceding paragraph, it is necessary to have some understanding of the response to heat treatment in 7*xxx* alloys. In general, the strengthening-precipitate precipitation/dissolution sequences are similar in many 7*xxx* alloys, and it is well established that the primary strengthening precipitate in 7050-T7451 alloy is the coherent $\eta'$ phase (Ref 11). Examination of the literature reveals the following regarding precipitate stability in the 7*xxx*-series alloys (Ref 17, 18):

- Dissolution of the strengthening $\eta'$ phase occurs at $T > 190$ °C (375 °F).
- The incoherent $\eta$ phase precipitates between approximately 215 to 250 °C (420 to 480 °F). This phase contributes much less to strengthening than does $\eta'$. Near 250 °C, $\eta$ begins to coarsen rapidly.
- $\eta$ phase begins to dissolve at $T > 320$ °C (610 °F).
- There is a maximum in the formation rate of the high-temperature, nonstrengthening, incoherent M phase at approximately 350 °C (660 °F). Hence, solute will be most rapidly depleted from the matrix at this temperature.

Further, there is pertinent information regarding the thermal conditions associated with HAZ formation in welding of 7075 (which, it is assumed, is similar to 7050 in this regard). Mahoney et al. (Ref 19) found the minimum HAZ hardness in a 7075 friction stir weld in a region where the maximum temperatures were in the range of 300 to 350 °C (570 to 660 °F). Hwang and Chou (Ref 20) performed weld simulation of alloy 7075 and found that the minimum strength resulted from a weld thermal cycle with a peak temperature of 377 °C (711 °F). This was not necessarily the temperature that would result in the absolute minimum hardness, because a continuum of peak temperatures was not examined (adjacent temperatures were 288 and 445 °C, or 550 and 833 °F). Hwang and Chou ascribed the low strength at 377 °C to rapid formation of coarse $\eta$. Temperatures above 377 °C were considered partial solution treatments, with subsequent natural aging leading to higher strength, while those below 377 °C resulted in less dissolution of $\eta'$ and hence higher strength. Based on the work of Archambault and Godard (Ref 18), it seems likely that the minimum hardness at a peak temperature of 377 °C may also be ascribed to rapid formation of the nonstrengthening M phase and concomitant solute depletion. Regardless, based on the work of Mahoney et al. and Hwang and Chou in welding of 7075, peak temperatures near 350 °C appear to be most effective in reducing the strength or hardness in the HAZ when HAZ temperatures greater than or equal to 350 °C are present.

As stated previously, a series of welds were made in the 7050 plate material. Table 4.1 lists all of the welding conditions. In order to correlate weld properties with the welding parameters, it is necessary to understand parameter effects on weld power, specific weld energy, and, ultimately, temperature history. Figures 4.6(a) and (b) illustrate the relationships between weld power and welding speed and specific weld energy and welding speed. It is important to keep in mind that weld power (like the torque) is not a controlled variable in FSW; it is a response variable. This is substantially different from, for example, arc welding, where weld power may be controlled to different levels by variation of the arc current and voltage. Figure 4.6(a) shows that the weld power increases with increasing welding speed in a nonlinear way. It should be borne in mind that the rpm is increasing with welding speed for a given APR as well. It is true, however, that for a given rpm, the required power increases with increasing welding speed. The increase in required power for increasing welding speed is

**Table 4.1 Friction stir welding process parameters**

| Spindle rotation rate, rpm | Welding speed, mm/s | *z*-axis load, kN |
|---|---|---|
| 90 | 0.85 | 28.9 |
| 135 | 1.27 | 30 |
| 180 | 1.7 | 27.8 |
| 270 | 2.54 | 37.8 |
| 315 | 2.96 | 37.8 |
| 405 | 3.81 | 45.6 |
| 120 | 0.85 | 24.5 |
| 180 | 1.27 | 24.5 |
| 240 | 1.7 | 24.5 |
| 360 | 2.54 | 30 |
| 540 | 3.81 | 41 |
| 720 | 5.1 | 37.8 |
| 180 | 0.85 | 20 |
| 270 | 1.27 | 22 |
| 360 | 1.7 | 24 |
| 540 | 2.54 | 33.5 |
| 630 | 2.96 | 36 |
| 810 | 3.81 | 39 |
| 900 | 4.2 | 36.5 |

intuitively reasonable, because more material is processed per unit time at higher welding speed. The energy per unit weld length (equal to the power divided by the welding speed) declines with increasing welding speed; essentially, the relationship between weld power and unit weld energy is inverse for this series of welds.

The peak temperatures for some of the welds (calculated using the input torque FEM simulation) (Ref 21) are plotted versus the weld power in Fig. 4.7. The peak temperature in the welds increases almost monotonically with the weld power. A similar plot of peak *T* versus weld energy reveals the opposite relationship for this set of welds; that is, peak *T* drops with increasing weld energy. This is an interesting note and worthy of a sidebar. In the early days of FSW, the terms *hot weld* and *cold weld* were typically applied to, respectively, slow welds and fast welds. This terminology came about because, very often, a slow weld would be relatively hot far from the weld line, while a fast weld would be

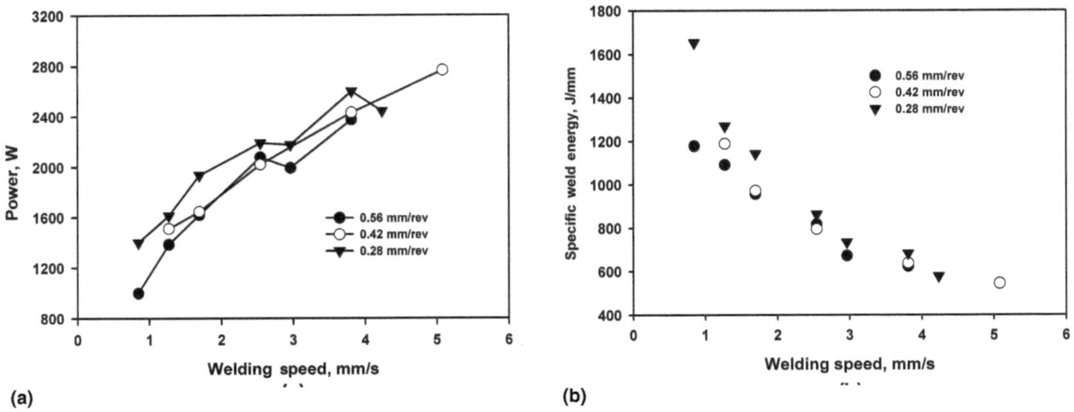


**Fig. 4.6** (a) Plot of weld power vs. welding speed for welds made using three different advances per revolution (APR). (b) Specific weld energy for welds made using three different APR

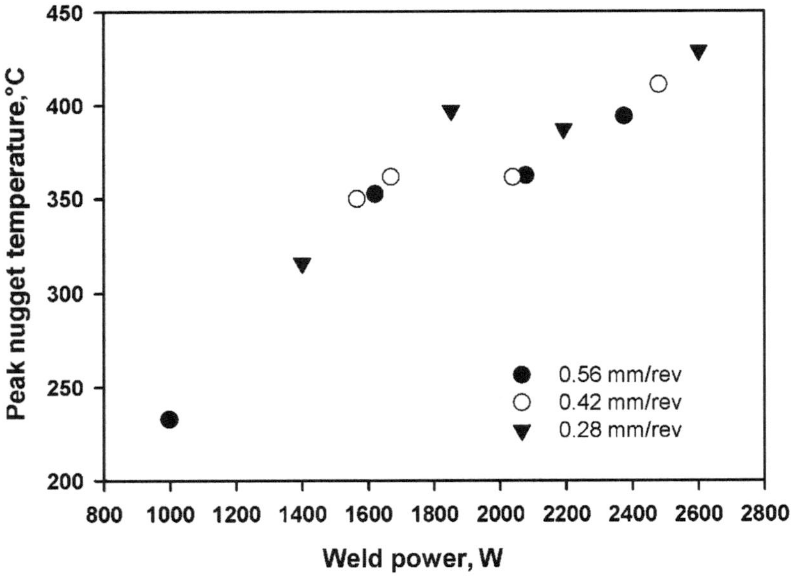


**Fig. 4.7** Calculated peak nugget temperature (from finite element modeling simulation) plotted vs. the weld power for a subset of the welds

relatively cool far from the weld line. The far-field temperature can be very misleading in relation to the peak temperature in a friction stir weld. The far-field temperature is more closely related to the unit weld energy than to the power, while the power is more important in determining the peak weld temperature. This is again different from the situation during fusion welding. In a fusion weld, the highest temperature in the solid metal (always at the fusion boundary) is the melting point. In a friction stir weld, the peak temperature can be substantially less than the melting temperature. So, in fusion welds, the peak temperature in the solid is independent of the weld parameters, but the far-field temperatures will be higher in welds made with high specific weld energy. In this series of welds, the fast welds are generally the hotter welds with respect to the peak temperature. One aspect of thermal history that is true of both FSWs and fusion welds is that the temporal length of the temperature transient experienced by the weldment is closely related to the welding speed. The higher the welding speed, the shorter the heatup and cooldown times. This transient time controls the time available for temperature-driven metallurgical processes in the weld nugget and HAZ as well as the quench rate (Ref 5, 21). Figure 4.8 illustrates the effect of welding speed on the peak temperature at the weld centerlines and the transient time for the welds made with an APR = 0.28 mm/rev. The heatup and quench rates vary directly with the welding speed, while the peak temperature is a nonlinear function of the weld power.

Figure 4.9 illustrates the effect of welding speed on the average nugget hardness in the as-welded condition. For each weld pitch, there is an initial relatively rapid increase in hardness with increasing welding speed (which, for a given weld pitch, also implies higher rpm). The rapid rise is followed by a hardness plateau; the plateau begins at a lower welding speed at lower APR, which again corresponds to a higher rpm. In Fig. 4.10, the change in average hardness of the nugget due to the PWHT is plotted versus the welding speed. Here, it is shown that a positive nugget hardness response to PWHT is observed for higher welding speeds but that lower speed coupled with higher rpm (smaller APR for a given welding speed) leads to a "better" response to PWHT. The implication in this case is that the higher-power welds (and the associated higher peak temperatures) lead to some solution treatment of the weld nuggets and hence some subsequent precipitation of strengthening precipitates during the PWHT. In those welds for which only particle coarsening has taken place during the weld thermal cycle, a negative response to PWHT is observed, probably due to additional coarsening.

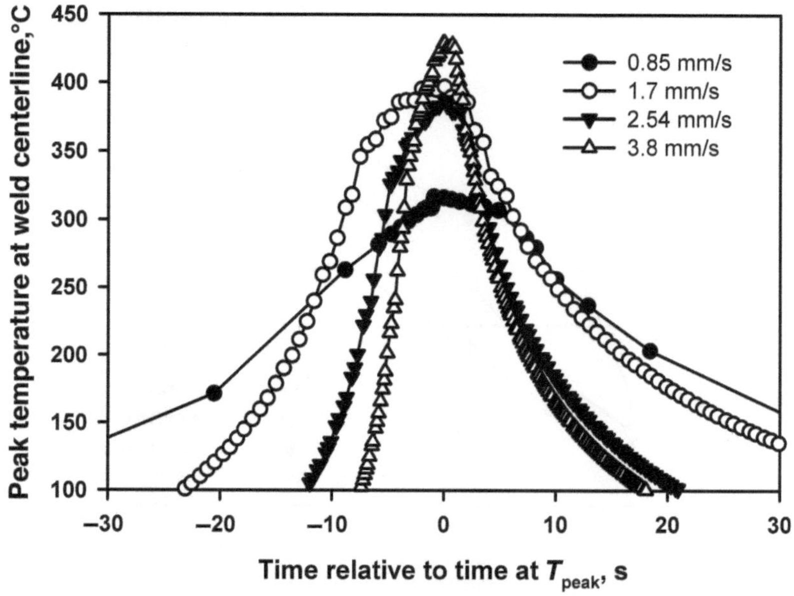


**Fig. 4.8** Plots of temperature ($T$) vs. time for 7050-T7451 welds made at different welding speeds

In the HAZ, the situation is somewhat different and overall less complicated. Figure 4.11 shows the effect of welding speed on HAZ minimum hardness for all three APRs. There is a general trend for increased HAZ hardness with increasing welding speed and no systematic variation with APR; that is, for a given welding speed, there does not seem to be an effect of rpm on the HAZ hardness. In addition, all of the HAZs exhibit a negative response to the postweld aging treatment.

Examination of the calculated temperature profiles shows that the HAZ minimum hardness location corresponds to a peak temperature of

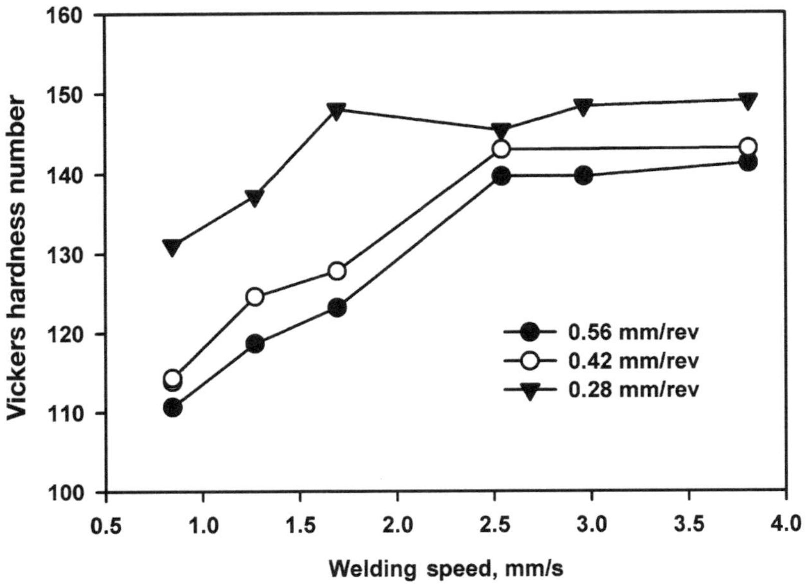


Fig. 4.9 Average weld nugget hardness in the as-welded condition

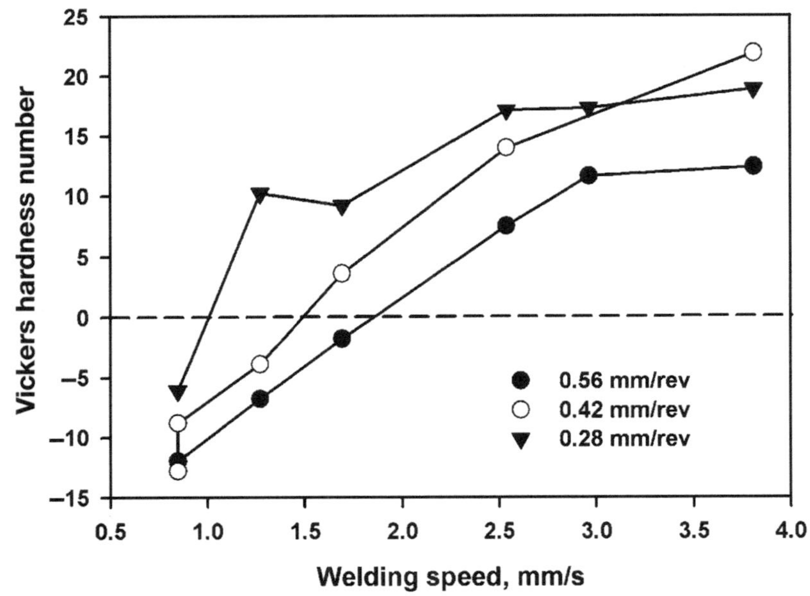


Fig. 4.10 Nugget response to postweld aging treatment

approximately 350 °C for all cases in which such a temperature is possible (that is, the peak weld temperature must be greater than 350 °C). On the other hand, a positive weld nugget response to PWHT is observed when the peak temperature in the nugget is greater than approximately 350 °C. These points are illustrated in Fig. 4.12, which shows nugget (closed symbols) and HAZ (open symbols) response to PWHT as the change in Vickers hardness number versus the calculated peak temperature at the pertinent locations. With very few exceptions, the peak temperature in the HAZ minimum hardness region is near 350 °C. The exceptions are for welds that had nugget peak temperatures less than 350 °C.

Data presented in this section indicate the following important points relative to FSW of 7050-T7451:

- For maximum nugget hardness, peak temperatures in the nugget must be high enough to provide some level of solution heat treatment.
- Peak temperature in the nugget depends primarily on weld power; higher power leads to higher temperature for a constant welding speed. Higher power at a constant welding speed is obtained by increasing the rpm.
- The hardness of the HAZ is dependent primarily on the welding speed; higher welding speed corresponds to higher HAZ hardness. This is likely due to the temporal length of the temperature transient; shorter time near 350 °C results in less overaging in the nugget.
- The HAZ minimum hardness is normally found where the peak temperature is near 350 °C; this temperature maximizes the kinetics of the overaging process in 7*xxx* alloys.

In order to maximize nugget hardness in alloys such as 7050-T7451, it is necessary to weld with sufficient power to achieve the solution treatment temperature in the weld nugget. In the nugget, there will likely be a secondary effect of welding speed that will influence the quench rate from the peak temperature, hence, the as-welded hardness and the response to PWHT. In order to maximize HAZ hardness, it is necessary to weld as fast as possible. If the peak $T$ in the nugget is above 350 °C (as it must be to achieve good nugget hardness), then at some distance from the weld centerline, the peak $T$ will be near 350 °C, and the time near this temperature must be minimized in order to limit overaging. The temporal length of the tem-

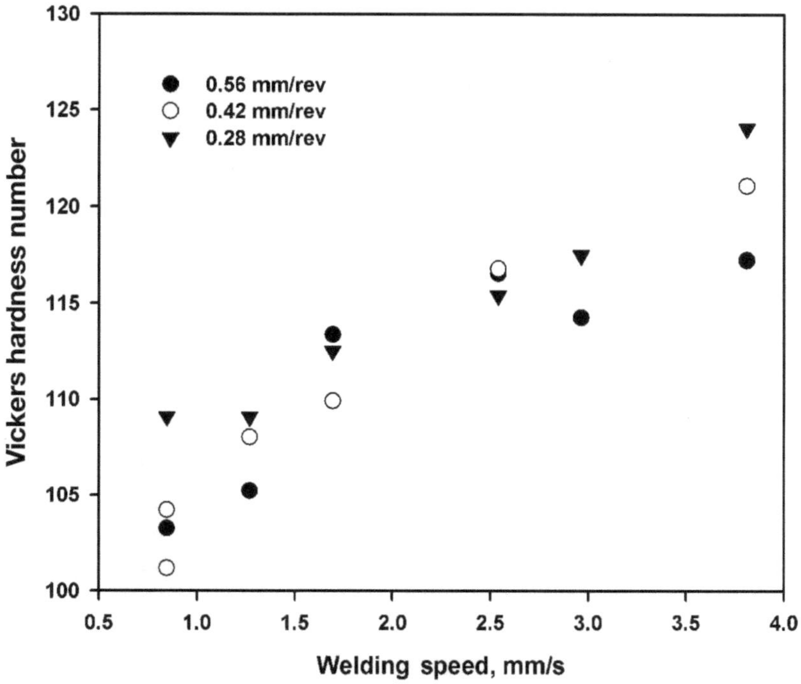


**Fig. 4.11** Heat-affected zone minimum hardness plotted vs. the welding speed

perature transient is minimized by welding at the highest possible speed.

**Naturally Aged Aluminum Alloys: 2524-T3.** Alloy 2524-T351 is a medium-strength, high-toughness aerospace alloy in a naturally aged condition. The nominal composition (weight percent) of the alloy is 4.2% Cu, 1.4% Mg, 0.6% Mn, 0.15% Zn, and 0.1% Ti, with traces of iron and silicon and the balance aluminum. The alloy is strengthened by Guinier-Preston-Bagaryatsky (GPB) zones in the solution-treated and naturally aged condition; artificially aged tempers are strengthened primarily by S′ phase. Both alloy 2524 and its older variant, 2024, are considered marginally weldable, at best, by fusion welding techniques. The alloys are in widespread use in the aerospace industry, and the advent of FSW spawned a substantial amount of effort in FSW research on these alloys. In the following, as for the 7050 discussed in the preceding section, an attempt is made to rationalize the response of 2524/2024 to variations in FSW parameters by reference to the metallurgy of the alloy.

One of the striking differences between FSWs in 2524-T351 (and 2024-T351) and FSWs in peak or overaged alloys is (as described briefly in a preceding section) the presence of inner and outer HAZ hardness minima. This phenomenon has been observed by several groups and has been well explained by Jones et al. (Ref 14). The advancing and retreating side inner HAZ hardness minima are normally separated by a local hardness maximum in the weld nugget. There is, of course, also a local maximum between the inner and outer minima on both sides of the weld. Jones et al. performed transmission electron microscopy in all of these regions and described the microstructure as follows:

- In the nugget, streaks consistent with the presence of very fine S-phase particles or GPB zones were observed.
- The inner hardness minimum contained coarse S-phase particles (overaged).
- The local maximum between the minima was strengthened by fine S-phase precipitates.
- The outer minimum was devoid of precipitates even after postweld natural aging.

Jones et al. speculate that the outer minimum results from GPB zone dissolution at relatively

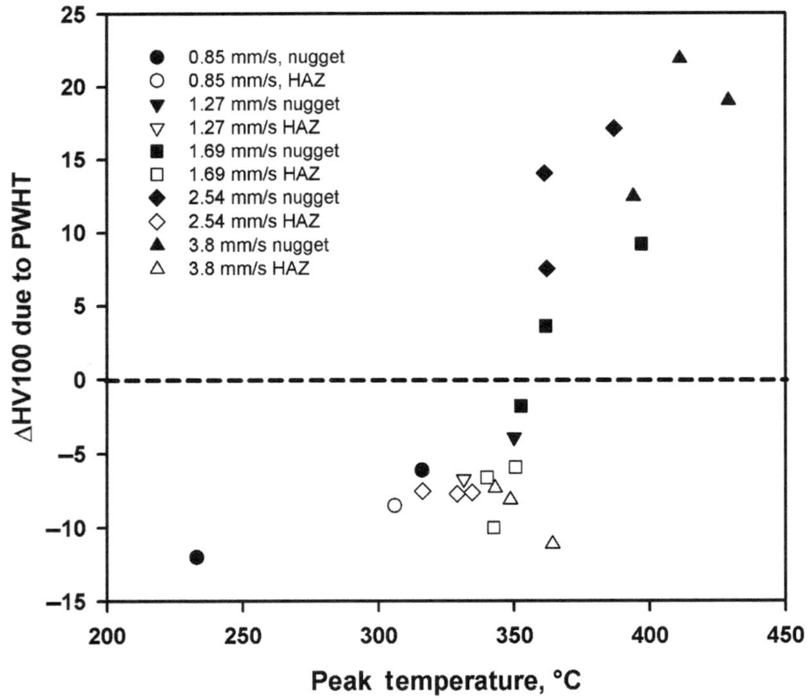


**Fig. 4.12** Nugget (closed symbols) and heat-affected zone (HAZ) (open symbols) minimum response to postweld heat treatment (PWHT) plotted vs. the peak temperature in the pertinent location

low temperature and that the zones do not reprecipitate due to the lack of quenched-in vacancies.

As for the heights and positions of the maxima and minima, these are functions of the weld process parameters; some aspects of this dependence are illustrated in Fig. 4.13. Figure 4.13 shows an example of the weld parameter dependence of the hardness distribution for two 2524-T351 welds referred to as fast and slow (Ref 22). Parameters for the fast weld were 480 rpm and 3.4 mm/s (0.13 in./s). Parameters for the slow weld were 120 rpm and 0.85 mm/s (0.033 in./s). Both welds were performed with the same tool APR. The two distributions have similarities but also differ substantially from each other. In the fast weld, nugget hardness is equivalent to the base-metal hardness, while the nugget hardness in the slow weld is quite low. The inner hardness minimum is just at the edge of the nugget in the slow weld and is only slightly lower than the slow weld nugget hardness. In the fast weld, the inner HAZ hardness minimum is somewhat removed from the nugget edge. Both welds exhibit local maxima at approximately 15 mm (0.6 in.) from the weld centerline and then local minima near 20 mm (0.8 in.) from the centerline. Beyond the second local minima, the hardness is nearly the same as the base-metal hardness. Both the inner and outer local hardness minima are much lower in the slow weld than in the fast weld.

The difference in nugget hardness between the fast and slow welds can be attributed to higher peak temperature in the fast weld, resulting in solution heat treatment of the fast weld nugget and overaging of the slow weld nugget; additionally, the fast weld nugget will have experienced a higher quench rate than the slow weld nugget. A higher peak temperature in the fast weld nugget is inferred from the nugget grain sizes of the two welds and the spindle torques required to make the two welds. In the nugget of the fast weld, the average grain size is 6 μm (0.24 mil); the grain size in the slow weld is not resolvable optically at 500×. Based on previous work, the larger grain size is indicative of a higher peak temperature, especially in light of the fact that the tool APR is the same in both welds. Also, the torque required for the fast weld is approximately half that needed for the slow weld. Assuming sticking conditions, or nearly sticking conditions, at the tool/workpiece interface, the torque should be a direct indicator of the flow stress of the material. Hence, low torque indicates high temperature through the relationship between temperature and flow stress.

In another study of 2524 FSW, relationships between nugget grain size, nugget hardness, and

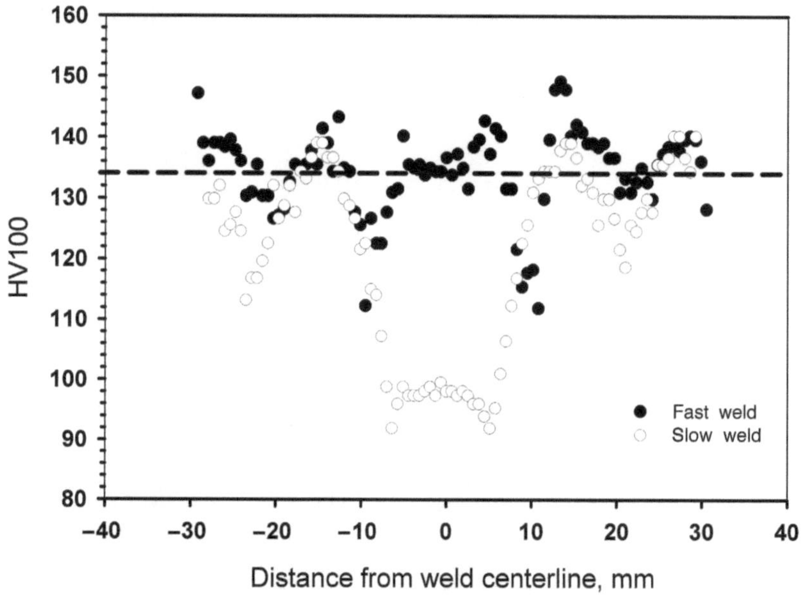


**Fig. 4.13** 2524-T351 friction stir weld transverse hardness distributions in a slow and a fast weld

HAZ hardness and various welding parameters have been studied and elucidated (Ref 13). Figure 4.14(a) shows the grain size in the weld nugget of 2524 FSWs as a function of rpm, with welding speed and *z*-force held constant. Figure 4.14(b) shows the nugget and inner HAZ hardness of the same welds also plotted against the rpm. Comparison of the two figures shows that the grain size and hardness have very similar relationships to the rpm. Both exhibit rapid increases with increasing rpm in the low-rpm range and then a plateau starting near 300 rpm and going up to 800 rpm. The measured nugget hardness becomes essentially flat above 300 rpm, while the grain size continues to grow slowly with increasing rpm above 300 rpm. Also shown in Fig. 4.14(b) is the inner HAZ minimum hardness, which is essentially unaffected by the rpm at constant welding speed. The combined behavior of the grain size and the nugget hardness may be explained by supposing that the solution heat treatment temperature is attained in the weld nugget at 300 rpm. At higher rpm, the peak temperature will continue to rise slowly, as attested to by the grain size; however, increased temperature above the solution heat treatment temperature has little effect on the nugget hardness. Figures 4.13 and 4.14 together indicate that the behavior of the 2524-T351 is similar to that observed in the 7050-T7451, with the exception of the presence of the outer HAZ hardness minimum in the 2524-T351. Specifically, maximum nugget hardness is obtained by welding at a sufficiently high peak temperature to enable solution heat treatment of the weld nugget (Fig. 4.14), and the HAZ minimum hardness is increased by welding at higher speeds (Fig. 4.13).

Another phenomenon that was observed in the study cited in the preceding paragraph is that of overheating. Although FSW is a nominally solid-state process, if there are low-melting regions embedded in the bulk higher-melting material, then local melting may occur. This phenomenon has also been observed in highly alloyed 7*xxx* alloys (Ref 12). Local melting in the 2524-T351 FSWs shown in Fig. 4.14 was discovered by performing tensile tests of all nugget material. Figures 4.15(a–d) show tensile properties for all nugget specimens produced by welding at one welding speed, 2.11 mm/s (0.083 in./s), and a range of rpm from 120 to 600. In each graph, data for full-thickness and root half-thickness (top half of the weld excluded) specimens are shown. Figure 4.15(a) shows the 0.2% offset yield strength for both specimen types. The yield strength exhibits a similar dependence on rpm as that shown by the nugget hardness (Fig. 4.14b), and the values are nearly identical for both the full-thickness and root-half specimens. Figure 4.15(b) shows the ultimate tensile strengths for the root-half and full-thickness specimens. The tensile strength of the root-half specimens has the same dependence on rpm as does the yield strength; however, the tensile strength of the full-thickness specimens declines sharply between 480 and 600 rpm. This decline in full-thickness specimen tensile strength is mirrored by the uniform

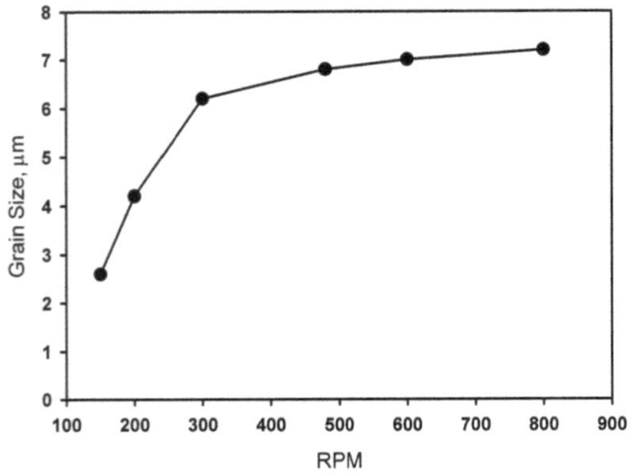


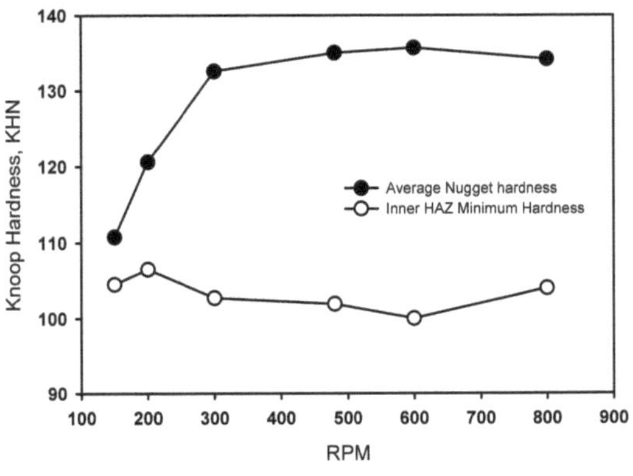


**Fig. 4.14** (a) Nugget grain size vs. rotation speed and (b) nugget center and heat-affected zone (HAZ) minimum hardness variation as a function of rotation speed with constant welding speed and *z*-axis force, Fz

and total elongations (Fig. 4.15c and d). Optical micrographs showing the changes in the near-crown nugget microstructure as tool rotation rate is changed from 480 to 800 rpm are shown in Fig. 4.16. The image in the top left corner of Fig. 4.16 is base metal. To the right of the base-metal image is an image from a 480 rpm weld. The 480 rpm weld shows constituent particle refinement via comminution. In the lower left (600 rpm), many grain boundaries are decorated with a second phase of unknown composition, and in the lower right (800 rpm), this decoration of the grain boundaries is even more complete than at 600 rpm. In Fig. 4.17, backscattered electron images of the same areas are shown. The brightly contrasting grain-boundary phases indicate that they are composed of relatively high-*z* elements, probably low-melting eutectic compositions. The decoration of the grain boundaries by high-*z* compounds is presumably responsible for the low ductility of the high-rpm welds (fracture surfaces indicated the presence of grain-boundary fracture). The morphology of the high-*z* compounds is consistent with a grain-boundary liquation process.

To summarize, as for the 7050 alloy described in the preceding section, to achieve maximum nugget strength, sufficient weld power must be used to produce a solution heat treated and subsequently naturally aged nugget. Unlike in the 7050, inner and outer HAZ hardness minima are produced. However, as for the precipitation-hardened 7*xxx* alloys, higher welding speed results in shallower hardness minima. Lastly, while it is desirable to weld with a sufficient peak temperature in the nugget to produce a solution heat treated condition, it is also important to keep the peak temperature below that which can result in local melting of eutectic phases.

## 4.4 Summary

In this chapter, some general guidelines for welding of various types of aluminum alloys have been presented. These guidelines were

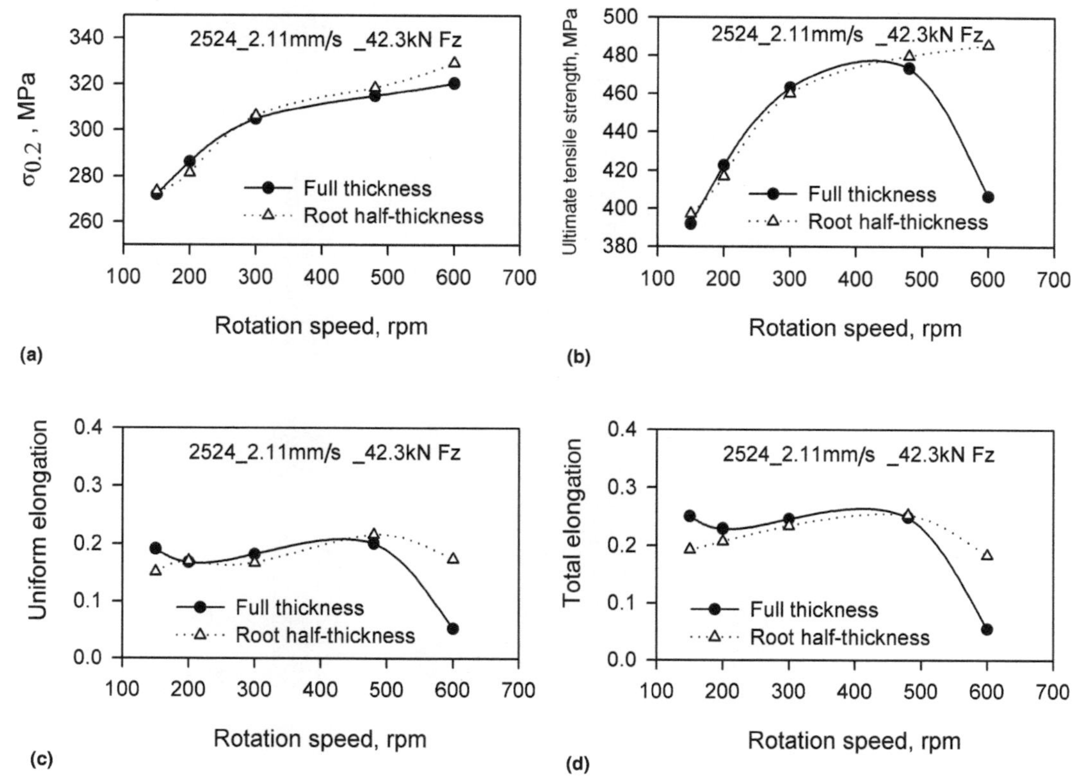


**Fig. 4.15** Tensile properties of all nugget material loaded in the longitudinal orientation (in the welding direction). Open symbols represent specimens taken from the root-half of the weld (excluding the crown region). Closed symbols are full-thickness specimens. Fz, *z*-axis force

developed based on the microstructure and properties of friction stir welds in the various classes of aluminum alloys; however, they are general guidelines only, and specific instances may require substantial deviation from these guidelines. Based on the foregoing, it has been shown that for precipitation-hardening alloys, maximum transverse tensile strength is normally obtained by welding at the highest possible welding speed. High welding speed minimizes the time available for overaging of the HAZ; hence, it results in the shallowest hardness minima in the HAZ. High welding speed generally requires relatively high weld power, which can result in high peak temperature in the weld nugget. Normally, a peak temperature in the weld nugget that is greater than the solution heat treatment temperature is desirable; however, if the nugget temperature exceeds that necessary to cause local melting (overheating), then the nugget may become brittle due to decoration of grain boundaries in the weld nugget with

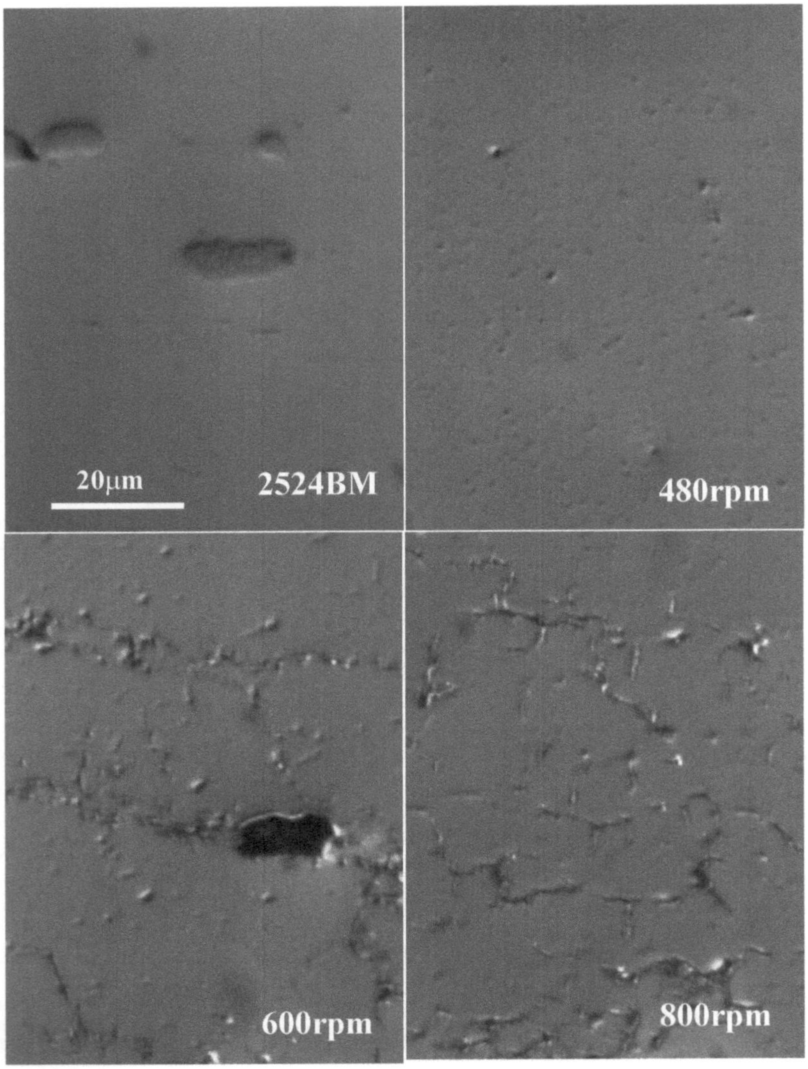


**Fig. 4.16** Optical micrographs of as-polished vertical-transverse sections (near the crown side) of 2524 base metal and friction stir welds made using 480, 600, and 800 rpm (welding speed, 2.11 mm/s, or 0.083 in./s; *z*-axis force, 42.3 kN, or 9500 lbf)

intermetallic phases. This overheating phenomenon is most likely in the highly alloyed high-strength alloys. For example, it is more likely to occur in 7075 than in 6061.

The properties of non-heat-treatable alloys are less sensitive to welding conditions. Alloys welded in the O temper will likely have weld nuggets that are slightly overmatched relative to the base metal. This overmatching of the weld nugget may be attributed to an increment of grain-boundary strengthening over that which is available in the base metal. Non-heat-treatable alloys that are welded in a strain-hardened temper will always be undermatched, because the recrystallization that occurs in the weld nugget eliminates all of the strengthening due to cold work. Increases in grain-boundary strengthening due to grain refinement in the nugget have not been demonstrated to be capable of compensating for the loss of cold work. Therefore, welding at high speed in non-heat-treatable alloys is more of a productivity issue than a property issue.

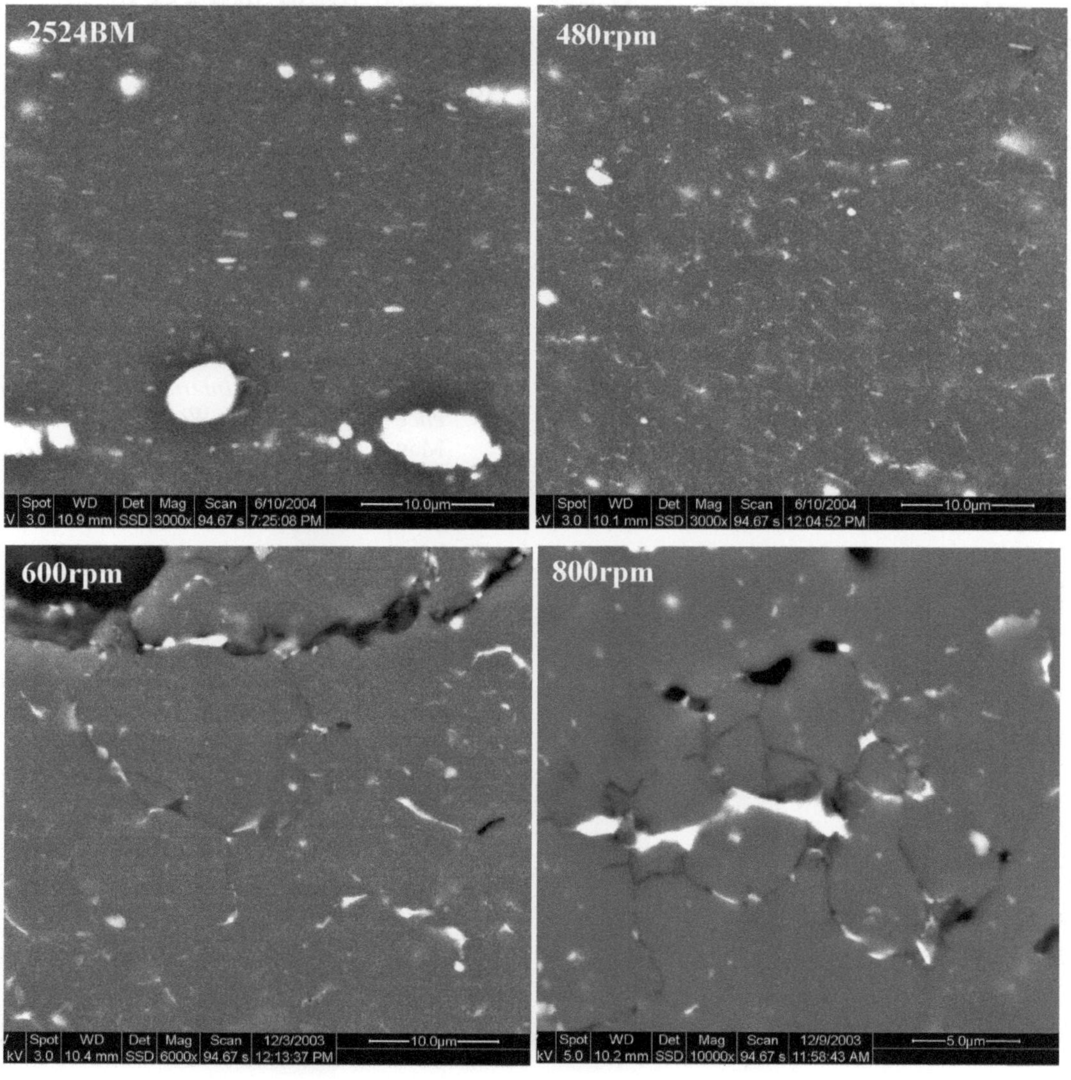


Fig. 4.17 Comparison of backscattered election SEM images from the as-polished vertical-transverse sections of 2524 base metal and friction stir welds produced with 480, 600, and 800 rpm

## REFERENCES


1. J.E. Hatch, Ed., *Aluminum: Properties and Physical Metallurgy,* American Society for Metals, 1984
2. S. Gourder, E.V. Konopleva, H.J. McQueen, and F. Montheillet, *Mater. Sci. Forum,* 1996, Vol 217–222, p 441
3. A.W. Bowen, *Mater. Sci. Technol.,* Vol 6, 1990, p 1058
4. R.D. Doherty, D.A. Hughes, F.J. Humphreys, J.J. Jonas, D.J. Jensen, M.E. Kassner, W.E. King, T.R. McNelley, H.J. McQueen, and A.D. Rollett, *Mater. Sci. Eng. A,* Vol 238, 1997, p 219–274
5. Y.S. Sato, M. Urata, and H. Kokawa, *Metall. Mater. Trans. A,* Vol 33, March 2002, p 625–635
6. A.P. Reynolds, W. Tang, Z. Khandakar, J.A. Khan, and K. Lindner, *Sci. Technol. Weld. Join.,* Vol 10 (No. 2), 2005, p 190–199
7. S. Gourdet and F. Montheillet, *Acta Mater.,* Vol 51, 2003, p 2685–2699
8. K. Jata and L. Semiatin, *Scr. Mater.,* Vol 43, 2000, p 743
9. B. Heinz and B. Skrotzki, *Metall. Mater. Trans. B,* Vol 33, June 2002, p 489–498
10. J.Q. Su, T.W. Nelson, and C.J. Sterling, *J. Mater. Res.,* Vol 18, 2003, p 1757
11. K.V. Jata, K.K. Sankaran, and J.J. Ruschau, *Metall. Mater. Trans. A,* Vol 31, 2000, p 2181
12. K.A.A. Hassan, P.B. Pragnell, A.F. Norman, D.A. Price, and S.W. Williams, *Sci. Technol. Weld. Join.,* Vol 8 (No. 4), Aug 2003, p 257–268
13. J. Yan, M.A. Sutton, and A.P. Reynolds, *Sci. Technol. Weld. Join.,* Vol 10 (No. 6), Dec 2005, p 725–736
14. M.J. Jones, P. Heutier, C. Dearayaud, D. Allehaux, and J.H. Driver, *Scr. Mater.,* Vol 52, 2005, p 693–697
15. A.P. Reynolds, unpublished research
16. A.P. Reynolds and F. Duvall, *Weld. J. Res. Suppl.,* Vol 78 (No. 10), Oct 1999, p 355-s–360-s
17. F. Viana, A.M.P. Pinto, H.M.C. Santos, and A.B. Lopes, *J. Mater. Process. Technol.,* Vol 92–93, 1999, p 54–59
18. P. Archambault and D. Godard, *Scr. Mater.,* Vol 42, 2000, p 675–680
19. M.W. Mahoney, C.G. Rhodes, J.G. Flintoff, R.A. Spurling, and W.H. Bingel, *Metall. Mater. Trans. A,* Vol 29, 1998, p 1955–1964
20. R.Y. Hwang and C.P. Chou, *Scr. Mater.,* Vol 38, 1998, p 215–221
21. M.Z.H. Khandkar, J.A. Khan and A.P. Reynolds, *Sci. Technol. Weld. Join.,* Vol 8 (No. 3), June 1, 2003, p 165–174
22. A.P. Reynolds and J. Pohlman, *Proceedings of the Seventh International Conference on Trends in Welding Research,* May 2005 (Calloway Gardens, GA), AWS/ASM International, in press

CHAPTER 5

# Mechanical Properties of Friction Stir Welded Aluminum Alloys

Murray W. Mahoney, Rockwell Scientific Company

FRICTION STIR WELDING (FSW) is a new solid-state welding process capable of welding all aluminum alloys, including the difficult-to-weld 2*xxx* and 7*xxx* aluminum alloys. Because there is no melting during FSW, that is, temperatures approach but remain below the solidus, friction stir welds can and most often do have superior properties compared to fusion welds. For example, some of the weld characteristics include a narrow heat-affected zone, a fine-grain wrought microstructure rather than a cast microstructure in the weld nugget, no filler material is needed, and there is no shrinkage porosity. Clearly, if the weld practice is performed properly, there is the potential for FSW to produce welds with high strength and ductility, increased fatigue life, and improved fracture toughness. During the early development of FSW, the process appeared simple, and indeed, it is simple compared to many conventional welding practices. However, as development continued, the complexity of FSW was realized. It is now known that properties following FSW are a function of both controlled and uncontrolled variables as well as external boundary conditions. For example, investigators have now illustrated that postweld properties can be a function of:

- *Tool travel speed:* influences total heat input
- *Tool rotation rate:* influences total heat input
- *Tool design:* shoulder diameter, scroll or concave shoulder, features on the pin, pin length
- *Tool tilt:* depends on the tool shoulder design but typically is 0 to 3°
- *Material thickness:* influences cooling rate and through-thickness temperature gradients
- *Alloy composition:* weld parameters not transferable from one aluminum alloy to another
- *Initial material temper:* influences alloy response
- *Cooling rate:* passive or active cooling
- *Heat sink:* thermal conductivity of materials in contact with the weld, for example, anvil and clamping system
- *Test sample size, location, and orientation:* where the sample is sectioned from the weld, especially through the thickness and longitudinal versus transverse orientation
- *Surface oxides:* potential for more or less of a continuous oxide within the weld
- *Joint design:* lap, butt, fillet
- *Postweld heat treatment:* dependent on alloy composition and preweld temper
- *FSW test system:* specific characteristics for each system, for example, spindle runout, heat dissipation through the spindle, anvil and clamps, and so on
- *Time between FSW and testing, that is, natural aging at room temperature:* For the 2*xxx* aluminum alloys, the weld zone stabilizes at room temperature within a few days. The 5*xxx* aluminum alloys do not naturally age. The 6*xxx* aluminum alloys naturally age slower than the 2*xxx* alloys, and more than 4 weeks may be necessary for welds to stabilize. For the 7*xxx* aluminum alloys, the weld zone does not stabilize without a postweld heat treatment.

This chapter presents properties for friction stir welded 2*xxx,* 5*xxx,* 6*xxx,* and 7*xxx* aluminum alloys as well as some results for aluminum-lithium alloys and aluminum metal-matrix composites. Not all variables and boundary conditions listed previously were reported by the different investigators, and indeed, it would not be reasonable to expect this considerable detail. Also, it is not possible within this document to identify all the variables reported by the different investigators. For experimental details, readers are encouraged to read the original manuscripts. Thus, properties presented herein are illustrative of what can be achieved using good FSW practices, that is, full-penetration butt welds with no detectable defects. Lap and fillet weld joints are not considered herein, because each of these weld joint geometries introduces issues specific to the joint geometry rather than inherent material properties. Further, results are not presented for FSW with a self-reacting or bobbin tool. There are insufficient properties data available at this time for this method of FSW. Where possible, property ranges are provided, illustrating the spread of results from different laboratories and facilities. At times, data are limited to one investigator, and thus, precaution should be exercised.

For testing of monolithic materials, test procedures and interpretation of test results are relatively straightforward. However, when welds are tested in the transverse orientation, a material with a composite of properties within the gage length is tested. That is, loads are applied across the weld nugget, thermomechanically affected zone (TMAZ), heat-affected zone (HAZ), and parent metal. For most aluminum alloys and temper conditions, each weld zone location will have different mechanical properties, and thus, strain localization will occur in the lowest-strength region. Because the gage length of this softer low-strength zone is not known, it is not realistic to measure transverse strain in the customary manner. However, if different weld locations are tested in the longitudinal orientation, and only one postweld microstructure is included within the gage diameter, then properties for each weld zone can be determined separately. Some investigators have isolated properties for the different weld zones, and, when available, these results are presented.

It is important to understand the FSW nomenclature to identify and recognize the different weld-zone microstructures and resultant properties. Early in the development of FSW, the term *thermomechanically affected zone* was used to identify the region between the weld nugget and the HAZ. This TMAZ region experienced both heat and deformation but the deformation was insufficient to facilitate full recrystallization. This nomenclature was convenient for aluminum alloys due to the existence of a distinct zone between the nugget and HAZ that met this definition. However, this definition of a TMAZ proved to be inappropriate for other alloy systems where a distinct TMAZ was not evident, for example, ferrous materials. Thus, the FSW licensees group (license holders with rights to use the TWI initial FSW patent) recommended the TMAZ be redefined to include all regions affected by both heat and deformation, with the weld nugget being a subset within the TMAZ. Unfortunately, the initial definition of the TMAZ has continued to be used in the recent literature and certainly in all of the early literature. It would be too confusing to attempt to change this nomenclature in a review document where reference is made to others' work when essentially all published data refer to the TMAZ as a region separate from the weld nugget. Accordingly, in this chapter, the initial nomenclature used for the TMAZ is followed for the different weld zones. Figure 5.1 illustrates the weld-zone nomenclature used in this chapter for aluminum alloys. Figure 5.1(a) illustrates a low-magnification view of a friction stir weld in an aluminum alloy, and Fig. 5.1(b) and (c) illustrate the uplifted grains on the retreating and advancing sides of the weld nugget.

## 5.1 2*xxx* Aluminum Alloys

The preponderance of research and reported data for the 2*xxx* aluminum alloys is concentrated on 2024 Al, an Al-Cu-Mg alloy (Ref 1–19). Thus, this section focuses on 2024 Al with reference to other 2*xxx* aluminum alloys, where data are available. In general, the weldability of 2024 Al by conventional fusion welding practices, that is, gas metal arc welding or gas tungsten arc welding, is limited. Aluminum 2024 can be welded with proper procedure and equipment, but except for resistance welding, weldability ratings for 2024 indicate limited weldability. Also, 2024 Al is more sensitive to cracking during conventional welding than other aluminum alloys, and the joint design and fixtures must be so proportioned as to put minimum strain on the joint during the cooling period (Ref 20). These precautions are not required during FSW, again due to the absence of melting associated with FSW. Basically, 2024 Al is easily friction stir welded without any special procedures, other than good FSW practices. Selected post-FSW properties are presented as follows, illustrating properties in the weld nugget, TMAZ, and HAZ.

**Hardness.** Many investigators use hardness data as an initial evaluation of variation in mechanical properties across the weld zone. First, it should be understood that 2024 Al will naturally age at room temperature following an excursion to a temperature above that where strengthening precipitates go into solution. In 2024, most of the strengthening occurs within a day at room temperature; the mechanical properties are essentially stable after four days (Ref 21). Figure 5.2 illustrates the change in hardness for a friction stir weld naturally aged at room temperature for >12,000 h (Ref 17). Most of the hardness change occurs in the first week. After this time, the hardness appears to stabilize, and the material reaches an equilibrium condition. Mechanical properties would be expected to increase in a corresponding manner to the increase in hardness. Although time between welding and testing is not usually reported, it is assumed that testing was performed at least one week after FSW, and thus, this temporary instability is of little concern when considering postweld properties in 2024 Al. However, as shown subsequently for the 7*xxx* aluminum alloys, because the 7*xxx* alloys do not stabilize in a reasonable time (if ever) after FSW, natural aging needs to be considered when evaluating mechanical properties.

The work of Bussu and Irving on 6.35 mm (0.25 in.) thick 2024-T351 Al sheet is illustrative of hardness variations following FSW in 2024-T351 Al (Ref 12). In their work (Fig. 5.3), hardness is illustrated both as a function of distance from the joint interface and depth from the top surface. As shown, a typical "W"-shaped hardness curve is created. Due to the close relationship between hardness profiles and tensile test results, this composite of hardness results has implications for resultant mechanical properties. The studies in this work show four distinct hardness zones:

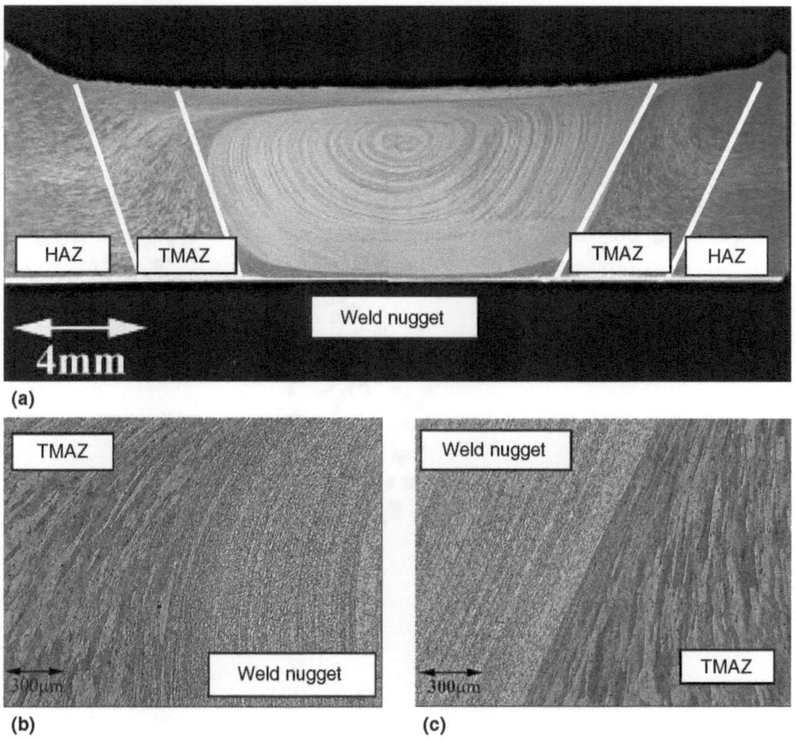


**Fig. 5.1** (a) Micrograph illustrating different zones in a friction stir welded aluminum alloy. (b) Retreating side. (c) Advancing side. HAZ, heat-affected zone; TMAZ, thermomechanically affected zone

- The weld nugget extending 5 to 6 mm (0.20 to 0.24 in.) from each side of the joint interface, where the hardness is nearly constant
- The remainder of the TMAZ extending an additional 5 to 6 mm from the weld nugget, where hardness decreases sharply
- The HAZ extending an additional 15 to 20 mm (0.6 to 0.8 in.) from the TMAZ, where hardness reaches a minimum and then increases as distance from the weld centerline increases, even achieving a hardness greater than the parent metal
- The hardness of the parent metal unaffected by FSW

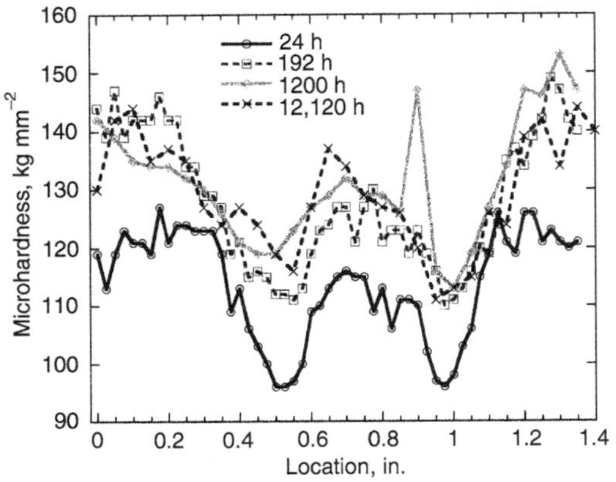


**Fig. 5.2** Hardness results for friction stir welded 2024 Al following natural aging for >7 months. Source: Ref 22

The work by Bussu and Irving also illustrates hardness differences from the crown surface to the root surface of the friction stir weld (Ref 12). During FSW, heat input and heat extraction are nonuniform through the material thickness. That is, the FSW tool generates heat from both the tool shoulder and the tool probe, but the influence of the shoulder is limited in depth, depending on shoulder design, for example, convex or scrolled, shoulder tilt, and applied axial force. In aluminum alloys, the shoulder influence is typically less than 2 mm (0.08 in.) deep (based on microstructural observations). Thus, more heat will be generated near the crown surface than the root. Further, the root surface is adjacent to an anvil, where heat is

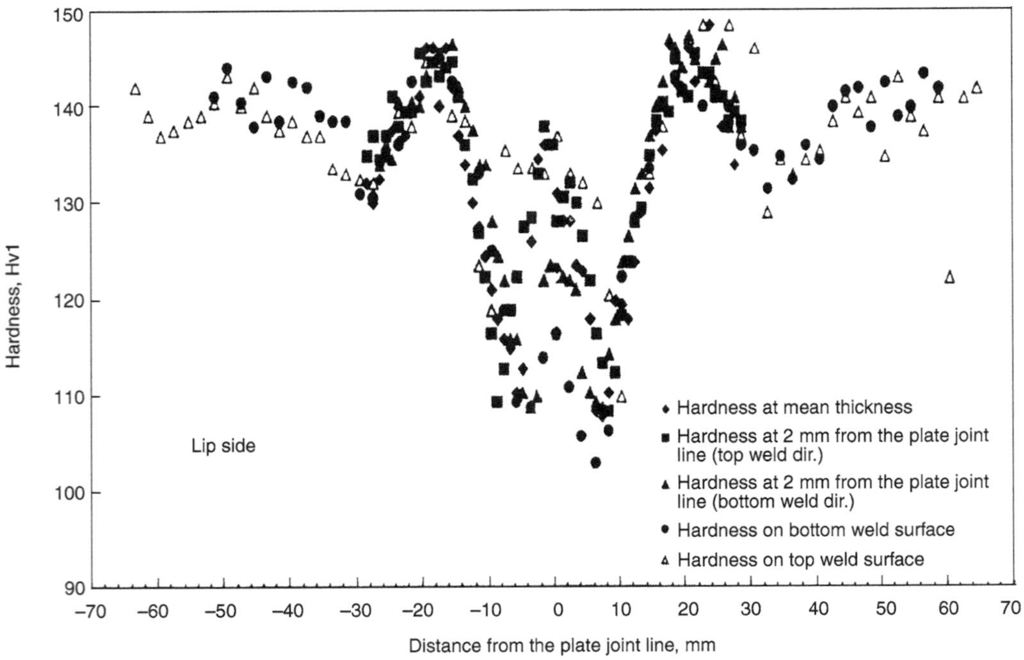


**Fig. 5.3** Microhardness traverse across the friction stir weld at various positions in the section. Source: Ref 12

extracted via conduction. Thus, a mild-to-severe temperature gradient can be expected through the thickness of a friction stir welded joint, depending on a number of boundary conditions, especially material thickness. Thus, through-thickness hardness variations would also be expected. However, due to the high thermal conductivity of aluminum alloys, this hardness difference is only evident within the regions where metal deformation occurs. This is clearly illustrated in Fig. 5.3, where through-thickness hardness varies within the weld nugget and TMAZ but does not vary within the HAZ or parent metal. As shown, hardness in the weld nugget is greatest near the crown surface and lowest near the root surface.

Microstructures within these different hardness zones are typical of what is achieved following good FSW practices. The nugget has a fine, recrystallized grain structure, with hardness values between 110 and 140 HV, again with hardness decreasing from crown to root. Immediately outside the nugget, the microstructure consists of highly elongated and deformed grains, with a sharp drop in hardness reaching a minimum either within the TMAZ or near the boundary between the TMAZ and HAZ. Outside the TMAZ is the HAZ with a parent-metal microstructure, where hardness increases until the parent metal, unaffected by either heat or deformation from FSW, is reached. For a more detailed explanation of FSW microstructures, including recrystallization, grain growth, particle coarsening, precipitate-free zones, and so on, see Chapter 4.

Hardness results for 2024 Al, including the T3 condition, are reported by additional authors illustrating results for different FSW parameters (Ref 1–3, 6, 8, 9, 15, 16, 19). For example, Kristensen et al. present hardness results at a fixed depth for 2024-T3 as a function of tool rotation rate and tool travel speed (Ref 15). Although differences were small, hardness in the weld zone was shown to be influenced by FSW parameters. In the work of Hashimoto et al. (Fig. 5.4), the hardness of 2024-T6 Al is compared for FSW and gas tungsten arc welding (GTAW) (Ref 1). This work illustrates the narrow HAZ associated with FSW compared to GTAW. Also, the hardness minimum was lower for GTAW compared to FSW. This is to be expected, due to potentially lower heat input and localized heat concentration associated with FSW. Biallas et al. illustrated hardness differences for different sheet thicknesses (Ref 2). Comparing hardness following FSW of 1.6 and 4 mm (0.06 and 0.15 in.) thick 2024-T3 sheet, higher maxima and lower minima were obtained for the 4 mm sheet, that is, higher hardness differences both through the sheet thickness and lateral from the weld centerline. This result was explained based on a critical cooling rate and partial reprecipitation of the hardening particles if a critical cooling rate is exceeded. Not only is the cooling rate higher in the thin sheet, the temperature gradients are also smaller, reducing the hardness minima.

Additional hardness data are available for friction stir welded 2524-T351 Al as a function of tool rotation speed (Ref 23). Alloy 2524 is a high-toughness aerospace alloy with improved plane isotropy and lower constitutive particle content relative to 2024. This work by Yan et al. illustrates an increase in hardness for rotation speeds from 150 to 300 rpm, reaching a plateau at 135 KHN and remaining constant from 300 to 800 rpm, the highest rotation speed evaluated. The nugget hardness values exhibit a trend that is nearly identical to that of the grain size relative to rotation speed but opposite of the typical Hall-Petch effect. The HAZ minimum hardness values (105 KHN) are nearly unaffected by changes in the rotation speed.

Surface hardness data for 2219-T8751, an aluminum-copper alloy, illustrate a similar "W"-curve response to FSW (Fig. 5.5) (Ref 24). As before, the hardness variation shown in Fig. 5.5 is presented as a function of distance from the weld centerline and indicates softened material in the weld zone, with the softest material at the edges of the stir weld boundary. In this case,

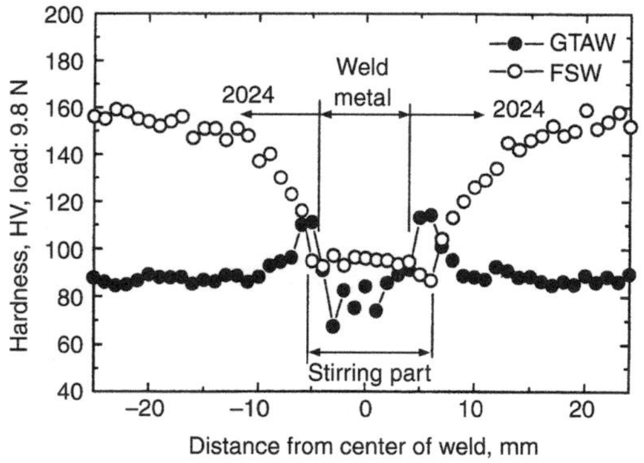


**Fig. 5.4** Hardness distribution in 2024 Al for both gas tungsten arc welding (GTAW) and friction stir welding (FSW). Source: Ref 1

hardness in the stirred region remains well below that of the parent metal.

**Mechanical Properties.** As noted previously, caution must be exercised when interpreting strain within weldments, due to the potential for strain localization in transverse tensile tests. However, yield and tensile strength results require no special consideration. Yield strength is often related to hardness, and based on the hardness curves typically obtained for 2024 Al, yield strength as well as fracture location should correlate with the lowest-hardness location in the "W"-hardness curve. Mechanical properties for 2024 Al have been reported in numerous publications as a function of FSW variables, including tool rotation rate, travel speed, and sheet thickness (Ref 2, 4–8, 11, 15).

Typical yield strength results for 2024-T3 Al are those reported by Biallas et al. as a function of process parameters and sample thickness (Ref 2). For 4 mm thick sheet, yield strength following FSW was 66 to 72% (280 to 305 MPa, or 41 to 44 ksi) of parent-metal yield strength (424 MPa, or 61.5 ksi), depending on FSW parameters. For the 1.6 mm sheet, yield strength was 93 to 100% (300 to 325 MPa, or 43.5 to 47 ksi) of parent-metal yield strength (325 MPa). For each material thickness, the highest yield strengths were obtained for the combination of higher tool rotation rate and travel speed. The increased strength with increasing lateral speed can be explained by a partial reprecipitation of the hardening particles, which takes place if a critical cooling rate is exceeded. The high yield strength for the friction stir welded 1.6 mm thick sheet is unusual and may be attributed to a lower heat input associated with both a high travel speed and a smaller tool shoulder diameter and an accompanying high cooling rate for this thin sheet. Most typical are the results for the 4 mm sheet, where yield strength is reduced approximately 30%. These results are comparable to those of von Strombeck et al., where, for 5 mm (0.2 in.) thick 2024-T351 Al, the yield strength of the welded sample is 77% (270 MPa, or 39 ksi) of the parent-metal yield strength of 350 MPa (51 ksi) (Ref 6). Similarly, the results of Magnusson et al. confirm the former conclusion of Biallas et al., where yield strength for thin sheet following FSW approaches 100% of parent-metal properties (302 and 310 MPa, or 43.8 and 45 ksi, respectively) (Ref 8). In the work of Magnusson et al., the sheet was 2 mm thick, very comparable to the 1.6 mm thick sheet used by Biallas et al. However, the rotation rate and travel speed used by Magnusson et al. was 1180 rpm and 110 mm/min (4.3 in./min), which were both approximately half that used by Biallas et al. (2400 rpm and 240 mm/min, or 9.5 in./min). From this comparison, it may be hypothesized that sheet thickness and possible tool design are more important than FSW parameters with regard to cooling rate, heat input, and resultant yield strength.

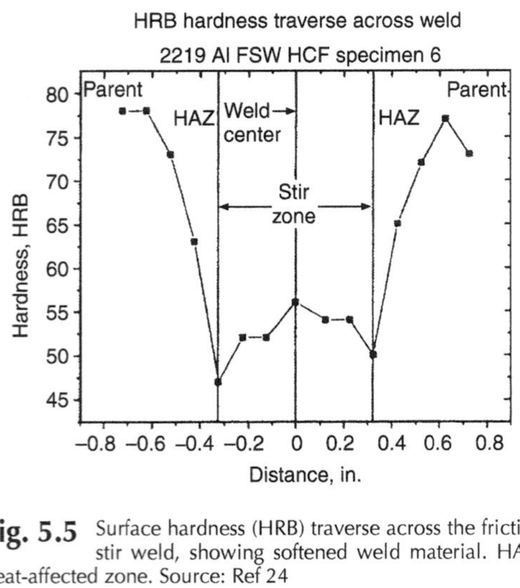


**Fig. 5.5** Surface hardness (HRB) traverse across the friction stir weld, showing softened weld material. HAZ, heat-affected zone. Source: Ref 24

Results by Biallas et al. are illustrative of transverse tensile strength as a function of sheet thickness and FSW parameters (Ref 2). For 4 mm thick 2024-T3 Al sheet, the tensile strength range is 82 to 87% (408 to 432 MPa, or 59 to 62.6 ksi) of the parent-metal strength of 497 MPa (72 ksi). For the 1.6 mm sheet, tensile strengths range from 90 to 98% (425 to 460 MPa, or 61.6 to 67 ksi) of the parent-metal tensile strength of 472 MPa (68.5 ksi). The same explanation for the high yield strength can be offered for these high tensile strength values. Similarly, von Strombeck et al. reported a tensile strength of 83% of parent-metal tensile strength for friction stir welded 5 mm thick 2024-T351 Al sheet (410 and 493 MPa, or 59.5 and 71.5 ksi, respectively) (Ref 6). Using 6 mm thick 2024-T3 Al, Kristensen illustrated tensile strength as a function of rotation rate and travel speed (Ref 15). When the weld travel speed was high (>400 mm/min, or 16 in./min), there was a significant variation in tensile strength, with lower tensile strength at the higher travel speeds (400 to 560 mm/min, or 16 to 22 in./min). These strength differences were attributed to fracture within the weld nugget as opposed to fracture in the HAZ or parent metal.

Although one may attribute the tensile strength differences to weld defects, the authors did not indicate this to be the cause, and indeed, the metallography illustrated defect-free welds. When the fracture was located within the HAZ, that is, low travel speeds (275 to 400 mm/min, or 11 to 16 in./min), tensile strength was both high (~440 MPa, or 64 ksi) and did not vary with travel speed. Hashimoto et al. evaluated 2024-T6 Al and illustrated a post-friction stir weld tensile strength 80% (440 MPa) of the parent-metal value (Ref 1). Further, Hashimoto et al. compared FSW to GTAW and showed the GTAW tensile strength to be 57% of that for the friction stir weld. Magnusson et al. evaluated postweld heat treatment for friction stir welded 2024-T3 Al and illustrated no change in tensile strength following a solution heat treatment and T3 age (Ref 8).

Russell et al. evaluated tensile strength with a 6.35 mm diameter hole located both in the center of the weld and in the HAZ (Ref 7). For 2.3 mm (0.09 in.) thick friction stir welded 2024-T3 Al, there was little change in the net section tensile strength when the hole was located within the weld zone (470 versus 428 MPa, or 66 versus 62 ksi, or ~83% of parent-metal tensile strength), but with the hole located within the HAZ, the tensile strength decreased further (325 MPa, or 47 ksi) to approximately 60% of parent-metal tensile strength. Apparently, the HAZ is notch-sensitive compared to the parent material, whereas the weld itself was not notch-sensitive.

As mentioned previously, transverse strain to failure is not meaningful, because the tensile gage length is a composite with variable strengths, thus resulting in strain localization at the strength minima. However, the hardness curves for 2024 Al can be relatively flat compared to some aluminum alloys following FSW, and although not completely accurate, transverse strain can provide some indication of ductility. Transverse strain to failure has been reported by some investigators (Ref 2, 6, 8, 10). Without consideration for other factors, the average transverse strain for a large number of samples was 8.3%, with a low of 5.1% and a high of 16.3%. Although, as expected, this is lower than the base material (15 to 21%), the weld zone is still ductile. Yan et al. evaluated mechanical properties of friction stir welded 2524-T351, including the influence of rotation speed on total elongation (Ref 23). Total elongation was relatively consistent up to 500 rpm but was significantly lower at 600 rpm. The authors attributed this reduced ductility to excessive rotation speeds, resulting in localized grain-boundary melting near the weld crown. In addition to Yan et al., investigators have only infrequently claimed melting during FSW (Ref 25, 26).

On a microscale, the microstructure in a friction stir nugget is inhomogeneous. One observation unique to FSW is the banded microstructure commonly seen within the weld nugget. This banding is associated with tool design and the tool advance per revolution. A banded microstructure in both 2024-T351 and 2524-T351 has been described, where the periodic bands have variations in grain size, band width, and particle distribution as a function of FSW process parameters (Ref 14, 27–29). Sutton et al. investigated local variations in the material response within the banded microstructure using minitension tests and digital image correlation (Ref 14). Periodic variations in strain response across the metallurgical bands indicated periodicity in particular features of the underlying banded microstructure. For example, Sutton et al. observed high-strain bands with a lower density of secondary particles and lower microhardness compared to the low-strain bands. Further, the bands had different hardening exponents but not different initial yielding behavior (Fig. 5.6). This suggests that the particles act as coarse aggregates with respect to the strain-hardening behavior of the weld nugget region.

**Fatigue, Fracture Toughness, Fatigue Crack Growth Rate.** Considerable data are available for fatigue, fracture toughness, and crack growth rate for friction stir welded 2024 Al (Ref 2–4, 6, 7, 9, 12, 13, 15, 16, 19). Recently published fatigue-life curves for 2024-T3 are shown in Fig. 5.7 as a function of material thickness (Ref 15). The denomination "as-welded" refers to the fact that the specimens were machined and polished only on the edges. The material flash and the rippled surface caused by the rotating shoulder were not removed. In comparison to polished specimens, lower fatigue strength is usually attributable to the as-welded surface. Results of Bussu et al. for a skimmed surface are included for comparison (Ref 3). The influence of sheet thickness is again evident for fatigue life. For the 1.6 mm (0.06 in.) thick 2024-T3 sample, fatigue strength, for the extent evaluated ($3 \times 10^5$ cycles), is unchanged compared to base material. For 4 and 6 mm (0.16 and 0.24 in.) thick friction stir welded samples, there is only a very small loss in fatigue strength. Further, the samples with a prepared smooth surface had lower fatigue strength. This is not usually the case, because crack initiation is most often associated with the spiral features created by the tool or a stress concentration associated with the weld flash. The authors attribute these higher fatigue strength results to a third-generation advanced FSW tool (not described) and to higher welding speeds (408 mm/min, or 16 in./min, at 340 rpm), resulting in lower overall heat input (Ref 15). Figure 5.8 illustrates the influence of specimen orientation on fatigue strength for 6.3 mm (0.25 in.) thick 2024-T351 Al in the as-welded condition with a stress ratio of $R = 0.1$ (Ref 3). A comparison between the parent plate and the weld data provides an indication of the potential degradation in fatigue properties due to FSW. Further, the loss in fatigue strength is greater for the transverse orientation compared to the longitudinal orientation. These samples were tested in the as-welded condition. These same authors evaluated the fatigue performance with machined surfaces (Ref 3). All the profile irregularities of the weld surface were removed, that is, tool marks, thickness variations, and the weld flash. Following surface machining, the results showed fatigue performance for both the longitudinal and transverse orientations to be improved, with fatigue life for each orientation nearly equivalent to parent-metal properties.

Hornbach et al. evaluated fatigue life for a variety of test conditions in friction stir welded 2219-T8751 Al (Ref 29). Test variables included a milled surface, a milled surface plus low-plasticity burnishing, a milled surface plus 100 h salt exposure, and a milled surface plus low-plasticity burnishing plus 100 h salt exposure. Low-plasticity burnishing introduces compressive residual stresses into the surface to a depth dependent on the applied burnishing load. For friction stir welded material with the flash and circular tool pattern removed by milling, the threshold stress was >230 MPa (33 ksi). When this same type sample was exposed to a salt solution for 100 h and subsequently fatigue tested, the threshold stress decreased to ~175 MPa (25 ksi). Burnishing the milled sample increased the threshold stress ~70 MPa to 300 MPa (~10 ksi to 43 ksi). Similarly, following burnishing, the threshold stress in the salt-exposed sample increased ~100 MPa to 275 MPa (~15 ksi to 40 ksi). (Ref 30).

Fatigue crack propagation data for 2024 Al are available for both compact tension and sur-

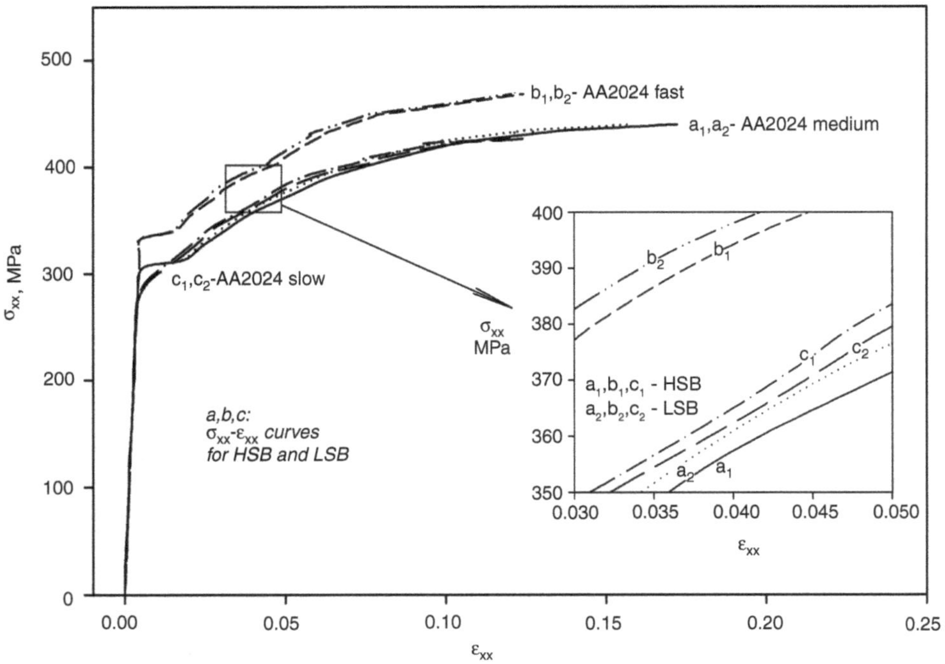


**Fig. 5.6** Comparison of stress-strain curves between high-strain bands (HSB) and low-strain bands (LSB) for fast, medium, and slow processing of AA2024-T351. Source: Ref 14

face crack tension specimens (Ref 9, 12, 13, 16). In the work of Christner et al. using 2024-T3 Al, compact tension specimens were used with the precrack located both within the weld nugget and within the HAZ/TMAZ zones. Crack propagation results with the crack oriented in the weld direction are illustrated in Fig. 5.9 (Ref 13). Testing showed crack growth rates (*da*/*dN*) in the HAZ/TMAZ to be equivalent to the base metal. Crack growth rate in the weld nugget was slightly faster than in the base metal, particularly at lower values of the stress-intensity range ($\Delta K$). A higher crack growth rate can be attributed to the fine-grain microstructure in the weld

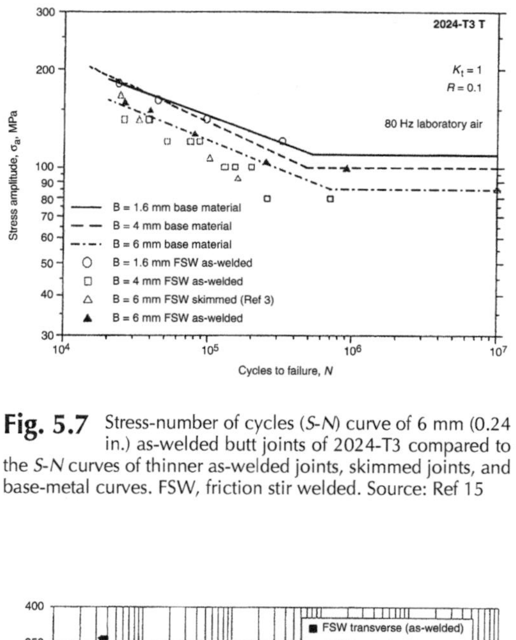


**Fig. 5.7** Stress-number of cycles (*S-N*) curve of 6 mm (0.24 in.) as-welded butt joints of 2024-T3 compared to the *S-N* curves of thinner as-welded joints, skimmed joints, and base-metal curves. FSW, friction stir welded. Source: Ref 15

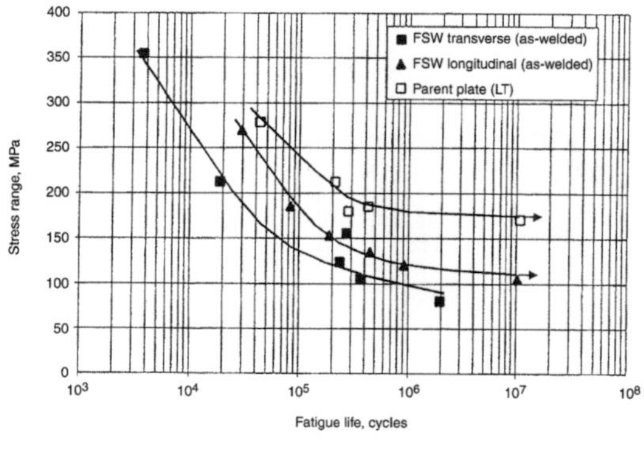


**Fig. 5.8** Stress-number of cycles (*S-N*) curves ($R$ = 0.1) of parent plate and friction stir welded joints in the as-welded condition. FSW, friction stir welded; LT, long transverse. Source: Ref 3

nugget. Similar studies using compact tension samples by Bussu et al. for 2024-T351 included cracks propagating as a function of distance from the plate joint line (Fig. 5.10) (Ref 12). The lowest threshold $\Delta K$ values and the highest growth rates were exhibited by cracks propagating at 28 mm (1.1 in.) from the plate joint line. At low $\Delta K$, cracks propagating in the weld nugget were slower than those of the unwelded plate. The largest threshold $\Delta K$ values, up to twice those of the unwelded plate, were observed for cracks originating 6 mm from the plate joint line. At this location, crack propagation rates were approximately 15 times less than in the unwelded plate. To investigate the effect of residual stress on the observed crack growth behavior, residual stresses were removed by mechanical stress relief. Stretching 2% decreased the residual stress orthogonal to the weld to 0. Figure 5.11 shows that following mechanical stress relief, crack growth rates are almost identical to those of the parent plate, regardless of location and orientation (Ref 12). This indicates that weld residual stress is responsible for the differences in fatigue crack growth rate and the observed crack growth threshold values ($\Delta K_{th}$). These observations are consistent with a crack closure-based model in which compressive residual-stress fields reduce the effective stress-intensity range ($\Delta K_{eff}$). Local hardness and microstructure changes appear to play a secondary role. In the work of Dalle Donne et al., studies evaluated the effects of pores within the weld nugget on crack propagation rates (Ref 9). After evaluating two different $R$ factors of 0.1 and 0.7, it was determined that pores had little influence on growth rates.

Crack-resistance curves were developed for 2024-T3 Al by Biallas et al. (Ref 2). Crack-resistance curves in terms of crack tip opening displacement ($\delta_5$) versus stable crack propagation ($\Delta a$) are shown in Fig. 5.12. A significant increase in fracture toughness is observed for the welded joints compared to the base material. This effect is mainly attributed to the large primary particles, which nucleate voids at relatively low loads and are therefore detrimental for fracture toughness (Ref 31). In the FSW joints, these primary particles have been fractured by the stirring process. Therefore, much smaller and rounder void-nucleating particles were present in the weld nugget than in the base material. Because higher stresses and strains were required to nucleate voids from these particles (Ref 31), fracture was retarded in the weld nugget, and a higher crack-resistance curve was obtained (Ref 2).

## 5.2 5*xxx* Aluminum Alloys

Similar to the 2*xxx* aluminum alloys, most investigations studied post-FSW properties of one commonly used 5*xxx* aluminum alloy, that is, 5083 Al (Ref 5, 15, 18, 32–40), with a few studies on a variety of other 5*xxx* aluminum alloys (Ref 6, 41–46). The 5*xxx* alloys are strengthened with magnesium additions from 1 to 5.5% and are non-heat-treatable, work-hardened alloys. Thus, the 5*xxx* aluminum alloys would be expected to behave differently than the heat treatable 2*xxx*, 6*xxx*, and 7*xxx* alloys following a thermal cycle associated with welding. Alloys in the 5*xxx* series possess good fusion welding characteristics.

**Hardness.** Post-FSW hardness has been reported by a number of investigators (Ref 6, 15, 32–35, 43, 46). Karlsson et al. and Kumagai et al. report similar trends for hardness following FSW (Ref 32, 34). Figure 5.13 shows the results of Karlsson et al. for annealed 5083-0 Al (4 to 4.9% Mg), illustrating an essentially horizontal line with no variation in hardness across the nugget into the HAZ following FSW. Sato et al. reported the same constant hardness results in transverse hardness measurements extending beyond the HAZ for 5083-0 from the weld root to the weld crown, but the scatter in the hardness data was considerable, varying from 60 to 80 HV (Ref 46). This is the expected response from a fully annealed work-hardenable alloy. Kumagai et al. show a slight increase in hardness across the weld nugget compared to the HAZ for a slightly hardened 5083-H112, but the increase is less than 6% (Ref 34). This small hardness increase may be due the very fine grain size created by FSW. Colligan et al. investigated 5083 hardened to the H131 temper and illustrated the change in hardness in a 25 mm (1 in.) thick friction stir weld from the crown to the root (Ref 35). Figure 5.14 illustrates these results, showing a modest decrease in hardness (~20%) for the weld nugget and the influence of heating from the shoulder at the crown surface; that is, softening extends beyond the nugget near the crown surface and to some depth (Ref 35). Additional hardness results can be found for 5005 Al, an alloy with a very low magnesium content (0.5 to 1.1%) (Ref 6) and for 5454 Al, an alloy with an intermediate magnesium content (2.4 to 3%) (Ref 40). In the work of Frankel et al., both the fully annealed and the H34 temper were evaluated and show the same hardness trends as that for annealed and hardened 5083 Al (Ref 43). In this same work, hardness was compared for FSW and GTAW 5454-H34 Al, illustrating the broader HAZ associated with GTAW.

**Mechanical properties** data for friction stir welded 5083 Al are limited (Ref 15, 34, 35). In

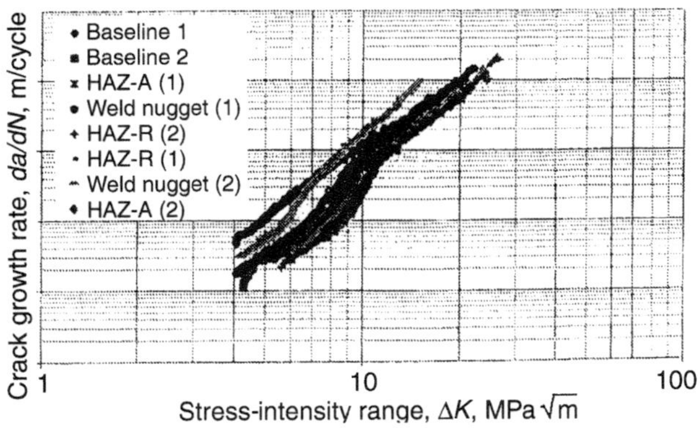


**Fig. 5.9** Crack growth rate in friction stir welded 2024-T351 compared to the parent metal. HAZ, heat-affected zone. Source: Ref 13

the work of Kristensen et al., tensile strength of friction stir welded 5083-H111 Al was investigated as a function of travel speed and rotation rate (Ref 15). All samples were tested in the transverse orientation. As is often the case, samples failed in the parent metal, and thus, no significant strength loss could be attributed to FSW or the welding parameters evaluated. Only by testing in the longitudinal orientation with all weld metal in the gage diameter can strengthening due to FSW be determined. However, based on the

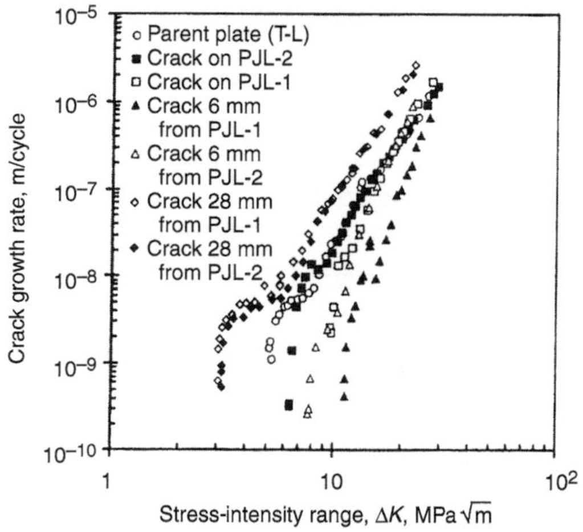


**Fig. 5.10** Crack growth data in 2024-T351 for cracks growing parallel to the weld in compact tension samples and cracks located at various distances from the plate joint line (PJL). Source: Ref 12

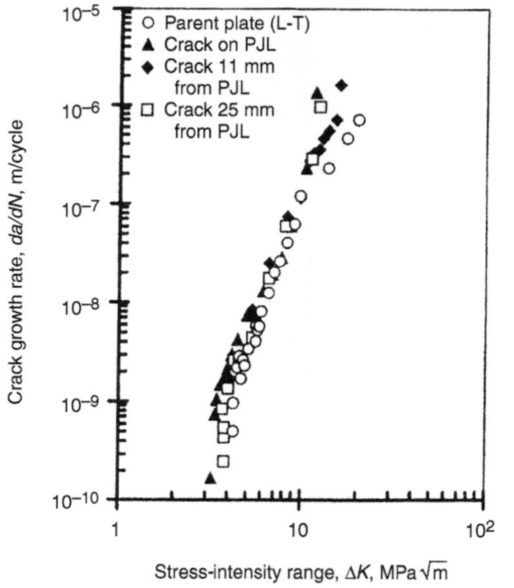


**Fig. 5.11** Crack growth data for friction stir welded 2024-T351 strained 2% parallel to the weld line, with surface cracks propagating orthogonal to the weld. Source: Ref 12

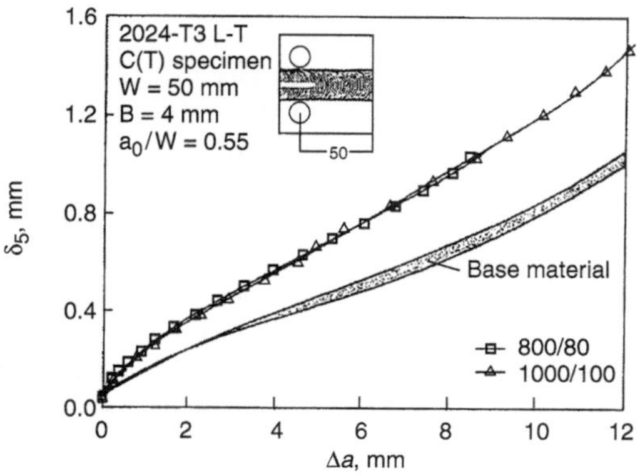


**Fig. 5.12** Crack tip opening displacement ($\delta_5$) versus stable crack growth ($\Delta a$). The numbers indicate tool rotational and lateral speed. Source: Ref 2

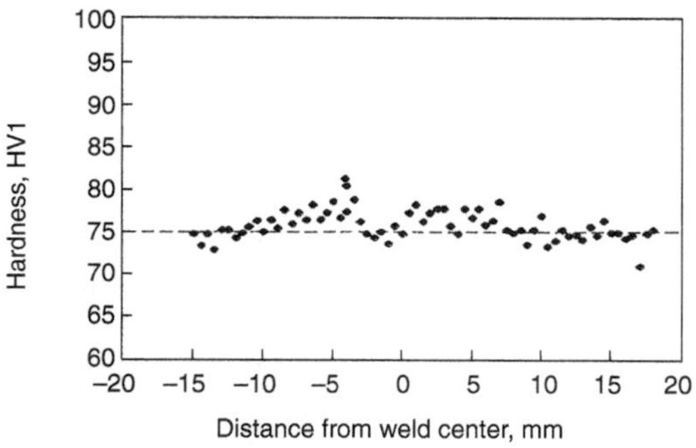


**Fig. 5.13** Horizontal hardness profile across a friction stir weld in AA5083 measured 1.7 mm (0.07 in.) from the root face. Source: Ref 32

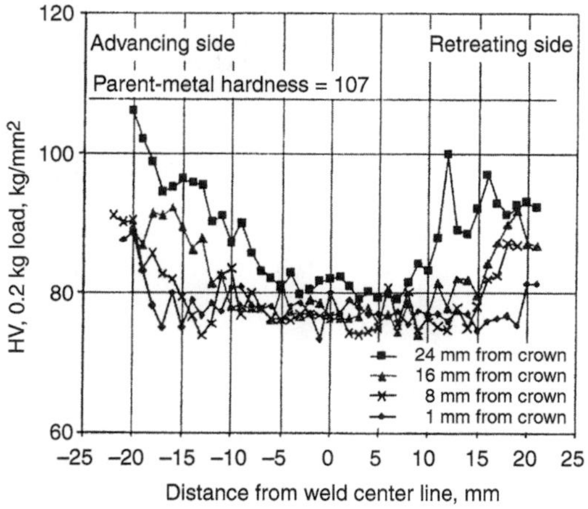


**Fig. 5.14** Microhardness traverse in 25.4 mm (1 in.) thick friction stir welded 5083-H131, 250 rpm at 127 mm/min (5.0 in./min) with zero tool axis tilt. Source: Ref 35

hardness results shown previously for 5083-H112, a slight strengthening could be predicted following FSW. Kumagai et al. compared tensile properties of 5083-H112 to metal inert gas (MIG) welding and base-metal properties (Ref 34). As shown in Fig. 5.15, they exhibit only slight differences in yield and tensile strength for the three material conditions. Due to the flat hardness curves of the friction stir and MIG welds, even the elongation measurements are relatively accurate and illustrate reasonable and comparable ductility for the three conditions. Colligan et al. show mechanical property results for 5083-H131 (Ref 35). However, in their work, tensile samples failed in the weld metal, providing weld-metal strength as opposed to parent-metal strength (Fig. 5.16) (Ref 35). This failure location is to be expected from the hardness curve (Fig. 5.14), where the more severely strain-hardened 5083-H131 showed softening in the weld nugget, especially on the crown surface. Also in this work, the yield strength was significantly reduced by ~44% (~155 MPa, or 22.5 ksi) compared to that of the parent-metal yield strength of 278 MPa (40.3 ksi). For all weld travel speeds evaluated (30 to 142 mm/min, or 1.2 to 5.6 in./min), strength in the friction stir welds was essentially constant with travel speed and compared well with gas metal arc welds.

In 5456-H116, an alloy very close in composition to 5083, Pao et al. evaluated properties in the longitudinal direction (parallel to the weld direction), with the gage diameter containing a constant microstructure (Ref 44). With this approach, Pao et al. were able to evaluate properties in the different weld zones (nugget, TMAZ, and HAZ). Figure 5.17 illustrates their results, revealing only a slight decrease in the tensile strength in the weld-affected region compared to the base material (~380 MPa, or 55 ksi). The one low data point was attributed to the presence of a layer of entrained oxides at that location. This entrained oxide layer has been called a "lazy S" and is the dispersed oxide associated with the

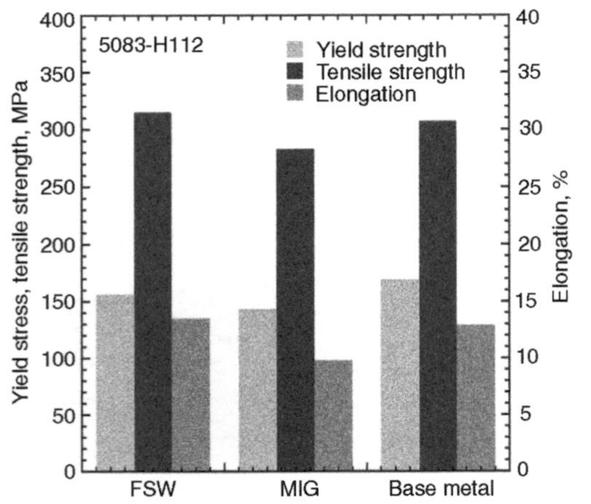


**Fig. 5.15** Tensile properties of welds in 5083-H112 Al. FSW, friction stir welding; MIG, metal inert gas welding. Source: Ref 34

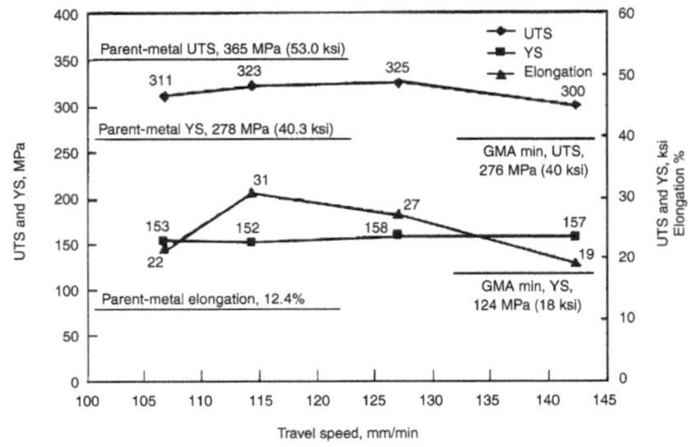


**Fig. 5.16** Transverse tensile properties versus travel speed for friction stir welded 25 mm (1 in.) thick 5083-H131 with zero tool tilt axis. UTS, ultimate tensile strength; YS, yield strength; GMA, gas metal arc. Source: Ref 35

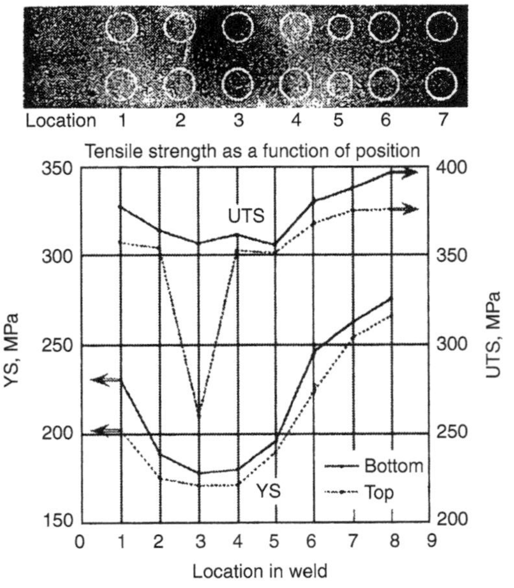


**Fig. 5.17** (top) Location and size of longitudinal tension test specimens and (bottom) graph of yield strength (YS) and ultimate tensile strength (UTS). Source: Ref 44

faying surfaces. Based on weld parameters and cleaning procedure, the oxide can remain after FSW in a semicontinuous "S"-shaped path from the root to the crown. Shifting the faying surfaces more to the advancing side of the tool, that is, biasing the tool to the retreating side, increases mixing of the weld interface and thus maximizes dispersion of the faying surface oxides. The yield strengths in the weld zones (~170 to 180 MPa, or ~24 to 26 ksi) are significantly lower than the base-plate yield strength of ~310 MPa (~45 ksi). The bottom of the weld retained higher strengths (both yield and tensile) than the top, due to the cooling effect of the anvil. The authors attribute this reduced yield strength to a lower dislocation density. It is difficult to reconcile the difference in results following FSW for the 5456-H116 (reduced yield strength) and 5083-H112 (yield strength approaching parent-metal strength) for alloys of comparable composition and with similar initial strain hardening.

**Fatigue, Fracture Toughness, Fatigue Crack Growth Rate.** Data available that illustrate fatigue and fatigue crack propagation rate in the 5*xxx* alloys are limited (Ref 33, 38–40, 44, 45). The results of James et al. in 8 mm (0.32 in.) thick single-pass butt joints of 5083-H321 sheet are the most complete, illustrating fatigue life as a function of weld travel speed for *S-N* testing performed in tension at 112 Hz and $R = -1$ (fully reversed loading) (Ref 38). Two specimen surface conditions were investigated: as-welded, with small burrs removed, but the tool shoulder ledges (~0.2 mm, or ~0.008 in.) remaining; and machined, where both burrs and ledges had been removed, leaving a smooth surface free of stress concentrations. This approach provides fatigue life data for as-welded samples, representing general engineering use, and inherent fatigue properties of the weld unaffected by surface artifacts induced by the welding process. Figure 5.18 presents the results of James et al. for the two surface conditions and four travel speeds (Ref 38). Data on cycles to failure ($N_f$) were obtained for $N_f$~$10^7$ cycles in all cases except for the 80 mm/min (3.15 in./min) travel speed as-welded case, where the curve apparently is asymptotic to the *x*-axis at approximately $10^6$ cycles. The authors attribute this asymptotic limit to initiation becoming controlled by surface notches. The FSW leaves circular arcs on the surface due to tool rotation and translation, which generally act as crack-initiation sites in as-welded specimens. Assuming an endurance limit of $10^7$ cycles, it is clear that the as-welded specimens have lower endurance-limit stress amplitudes than the polished specimens. It is difficult to identify a rela-

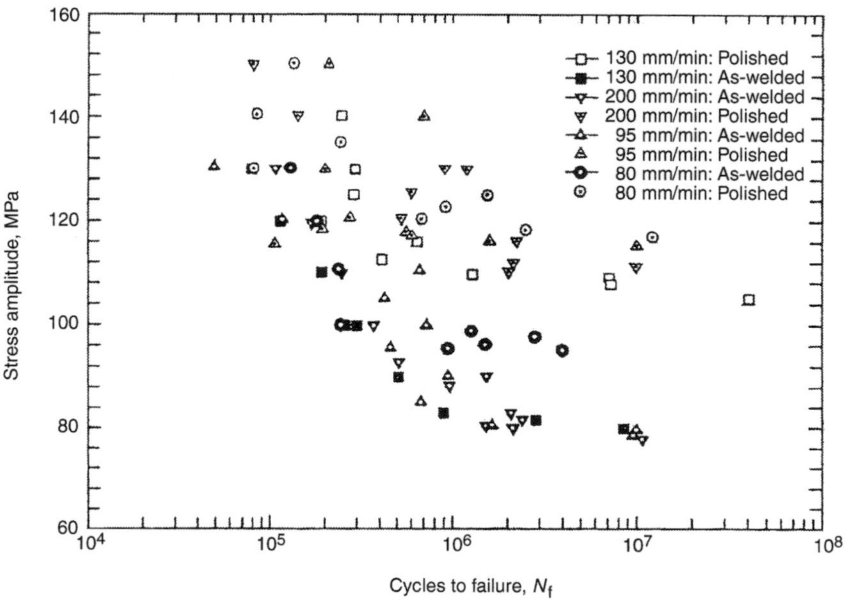


**Fig. 5.18** Fatigue results at $R = -1$ for friction stir welded 5083-H321 Al. Four travel speeds (80 to 200 mm/min, or 3.15 to 7.9 in./min) and two surface conditions (as-welded and polished) were considered. Source: Ref 38

tionship between fatigue life and weld travel speed. For the as-welded samples, this is even more difficult, because the surface condition varies for the different travel speeds, contributing more to the scatter.

Additional fatigue life results are available for lap welds in 5083-H111 (Ref 40). In this work, Thomas et al. evaluated fatigue life performance following FSW using the Skew-Stir tool design (TWI Ltd.). The Skew-Stir tool is described in detail in Chapter 2 within this book.

Lap joints are considered more difficult to friction stir weld than butt joints. Issues of joint fit-up, oxide dispersion across the interface, and uplift or subduction at the nugget/parent-metal interface are all critical to eventual weld joint performance. These weld quality issues are significantly influenced by the FSW tool design. As evidenced by the work of Thomas et al., unconventional tool designs (compared to those used for butt welds) are necessary to impart dispersion of faying surface oxides across a horizontal lap joint interface without severe uplift and thinning of the upper sheet. Unfortunately, friction stir lap welding requires considerable attention to weld-procedure detail, with interpretation of results also highly dependent on the postweld test method. Thus, it is not adequate to briefly summarize results of different investigators for friction stir welded lap joints. It is best to refer directly to the primary work to evaluate all weld-boundary conditions and postweld test methods. Suffice it to say, the results of Thomas et al. do highlight increased lap joint strength using the Skew-Stir tool design (Ref 40).

Work by Fuller et al. evaluated fatigue life of 5083-H321 of friction stir processed fusion welds (Ref 33). In this work, only the surface of the fusion weld was penetrated by the FSW tool, creating a forged microstructure on the surface and a substantial increase in fatigue life. This subject is addressed in greater detail in Chapter 14 within this book.

Limited work has been directed to fracture toughness testing of friction stir welded 5*xxx* alloys. However, Dawes et al. tested 5083-0 using the unloading compliance method to determine crack tip opening displacement and crack growth energy release rate curves (Ref 39). Using single-edge-notched three-point-bend fracture toughness specimens with the notch centered in the weld nugget, it was concluded that friction stir welds in 5083-0 had a higher fracture toughness than the corresponding parent metal.

Crack propagation results were reported by Pao et al. for butt-welded 5456-H116 12.7 mm (0.5 in.) plate (Ref 44). For the fatigue crack growth studies, wedge-opening-load fracture mechanics specimens were used, with the crack propagation direction parallel to the welding direction. Starting cracks were located in the center of the weld nugget, in the middle of the TMAZ, and in the base plate. Testing was performed with $R = 0.1$. Figure 5.19 shows fatigue crack growth rates as a function of $\Delta K$ through the base plate, weld nugget, and the advancing-side TMAZ (Ref 44). Even with quite different microstructures, fatigue crack growth rates of the base plate and nugget region are comparable and are significantly higher than those in the TMAZ. The differences in fatigue crack growth rates are most pronounced within the low-to-moderate $\Delta K$ regions. Also, the fatigue crack growth threshold stress-intensity range for the TMAZ is substantially higher than those of the base plate and weld nugget. The superior fatigue crack growth resistance in the TMAZ is believed to be associated with the presence of compressive residual stresses.

## 5.3 6*xxx* Aluminum Alloys

Extensive research is available presenting properties of friction stir welded 6*xxx* aluminum alloys (Ref 8–10, 18, 32, 33, 41, 42, 47–71). For the 6*xxx* family of alloys, properties are available for a number of specific alloys, including 6013, 6056, 6061, 6063, 6082, and the Japanese

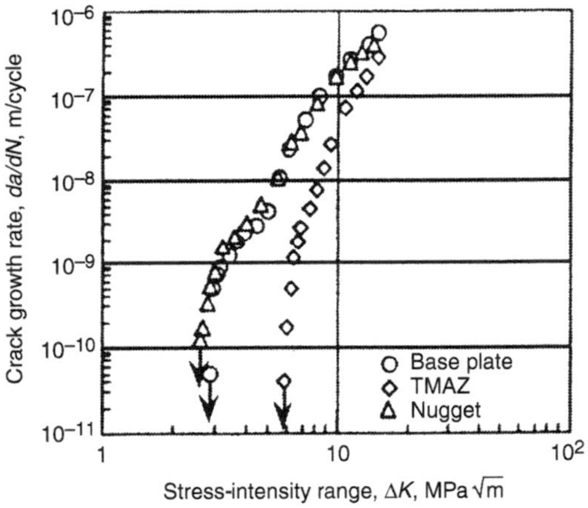


**Fig. 5.19** Fatigue crack growth kinetics in air for friction stir welded 5456-H116 12.7 mm (0.5 in.) thick plate. Source: Ref 44

extrusion alloy JIS6N01. Accordingly, when available, hardness, mechanical properties, and fatigue properties are presented for each of these alloys. The 6*xxx* alloys naturally age more slowly than 2024 Al, but strength changes are not as dramatic as those for either the 2024 or 7*xxx* alloys. Although strength increases are less, properties will change for friction stir welded 6*xxx* alloys based on the time between welding and testing. This alone can result in scatter in the results between investigators when comparing as-welded properties. Unfortunately, the time between welding and testing is seldom reported.

### 5.3.1 6013 Aluminum

Aluminum alloy 6013 is a relatively new heat treatable alloy of medium strength that derives its heat treat response from the precipitation of magnesium-silicon and an Mg-Si-Cu-Al phase. Alloy 6013 is weldable by GTA, MGA, and resistance methods. It has similar weldability to 6061 when welded by arc methods using 4043 or 4063 fillers and has weld strengths typically 27 to 40 MPa (3.9 to 18 ksi) higher than 6061.

**Hardness** data are limited, but results reported by Juričić et al. for 6013-T6, naturally aged following FSW, illustrate a shallow "W" curve, with hardness minima in both HAZs of 84 HV, ~35% lower than the parent metal (130 HV) (Ref 47). Hardness reduction in the nugget is only slightly less, at ~26% of parent-metal hardness. In addition, these investigators evaluated preweld heat treatment conditions of T4 and T6, followed by a postweld age to T6. Following each of these postweld heat treatments, the nugget hardness approached parent-metal hardness. Although the hardness minima were less, neither of the heat treatment conditions was able to prevent the hardness decrease in the HAZ, where the lowest hardness values were reached. The best result was attained by welding in the T4 temper, followed by a postweld T6 age. This increased the HAZ minimum hardness to 120 HV. Welding in the T6 condition and subsequently reaging to T6 resulted in a minimum hardness of 104 HV and a very narrow low-hardness region in the HAZ. This narrow low-hardness region could be detrimental to fracture toughness if a crack is located in this low-hardness band.

Hardness results are available for 6013-T4, 6013-T4 with a postweld heat treatment of 190 °C (375 °F) for 4 h, and 6013-T6 (Ref 71). The hardness decrease compared to the parent-metal hardness, that is, the minimum hardness in the HAZs, following FSW is approximately 17% for the T4 temper, 29% following the 190 °C for 4 h postweld heat treatment, and 21% for the T6 temper (Ref 71). In addition, the results by Heinz and Skrotzki illustrate hardness as a function of through thickness. As expected, the weld zone is considerably softer than the base metal for all three heat treat conditions, with the soft zone on the surface extending 10 mm (0.4 in.) on both sides of the weld on the crown surface and approximately 6 mm away from the weld centerline near the root surface. This result illustrates the anisotropy in through-thickness hardness (and other properties) that can occur as a result of FSW.

**Mechanical Properties.** Postweld mechanical properties of 6013 Al have been evaluated as a function of initial temper (T4 and T6) with subsequent aging to the T6 condition (Ref 8, 10, 47–49). For reference, base-material yield strength is 226 and 351 MPa (33 and 51 ksi) for the T4 and T6 tempers, respectively, with corresponding tensile strengths of 346 and 396 MPa (50 and 57.5 ksi) (Ref 47). The only available data for 6013-T4 with a natural age following FSW are those reported by Heinz et al. (Ref 48). Transverse yield and tensile strengths were 160 and 300 MPa (23 and 43.5 ksi), respectively, that is, 75 and 85% of parent-metal strength values. Both Juričić et al. and Lohwasser report strength results for friction stir welded 6013-T4 followed by a postweld age to T6, showing a yield strength of 340 MPa (49 ksi) and a tensile strength of 370 MPa (54 ksi) (Ref 47, 49). These strength levels are almost equivalent to parent-metal properties. Heinz et al. reported results for the same heat treat conditions, but even following a postweld T6 age, the strengths were still comparatively low, that is, ~250 MPa (36 ksi) yield and ~325 MPa (47 ksi) tensile strengths (Ref 48). Mechanical properties for friction stir welded 6013-T6 followed by natural aging are somewhat inconsistent. Again, Heinz et al. (Ref 48) reported a very low yield strength value of 165 MPa (24 ksi), compared to as-welded yield strengths of 215 MPa (31 ksi) for Lohwasser (Ref 49) and 228 MPa (33 ksi) by Juričić et al. (Ref 47). Tensile strengths reported by these same investigators were relatively consistent, ranging from 295 to 320 MPa (43 to 46 ksi). For each of these three investigators, the material thickness was 4 mm (0.16 in.) (Ref 47–49). The large variation in properties between investigators may be symptomatic of a newly emerging technology with continual advancements in tool design, process control, and other boundary conditions. The work of Heinz et al. was performed at the earliest date, and it is possible that advancements in FSW since that time have led to the improved results obtained by other investigators at a later date. Chapter 2 illustrates the evolution of tool design, leading to FSW at higher travel speeds and concurrent higher properties.

**Fatigue, Fracture Toughness, Fatigue Crack Growth Rate.** Fatigue life results for 6013-T4 followed by a postweld T6 age are illustrated in Fig. 5.20 (Ref 8). The specimens were tested in both the as-welded condition and after flush milling the weld crown and root surfaces. Parent-metal fatigue tests at $R = 0.1$ were performed for both unnotched specimens and for specimens with a 5 mm (0.2 in.) hole, creating a stress concentration ($K_t$) of 2.5. Friction stir welding does reduce the fatigue life compared to the parent metal. However, in the as-welded condition plus postweld T6 age, the fatigue life curve is above the reference curve for the open-hole specimens of the parent material ($K_t = 2.5$). Surface milling of the welds completely restores the apparent applied threshold stress to a level equivalent to that of the unnotched parent material (~200 MPa, or 29 ksi). Additional fatigue life studies in 6013 have been performed as a function of different pre- and postheat treat conditions (Ref 49). In this work, fatigue life results are compared to riveted joints, and all welded results show a great improvement over riveted structure.

Crack propagation rate studies for friction stir welded 6013 were investigated by Juričić et al. (Ref 47). Material conditions included:

- *T6 FSW:* welding in the T6 condition plus naturally aged for 4 weeks
- *T4 FSW T6:* welding in the T4 condition (solution heat treated), subsequent T6 aging
- *T6 FSW T6:* welding in the T6 condition, subsequent reaging to T6 (190 °C for 4 h)

Figure 5.21 illustrates fatigue crack propagation curves for these three heat treat conditions at $R = 0.1$ and $R = 0.7$. The fatigue crack propagation specimen geometry was a center-cracked [M(T)] specimen with the slot introduced in the center of the weld nugget, parallel to the weld direction. Base-material results are illustrated by the solid line. At $R = 0.1$, crack growth rates are faster compared to the parent metal, with the samples artificially aged to the T6 condition exhibiting the highest crack growth rate. However, at $R = 0.7$, there is little difference in crack growth rate between the welded samples and the parent metal. At high-$R$ ratios, closure is less influential. Closure stresses can be influenced by both residual stresses and the change in grain size. The effect of intentionally induced porosity on crack growth rate in 6013-T6 was investigated

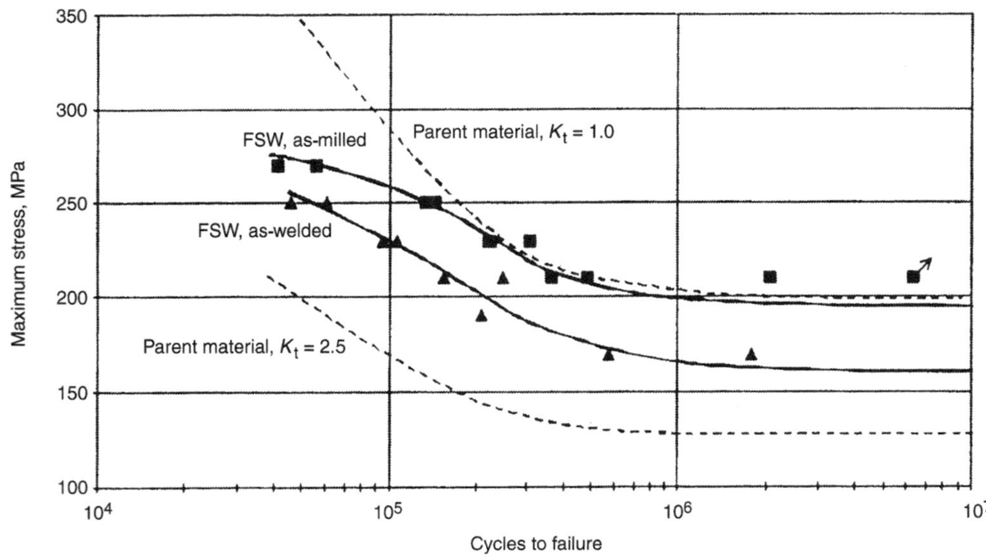


**Fig. 5.20** Fatigue results for friction stir welded (as-welded and milled) and parent material (stress concentration, $K_t$ = 1 and 2.5) 6013-T6 at $R = 0.1$. Source: Ref 8

by Dalle Donne et al. (Ref 9). Similar conclusions were found for *R*-ratio and effective stress intensity to those mentioned previously. In this work, the authors investigated the influence of specimen geometry and thus a different residual-stress distribution on *da*/*dN*-$\Delta K$ curves. Comparing crack propagation rates from center-cracked specimens versus edge-cracked specimens, large discrepancies in crack growth rate were found in the welded samples, whereas the base-material curves remained in a common scatter band (Ref 9). In this example, at low D*K* (<12 MPa $\sqrt{\text{m}}$), crack growth rates were significantly higher in the center-cracked samples.

Limited fracture toughness data are available for 6013. Fracture toughness data are reported by Juričić et al. where the highest fracture toughness is recorded when the crack is located in the center of the joint of the sheet welded in the T6 temper and subsequently naturally aged (Ref 47). The T6 heat treatment after welding increased the joint strength and had a detrimental effect on fracture toughness but was still comparable to fracture toughness of the base material. Even with this postweld heat treat condition, the toughness of the nugget material (where the crack was located) was so high that loading could be increased until the local ultimate strength was reached. The specimen then fractured by necking and local plastic deformation. Fracture toughness measurements were not made with the crack located in the narrow hardness regions of the HAZ.

### *5.3.2 6056 Aluminum*

Nominal composition for 6056 Al is 1 Si, 0.9 Mg, 0.8 Cu, 0.7 Mn, <0.5 Fe, 0.4 Zn, and 0.14 Zr (wt%).

**Hardness.** Microhardness results are reported by Denquin et al. (Ref 50) for 6056-T78 as-welded and for 6056-T4 with a postweld age to T78 (Ref 52). Typical "W" hardness curves were shown for both conditions. For the as-welded 6056-T78, the hardness minima occur in each HAZ, with a reduction of 35 to 40% in hardness compared to the parent metal (70 versus 105 HV, respectively) (Ref 50). Weld nugget hardness is only moderately less than the parent metal (95 versus 105 HV). For the 6056-T4 postweld aged to T78, the hardness mini-

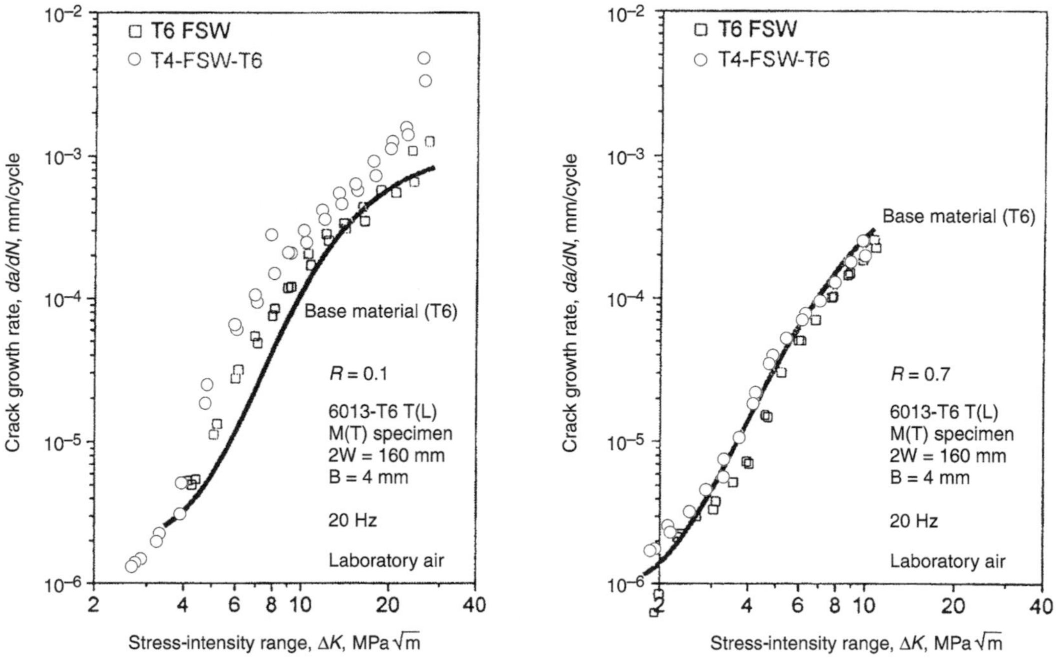


**Fig. 5.21** Fatigue crack propagation curves of friction stir welded specimens (points) compared to the base-material data (line) for $R = 0.1$ and $R = 0.7$. Source: Ref 47

mum is still 70 HV in the HAZ troughs, but the nugget hardness increases and is essentially the same as the parent metal (Ref 52).

**Mechanical Properties.** Denquin et al. evaluated mechanical properties of different regions of the weld zone in 6056-T4 postweld aged to T78 using microtensile samples (Ref 52). Figure 5.22 is a schematic illustration of the testing approach showing where eighteen 2 mm thick microtensile bars were removed from the weld zone. These samples are machined in the longitudinal direction and thus contain microstructures unique to each weld zone, that is, weld nugget, TMAZ, lowest-hardness zone, HAZ, and parent metal. This microtensile approach allows for more realistic ductility measurements as compared to transverse tensile tests. Figure 5.23 presents the results of Denquin et al. for tensile tests performed on microspecimens following a postweld aging heat treatment to T78 (Ref 52). These results are in direct agreement with the microhardness results. The weakest zone in tension corresponds to the low-hardness zone of the FSW joint. Decreases in reference to the base metal of 41% for yield strength (297 MPa, or 43 ksi), 25% for tensile strength (332 MPa, or 48 ksi), and 6% for fracture elongation (12%) are shown. As shown, the ductility is constant for the weld zones and is close to that of the base metal. In comparison, ductility in a transverse tensile test for the same material temper was 2.2%, illustrating strain localization and an unrealistically low reported ductility. Yield and tensile strengths in the weld nugget are comparable to parent-metal properties.

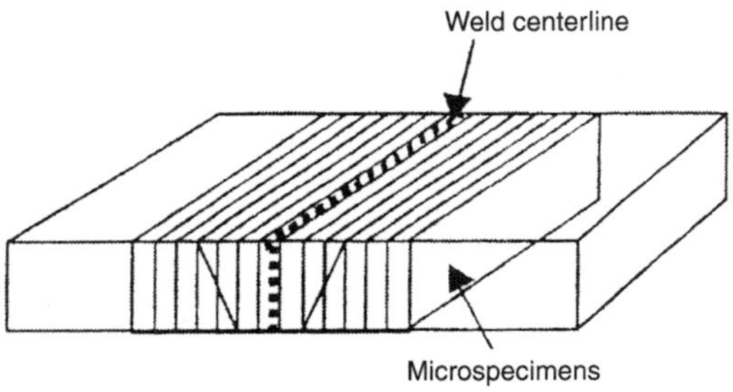


Fig. 5.22 Schematic illustration of microspecimens extracted from a friction stir weld. Source: Ref 52

**Fatigue, Fracture Toughness, Fatigue Crack Growth Rate.** Fatigue life and fatigue crack growth rate data for friction stir welded 6056-T4 artificially aged to T6 were established by Lohwasser (Ref 49). These fatigue life results, following a postweld T6 age, showed a drop of approximately 10% compared to the base material. The fatigue crack propagation behavior is better or equivalent to base material, even in the TMAZ. Fracture toughness results are in the range of the base material.

### 5.3.3 6061 and 6063 Aluminum

Alloy 6061 Al is the most used of the 6000-series aluminum alloys and possesses superior weldability as compared to other heat treatable

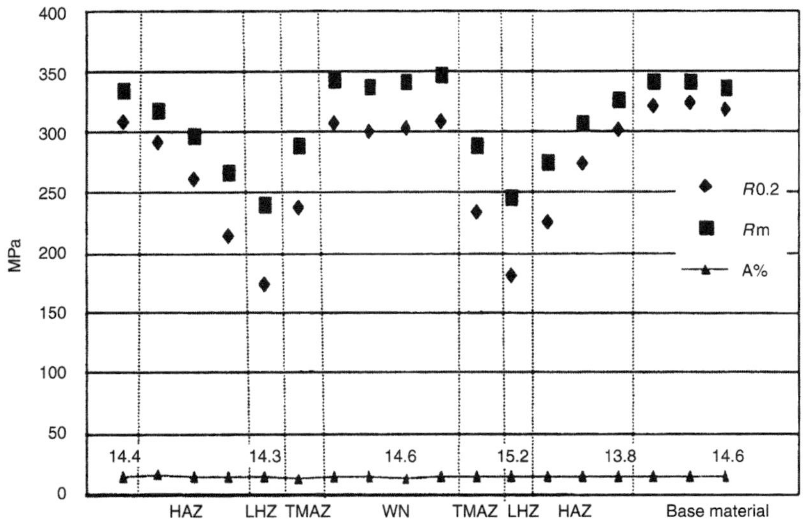


Fig. 5.23 Yield and ultimate strengths and fracture elongation profiles across the 6056 friction stir weld following a postweld aging heat treatment to T78. HAZ, heat-affected zone; LHZ, lowest-hardness zone; TMAZ, thermomechanically affected zone; WN, weld nugget; *R*0.2, yield strength; *Rm*, ultimate strength; A%, fracture elongation. Source: Ref 52

alloys. The alloy exhibits excellent welding characteristics in all tempers when welded by any of the commonly used fusion and resistance welding procedures. The strength properties of 6061 are not as high as the 2024 or 7075 heat treatable aluminum alloys. However, 6061 possesses excellent corrosion resistance, good machinability, and good formability. Alloy 6061 is the most popular aluminum alloy extrusion. Alloy 6063 is similar in composition to 6061 but possesses a superior surface appearance after extrusion. Because of their similarity, properties of friction stir welded 6061 and 6063 are presented together.

**Hardness** results for friction stir welded 6061 have been reported by a number of investigators (Ref 23, 33, 41, 53–56), with more limited results reported for 6063 (Ref 58, 59). Malin, in a gas metal arc weld, evaluated the effect of weld storage period on HAZ hardness following welding of 6061-T6 (Ref 57). In heat treatable magnesium-silicon alloys, hardness is reduced in the HAZ following welding, and natural aging can restore some of the hardness loss. Following gas metal arc welding, four samples sectioned from the HAZ were naturally aged at room temperature for times of 4 h, 7, 14, and 28 days (Ref 57). The minimum hardness in the HAZ recovered 16, 21, and 28% of as-welded hardness after times of 1, 2, and 4 weeks, respectively (Ref 57). Unfortunately, investigators seldom report time at room temperature prior to hardness or mechanical testing, even for heat treatable aluminum alloys. Although it is believed that FSW has a lower heat input than conventional welding practices, a similar natural aging response will occur in the HAZ and weld nugget of a friction stir weld. Thus, caution should be exercised when interpreting or comparing hardness or mechanical properties data between different investigators.

Chang et al. and Lim et al. illustrate the typical "W"-shaped hardness curves for as-welded 6061-T6 as a function of weld process parameters of travel speed and rotation rate (Ref 53, 55). Figure 5.24 illustrates typical hardness results for 6061-T6 Al as a function of weld travel speed (Ref 53). Slight hardness differences are evident with changing weld parameters, but differences are not significant. Depending on the starting temper, hardness in the minimum HAZ troughs can be 40% less than the parent metal, with the weld nugget showing an approximately 20 to 30% decrease in hardness. Time between FSW and testing was not reported. Hori et al. illustrated hardness in 6061-T6 compared to conventional fusion welding methods (Ref 54). Compared to tungsten inert gas (TIG) welding, hardness in the weld nugget following FSW is higher, and the HAZ is much smaller. However, hardness curves for MIG and FSW were identical, including the breadth of the HAZ. A laser weld showed the same minimum hardness as FSW, but the extent of the HAZ was significantly less in the laser weld. Results from Reynolds comparing FSW to TIG show the same trends as those reported by Hori et al., that is, an approximately 25% increase in hardness in the friction stir weld compared to a TIG weld (Ref 41).

Hardness results are available for 6063-T5 (Ref 59) and 6063-T561 (Ref 58). Results by Sato and Kokawa illustrate the ability to achieve near-parent-metal hardness when the initial temper prior to welding is T5 and a postweld age is applied (Ref 59). Complete recovery of hardness is attained if a postweld solution heat treatment is followed by an artificial age. Sato and Kokawa also show yield strength to be roughly proportional to hardness (Fig. 5.25), establishing a relationship between minimum hardness and yield strength of HV ~ $2.85\sigma_y + 199$ (MPa) for 6063 Al (Ref 59).

**Mechanical Properties.** Results on mechanical properties for friction stir welded 6061 and 6063 are relatively limited but are adequate to illustrate the range of strengths possible for friction stir welded 6061-T6 and 6063-T5 aluminum alloys (Ref 11, 23, 41, 53, 55, 57–59, 72). Most investigators evaluated the T6 temper as-welded, but again, time delay from welding to testing was not reported. A summary of

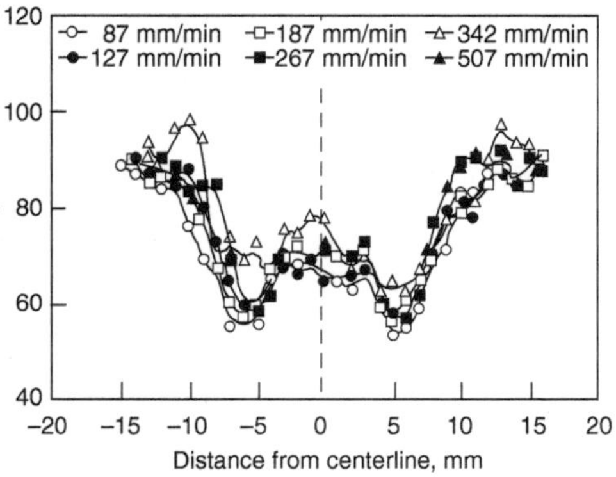


**Fig. 5.24** Hardness profiles across the weld zone in 6061-T6 Al as a function of travel speed. Source: Ref 53

mechanical property results from these studies for both 6061 and 6063 Al with a preweld T6 temper is presented in Table 5.1.

The variations in yield and tensile strengths shown in Table 5.1 are considerable but perhaps not unexpected. As noted, time between welding and testing is not reported, and this alone can contribute to variability in the mechanical property results. Also, as presented at the beginning of this chapter, there are a considerable number of variables between investigators, both reported and not reported, such as material thickness, cooling rate, weld parameters, tool design, and so on. Each of these weld variables can also influence resultant properties. When all variables are considered, property differences should be expected, and these results for 6061-T6 reflect the range of properties that may be obtained.

Hori et al. illustrated the influence of cooling rate, that is, air cool versus water cooling, on postweld mechanical properties of 6061-T6 (Ref 54). Specimens were longitudinal, that is, the gage diameter contained material only from the weld nugget, so the HAZ was not included. The tensile strength of 330 MPa (48 ksi) and yield strength of 298 MPa (43 ksi) of the water-cooled and aged FSW joint were higher than the tensile (302 MPa, or 44 ksi) and yield (272 MPa, or 39 ksi) strengths of the air-cooled FSW joint and even higher than those of the parent material (317 and 286 MPa, or 46 and 41.5 ksi, respectively). This illustrates the quench sensitivity of 6061 Al.

Additional mechanical property results are reported by Sato et al. for 6063-T5, illustrating mechanical properties for as-welded, welded then aged at 175 °C (347 °F) for 12 h, and welded plus solution heat treated at 530 °C (985 °F) for 1 h and subsequently aged 175 °C for 12 h (Ref 59). Sato's results are tabulated in Table 5.2, illustrating the ability to fully recover strength compared to base-metal properties. Sato's results are confirmed by the work of Luan et al., where 100% weld joint efficiency can be obtained with T5-treated 6063 (Ref 58). Mechanical properties from Heinz and Skrotzki are shown in Table 5.3 for 6063 Al for various temper conditions (Ref 71). The decrease in yield strength following FSW for the T4 temper is 28%, whereas the loss for the T6 temper is considerably more at 54%. The ability to recover strength for a postweld age of 190 °C (375 °F) for 4 h is illustrated by the increase in yield strength to 247 MPa (35.8 ksi) compared to the as-welded yield strength of 160 MPa (23 ksi) for the T4 temper. The low transverse strain values shown in Table 5.3 are attributed to strain localization in the minimum hardness location in the HAZs. Local strain measurements illustrated strain concentration in the

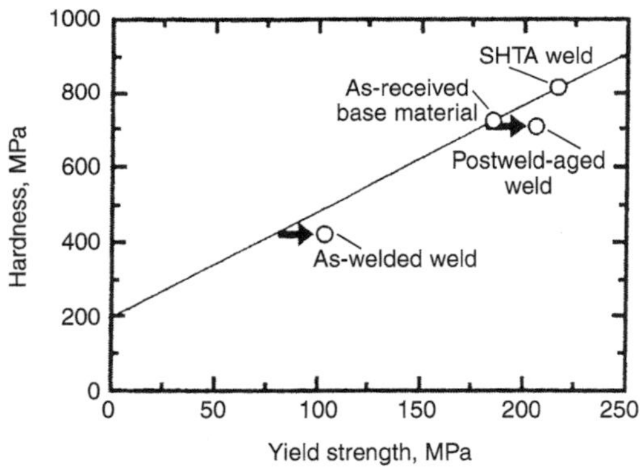


**Fig. 5.25** Relationship between the yield strength and the minimum hardness in the base material and the welds. SHTA, solution heat treated and aged. Source: Ref 59

**Table 5.1 Tensile properties of friction stir welded 6061-T6 and 6063-T6 Al alloys**

| Yield strength | | Tensile strength | | | | |
|---|---|---|---|---|---|---|
| MPa | ksi | MPa | ksi | Elongation, % | Variable | Reference |
| 155 | 22.5 | 260 | 37.7 | ... | ... | 11 |
| 143 | 20.7 | 230 | 33.4 | 6.4 | ... | 23 |
| ... | ... | 190–205 | 27.6–29.7 | ... | Travel speed | 53 |
| 135–150 | 19.6–21.8 | 210–240 | 30.5–34.8 | 10–18 | Travel speed | 55 |
| 100 | 14.5 | 249 | 36.1 | 6.7 | ... | 57 |
| ... | ... | 220–240 | 31.9–34.8 | ... | Travel speed | 58(a) |

(a) 6063-T6 as-welded

low-hardness regions, with the stronger regions within the test sample gage length, that is, base metal and weld nugget, resisting strain. Maximum local strains up to 40% were measured, illustrating the ductile response of the friction stir welds.

**Fatigue, Fracture Toughness, Fatigue Crack Growth Rate.** At this time, the available data characterizing fatigue and fracture of 6061 and 6063 Al alloys are very limited (Ref 57, 58, 72). Brinckmann et al. evaluated 6061-T6 Al following both FSW repair of intentionally defective friction stir welds and defect-free production friction stir welds (Ref 72). The repair procedure added an additional thermal cycle to the weld-zone material. Using standard compact tension samples, precracks were located in the nugget and the HAZ to determine fracture toughness properties. As shown in Fig. 5.26, the crack tip opening displacement ($\delta_{5m}$) for the welded specimens, in both the nugget and HAZ, is far superior (40 to 69%) to those measured in the base material (Ref 72). The additional thermal cycle and deformation imposed by the repair weld further improved toughness in the nugget area without causing any deterioration of the HAZ properties. These same investigators also evaluated fracture toughness using wide-plate through-thickness fatigue [M(T)] prerack specimens, with cracks centered in the weld nugget. For these through-thickness samples, the base-material toughness results are ~55% lower than those in both the production and repair welds. The superior toughness behavior was attributed to the strength undermatch of the nugget area compared to the base-material strength.

Nagano et al. compared Charpy impact results for friction stir, yttrium-aluminum garnet laser, and gas tungsten arc welds in 6061-T6 Al (Ref 56). Samples were machined with the Charpy V-notch on the weld centerline. Impact properties of the friction stir welded 6061-T6 were twice that of fusion welds (20 versus 10 J/cm$^2$). The higher impact strength was attributed to the fine, recrystallized microstructure in the friction stir weld compared to the cast microstructure created by the fusion welds. Further, the high silicon content in the filler materials used in the fusion welds likely contributed to the lower impact strength. Luan et al. illustrated excellent impact values for friction stir welded 6063 Al in the T561 temper (Ref 58). Impact

**Table 5.2 Tensile properties of 6063-T5 including base metal, as-friction stir welded, an aged weld, and welded plus solution heat treated (ST) and aged**

| Material condition | Yield strength MPa | Yield strength ksi | Tensile strength MPa | Tensile strength ksi | Elongation, % |
|---|---|---|---|---|---|
| Base material (T-5) | 185 | 26.8 | 215 | 31.2 | 19 |
| As-welded | 105 | 15.2 | 155 | 22.5 | 10 |
| Aged 175 °C (347 °F) for 12 h | 210 | 30.5 | 225 | 32.6 | 13 |
| ST + aged 175 °C (347 °F) for 12 h | 215 | 31.2 | 235 | 34.1 | 18 |

Source: Ref 58

**Table 5.3 Tensile properties of 6063 for a variety of pre- and postweld tempers**

| Material condition | Yield strength MPa | Yield strength ksi | Tensile strength MPa | Tensile strength ksi | Elongation, % |
|---|---|---|---|---|---|
| T4 - base metal | 222 | 32.2 | 320 | 46.4 | 20.5 |
| T6 - base metal | 357 | 51.8 | 394 | 57.1 | 11.5 |
| T4 + FSW | 160 | 23.2 | 300 | 43.5 | 8.7 |
| T4 + FSW + 190 °C (374 °F) for 4 h | 247 | 35.8 | 323 | 46.8 | 1.2 |
| T6 + FSW | 165 | 23.9 | 295 | 42.8 | 4.5 |

FSW, friction stir welding. Source: Ref 71

values of weld joints were shown to be 60% higher than that of the base material.

### *5.3.4 6082 Aluminum*

Alloy 6082 is a precipitation-strengthened alloy of nominal composition 1Si-0.65Mg-0.2Fe-0.52Mn, with the relatively high manganese content added to increase ductility. Fusion welding results in a significant loss of mechanical properties (Ref 60). Alloy 6082 is a common, strong, general alloy in the United Kingdom.

**Hardness** results for friction stir welded 6082 Al have been reported by a number of investigators, with welding performed in the T4, T5, and T6 tempers (Ref 32, 61–64). Starting with 6082-T6, a horizontal hardness profile 2.5 mm (0.1 in.) from the weld root surface shows the characteristic "W"-shaped curve, with hardness minima in the HAZ 45% less than parent-metal hardness (Fig. 5.27) (Ref 32). Hardness in the weld nugget is slightly higher, with a 36% reduction compared to parent-metal hardness. Not seen in other alloys is a hardness shelf at 85 HV, which the authors correlate with the tool shoulder diameter. Microhardness measurements across the weld nugget did not reveal any systematic variations that could be correlated to the microstructurally observed ring pattern (Ref 32). Results from Backlund et al. (Ref 61) illustrate hardness for a variety of pre- and postweld heat conditions, including the following:

- FSW-T6
- FSW-T6 + aged at 185 °C (365 °F) for 3 h
- FSW-T4
- FSW-T4 + aged at 185 °C for 3 h

Figure 5.28(b) illustrates the ability to fully recover hardness when welding in the T4 temper followed by an artificial age, whereas when welding in the T6 temper (Fig. 5.28a), small hardness minima still are evident in the HAZ. Note the similarity in hardness drop for the T6 temper between Fig. 5.27 and 5.28(a), except for the hardness isotherm in Fig. 5.27 attributable to the tool shoulder. This one difference may be due to location of the hardness trace, sheet thickness, and/or weld parameters. However, with potentially different welding practices, that is, different heat inputs between these two research studies, it is interesting that the hardness curves are nearly identical.

**Mechanical properties** for friction stir welded 6082 have been established for a variety of tempers and material thickness (Ref 32, 60, 61, 64–66). Not all weld variables can be reported for the different studies, but Table 5.4 summarizes mechanical properties for some variables. Again, considering different weld procedures, tool designs, thermal-boundary conditions, natural aging times, and different thicknesses of the workpieces, mechanical property results are remarkably similar between investigators. All specimens were tested transverse to the weld. Thus, elongation values are not always realistic, due to strain localization in the softer HAZ, but for the postweld-aged specimens, where parent-metal properties are fully recovered, transverse ductility can be meaningful. For the T4 temper, base-metal and postweld

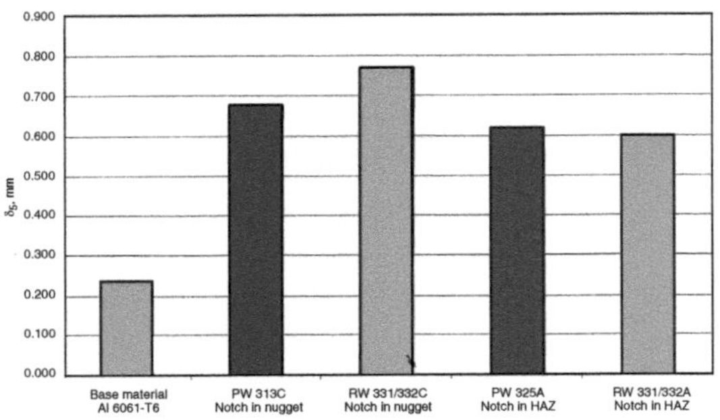


**Fig. 5.26** Crack tip opening displacement ($\delta_{5m}$) in friction stir welded 6061-T6 Al in both the nugget and heat-affected zone (HAZ) compared to the base material. RW repair weld; PW, production weld. Source: Ref 72

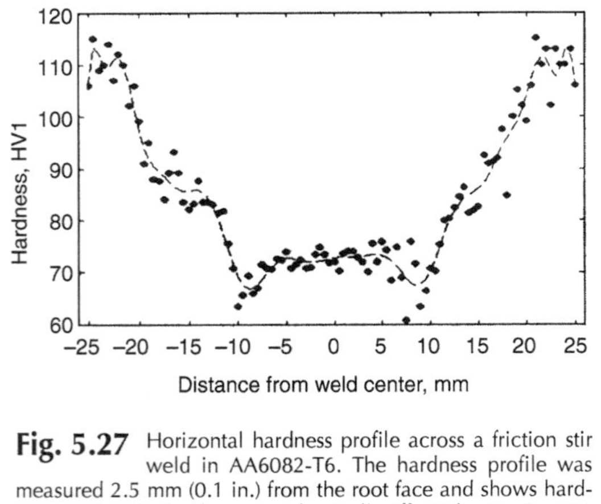


**Fig. 5.27** Horizontal hardness profile across a friction stir weld in AA6082-T6. The hardness profile was measured 2.5 mm (0.1 in.) from the root face and shows hardness minima in the thermomechanically affected zone. Source: Ref 32

strengths are nearly comparable, whereas for the T6 temper, both yield and tensile strengths are reduced considerably (53% and up to 25%, respectively). Results vary, but strengths are nearly completely recovered by postweld aging from either the T4 or T6 initial temper.

**Fatigue, Fracture Toughness, Fatigue Crack Growth Rate.** Fatigue strength of friction stir welded 6082 T4, T6, and T4 + 185 °C for 5 h with defect-free welds has been established (Ref 61, 64). The number of cycles to fracture at different stress levels for the T6 and T4 + 185 °C for 5 h conditions is illustrated in Fig. 5.29 (Ref 64). For the postweld heat treated T4 specimens, the number of cycles to failure is slightly below that of T6. This was an unexpected result, because the T4 + 185 °C for 5 h condition has higher yield and tensile strengths.

Backlund et al. compare fatigue life for FSW, MIG, and plasma keyhole welding (Ref 61). In all cases, fatigue life for the friction stir weld is far superior to that of the fusion welds. Care should be exercised to be sure lack of penetration defects do not occur at the root of a friction stir weld. This type of defect is very difficult to detect. In the work of Haagensen et al., the presence of root-notch lack of bonding was shown to influence fatigue life results (Ref 65). In this study, all fatigue cracks in the friction stir welds initiated in defects located in the lower part of the weld. However, fatigue life of the friction

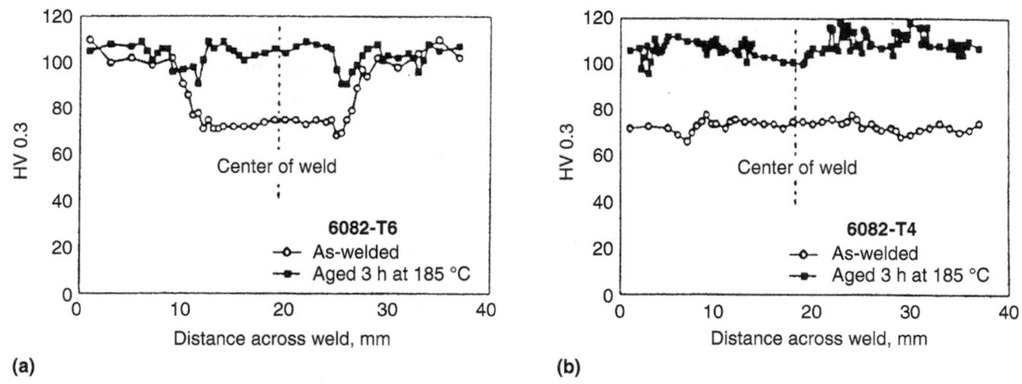


**Fig. 5.28** Hardness distribution across friction stir welds in AA6082. (a) Welded in the T6 temper and welded in the T6 temper and aged. (b) Welded in the T4 temper and welded in the T4 temper and aged. Source: Ref 61

**Table 5.4 Tensile properties of friction stir welded 6082 for a variety of pre- and postweld tempers**

| | | Yield strength | | Tensile strength | | | Thickness | | |
|---|---|---|---|---|---|---|---|---|---|
| Temper | Postweld age | MPa | ksi | MPa | ksi | Elongation, % | mm | in. | Ref |
| Base T4 | ... | 149 | 21.6 | 260 | 37.7 | 22.9 | 4 | 0.16 | 61 |
| Base T6 | ... | 291 | 42.2 | 303 | 43.9 | 11.3 | 4 | 0.16 | 61 |
| MIG T4(a) | ... | 129 | 18.7 | 163 | 23.6 | 3 | 5 | 0.20 | 65 |
| **Friction stir welded** | | | | | | | | | |
| T6 | ... | ... | ... | 226 | 32.8 | ... | 10 | 0.39 | 32 |
| T6 | ... | ... | ... | 254 | 36.8 | ... | 5 | 0.20 | 32 |
| T6 | ... | 135 | 19.6 | 220 | 31.9 | ... | 5.8 | 0.23 | 64 |
| T6 | ... | 160 | 23.2 | 254 | 36.8 | 4.9 | 4 | 0.16 | 61 |
| T6 | 185 °C (365 °F) for 3 h | 274 | 39.7 | 300 | 43.5 | 6.4 | 4 | 0.16 | 61 |
| T5 | ... | 125 | 18.1 | 198 | 28.7 | 7.8 | 3.5 | 0.14 | 60 |
| T5 | ... | 125 | 18.1 | 196 | 28.4 | 9.8 | 3 | 0.12 | 62 |
| T4 | ... | 144 | 20.9 | 239 | 34.7 | ... | 5 | 0.20 | 65 |
| T4 | ... | 138 | 20.0 | 244 | 35.4 | 18.8 | 4 | 0.16 | 61 |
| T4 | 185 °C (365 °F) 3 h | 221 | 32.1 | 246 | 35.7 | 5.7 | 3.5 | 0.14 | 60 |
| T4 | 185 °C (365 °F) 3 h | 285 | 41.3 | 310 | 45.0 | 9.9 | 4 | 0.16 | 61 |
| T4 | 185 °C (365 °F) 3 h | 260 | 37.7 | 289 | 41.9 | ... | 5.8 | 0.23 | 64 |
| T4 | 185 °C (365 °F) 3 h | 227 | 32.9 | 250 | 36.3 | 7.6 | 3 | 0.12 | 62 |

(a) MIG, metal inert gas

stir welds containing surface defects was still significantly greater than that of a MIG weld.

Unfortunately, fracture toughness and fatigue crack propagation data for 6082 Al are very limited. Haagensen et al. refer to the work of others where crack growth rates in the friction stir weld and HAZ of 6082 are lower than in the base material (Ref 65). Also, Ossterkamp et al. present fracture toughness under rapid loading and come to the conclusion that "friction stir welds have very high fracture toughness compared to the nonwelded material" (Ref 66).

### 5.3.5 JIS6N01 Aluminum

In Japan, FSW is in use in the transportation industries to fabricate rolling stock and aluminum decking in order to maintain high postweld strength compared to conventional fusion welding. The aluminum alloy of choice is JIS6N01, an easily extrudable aluminum alloy of nominal composition 0.6Si-<0.35Fe-<0.35Cu-<0.5Mn-0.6Mg-<0.3Cr. This alloy composition is very similar to aluminum alloy 6005. Friction stir welding studies have been completed on JIS6N01 in both the T5 and T6 tempers (Ref 54, 67–69).

**Hardness.** Post-friction stir weld hardness results for JIS6N01-T5 show the typical "W"-shaped curve, as previously seen for other age-hardenable aluminum alloys (Ref 54, 67, 68). For the as-welded condition, postweld hardness decreases from ~105 to ~75 HV, a decrease of 28%, with hardness in the nugget ~5 HV higher than the HAZ. Compared to MIG welding, the minimum hardness in the friction stir welded nugget is higher (78 versus 62 HV), and the size of the HAZ is considerably less (Ref 67). This would be expected due to the lower and more localized heat input associated with FSW. This same result was not demonstrated by Hori et al. where hardness traverses were equivalent for FSW and MIG, but TIG welds showed a much more extensive HAZ and a further decrease in hardness (Ref 54). Hori et al. also demonstrated the influence of cooling rate, that is, air cooled and water cooled, and postweld aging (180 °C for 6 h) on hardness in friction stir welded JIS6N01-T6 (Ref 54). Figure 5.30 illustrates hardness profiles for these different cooling and aging conditions. There is little difference in the hardness minima for air versus water cooling, but the size of the HAZ is reduced with water cooling. Aging completely recovered hardness in the nugget, but significant hardness minima remained in the HAZ.

**Mechanical Properties.** Strengths in friction stir welded JIS6N01 mirror the hardness results. Okura et al. tensile tested both transverse and longitudinal samples (the gage diameter contained only weld nugget material) (Ref 68). Tensile strengths were comparable for the different orientations (217 MPa, or 31.5 ksi) and were ~20% lower than parent-metal tensile strength of 270 MPa (39 ksi). However, the longitudinal orientation showed higher yield strength, (127.5 MPa, or 18.5 ksi) compared to the transverse orientation (111.2 MPa, or 16 ksi), both lower than the parent-metal yield strength of 245.7 MPa (35.6 ksi). An elongation of 29.8% was reported for the longitudinal orientation, illustrating the high ductility associated with the fine-grain weld nugget microstructure.

**Fatigue Strength.** Figure 5.31 shows the results of transverse fatigue tests for the parent material and friction stir welded JIS6N01-T5 for $R = 0.1$ and $R = -1$ (Ref 68). Friction stir welding reduces fatigue life at a given stress, but the decrease is small. Most samples failed in

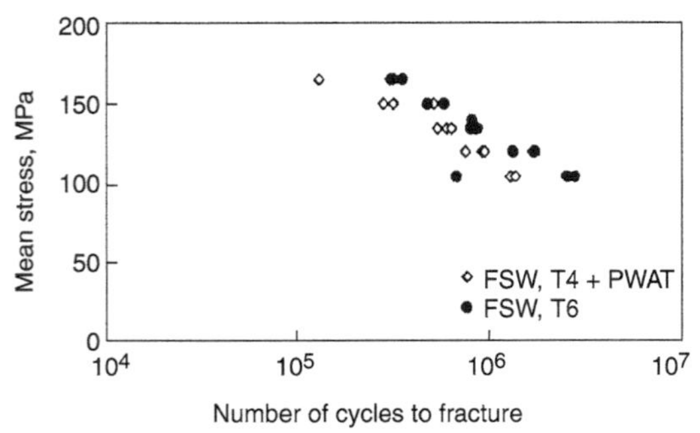


**Fig. 5.29** Fatigue test results for friction stir welded (FSW) 6082 for different tempers. PWAT, postweld heat treated. Source: Ref 64

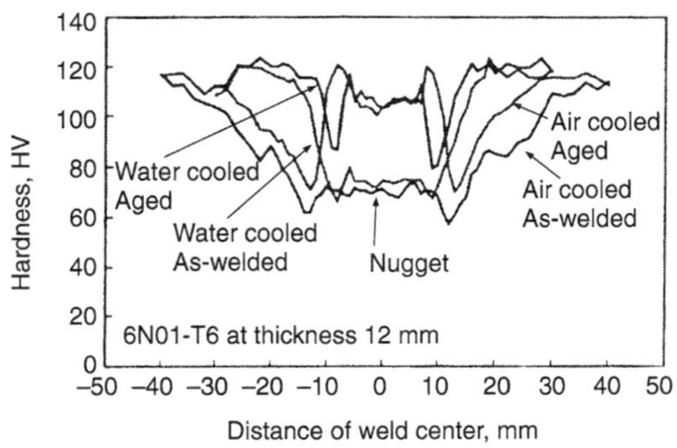


**Fig. 5.30** Effect of cooling conditions and aging on the hardness profiles of friction stir welds in JIS6N01-T6. Source: Ref 54

the HAZ, with a small number failing in the weld nugget. Other investigators have compared fatigue life for friction stir welds to MIG and laser welds (Ref 67, 69). In all cases, fatigue strength in the friction stir weld is greater. Weld speed did not influence fatigue life when defect-free welds were produced (Ref 69).

## 5.4 7*xxx* Aluminum Alloys

The 7*xxx* aluminum alloys are age hardenable, with a good combination of strength, fracture toughness, and corrosion resistance in both thick and thin wrought sections. The addition of zinc with other elements, notably copper, magnesium, and chromium, produces very high strength, including the highest strength available in any wrought aluminum alloy. In general, weldability of the high-strength 7*xxx* aluminum alloys by conventional fusion welding techniques is not good in any temper. However, because of the considerable interest in the high-strength 7*xxx* aluminum alloys in the aerospace industry and due to the inability to join these alloys by fusion welding, there has been considerable research into the ability to join 7*xxx* alloys by using the solid-state friction stir weld technique (Ref 6, 7, 17, 18, 50, 60, 66, 73–88).

**Hardness.** Following exposure to elevated temperature, the high-strength 7*xxx* alloys (7075, 7050, etc.) are in an unstable temper designated as W. For example, either a solution heat treatment or FSW, where the weld nugget experiences temperatures sufficient to dissolve the strengthening precipitates, is necessary. In the W temper, these alloys spontaneously age at room temperature, continuing to harden essentially forever, albeit at a decreasing rate. Hardness data for friction stir welded 7050 and 7075 Al alloys, naturally aged for >6 years, are illustrated in Fig. 5.32 and 5.33, respectively (Ref 73). These data illustrate the caution necessary when evaluating hardness (or mechanical properties) data for the 7*xxx* aluminum alloys. Most often, investigators do not report the time between welding and testing. However, as shown in Fig. 5.32 and 5.33, after >6 years of natural aging at room temperature, the hardness increased 55 to 65% in the weld nugget and 59 to 62% in the HAZ. Hardness increased slightly more in the 7050 alloy compared to 7075 Al. Also, the HAZ narrows considerably, and the minimum hardness zone moves outward from the weld nugget. Further, transverse weld failures corresponded directly with the hardness minima where failure location is shown to move further into the HAZ with increasing aging time.

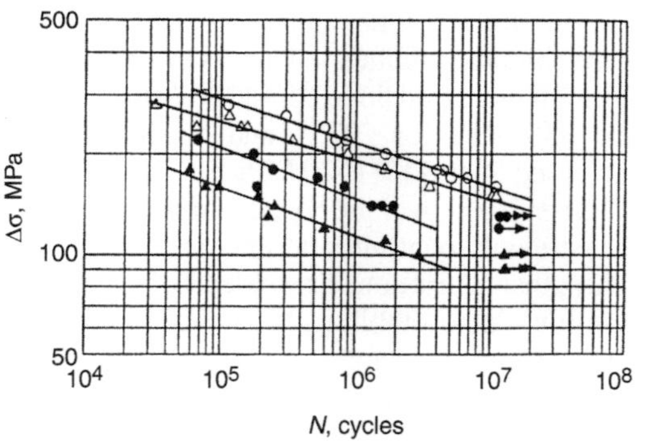


**Fig. 5.31** Fatigue life curves for friction stir welded JIS6N01-T5 for $R = 0.1$ and $R = -1$. Source: Ref 68

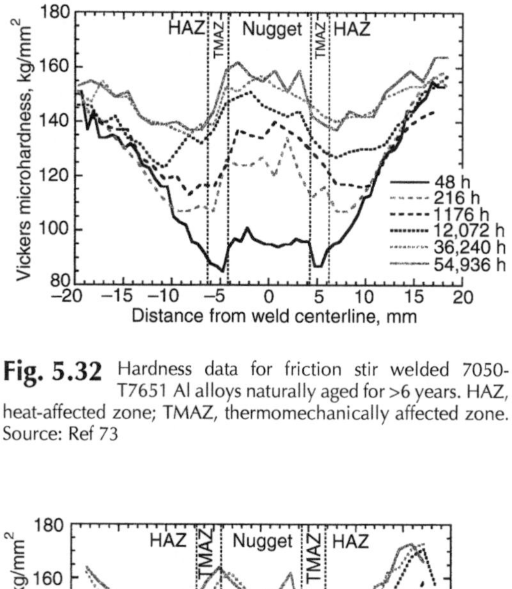


**Fig. 5.32** Hardness data for friction stir welded 7050-T7651 Al alloys naturally aged for >6 years. HAZ, heat-affected zone; TMAZ, thermomechanically affected zone. Source: Ref 73

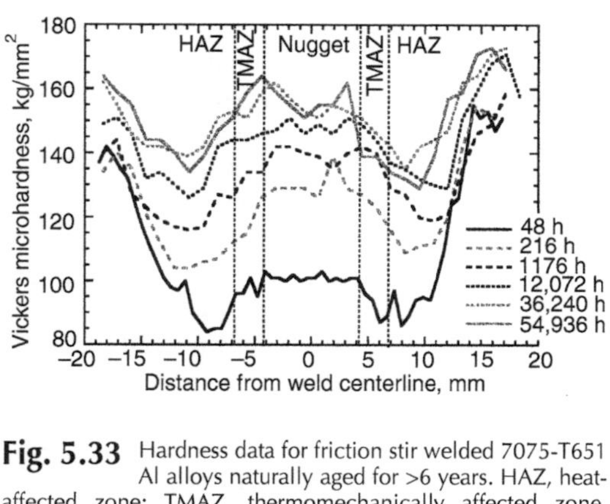


**Fig. 5.33** Hardness data for friction stir welded 7075-T651 Al alloys naturally aged for >6 years. HAZ, heat-affected zone; TMAZ, thermomechanically affected zone. Source: Ref 73

As shown, the weld zone continues to harden even after 6 years of natural aging. Similar hardness results are reported by Merati et al., where minimum hardness in the HAZ of friction stir welded 7050-T7651 increased by 9% after natural aging for 10 months (Ref 74). Time between FSW and the initial hardness measurements was not reported but is believed to be more than 2 months.

The first hardness data in Fig 5.32 and 5.33 were taken 48 h after FSW. Likely, there was significant hardening during this time. In the work of Nelson et al., the immediate recovery of hardness was noted where, after even just 5 h of natural aging, considerable hardness was recovered (Ref 75). In Fuller et al., the hardening rate is most dramatic during the time between 48 and 216 h (9 days), accounting for approximately half the hardness increase in these 6 years (Ref 73). These results emphasize the need to exercise caution when evaluating postweld properties in some friction stir welded 7*xxx* aluminum alloys. Likely, most investigators performed their mechanical testing sometime during the time of most rapid hardness change. Not only does the hardness increase via natural aging, but all other mechanical properties—fatigue, fracture toughness, and corrosion resistance—also change with time.

Additional hardness data are available for a number of friction stir welded 7*xxx* aluminum alloys, including 7010 (Ref 78, 81), 7017 (Ref 77), 7249 (Ref 82), 7349 (Ref 50, 83), 7050 (Ref 76, 80), 7075 (Ref 17, 79), and 7475 (Ref 8, 80). Time between welding and when hardness measurements were taken was not reported. However, these hardness results can illustrate trends with FSW variables. The results of Hassan et al. on 7010-T651 illustrate how hardness in the weld zone is influenced by different FSW parameters and, in addition, highlight unique features associated with FSW (Ref 78). Figure 5.34 illustrates hardness as a function of spindle speed and distance from the weld center for the top, center, and weld root. These are the similar "W"-shaped hardness profiles reported for other friction stir welded aluminum alloys. Each profile consists of a central uniform plateau that corresponds to the width of the nugget zone. Moving outward from the center, the profile then falls through the TMAZ, reaches a minimum (~110 HV) in the HAZ, and then gradually recovers to the level of the parent plate (~170 HV). Overall, the hardness of the plateau region is lower than the parent alloy and lies in the range of 130 to 155 HV. Although for the range of conditions investigated, the hardness minima did not vary, it can be seen that the hardness minima in the HAZ troughs shift out from the weld centerline with increasing rotation rate. Also, the different process parameters do have a significant effect on the hardness of the nugget zone, which changes significantly with spindle speed and depth within the weld (Ref 78). Unique to FSW, the width of the hardness plateau is largely independent of the spindle speed when it is controlled by the width of the tool shoulder, that is, near the top surface (Fig. 5.34a). However, this is not the case for the center and weld root, where the width of the hardness plateau increases with higher spindle speeds due to the higher heat input (Fig. 5.34b, c). This asymmetric hardness example illustrates the inhomogeneous behavior of FSW.

Other hardness observations of interest include those of Bassett et al. on 7017, where FSW was compared to MIG welding, illustrating a larger HAZ in the friction stir weld zone (Ref 77); Li et al. (Ref 82), where direct correlations are made between hardness values and conductivity profiles for a variety of postweld heat treatments following FSW in 7249; and the work of Jata et al., where postweld heat treatments in a 7050-T7451 alloy illustrate the ability to restore the weld nugget to near-parent-metal hardness, but the hardness minima troughs are unaffected in the HAZ (Ref 76).

**Mechanical Properties.** As shown previously, over time, hardness increases dramatically in the weld zone at room temperature following FSW. Similarly, transverse yield and tensile strengths in the weld zone also show significant increases (Ref 73, 74). Figures 5.35 and 5.36 illustrate strength increases in friction stir welded 7050-T7651 and 7075-T651 Al alloys following nearly 8 years of natural aging (Ref 73). In these results, initial tensile properties were obtained within a few hours of FSW. Yield strength increased 56 to 63% (385 MPa for 7050 and 355 MPa for 7075), and tensile strength increased 38 to 41% (520 MPa for 7050 and 500 MPa for 7075) over the 8 years. More important, the strength continues to increase after long times; that is, friction stir welded 7050 and 7075 Al alloys do not stabilize. Mertati et al. used longitudinal subsize samples to evaluate mechanical properties in the different weld zones of friction stir welded 7050-T7651 following natural aging for ~12 months (Ref 74). Their results illustrated the agreement between hardness results and tensile tests and also between longitudinal and transverse tensile tests. For example, comparing results of transverse and longitudinal samples, the yield and tensile strength of transverse specimens matched the minimum data for the longitudinal samples. These effects are expected, because specimens fail in the weakest point and the softest location.

Selected postweld transverse tensile properties for a variety of 7*xxx* aluminum alloys are shown in Table 5.5. Natural aging times were not reported. Although not always identified, the failure location is most often in the soft HAZ. Transverse weld failures correspond directly with hardness minima where failure location is shown to move further into the HAZ

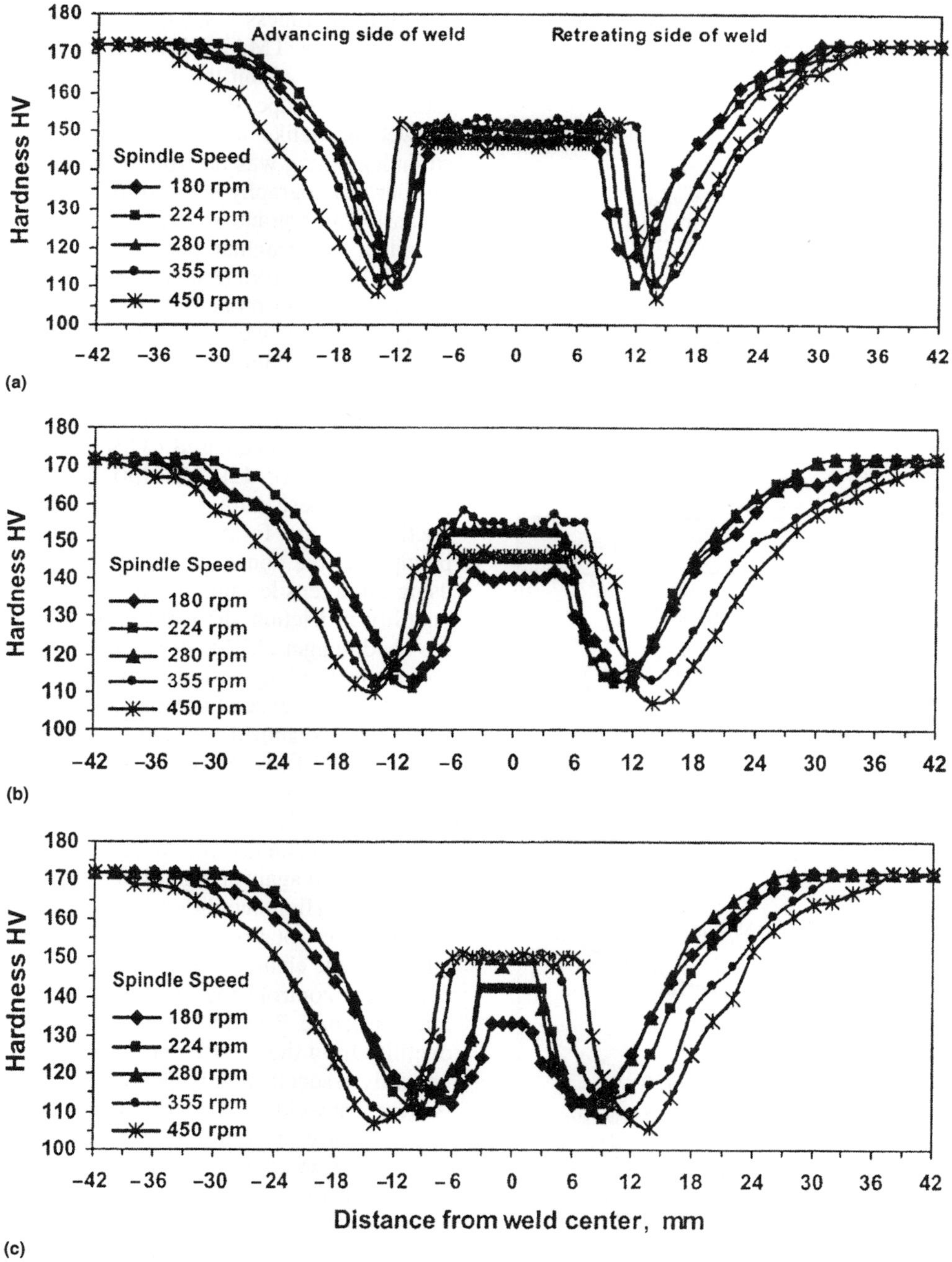


**Fig. 5.34** Hardness curves for friction stir welded 7010-T651 for a travel speed of 95 mm/min (3.7 in./min). (a) 0.1 mm (0.004 in.) below the top. (b) Center (c) 0.1 mm above the weld root. Source: Ref 78

with increasing aging time. Thus, properties in Table 5.5 are illustrative of HAZ yield and tensile strengths. As would be expected from a precipitation-hardenable alloy following FSW, the HAZ is overaged during FSW. For 7050 and 7075 in the as-welded condition, yield strength is reduced to 60 to 70% of base-metal properties, and tensile strength reduced to 70 to 80% of base-metal properties. However, as discussed previously, these alloys naturally age at room temperature, resulting in continued increases in strength.

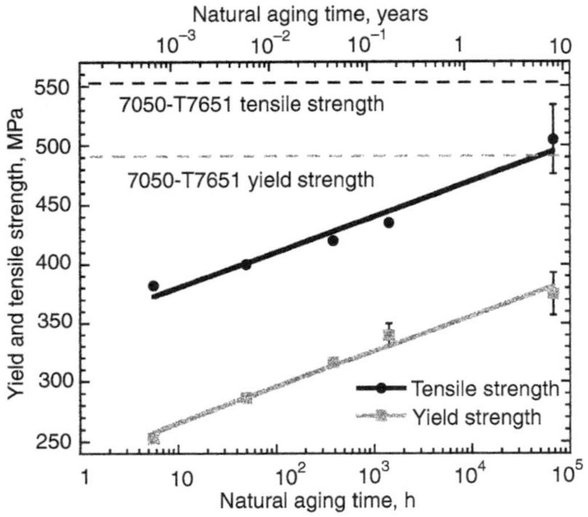


**Fig. 5.35** Yield and tensile strength for friction stir welded 7050-T7651 Al as a function of natural aging time. Source: Ref 73

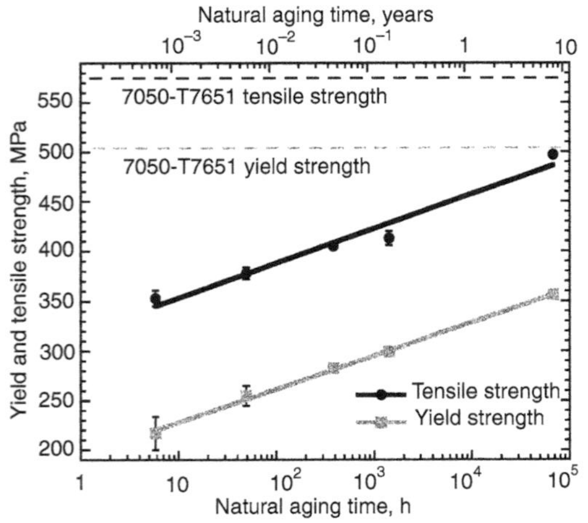


**Fig. 5.36** Yield and tensile strength for friction stir welded 7075-T651 Al as a function of natural aging time. Source: Ref 73

Allehaux et al. used the microspecimen technique to evaluate mechanical properties in the different regions of the weld zone for friction stir welded 7349-T6 (Ref 83). Figure 5.22 illustrates the microsample sectioning method. Figure 5.37 shows mechanical properties for microtensile samples across the different weld zones (Ref 83). As is customary, profiles of the yield and tensile strengths are in accordance with microhardness results, illustrating lowest strengths in the TMAZ. However, the low ductility obtained within the weld nugget is interesting. The shape of the stress-strain curve showed brittlelike behavior, no necking, and strain hardening was far from being exhausted. Further, fractography showed that rupture was primarily intergranular and associated with abundant intergranular precipitates within the nugget. Results from Mahoney et al. also evaluated weld nugget properties in the longitudinal direction (only weld nugget microstructure in the gage section) for friction stir welded 7075-T651 (Ref 85). In the as-welded condition, the ductility in the weld nugget was high (15%) and only decreased (3.5%) following a post-weld age of 120 °C (150 °F) for 24 h (Ref 85). These authors attributed decreased ductility to the formation of grain-boundary precipitate-free zones. Similarly, Paglia et al., using microtensile samples, demonstrated high ductility in friction stir welded 7075-T6 Al in the weld nugget (15%) (Ref 79). Hassan et al., also evaluating nugget-only properties, demonstrated the influence of travel speed and spindle speed on mechanical properties in friction stir welded 7010-T651 Al (Ref 78). In their work (Fig. 5.38), ductility was significantly influenced by spindle speed (Ref 78). Low ductility was observed at a low rotation speed (180 rpm) and decreased again at high (450 rpm) tool rotation speeds. High ductility was observed for the intermediate rotation speeds. These authors attributed the change in ductility to the thermal cycle that controls the eventual weld nugget microstructure. The differences in weld nugget ductility from these different investigators may be partly associated with alloy chemistry, but it is clear that weld parameters can have a significant influence on resulting mechanical properties in the 7*xxx* aluminum alloys.

Strength following postweld aging has been investigated by a number of investigators for friction stir welded 7*xxx* aluminum alloys (Ref 8, 76, 77, 82, 85). Solution treatment followed by artificial aging nearly completely restored

**Table 5.5 Transverse tensile properties of friction stir welded 7*xxx* aluminum alloys**

| Alloy temper | Postweld age | Yield strength | | Tensile strength | | Ref |
|---|---|---|---|---|---|---|
| | | MPa | ksi | MPa | ksi | |
| **Base-material properties** | | | | | | |
| 7108-T79 | ... | 295 | 42.8 | 370 | 53.7 | 60 |
| 7050-T7451 | ... | 489 | 70.9 | 555 | 80.5 | 76 |
| 7050-T73 | ... | ... | ... | 546 | 79.2 | 18 |
| 7075-T73 | ... | ... | ... | 515 | 74.7 | 18 |
| 7075-T6 | ... | 486 | 70.5 | 553 | 80.2 | 86 |
| 7349-T6 | ... | 586 | 85.0 | 636 | 92.2 | 83 |
| 7475-T76 | ... | ... | ... | 528 | 76.6 | 8 |
| **Friction stir welded properties** | | | | | | |
| 7017 | AW(a) | 256 | 37.1 | 376 | 54.5 | 77 |
| 7017 | 8 h at 150 °C (302 °F) | 343 | 49.7 | 261 | 37.9 | 77 |
| 7108 | AW | 275 | 39.9 | 380 | 55.1 | 66 |
| 7108-T79 | AW | 210 | 30.5 | 320 | 46.4 | 60 |
| 7108-T79 | Natural age | 245 | 35.5 | 350 | 50.8 | 60 |
| 7050-T73 | AW | ... | ... | 436 | 63.2 | 80 |
| 7050-T7451 | AW | 304 | 44.1 | 429 | 62.2 | 76 |
| 7050-T7451 | T7 | 287 | 41.6 | 371 | 53.8 | 76 |
| 7050-T7451 | T6 | 291 | 42.2 | 417 | 60.5 | 76 |
| 7075-T6 | AW | 333 | 48.3 | 410 | 59.5 | 86 |
| 7075-T651 | AW | 312 | 45.3 | 468 | 67.9 | 85 |
| 7075-T651 | T6 | 312 | 45.3 | 447 | 64.8 | 85 |
| 7075-T73 | AW | ... | ... | 416 | 60.3 | 18 |
| 7075-T651 | AW | 340 | 49.3 | 485 | 70.3 | 84 |
| 7249-W511 | AW | 367 | 53.2 | 520 | 75.4 | 82 |
| 7249-W511 | T6 + T76 | 434 | 62.9 | 500 | 72.5 | 82 |
| 7249-T6511 | AW | 374 | 54.2 | 490 | 71.1 | 82 |
| 7249-T6511 | T76 | 379 | 55.0 | 454 | 65.8 | 82 |
| 7249-T6511 | T6 + T76 | 394 | 57.1 | 470 | 68.2 | 82 |
| 7249-T76511 | AW | 378 | 54.8 | 511 | 74.1 | 82 |
| 7249-T76511 | T6 | 405 | 58.7 | 503 | 73.0 | 82 |
| 7349-T6 | AW | 368 | 53.4 | 515 | 74.7 | 83 |
| 7475-T76 | AW | 381 | 55.3 | 465 | 67.4 | 8 |

(a) AW, as-welded

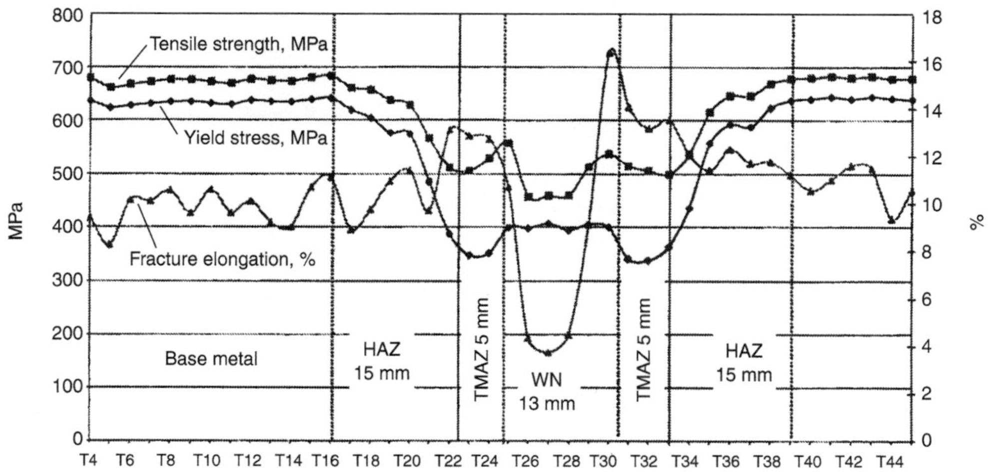


**Fig. 5.37** Tensile test results for longitudinal microspecimens in friction stir welded 7349-T6. HAZ, heat-affected zone; TMAZ, thermomechanically affected zone; WN, weld nugget. Source: Ref 83

parent-metal tensile strength (97%) in 7475-T76 Al (Ref 8). However, postweld aging without first solution treating can reduce strength when testing transverse to the weld direction. The results of Jata et al. for friction stir welded 7050-T7451 illustrate decreases in both yield and tensile strength following conventional postweld aging treatments to either T6 (121 °C, or 250 °F, for 24 h) or T7 (121 °C for 24 h + 175 °C, or 347 °F, for 8 h) (Ref 76). Their results, tabulated in Table 5.5, illustrate significant decreases in strength following FSW, with a further decrease following postweld aging. Similar results were obtained by Bassett et al. for 7017 where, following a postweld age of 8 h at 150 °C (300 °F), tensile strength decreased from the as-welded condition by an additional 10%, but yield strength was essentially unaffected (Ref 77). Li et al. evaluated a variety of postweld heat treatments for friction stir welded 7249-W511, obtaining similar results to others (Ref 82). In general, postweld aging had very little influence on yield strength, increasing from 2 to 10% for most heat treatments, while tensile strength decreased from 1.3 to 9.7% for all age treatments (Ref 82). The one exception was an 18% increase in yield strength for a postweld age of 121 °C for 24 h + 163 °C (325 °F) for 8 h. As one would expect, because failures are located in the soft, overaged HAZ, without first solution treatment, postweld aging simply overages the HAZ even more. This results in only slight changes in transverse weld strength and, more often than not, slight decreases.

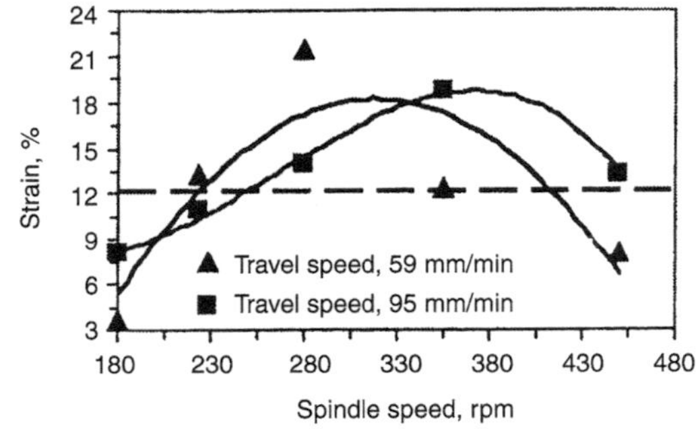


**Fig. 5.38** Tensile elongation in the weld nugget zone for friction stir welded 7010-T651. Source: Ref 78

**Fatigue and Fatigue Crack Growth Rate.** Fatigue life results for friction stir welded 7475-T76, 7475-T7351, and 7050-T7451 have been reported (Ref 8, 80). Magnusson et al. illustrate fatigue life for both the as-welded and surface-milled conditions (Ref 8). As shown in Fig. 5.39, FSW does reduce the fatigue threshold stress ~40 MPa (5.8 ksi) compared to parent metal when the weld bead is not removed. However, these results are more a reflection of surface-initiated fatigue failure, and the surface roughness of the weld bead can be considerably different, depending on the FSW tool design used. For example, the scroll shoulder tool used with zero tilt results in a relatively smooth surface and very little flash compared to the older concave tool

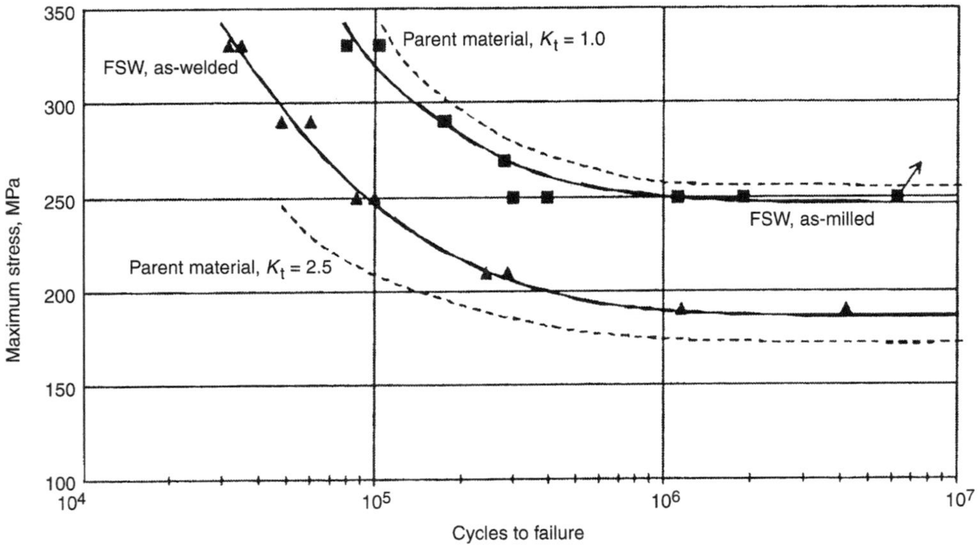


**Fig. 5.39** Fatigue results for friction stir welded (FSW) 7475-T76 at $R = 0.1$ for as-welded and milled surfaces compared to parent metal for $K_t = 1.0$ and 2.5. Source: Ref 8

design commonly tilted 2.5 to 3° in the travel direction. When the flash is removed from the weld crown, that is, removed by milling, fatigue life of the friction stir welded 7475-T76 approaches that of the parent metal. Results from Kumagai et al. are comparable, again illustrating fatigue life equivalent to parent metal when the crown surface is machined smooth (Ref 80). Additional fatigue life data are available for 7075-T6 Al from Talwar et al., but these results are for a lap weld geometry, which introduces other considerations not addressed in this chapter (Ref 86).

Crack growth rate data are available for friction stir welded 7050-T7451 from several investigators (Ref 76, 87, 88). In the work of John et al., crack growth rates were compared for different specimen geometries, compact and middle tension specimens, for two *R* ratios of 0.05 and 0.8 (Ref 87). In each case, the notch was placed along the weld direction but away from the weld centerline in the HAZ. Crack propagation results are illustrated in Fig. 5.40, with results compared to crack growth rate data for 7050-T7451 using curve fits from the AFGROW program (shown to be similar to parent-metal properties) (Ref 89). The crack growth rate results in Fig. 5.40 show a geometry dependency for crack growth behavior in the HAZ at low *R*. That is, the compact tension specimen has a much lower crack growth rate and significantly higher threshold stress-intensity factor than the middle crack tension sample at $R = 0.05$. However, the geometry dependency nearly vanished at high *R*. The implication of these results is that compressive residual stresses are present in the HAZ, and they can influence crack propagation rates. Jata et al. evaluated the effects of *R* ratio and crack location (nugget or HAZ) on crack growth rates for friction stir welded 7050-T7451 with a post-weld T6 heat treatment (121 °C for 24 h), as shown in Fig. 5.41 (Ref 76). No difference in fatigue crack growth rate was seen in the weld nugget for different *R* ratios. However, crack growth rates were the highest when the crack was centered in the weld nugget, even higher than in the parent metal. In the weld nugget, the fine-grain microstructure and intergranular failure dominated the fatigue crack growth rate. In the HAZ, as before, a low-*R* ratio results in considerably lower fatigue crack growth rates. Again, compressive residual stresses dominate fatigue crack growth rate in the HAZ. However, fatigue crack growth rates in the HAZ, for either *R* ratio, are lower than for the parent material.

## 5.5 Aluminum-Lithium Alloys

The aluminum-lithium alloys are of interest, especially for space applications, due to their high specific strength, that is, strength-to-weight ratio. Although there are many advantages associated with aluminum-lithium alloys, fusion welding is difficult. Thus, some investigators have evaluated the ability to friction stir weld both the 2195 and 2095 alloy compositions (Ref 11, 75, 90–95). Data for friction stir welded aluminum-lithium alloys are limited,

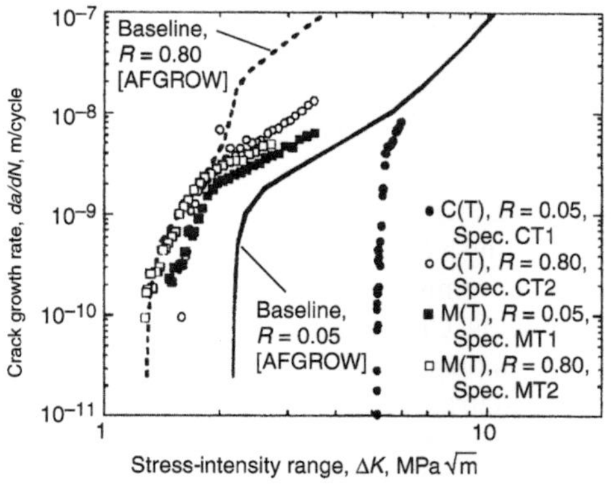


**Fig. 5.40** Fatigue crack growth rates in friction stir welded 7050-T7451 comparing compact tension (CT) and center-cracked (MT) specimens at $R = 0.05$ and $R = 0.8$. Source: Ref 87

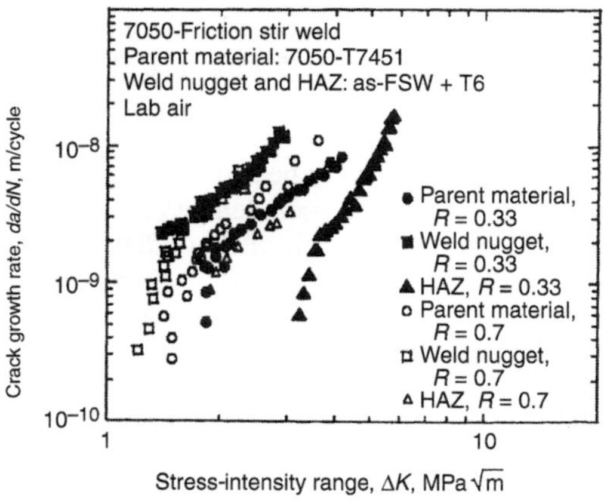


**Fig. 5.41** Comparison of fatigue crack growth rates between the weld nugget and heat-affected zone (HAZ) at $R = 0.33$ and $R = 0.7$ for friction stir welded (FSW) 7050. Fatigue crack growth rates were evaluated in the as-FSW +T6 condition. Source: Ref 76

and thus, variations in properties for different FSW conditions are not available. Accordingly, data presented herein should be considered with caution until more information is available.

**Hardness** contour maps for 2095 were developed by Attallah and Salem for two different travel speeds and two different rotation rates (Ref. 94). There was an increase in the hardness throughout the weld as the travel speed increased. On increasing the rotation rate while maintaining the same travel speed, the heterogeneity in the hardness distribution was significantly minimized (Ref 94). Unfortunately, some hardness results available for 2195 are for a bialloy weld, wherein 2195 was welded to 2219 Al, with the 2195 on the advancing side in one case and on the retreating side in another (Ref 90). A hardness reduction from the parent material (150 HV) to a hardness minimum in the HAZ (100 HV) was observed. Within the weld nugget itself, hardness varied considerably due to the mechanical intermixing of the two alloys.

Nelson et al. evaluated hardness in friction stir welded 2195-T8 for different weld conditions, that is, active cooling and active heating, followed by natural aging for 96 h (Ref 75). Hardness results from their studies are shown in Fig. 5.42, illustrating a number of findings. At 0 h, the hardness curve for the actively heated sample exhibits a more uniform hardness profile across the weld nugget compared to the actively cooled sample (Fig. 5.42a).The authors attribute this difference to "quenching-in" higher vacancy and solute concentrations during active cooling. The influence of active cooling is even more evident when considering the hardness curves following natural aging for 96 h (Fig. 5.42b). Comparing Fig. 5.42(a) and (b), hardness in the weld nugget of the actively cooled sample is shown to increase by >20%, whereas hardness in the actively heated sample changes very little with time at room temperature. This is an interesting illustration of how FSW boundary conditions can significantly change postweld properties, even in a solid-state weld.

**Mechanical Properties.** Mechanical property results have been reported for 2195 aluminum-lithium over a wide range of material thicknesses (Ref 11, 90–93). A summary of mechanical properties is presented in Table 5.6. Following FSW and combining results for different investigators, yield strength decreased from 53 to 63%, and tensile strength decreased from 32 to 40%. Although this is a significant decrease over parent-metal properties, FSW is still an improvement over variable polarity plasma arc (VPPA) welding (319 MPa, or 46 ksi), where the decrease in tensile strength is 47% (Ref 95). Mechanical properties at cryogenic temperatures (–253 °C, or –423 °F) are reported by Kinchen et al. and Loftus et al. (Ref 91, 92). Both investigators report significant strength increases for friction stir welds over room-temperature properties, with tensile strength ~630 MPa (91 ksi) and yield strength ~408 MPa (59 ksi). Again, this is considerably higher than the tensile strength of a VPPA weld tested at –253 °C (435 MPa, or 63 ksi) (Ref 92).

**Fracture Toughness.** Fracture behavior for 2195-T8 has been reported by Kroninger and Reynolds (Ref 93). *R*-curves were produced with compact-type specimens, using the single-specimen unloading compliance method. The base-metal and the weld-metal specimens all exhibited rising *R*-curve behavior and substantial crack extension before the onset of instability. The friction stir welded specimens exhibited higher crack resistance than the base metal at both large and small crack extensions. Further, the toughness of the friction stir welds, the base metal, and VPPA welds were compared. Toughness in the friction stir weld was greater than that of the base metal, while the VPPA toughness was substantially worse in the initiation region and, on average, worse at large crack extensions as well, compared to friction stir welds (Ref 93).

## 5.6 Aluminum Metal-Matrix Composites

Fusion welding has been applied to particulate-reinforced composites since 1985. However, during welding, the liquid aluminum reacts with SiC particles and results in the formation of aluminum carbide along with an increase in silicon in the matrix alloy. The use of low-power TIG welding along with the concentration of heat on the unreinforced filler metal can produce sound welds. Unfortunately, this technique relies heavily on operator skill and still results in some level of matrix/reinforcement reaction. The use of aluminum oxide as the reinforcement minimizes the severity of the reaction of the molten aluminum with the ceramic phase. However, even with this composite, the reaction that occurs decreases the strength of the matrix in the weld region. Conventional inertia or friction welding

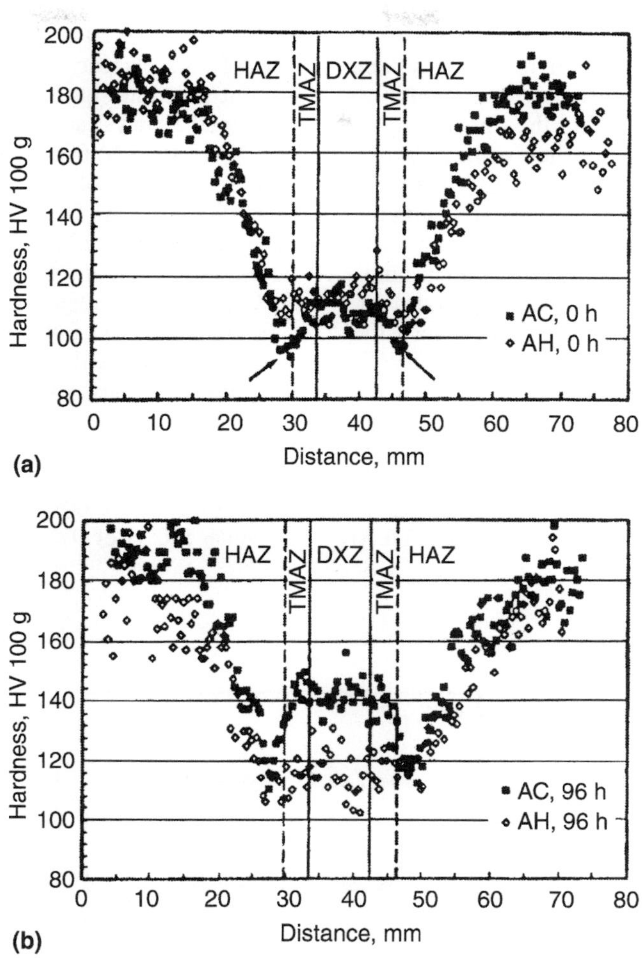


**Fig. 5.42** Microhardness data for actively cooled (AC) and heated (AH) conditions for friction stir welded 2195-T8 Al. HAZ, heat-affected zone; TMAZ, thermomechanically affected zone; DXZ, dynamically recrystallized zone. Source: Ref 75

**Table 5.6 Tensile properties of friction stir welded (FSW) 2195 for different material thicknesses**

| | Yield strength | | Tensile strength | | Thickness | | |
|---|---|---|---|---|---|---|---|
| 2195 Al-Li | MPa | ksi | MPa | ksi | mm | in. | Ref |
| Base-T8 | 570 | 82.7 | 600 | 87.0 | 8.1 | 0.319 | 93 |
| FSW-1424 | 225 | 32.6 | 390 | 56.6 | 4 | 0.157 | 11 |
| FSW-T8 | 270 | 39.2 | 410 | 59.5 | 8.1 | 0.319 | 93 |
| FSW | 251 | 36.4 | 401 | 58.2 | ... | ... | 92 |
| FSW-T8 | 249 | 36.1 | 399 | 57.9 | 8.1 | 0.319 | 91 |
| FSW-T8 | 209 | 30.3 | 357 | 51.8 | 16.5 | 0.650 | 91 |
| FSW | 217 | 31.5 | 368 | 53.4 | 25.4 | 1 | 90 |

produces sound welds with good mechanical properties but is limited to relatively simple geometries, typically, rod or tube configurations. Friction stir welding offers the opportunity to weld metal-matrix composites without limitation to the geometric shape. Hardness and property results are reported subsequently, summarizing the small amount of available data. Although the mechanical properties are promising, the particulate reinforcement acts as an abrasive on the FSW tool. Not only is tool life severely limited, but debris from the FSW is deposited in the weld joint.

**Hardness.** A limited number of investigations have evaluated FSW of discontinuously reinforced aluminum alloys (Ref 25, 96, 97). In the work of Nakata et al., hardness and mechanical properties were established for 6061-T6 Al reinforced with 10 and 20 vol% $Al_2O_3$ and 6092-T6 Al. Hardness results by Nakata et al. are illustrated in Fig. 5.43 for as-welded (naturally aged 20 days) and postweld aged (175 °C, or 347 °F, for 8 h) 6061-T6 Al reinforced with both 10 and 20 vol% $Al_2O_3$ particulate additions. For 10% $Al_2O_3$, it is difficult to distinguish a hardness difference between the weld nugget (stir zone) and HAZ. However, for 20% $Al_2O_3$, the weld nugget is substantially higher (90 versus 115 HV). For each volume loading of $Al_2O_3$, the postweld heat treatment increased the nugget hardness to a level greater than the parent metal, while the HAZ was near parent-metal strength.

Similar hardness behavior is reported by Mahoney et al. for 6092-T6 Al reinforced with 17 vol% SiC (Ref 97). Figure 5.44 illustrates

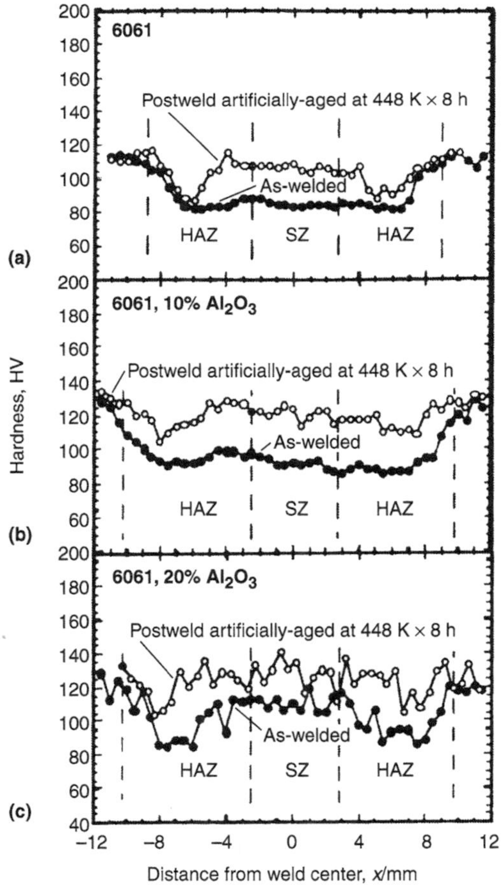


**Fig. 5.43** Hardness in friction stir welded 6061-T6 Al reinforced with 10 and 20 vol% $Al_2O_3$ particulate, naturally aged 20 days and postweld aged (175 °C, or 347 °F, for 8 h). HAZ, heat-affected zone; SZ, stir zone. Source: Ref 96

transverse hardness traces for this composite alloy as a function of postweld heat treatment. Following solution treatment and aging, full parent-metal hardness is recovered (RB 85) in the HAZ, while the weld nugget hardness is slightly higher. This hardness increase could be associated with either tool debris in the weld or the breakup of the particulate reinforcement, resulting in a more homogeneous distribution of SiC. Within the weld nugget, Mahoney et al. showed a very fine dispersion of reinforcement particles between the larger particles. It appears that the small particle dispersion is created during FSW; that is, the sharp edges of the larger particles are sheared off, leaving behind the very small particles dispersed among the larger particles with rounded edges. In addition to the fine SiC dispersion, x-ray results identified a very fine dispersion of debris from the FSW tool.

**Mechanical Properties.** Following FSW, the tensile strengths of the 6061-T6 base metal, 6061 + 10% $Al_2O_3$, and 6061 + 20% $Al_2O_3$ are essentially the same (260 MPa, or 37.7 ksi) and are approximately 20% less than unwelded tensile strengths (320 MPa, or 46.4 ksi) (Ref 96). Following a postweld age of 175 °C for 8 h, tensile strength is fully restored to base-metal strength. Similar postweld results are reported by Mahoney et al. where friction stir welding reduces the yield strength by 45% (220 versus 396 MPa, or 31.9 versus 57.4 ksi) and the tensile strength by 33% (303 versus 449 MPa, or 43.9 versus 65 ksi) for 6092-T6 Al reinforced with 17 vol% SiC (Ref 97). In a qualitative measure of postweld ductility, Mahoney et al. performed bend tests and illustrated an 18° bend for the welded sample versus a 10° for the parent metal prior to fracture. In the welded sample, failure occurred in the lower-strength HAZ, where overaging of strengthening precipitates reduced the yield strength.

## REFERENCES


1. T. Hashimoto, N. Nishikawa, S. Tazaki, and M. Enomoto, Mechanical Properties of Joints for Aluminum Alloys with Friction Stir Welding Process, *Joints in Aluminum, INALCO, 1998: Seventh International Conference,* Vol 2, Abington Publishing, 1999
2. G. Biallas, R. Braun, C. Dalle Donne, G. Staniek, and W.A. Kaysser, "Mechanical Properties and Corrosion Behavior of Friction Stir Welded 2024-T3," First International Symposium on Friction Stir Welding, June 1999 (Thousand Oaks, CA), TWI
3. G. Bussu and P.E. Irving, "Fatigue Performance of Friction Stir Welded 2024-T351," First International Symposium on Friction Stir Welding, June 1999 (Thousand Oaks, CA), TWI
4. R. Pedwell, H. Davies, and A. Jefferson, "The Application of Friction Stir Welding to Wing Structures," First International Symposium on Friction Stir Welding, June 1999 (Thousand Oaks, CA), TWI
5. T. Hashimoto, S. Jyogan, K. Nakata, Y. Kim, and M. Ushio, "FSW Joints of High Strength Aluminum Alloys," First Inter-


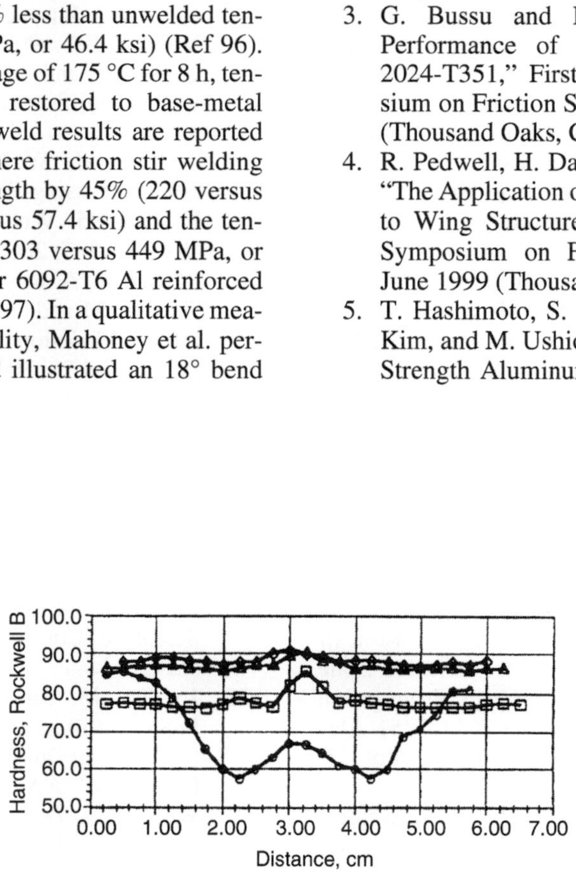


**Fig. 5.44** Hardness behavior for friction stir welded 6092-T6 Al reinforced with 17 vol% SiC as a function of postweld heat treatment. Source: Ref 97

national Symposium on Friction Stir Welding, June 1999 (Thousand Oaks, CA), TWI
6. A. von Strombeck, J. dos Santos, F. Torster, P. Laureano, and M. Kocak, "Fracture Toughness Behavior of FSW Joints on Aluminum Alloys," First International Symposium on Friction Stir Welding, June 1999 (Thousand Oaks, CA), TWI
7. S. Russell, M. Tester, E. Nichols, A. Cleaver, and J. Maynor, Static, Fatigue and Crack Growth Behavior of Friction Stir Welded 7075-T6 and 2024-T3 Aluminum Alloys, *Friction Stir Welding and Processing,* K.V. Jata, M.W. Mahoney, R.S. Mishra, S.L. Semiatin, and D.P. Field, Ed., TMS, 2001
8. L. Magnusson and L. Kallman, "Mechanical Properties of Friction Stir Welds in Thin Sheet of Aluminum 2024, 6013, and 7475," Second International Symposium on Friction Stir Welding, June 26–28, 2000, (Gothenburg, Sweden), TWI
9. C. Dalle Donne, G. Biallas, T. Ghidini, and G. Raimbeaux, "Effect of Weld Imperfections and Residual Stresses on the Fatigue Crack Propagation in Friction Stir Welded Joints," Second International Symposium on Friction Stir Welding, June 26–28, 2000 (Gothenburg, Sweden), TWI
10. D. Lohwasser, "Application of Friction Stir Welding for Aircraft Industry," Second International Symposium on Friction Stir Welding, June 26–28, 2000 (Gothenburg, Sweden), TWI
11. D. Lohwasser, Thin Section Airframe Alloy Welding within WAFS, THERMEC 2003, International Conference on Processing and Manufacturing of Advanced Materials (Leganes, Madrid, Spain), *Mater. Sci. Forum,* Vol 426–432, 2003, p 2879–2884
12. G. Bussu and P.E. Irving, The Role of Residual Stress and Heat Affected Zone Properties on Fatigue Crack Propagation in Friction Stir Welded 2024-T351 Aluminum Joints, *Int. J. Fatigue,* Vol 25, 2003, p 77–88
13. B. Christner, J. McCoury, and S. Higgins, "Development and Testing of Friction Stir Welding (FSW) as a Joining Method for Primary Aircraft Structure," Fourth International Symposium on Friction Stir Welding, May 14–16, 2003 (Park City, UT), TWI
14. M. Sutton, B. Yang, A. Reynolds, and J. Yan, "Banded Microstructure in 2024-T351 and 2524-T351 Aluminum Friction Stir Welds, Part II: Mechanical Characterization, *Mater. Sci. Eng. A,* Vol 364, 2004, p 66–74
15. J. Kristensen, C. Dalle Donne, T. Ghidini, J. Mononen, A. Norman, A. Pietras, M. Russell, and S. Slater, "Properties of Friction Stir Welded Joints in the Aluminum Alloys of 2024, 5083, 6082/6060, and 7075," Fifth Int. Friction Stir Welding Conf., Sept 14–16, 2004 (Metz, France), TWI
16. C. Dalle Donne and G. Biallas, Fatigue and Fracture Performance of Friction Stir Welded 2024-T3 Joints, *Proceedings European Conference on Spacecraft Structures, Materials and Mechanical Testing,* Nov 4–6, 1998 (Braunschweig, Germany) European Space Agency, SP Vol 428, Feb 1999, p 309–314
17. C. Fuller and M.W. Mahoney, Rockwell Scientific Corporation, unpublished research
18. A. Reynolds, W. Lockwood, and T. Seidel, Processing-Property Correlation in Friction Stir Welds, *Mater. Sci. Forum,* Vol 331–337, 2000, p 1719–1724
19. M. Sutton, A. Reynolds, B. Yang, and R. Taylor, Mixed Mode I/II Fracture of 2024-T3 Friction Stir Welds, *Eng. Fract. Mech.,* Vol 70, 2003, p 2215–2234
20. *Alloy Digest, Aluminum 2024,* Engineering Alloys Digest, Inc., Upper Montclair, NJ, Aug 1973
21. J.E. Hatch, Ed., *Aluminum Properties and Physical Metallurgy,* American Society for Metals, Metals Park, OH, May 1984, p 175–177
22. C. Fuller and M.W. Mahoney, Rockwell Scientific Corporation, unpublished research
23. J. Yan, M. Sutton, and A. Reynolds, Process-Structure-Property Relationships for Nugget and Heat Affected Zone Regions of AA2524-T351 Friction Stir Welds, *Sci. Technol. Weld. Joining,* Vol 10 (No. 6), Dec 2005, p 725–736
24. P. Prevey and M. Mahoney, Improved Fatigue Performance of Friction Stir

Welds with Low Plasticity Burnishing: Residual Stress Design and Fatigue Performance Assessment, THERMEC 2003, International Conference on Processing and Manufacturing of Advanced Materials (Leganes, Madrid, Spain), *Mater. Sci. Forum,* Vol 426–432, 2003, p 2933–2940

25. J. Wert, Microstructure of Friction Stir Weld Joints between an Aluminum-Base Metal Matrix Composite and a Monolithic Aluminum Alloy, *Scr. Mater.,* Vol 49, 2003, p 607–612
26. M. Mahoney, unpublished research
27. M. Sutton, B. Yang, A. Reynolds, and R. Taylor, Microstructural Studies of Friction Stir Welds in 2024-T3, *Mater. Sci. Eng. A,* Vol 323 (No. 1–2), 2002, p 160–166
28. M. Sutton, A. Reynolds, B. Yang, and R. Taylor, Model I Fracture and Microstructure for 2024-T3 Friction Stir Welds, *Mater. Sci. Eng. A,* Vol 354, 2003, p 6–16
29. D. Hornbach, M. Mahoney, P. Prevey, D. Waldron, and J. Cammett, Low Plasticity Burnishing of Friction Stir Welds in 2219 Aluminum to Increase Corrosion Fatigue Life, *Sixth International Trends in Welding Research Conf. Proceedings,* April 15–19, 2002 (Pine Mountain, GA), ASM International, Materials Park, OH, 2003, p 302–306
30. B. Yang, J. Yan, M. Sutton, and A. Reynolds, Banded Microstructure in AA2024-T351 and AA2524-T351 Aluminum Friction Stir Welds, Part 1: Metallurgical Studies, *Mater. Sci. Eng. A,* Vol 364, 2004, p 55–65
31. R. Van Stone, R. Merchant, and J. Low, Jr., *Fatigue and Fracture Toughness—Cryogenic Behavior,* STP 556, ASTM, 1974, p 93–124
32. L. Karlsson, L. Svensson, and H. Larsson, Characteristics of Friction Stir Welded Aluminum Alloys, *Trends in Welding Research, Proc. of the Fifth Int. Conf.,* June 1–5, 1998 (Pine Mountain, GA), J. Vitek, S. David, J. Johnson, H. Smartt, and T. DebRoy, Ed., ASM International, Materials Park, OH, 1999, p 574–579
33. C. Fuller, M. Mahoney, and W. Bingel, "A Study of Friction Stir Processing Tool Designs for Microstructural Modifications as Demonstrated in Aluminum Fusion Welds," Fifth Int. Friction Stir Welding Conf., Sept 14–16, 2004 (Metz, France), TWI
34. M. Kumagai and S. Tanaka, "Properties of Aluminum Wide Panels by Friction Stir Welding," First International Symposium on Friction Stir Welding, June 1999 (Thousand Oaks, CA), TWI
35. K. Colligan, I. Ucok, K. McTernan, P. Konkel, and J. Pickens, "Friction Stir Welding of Thick Section 5083-H131 and 2195-T8P4 Aluminum Plates," Third International Symposium on Friction Stir Welding, Sept 27–28, 2001 (Kobe, Japan), TWI
36. M. Hayashi and K. Oyama, "Mechanical Properties of Friction Stir Welded 5083 Aluminum Alloy at Cryogenic Temperatures," Third International Symposium on Friction Stir Welding, Sept 27–28, 2001 (Kobe, Japan), TWI
37. C. Fuller, M. Mahoney, and W. Bingel, "Friction Stir Processing of Aluminum Fusion Welds," Fourth International Symposium on Friction Stir Welding, May 14–16, 2003 (Park City, UT), TWI
38. M. James, D. Hattingh, and G. Bradley, Influence of Travel Speed on Fatigue Life of Friction Stir Welds in 5083 Aluminum, *Fatigue 2002,* International Fatigue Congress, p 413–420
39. M. Dawes, S. Karger, and T. Dickerson, "Strength and Fracture Toughness of Friction Stir Welds in Aluminum Alloys," Second International Symposium on Friction Stir Welding, June 26–28, 2000, (Gothenburg, Sweden), TWI
40. W. Thomas, D. Staines, P. Tubby, M. Dore, and M. Gittos, "Friction Skew-Stir Welding—Fatigue Performance," INALCO 2004, Ninth Int. Conf. on Aluminum Structural Design, June 2–4, 2004 (Cleveland, OH)
41. A. Reynolds, Mechanical and Corrosion Performance of TGA and Friction Stir Welded Aluminum for Tailor Welded Blanks: Alloys 5454 and 6061, *Trends in Welding Research, Proc. of the Fifth Int. Conf.,* June 1–5, 1998 (Pine Mountain, GA), J. Vitek, S. David, J. Johnson, H. Smartt, and T. DebRoy, Ed., ASM International, Materials Park, OH, 1999 p 563–567
42. S. Sheikhi, J. dos Santos, and S. Losch,

"On the Formability of Friction Stir Welded Aluminum Tailored Blanks," Fifth Int. Friction Stir Welding Conf., Sept 14–16, 2004 (Metz, France), TWI
43. G. Frankel and Z. Xia, Localized Corrosion and Stress Corrosion Cracking Resistance of Friction Stir Welded Aluminum Alloy 5454, *Corrosion,* Vol 55 (No. 2), Feb 1999, p 139–150
44. P. Pao, R. Fonda, H. Jones, C. Feng, B. Connolly, and A. Davenport, Microstructure, Fatigue Crack Growth, and Corrosion in Friction Stir Welded Al 5456, *Friction Stir Welding and Processing III,* K. Jata, M. Mahoney, R. Mishra, and T. Lienert, Ed., 2005 TMS Annual Meeting, Feb 13–17, 2005 (San Francisco, CA), p 27–34
45. P. Nerman and J. Andersson, "Fatigue Strength of Mixed Al-Joints Performed with FSW," Fourth International Symposium on Friction Stir Welding, May 14–16, 2003 (Park City, UT), TWI
46. Y. Sato, S. Hwan, C. Park, and H. Kokawa, Microstructural Factors Governing Hardness in Friction-Stir Welds of Solid Solution Hardened Al Alloys, *Metall. Mater. Trans. A,* Vol 32, Dec 2001, p 3033–3042
47. C. Juričić, C. Dalle Donne, and U. Drebler, "Effect of Heat Treatments on Mechanical Properties of Friction Stir Welded 6013," Third International Symposium on Friction Stir Welding, Sept 27–28, 2001 (Kobe, Japan), TWI
48. B. Heinz, B. Skrotzki, and G. Eggeler, Microstructure and Mechanical Characterization of a Friction Stir Welded Al-Alloy, *Mater. Sci. Forum,* Vol 331–337, 2000, p 1757–1762
49. D. Lohwasser, "Friction Stir Welding of Aerospace Alloys," Fourth International Symposium on Friction Stir Welding, May 14–16, 2003 (Park City, UT), TWI
50. A. Denquin, D. Allehaux, M. Campagnac, and G. Lapasset, Microstructure and Mechanical Evolutions within Friction Stir Welds of Precipitation Hardened Aluminum Alloys, *Mater. Sci. Forum,* Vol 426–432, 2003, p 2921–2926
51. H. Steiger, M. Schwalm, and F. Palm, "Highest FSW Joints Properties versus FSW Process Stability," Fourth International Symposium on Friction Stir Welding, May 14–16, 2003 (Park City, UT), TWI
52. A. Denquin, D. Allehaux, M. Campagnac, and G. Lapasset, "Microstructural Evolution and Strength Mismatch within a Friction Stir Welded 6056 Aluminum Alloy," Third International Symposium on Friction Stir Welding, Sept 27–28, 2001 (Kobe, Japan), TWI
53. W. Chang, H. Bang, S. Jung, Y. Yeon, H. Kim, and W. Lee, Joint Properties and Thermal Behaviors of Friction Stir Welded Age Hardenable 6061Al Alloy, *Mater. Sci. Forum,* Vol 426–432, 2003, p 2953–2958
54. H. Hori, S. Makita, T. Minamida, S. Watanabe, E. Anzai, and H. Hino, "Joint Strength of Thick Sheet Welded by Friction Stir Welding," Third International Symposium on Friction Stir Welding, Sept 27–28, 2001 (Kobe, Japan), TWI
55. S. Lim, S.S. Kim, C.-G. Lee, and S.J. Kim, Mechanical Properties of Friction Stir Welded Al 6061-T651, *LiMAT-2003,* W. Frazier, Y. Han, N. Kim, and E. Lee, Ed., Center for Advanced Aerospace Materials, POSTECH, 2004, p 145–152
56. Y. Nagano, S. Jogan, and T. Hashimoto, "Mechanical Properties of Aluminum Die Casting Joined by FSW," Third International Symposium on Friction Stir Welding, Sept 27–28, 2001 (Kobe, Japan), TWI
57. V. Malin, Study of Metallurgical Phenomena in the HAZ of 6061-T6 Aluminum Welded Joints, *Weld. Res. Suppl.,* Sept 1995, p 305S–318S
58. G. Luan, S. Lin, P. Chai, and H. Li, "Friction Stir Welding in Large 6063Al Extrusions," Fifth Int. Friction Stir Welding Conf., Sept 14–16, 2004 (Metz, France), TWI
59. Y. Sato and H. Kokawa, Distribution of Tensile Property and Microstructure in Friction Stir Weld of 6063 Aluminum, *Metall. Mater. Trans. A,* Vol 32, Dec 2001, p 3023–3031
60. O. Midling, L. Oosterkamp, and J. Bersaas, "Friction Stir Welding Aluminum—Process and Applications," INALCO '98, Seventh Int. Conf. Joints in Aluminum (Cambridge, U.K.)
61. J. Backlund, A. Norlin, and A. Anderson, "Friction Stir Welding—Weld Prop-

erties and Manufacturing Techniques," INALCO '98, Seventh Int. Conf. Joints in Aluminum (Cambridge, U.K.)
62. O. Midling, Material Flow Behavior and Microstructural Integrity of Friction Stir Butt Welds, *Fourth Int. Conf. on Aluminum Alloys,* Vol 1, Sept 11–16, 1994
63. J. Karlson, B. Karlsson, H. Larsson, L. Karlsson, and L. Svensson, "Microstructure and Properties of Friction Stir Welded Aluminum Alloys," INALCO '98, Seventh Int. Conf. Joints in Aluminum (Cambridge, U.K.)
64. M. Ericsson and R. Sandstrom, Fatigue of Friction Stir Welded AlMgSi-Alloy 6082, *Mater. Sci. Forum,* Vol 331–337, 2000, p 1787–1792
65. P. Haagensen, O. Midling, and M. Ranes, "Fatigue Performance of Friction Stir Butt Welds in a 6000-Series Aluminum Alloy," Surface Treatment Effects II, June 7–9, 1995 (Milan, Italy)
66. L. Ossterkamp, A. Ivankovic, and A. Oosterkamp, "Initiation Fracture Toughness of Friction Stir Welds in Commercial Aluminum Alloys under Rapid Loading," Second International Symposium on Friction Stir Welding, June 26–28, 2000 (Gothenberg, Sweden), TWI
67. T. Kawasaki, T. Makino, S. Todori, H. Takai, M. Ezumi, and Y. Ina, "Application of Friction Stir Welding to the Manufacturing of Next-Generation A-Train Type Rolling Stock," Second International Symposium on Friction Stir Welding, June 26–28, 2000 (Gothenburg, Sweden), TWI
68. I. Okura, M. Naruo, L. Vigh, N. Hagasawa, and H. Toda, "Fatigue of Aluminum Deck Fabricated by Friction Stir Welding," Eighth INALCO 2001, March 2001 (Munich, Germany)
69. H. Hori, S. Makita, and H. Hino, "Friction Stir Welding of Rolling Stock for Subway," First International Symposium on Friction Stir Welding, June 1999 (Thousand Oaks, CA), TWI
70. P. Haagensen, M. Ranes, A. Kluken, and I. Kvale, Fatigue Performance of Welded Aluminum Deck Structures, *1996 OMAE,* Vol III, *Mater. Eng.,* ASME, 1996, p 505–512
71. B. Heinz and B. Skrotzki, Characterization of a Friction-Stir-Welded Aluminum Alloy 6013, *Metall. Mater. Trans. B,* Vol 33, June 2002, p 489–498
72. S. Brinkmann, A. von Strombeck, C. Schilling, J. dos Santos, D. Lohwasser, and M. Kocak, "Mechanical and Toughness Properties of Robotic-FSW Repair Welds in 6061-T6 Aluminum Alloys," Second International Symposium on Friction Stir Welding, June 26–28, 2000 (Gothenburg, Sweden), TWI
73. C. Fuller and M. Mahoney, Rockwell Scientific Corp., unpublished research
74. A. Merati, K. Sarda, D. Raizenne, and C. Dalle Donne, Improving Corrosion Properties of Friction Stir Welded Aluminum Alloys by Localized Heat Treatment, *Friction Stir Welding and Processing II,* K. Jata, M. Mahoney, R. Mishra, S. Semiatin, and T. Lienert, Ed., 2003 TMS Annual Meeting, March 2–6, 2003 (San Diego, CA) p 65–75
75. T. Nelson, R. Steel, and W. Arbegast, In Situ Thermal Studies and Post-Weld Mechanical Properties of Friction Stir Welds in Age Hardenable Aluminum Alloys, *Sci. Technol. Weld. Joining,* Vol 8 (No. 4), 2003, p 283–288
76. K. Jata, K. Sankaran, and J. Ruschau, Friction-Stir Welding Effects on Microstructure and Fatigue of Aluminum Alloy 7050-T7451, *Metall. Mater. Trans. A,* Vol 31, Sept 2000, p 2181–2192
77. J. Bassett and S. Birley, "Friction Stir Welding of Aluminum Armor," Second International Symposium on Friction Stir Welding, June 26–28, 2000 (Gothenburg, Sweden), TWI
78. A. Hassan, A. Norman, and P. Prangnell, "The Effect of Welding Conditions on the Microstructure and Mechanical Properties of the Nugget Zone in AA7010 Alloy Friction Stir Welds," Third International Symposium on Friction Stir Welding, Sept 27–28, 2001 (Kobe, Japan), TWI
79. C. Paglia, M. Carroll, B. Pitts, T. Reynolds, and R. Buchheit, Strength, Corrosion, and Environmentally Assisted Cracking of a 7075-T6 Friction Stir Weld, *Friction Stir Welding and Processing II,* 2003 TMS Annual Meeting, March 2–6, 2003 (San Diego, CA), p 65–75
80. M. Kumagai, S. Tanaka, H. Hatta, and H. Yoshida, "Integral Wing Panel for Airplanes Produced by Friction Stir Welded

Extrusion," Third International Symposium on Friction Stir Welding, Sept 27–28, 2001 (Kobe, Japan), TWI
81. A. Hassan, A. Norman, and P. Prangnell, The Effect of the Welding Conditions on the Nugget Zone in Friction Stir Welds in an AA7010 Alloy, *Sixth International Trends in Welding Research Conf. Proceedings,* April 15–19, 2002 (Pine Mountain, GA), ASM International, Materials Park, OH, 2003, p 287–292
82. Z. Li, W. Arbegast, A. Wilson, J. Moran, and J. Liu, Post-Weld Aging of Friction Stir Welded Al 7249 Extrusions, *Sixth International Trends in Welding Research Conf. Proceedings,* April 15–19, 2002 (Pine Mountain, GA), ASM International, Materials Park, OH, 2003, p 312–317
83. D. Allehaux, G. Petit, M. Campagnac, G. Lapasset, and A. Denquin, "Microstructure and Properties of a Friction Stir Welded 7349-T6 Aluminum Alloy," Fourth International Symposium on Friction Stir Welding, May 14–16, 2003 (Park City, UT), TWI
84. M. Salagaras, P. Bushell, and B. Hinton, Effects of Friction Stir Welding on the General Corrosion and Stress Corrosion Cracking Resistance of Aluminum Alloy 7075-T651, Paper 9, *Proc. Int. Conf. Tech. Dev. Weld. Def. Equip.,* March 18–19, 2002
85. M. Mahoney, C. Rhodes, J. Flintoff, R. Spurling, and W. Bingel, Properties of Friction-Stir-Welded 7075 T651 Aluminum, *Metall. Mater. Trans. A,* Vol 29, July 1998, p 1955–1964
86. R. Talwar, D. Bolser, R. Lederich, and J. Baumann, "Friction Stir Welding of Airframe Structures," Second International Symposium on Friction Stir Welding, June 26–28, 2000 (Gothenburg, Sweden), TWI
87. R. John and K. Jata, Residual Stress Effects on Near-Threshold Fatigue Crack Growth in Friction Stir Welds, *Friction Stir Welding and Processing,* K. Jata, M. Mahoney, R. Mishra, S. Semiatin, and D. Field, Ed., TMS, 2001, p 57–69
88. K. Jata, Friction Stir Welding of High Strength Aluminum Alloys, *Mater. Sci. Forum,* Vol 331–337, 2000, p 1701–1712
89. J.A. Harter, "AFGROW Users Guide and Technical Manual," AFRL-VA-WP-TR-1999-3016, Air Force Research Laboratory, Air Vehicles Directorate (AFML/VASM), Wright Patterson AFB, OH, Feb 1999
90. M. Skinner and R. Edwards, Improvements to the FSW Process Using the Self-Reacting Technology, THERMEC 2003, *Mater. Sci. Forum,* Vol 426–432, 2003, p 2849–2854
91. D. Kinchen, Z. Li, and G. Adams, "Mechanical Properties of Friction Stir Welds in Al-Li 2195-T8," First International Symposium on Friction Stir Welding, June 1999 (Thousand Oaks, CA), TWI
92. Z.S. Loftus, W.J. Arbegast, and P.J. Hartley, Friction Stir Weld Tooling Development for Application on the 2195 Al-Li-Cu Space Transportation System External Tank, *Trends in Welding Research, Proc. of the Fifth Int. Conf.,* J. Vitek, S. David, J. Johnson, H. Smartt, and T. DebRoy, Ed., June 1–5, 1998 (Pine Mountain, GA), ASM International, Materials Park, OH, 1999, p 580–584
93. H. Kroninger and A. Reynolds, R-Curve Behavior of Friction Stir Welds in Aluminum-Lithium Alloy 2195, *Fatigue Fract. Eng. Mater. Struct.,* Vol 25, 2002, p 283–290
94. M. Attallah and H. Salem, "Effect of Friction Stir Welding Process Parameters on the Mechanical Properties of the As-Welded and Post-Weld Heat Treated AA2095," Fifth Int. Friction Stir Welding Conf., Sept 14–16, 2004 (Metz, France), TWI
95. A. Reynolds, T. Seidel, and M. Simonsen, "Visualization of Material Flow in an Autogenous Friction Stir Weld," First International Symposium on Friction Stir Welding, June 1999 (Thousand Oaks, CA), TWI
96. K. Nakata, S. Inoki, Y. Nagano, and M. Ushio, Friction Stir Welding of $Al_2O_3$ Particulate 6061 Al Alloy Composite, THERMEC 2003, *Mater. Sci. Forum,* Vol 426–432, 2003, p 2873–2878
97. M. Mahoney, W. Harrigan, and J. Wert, "Friction Stir Welding SiC Discontinuously Reinforced Aluminum," INALCO '98, Seventh Int. Conf. Joints in Aluminum (Cambridge, U.K.)

# CHAPTER 6

# Friction Stir Welding of Ferrous and Nickel Alloys

Carl D. Sorensen and Tracy W. Nelson
Department of Mechanical Engineering, Brigham Young University

FRICTION STIR WELDING (FSW) is a solid-state joining process invented by The Welding Institute of Cambridge, England (Ref 1). In the FSW process, a rotating tool containing a pin and a shoulder is plunged into the joint between two workpieces, generating heat by friction. Once the heat has built up to the desired level, the tool is translated along the joint. Plasticized base material passes around the tool, where it is consolidated due to force applied by the shoulder of the tool.

Friction stir welding has been applied to metals with moderate melting points. Initially, FSW was applied primarily to aluminum alloys, which could be easily welded due to the relatively low softening temperatures of these alloys. Other relatively soft metals, such as copper, lead, zinc, and magnesium, have also been welded. In contrast, for a number of years it was difficult to weld ferrous alloys and other high-softening-temperature metals due to the lack of suitable tool materials.

Until recently, there were no tool materials that would stand up to the high stresses and temperatures necessary for FSW of materials with higher melting points, such as steels, stainless steels, and nickel-base alloys. In 1998, tungsten alloys and polycrystalline cubic boron nitride (PCBN) were developed to create FSW tools for use in steel, stainless steel, titanium alloys, and nickel-base alloys. Properties of the resultant welds have been shown to be outstanding. Although some issues remain (primarily limited tool life with tungsten-base tools), FSW has been demonstrated as a technically and economically feasible process in high-temperature materials. This chapter summarizes research work performed at a number of different laboratories to make FSW of high-temperature materials a reality. It covers the development of suitable tools, welding equipment, and welding procedures, describes the characteristics of the resulting weldments, and describes the variety of materials that have been tested with the FSW process.

## 6.1 Tool Materials

The requirements for an FSW tool in high-temperature materials (HTM) are significant. Obviously, the tool must maintain sufficient strength to constrain the weld material at softening temperatures in excess of 1000 °C (1830 °F). Perhaps less apparent, the tool must also be resistant to fatigue, fracture, mechanical wear, and chemical reactions with both the atmosphere and the weld material. To date, two classes of materials have been found that meet these requirements: refractory metal tools and superabrasive tools.

**Refractory Metal Tools.** The first class of tool materials to be used for FSW of HTM were refractory metal tools. Initially, the tool materials were considered proprietary. Eventually, however, the composition of the tools was revealed.

Tungsten was used as a tool material in many of the early welds performed (Ref 2). Tungsten appeared to have sufficient hot strength to serve as an FSW tool but suffered problems on the plunge due to its high ductile-to-brittle transition temperature. This necessitated preheating of the tool to temperatures above 300 °C (570 °F) and the drilling of a pilot hole for the tool (Ref 3).

Later tool materials included additions of up to 25% Re to tungsten, which lowered the transition temperature to below room temperature. Tungsten-rhenium tools show increased fracture resistance and improved wear resistance compared to pure tungsten and appear to have become the most widely used refractory metal. Development of production processes continues to improve the tool life of tungsten-rhenium tools. Molybdenum has been used on at least one occasion as a tool material for FSW of steel (Ref 4).

Early tungsten and tungsten-rhenium tools showed a tendency to wear rapidly in the weld, leading to macroscopic inclusions of tool material in the weld zone. Later tools were much more resistant to this problem, but the tool material often continues to dissolve in the weld, leaving a tungsten-enriched stir zone. Furthermore, some researchers report that microstructural changes in the tool indicate ongoing deformation during welding.

Refractory metal tools have been used to weld low-carbon steels, carbon-manganese steels, austenitic stainless steels, and ferritic stainless steels.

Tungsten-rhenium tools show good fracture toughness and can be used for relatively thick welds (up to 13 mm in a single pass). Reported tool life ranges from a quarter meter (tens of inches) up to approximately 4 m (over 10 ft).

**Superabrasive Tools.** The second class of tool materials used for FSW of HTM is superabrasives. Superabrasives are materials that are formed in presses under extreme temperature and pressure. The two superabrasives that have been used in FSW are polycrystalline diamond (PCD) and PCBN. Both materials consist of small crystals of ultrahard material (diamond or CBN) bonded together in a skeletal matrix with a second-phase material that serves as a catalyst for the formation of the matrix. Reference 5 gives a summary of the characteristics of superabrasive materials.

Polycrystalline diamond has been used for aluminum-matrix composites reinforced with particulate silicon carbide, boron carbide, or alumina. It also shows promise as a tool material for welding titanium, although this work is only in a preliminary stage.

Polycrystalline cubic boron nitride has been used to weld carbon steels, carbon-manganese steels, high-strength, low-alloy (HSLA) steels, high-strength pipeline steels, austenitic stainless steels, duplex stainless steels, dual-phase steels, nickel-base alloys, and other exotic alloys. It has been tested in titanium alloys, with inconsistent results. At times, it performs well; at others, chemical reactions with the workpiece cause rapid wear.

Superabrasive materials can be made only in relatively small pieces, due to the high pressure required for manufacturing. Furthermore, these materials are very difficult or impossible to braze. Therefore, superabrasives are used in a composite tool design, as described by Ref 5.

Early trials of PCBN tools in 316L stainless steel showed tool life of 1 to 4 m (3 to 13 ft), with life limited by fracture. Continued efforts at improving the design of the composite tools, together with improvements in the grade of the PCBN, have greatly reduced the tendency of the tool to fracture and have increased its life significantly (Ref 6). The most recent tool life test on PCBN tools showed a tool life of 80 m (260 ft) in 1018 steel.

Polycrystalline cubic boron nitride tools produce an exceptionally smooth surface on the weld. This is thought to be due to the low coefficient of friction between PCBN and the weld metal.

The major limitation in PCBN tools is the maximum depth of the weld. Although a pin 13 mm (0.5 in.) in length has been tested, for practical purposes, the maximum depth of welding at the present time is 10 mm (0.4 in.). Ongoing efforts in the design of PCBN tools should lead to increases in pin length. MegaStir Technologies, the provider of PCBN tools, has plans to achieve a 13 mm weld depth within a year.

Over the past several years, significant efforts were expended on developing tougher, more wear-resistant grades of PCBN (Ref 6). Efforts to understand the effects of different binder phases, ratio of CBN to binder phase, and grain size distributions of CBN on performance were investigated. Performance was evaluated via a turning test on 304L stainless steel. Those grades exhibiting greater wear resistance in the turning tests were subsequently evaluated via FSW in 304L to compare wear results and evaluate toughness.

The PCBN grade-development program was quite successful in that tougher, more wear-resistant grades of PCBN were developed. In addition to improved wear resistance, the improved toughness of the new grades has enabled both deeper weld penetration (up to 12 mm, or 0.47 in.) and threaded-type features to be incorporated into the tool design. These features are illustrated in Fig. 6.1.

## 6.2 FSW Equipment

The FSW equipment for high-temperature materials requires improved cooling, higher-precision spindles, and increased machine stiffness compared to that required for aluminum.

**Tool Cooling.** The welding zone temperatures frequently reach 900 to 1200 °C (1650 to 2190 °F). Further, the materials used for the tool (either tungsten alloys or tungsten carbide shanks) have high thermal conductivity relative to the tool steel commonly used for aluminum. To prevent damage to the spindle bearings and to establish a consistent thermal environment for the tool, cooling of the tool shank is required.

Two different methods for cooling the tool have been used. In the first, a hollow drawbar is used to conduct coolant directly onto the back end of the tool shank. This method provides the highest cooling rate but sometimes provides a machine-specific thermal environment that can make it difficult to transfer operating parameters between machines. There can also be difficulties in establishing a consistent seal between the tool holder and the shank.

The second method used for cooling the tool is to mount a cooled tool holder in the machine spindle. The holder can be designed for any spindle configuration, and the cooling is consistent from machine to machine. The major disadvantage of this cooling method is that the cooled tool holder is generally less stiff than the machine spindle.

**Precision Spindles.** Strengths of metallic tools at process temperatures are substantially higher than the aluminum alloys being welded. In contrast, for high-temperature materials, the tool strengths are only marginally higher than the alloys being welded. Thus, tool deformation for metallic tools and fracture for PCBN tools are common.

Spindle runout has been demonstrated to be a significant factor limiting the life of PCBN tools. Many FSW machines built for aluminum alloys have relatively high spindle runout, because they were designed primarily to accommodate high process loads. Producers of PCBN tools have recognized the importance of precise spindles and specify a maximum spindle runout of 0.01 mm (0.0004 in.) (Ref 7). Failure to meet this spindle runout requirement has led to premature tool fracture.

**Stiff Machines.** Cyclic process loads in FSW tend to be higher in many high-melting-temperature alloys than in aluminum. Deflections under load can lead to problems with fatigue failure, particularly with PCBN tools. To minimize these problems, the stiffness for the machine is specified. A deflection of 0.75 mm (0.030 in.) under a load of 45 kN (10 kip) is suggested by Ref 7.

## 6.3 Weld Metal Properties

A few studies have carefully examined the metallurgy of welds produced in a variety of HTM by FSW. This section summarizes the detailed property and structure results.

### *6.3.1 Ferritic Steels*

Reference 3 reported on welds in low-carbon and Fe-12%Cr steels, using a tool that was later reported to be tungsten-base. The weld zone

(a) (b) (c)

**Fig. 6.1** Pin features produced on polycrystalline cubic boron nitride friction stir processing tools, including (a) flats, (b) helical threads, and (c) a combination of convex scrolled shoulder and helical threaded pin was shown to contain a range of martensite, bainite, and ferrite structures, along with tool debris. A unique feature of this study is a preliminary look at the typical costs of welding, showing that FSW could easily be superior to a variety of other welding processes.

Reference 8 reported on welds made in DH-36 using a tungsten alloy tool. Radiographic inspection showed full-penetration, sound welds. However, there were indications in the radiograph that tool material was being mixed into the stir zone. Transverse tensile tests showed overmatching of the weld, with failures occurring in the base metal. All-stir-zone tensile tests showed yield strength approximately 50% higher than the base metal, and tensile strength approximately 33% higher than the base metal.

Reference 9 evaluated the feasibility of welding 1018 steel using tungsten- and molybdenum-base alloy tools. Observations of the peak temperature seen during the weld were extrapolated to give a probable maximum weld temperature of 1200 °C. The thermomechanically affected zone was not readily observable in the microstructure of the weld, likely due to the allotropic transformation on cooling. Evidence of microalloying between the tool and the workpiece was found. Stir-zone microstructure was found to consist of ferrite, grain-boundary ferrite, and fine pearlite. In the stir zone, the structure was found to be finer near the shoulder and coarser away from the shoulder. Tensile properties of the resulting welds were found to be acceptable.

Reference 10 reported on the welding of S355 carbon-manganese steel plates, using tungsten-rhenium tools. The welds were made in 12 mm (0.5 in.) thick plate using tools with a pin length of 7.5 mm (0.3 in.). Welds were made from both sides of the plate in order to achieve full penetration. Tool wear was a significant issue. One significant microstructural observation was the tempering of the first pass by the heat from the second pass. Hardness was shown to be higher in the weld zone than in the base material. Longitudinal microtensile specimens were taken from the various regions of the weld, and yield and tensile strengths were consistent with microhardness results. Charpy impact testing revealed that the toughness at –40 °C (–40 °F) was equivalent for the weld material and the base plate. At higher temperatures, toughness for the weld material was significantly lower than the base metal, with the lowest toughness in the heat-affected zone (HAZ). No comparison was made between the toughness of friction stir welds and those produced by fusion welding processes.

Reference 11 reported on FSW of DH-36 steel with W-25%Re tools. No measurable change in tool dimension was found after a welding distance of approximately 1.8 m (5.9 ft). Tensile properties were found to be acceptable, in spite of some defects in the weld zone.

Scientists (Ref 12) welded HSLA-65 using tungsten-base tools. Subjected to bend tests, a 10 mm (0.4 in.) thick weld passed, and a 6 mm (0.24 in.) thick weld failed when bent with the root in tension, due to the formation of surface cracks. Tensile properties of the 10 mm thick welds exceeded the specifications for the base metal. Some 6 mm thick welds exceeded the plate specifications, while others were approximately 10% below the plate specifications. Charpy V-notch (CVN) toughness at both –29 and –40 °C (–20 and –40 °F) were below the base material toughness but exceeded the minimum specification of the plate. The surface of the welded material was found to have small defects due to the roughness caused by the interaction between the shoulder and the surface of the plate. Salt spray corrosion tests indicated no preference for corrosion in the weld zone.

Reference 13 reported on welds in 6.4 and 12.7 mm (0.25 and 0.5 in.) thick HSLA-65 using tungsten-rhenium tools. Radiographic inspection showed traces that may indicate the formation of a wormhole defect at the start of the weld. Postweld distortion of the 12.7 mm plate was measured to be less than that in submerged arc welded (SAW) or gas metal arc welded (GMAW) plates. The welded plates were tested by an underwater explosion test known as shock-holing; the welded specimen met the shock-hole requirements in spite of the radiographic indications and pieces of broken tool material that remained in the weld. Tensile strength of the weld zone was slightly less than the base material. Charpy toughness of the weld zone was significantly less than the base material and showed extreme variability, which was unexplained. Average Charpy values exceeded the specification for HSLA-65 welds.

Reference 2 reported on welds in 0.29C-Mn-Si-Mo-B quenched and tempered steel using a PCBN tool. Weld thicknesses included both 6.4 and 12.7 mm. Microhardness of the stir zone was found to approximately equal that of the base metal. Significant softening was observed in the HAZ. Transverse tensile properties of friction stir weldments were found to be less than the base metal but greater than comparison GMAW welds. The CVN toughness in the weld zone was found to be at or above the base metal but below the toughness of the GMAW welds, except in the case of the HAZ in the 6.4 mm welds, where the FSW toughness was more than twice the GMAW toughness. In this study, the filler material for the GMAW was carbon steel, so it is expected that the weld material will be both softer and tougher than the weld material with the same composition as the base metal in the friction stir weld.

### *6.3.2 Austenitic Stainless Steels*

Researchers (Ref 14) welded 304L stainless using a tungsten alloy tool. They reported extrapolated peak temperatures in the weld zone of approximately 1200 °C. They reported equiaxed grains in the stir zone, with a grain size slightly reduced from the base metal. They also noticed narrow bands in the stir zone but made no determination as to the origin or detailed structure of the bands. The weld material was found to be stronger than the base metal and to exhibit excellent ductility, with elongation to fracture of more than 50%. Longitudinal residual stresses were found to be close to the base material yield stress.

Researchers (Ref 15) reported on welding of 304L and AL-6XN stainless steels. They found a highly refined stir-zone microstructure, with an unidentified dark banded structure in the stir zone. They reported increased microhardness in the weld zone and excellent ductility for both 304L and AL-6XN. They also described the difficulty of achieving sound welds in AL-6XN, because a number of pores were found in the resulting weld. A later report (Ref 16) gave properties of friction stir welds and AL-6XN base metal. Weld metal was higher in yield strength (700 MPa compared to 430, or 102 ksi compared to 62) and ultimate strength (930 MPa compared to 780, or 135 ksi compared to 113) but lower in ductility (50 to 60% reduction in area compared to 75%; 28% elongation compared to 46%). The elongation of the friction stir welds was only slightly below the 30% minimum elongation specified for the base metal.

Scientists (Ref 17) analyzed friction stir welds made in 304 stainless steel. They found a banded structure similar to that identified by Reynolds et al. The dark bands were found to be narrow regions of ultrafine grains. The advancing side of the stir zone was found to contain fine sigma particles as well as even finer carbide precipitates.

Researchers (Ref 18) investigated sigma formation in FSW of various stainless alloys with compositions at various distances from the sigma + austenite region of the Fe-Ni-Cr ternary diagram. They were able to predict the propensity for sigma formation and hypothesized that sigma formation was a marker for recrystallization in 304L. They also demonstrated that welding parameter changes affected the amount and location of sigma.

Later studies (Ref 19) with a convex shoulder, step spiral (CS4) pin tool showed dramatically reduced sigma formation in 304L with the new tool design. No sigma has yet been identified in welds with the new tool.

Because the temperature of the weld zone exceeds 800 °C (1470 °F), the possibility of sensitization exists. A scientist (Ref 20) explored both sensitization and stress-corrosion cracking (SCC) in FSW 304L. The welds analyzed qualified as nonsensitized during an oxalic acid etch test. Double-loop electrochemical potentiokinetic reactivation testing showed regions of increased corrosion susceptibility away from the surface of the specimen. U-bend specimens in boiling 25% NaCl showed no increased SCC susceptibility compared with the base metal.

## 6.4 Materials Welded with PCBN

As part of the evaluation of PCBN as a tool material for FSW of high-temperature materials, a variety of different alloys have been tested. The materials that have been tested, along with results of preliminary mechanical testing, are given in this section. Table 6.1 summarizes the results of this testing.

### *6.4.1 Ferritic Steels*

**A-36.** Almost 200 m (over 200 yd) of A-36 have been welded using PCBN tools. A wide range of weld parameters has been found to give fully consolidated welds. Surface quality is excellent. No mechanical property data are available.

**Quenched and Tempered Carbon-Manganese Steel.** Scientists (Ref 21) welded 6.4 mm (0.25 in.) thick quenched and tempered carbon-manganese steel using PCBN. Tool wear was very low but not measured quantitatively. Greatly refined grain structures in the stir zone were observed, both in the prior-austenite grains and in the transformation product. The microhardness in the weld zone was approximately the same as that of the base metal. The HAZ showed a hardness reduction from 550 to 350 HV. Transverse tensile specimens exhibited a strength approximately 70% of the base metal, with failure in the HAZ. Elongation as measured in the transverse tensile test was reduced from 9.5% in the base metal to 2.6%. However, because of the reduction in strength in the HAZ, it is likely that this elongation is nonuniform and hence greatly underestimates the ductility of the weldment.

**DH-36** steel has been test welded with PCBN tools. It appears to weld at approximately the same parameters as A-36. Fully consolidated welds at travel speeds of up to 250 mm/min (10 in./min) have been achieved. No mechanical properties are presently available.

**HSLA-65** steel has been welded at travel speeds of up to 200 mm/min (8 in./min). The resulting welds are of excellent quality. Surface appearance is excellent. The yield and ultimate strengths of all-weld-material specimens are 597 and 788 MPa (86.6 and 114 ksi), respectively, compared with 605 and 673 MPa (87.7 and 97.6 ksi) in the base metal. Elongation and reduction in area are 14.5 and 77% for the weld material, compared with 18.7 and 81% for the base metal. Tool life in HSLA-65 appears to be excellent, although it has not been quantified due to the lack of available metal for carrying out the life test.

**X-65.** Reference 22 reported postweld mechanical properties in 6 mm (0.25 in.) thick FSW X-65 pipe. Transverse tensile strengths were equivalent to the base metal. All tensile samples fractured in the base metal well removed from the weld or HAZ. Charpy impact results in the weld nugget and HAZ exceeded that of the base metal at –50, 0, and 20 °C (–58, 32, and 68 °F). These results are shown in Fig. 6.2.

**L-80, X-80, and X-120.** These pipeline steels were welded using PCBN tools. All of these alloys appear to be readily weldable by FSW. An in-depth examination of these alloys is presented by Ref 23. Welding parameters for X-80 were 550 rpm and 100 mm/min (4 in./min) with argon shielding gas. No transverse tensile tests were done on this weld, but the HAZ and stir-zone microhardnesses were higher than the base material. Welds were fully consolidated. A small region on the advancing side of the stir zone appears to have higher hardness than the rest of the weld.

**Dual-Phase Steel.** Dual Ten 590 dual-phase steel (United States Steel Corporation)

**Table 6.1 Results of preliminary friction stir welding testing with polycrystalline cubic boron nitride tools**

| | Yield strength (weld/base metal) | | Ultimate strength (weld/base metal) | | rpm/travel | | |
|---|---|---|---|---|---|---|---|
| Material | MPa | ksi | MPa | ksi | mm/min | in./min | Comments |
| A-36 | N/A | | N/A | | 600/150 | 24/6 | 80 m (260 ft), 79 plunges tool life, 7 |
| Quenched and tempered C-Mn steel | 1040/1400 | 151/203 | 1230/1710 | 178/248 | 545/130 | 21/5 | ... |
| DH-36 | N/A | | N/A | | 500/200 | 20/8 | ... |
| HSLA-65 | 597/605 | 87/88 | 788/673 | 114/98 | 500/200 | 20/8 | ... |
| L-80 | N/A | | N/A | | 550/100 | 22/4 | ... |
| X-80 | N/A | | N/A | | 550/100 | 22/4 | ... |
| X-120 | N/A | | N/A | | 550/100 | 22/4 | ... |
| Dual Ten 590 dual phase | 496/340 | 72/49 | 710/590 | 103/86 | 450/240 | 18/9.5 | ... |
| 304L | 51/55 | 7.4/8.0 | 95/98 | 13.8/14 | 400/75 | 16/3.0 | ... |
| 316L | 434/338 | 63/49 | 641/674 | 93/98 | 550/80 | 22/3.2 | ... |
| AL-6XN | N/A | | N/A | | 350/25 | 14/1.0 | ... |
| 301L | N/A | | N/A | | 600/300 | 24/12 | Lap weld, small-diameter tool |
| 430 | N/A | | N/A | | 550/80 | 22/3.2 | ... |
| 2507 super duplex | 762/705 | 110/102 | 845/886 | 123/128 | 450/60 | 18/2.4 | ... |
| 201 | 193/103 | 28/15 | 448/406 | 65/59 | 1000/100 | 39/4.0 | 16 mm (0.6 in.) tool |
| 600 | 374/263 | 54/38 | 719/631 | 104/91 | 450/56 | 18/2.2 | ... |
| 718 | 668/1172 | 97/170 | 986/1392 | 143/202 | 500/50 | 20/2.0 | 16 mm (0.6 in.) tool |
| Narloy-Z | N/A | | N/A | | 450/100 | 18/4.0 | demonstration only |
| Invar | N/A | | N/A | | 600/150 | 24/6.0 | demonstration only |
| Ni-Al bronze | 420/193 | 61/28 | 703/421 | 102/61 | 1000/102 | 39/4.0 | ... |

has been welded in a variety of geometries, including automotive sheet. Spindle speeds were 450 to 550 rpm, with travel speeds varying from 150 to 340 mm/min (6 to 13 in./min). Argon was used as a shielding gas. Welds were fully consolidated. Microhardness in the weld zone is higher than the base material. Transverse yield and tensile strengths of 71 and 103 MPa (10 and 15 ksi) are higher than that of the base material (49 and 85 MPa, or 7.1 and 12 ksi). Elongation is only slightly lower than the base metal (22% compared with 25%). Preliminary forming studies have indicated that the weld zone forms about as well as the base metal.

### 6.4.2 Austenitic Stainless Steels

**304L.** A 6mm (0.25 in.) thick 304L plate was welded using PCBN tools. Spindle speed was 400 rpm; travel speed was 75 mm/min (3 in./min). A variety of welding parameters were tried. Different parameters were found to lead to widely varying microstructures. Under some conditions, sigma phase was found to be present in the stir zone (Ref 17). Yield strength, tensile strength, and ductility were almost identical in the weld and base metal. Tool life in 304 exceeded 30 m (98 ft). Tool wear in austenitic stainless steels appears to be higher than in ferritic alloys, possibly due to chemical interactions between the tool and weld material.

**316L.** Reference 24 reported on the welding of 316L using PCBN tools. Welds had full consolidation and good surface appearance. Transverse yield and tensile strength of the weld were essentially the same as the base metal. No significant softening was reported in the HAZ. Ductility of the resulting welds is excellent.

**301L.** Alloy 301L was welded in a lap weld configuration. Sheet thickness was 1.5 mm (0.06 in.). To avoid wrinkling on the free surface of the lap, a small-diameter tool (10 mm shoulder, 3 mm pin) was used. The small-diameter tool required correspondingly higher rotation speeds to achieve welding temperatures. The joint appeared to be fully consolidated and defect-free under optical inspection. The joint was tested for corrosion in a salt spray environment. Slight corrosion appeared in the HAZ. Significant corrosion appeared in the crevice between the flash and the top surface. Better control of the flash or mechanical removal of the flash following welding are expected to improve the corrosion performance of the lap weld.

**AL-6XN** has been welded with PCBN tools. Microhardness values look appropriate. It is very difficult to fully consolidate the advancing side of the weld. No mechanical properties data are available. Further weld development on this alloy is dependent on improved PCBN grades.

### 6.4.3 Type 430 Stainless Steel

Reference 24 reported on the welding of type 430 stainless steel using PCBN tools. The weld was performed at 550 rpm, with a travel speed of 80 mm/min (3.15 in./min). The weld was a partial penetration bead-on-plate weld. No mechanical property data were obtained. The weld appeared to be fully consolidated. Surface

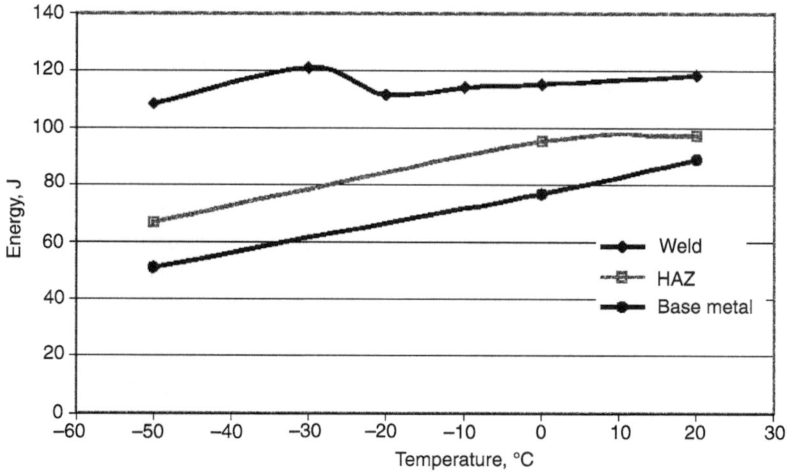


**Fig. 6.2** Charpy impact results. HAZ, heat-affected zone. Courtesy of Z. Feng, Oak Ridge National Laboratory

quality was excellent. No HAZ softening was observed. The weld zone had higher microhardness than the base material.

### 6.4.4 Super Duplex Stainless Steel (2507)

The SAF 2507 (UNS S32750) super duplex stainless steel was welded with a 25 mm (1.0 in.) diameter PCBN tool (Ref 25). Welding parameters of 450 rpm and 60 mm/min (2.4 in./min) produced sound welds with an excellent surface finish. The resulting microstructure was fine-grained (average 4 μm in the stir zone) and equiaxed. Ferrite content varied from 40 to 50% across the weld zone, compared with 45% in the base metal. Corrosion resistance of the weld was determined by ASTM G-48C, which measures the critical pitting temperature (CPT). The CPT for the FSW joints was 65 °C (150 °F compared to 40 to 55 °C (100 to 130 °F) for typical arc welding processes. The yield and ultimate strengths of the welds were 846 and 1045 MPa (122.7 and 151.5 ksi), which were higher than the base metal (705 and 886 MPa, or 102 and 128 ksi). Transverse elongation of the weld was 18%, compared with 30% elongation in the base material.

### 6.4.5 Nickel-Base Alloys

**Alloy 201.** A 3.2 mm (0.125 in.) thick alloy 201 sheet was welded in a butt joint configuration using a tool with a 16 mm (0.63 in.) diameter shoulder. The weld was a partial penetration weld, to avoid complications associated with getting the pin close to the backing plate. Yield and tensile strengths of the weld metal were 193 and 448 MPa (28 and 65 ksi), respectively, compared with 103 and 406 MPa (15 and 60 ksi) for the base material. Elongation was 34% for the transverse specimen, compared to 50% for the base material. Very little tool wear was observed in this weld.

**Alloy 600 plates** (~6 mm thick) were butt welded using a PCBN tool. Spindle speed was 450 rpm, and travel rate was 56 mm/min (2.2 in./min). Substantial grain refinement was observed in the stir zone. Mechanical properties were excellent. Yield strength and ultimate strength were 370 and 720 MPa (54 and 104 ksi), respectively, compared with 265 and 630 MPa (38 and 92 ksi) for the base metal. Elongation was reduced from 50% in the base metal to 27% in the transverse weld specimen. However, it is important to recognize that the nonuniform deformation in transverse weld specimens generally results in reduced elongation measurements.

**Alloy 718 sheets** (3.2 mm thick) were butt welded using a tool with a 16 mm diameter shoulder. Spindle speed was 500 rpm, and travel speed was 50 mm/min (2.0 in./min). The weld was fully consolidated and exhibited substantial grain refinement as compared with the base material. Yield and ultimate strengths of the transverse weld specimens were 670 and 985 MPa (97 and 143 ksi), respectively. There was not enough material available to make a base-metal measurement. However, for comparison purposes, typical yield and tensile strengths are 460 and 895 MPa (67 and 130 ksi) for alloy 718 in the annealed condition and 1170 and 1390 MPa (170 and 202 ksi) in the precipitation-hardened condition.

### 6.4.6 Specialty Alloys

**Narloy-Z plates** (~6 mm thick) were welded at Boeing's Huntington Beach facility using the ESAB SuperStir machine. The weld was made at 450 rpm and 100 mm/min (4.0 in./min). The surface finish of the resulting weld appeared excellent. There was no visible tool wear. Microstructural and tensile data are not available.

**Ni-Al Bronze.** Cast Ni-Al bronze has been friction stir processed (FSP), as reported by Ref 26. Yield strength of the FSP material (420 MPa, or 61 ksi) was more than double that of the cast alloy (193 MPa, or 28 ksi). Tensile strength also increased substantially due to the processing (700 MPa compared with 420, or 102 ksi compared with 61). However, elongation dropped to 14%, compared with 20% in the as-cast material. Surface finish and tool life were both excellent. In addition to the improvement in as-cast properties, FSP was demonstrated to reduce or eliminate internal porosity due to casting defects.

**Invar** has been welded in a variety of thicknesses with different welding parameters. Surface finish has been excellent. Weld distortion has been low. Welds are fully consolidated. No mechanical properties are available at this time.

## 6.5 Additional Benefits

One of the mounting obstacles facing welded fabrication is the mandated restriction of hazardous fumes from arc welding processes. Both hexavalent chromium and manganese are under heavy scrutiny in the United States and European communities. It is anticipated that the new Occupational Safety and Health Administration (OSHA) restrictions on permissible exposure limits of hexavalent chromium will dramatically increase the cost of welded fabrication in the United States.

Generally, solid-state welding processes are not known for hazardous fume generation. Although it was assumed that FSW would fall into this category, FSW had never been evaluated specifically for hazardous fumes. Reference 27 compared both gas tungsten arc welding (GTAW) and FSW to evaluate the two in a side-by-side evaluation. Both processes were completely enclosed in sealed containers with both inlet and outlet filters. Over the same duration of weld time, the GTAW process generated 1.88 and 0.02 mg/m$^3$ of manganese and hexavalent chromium, respectively. In contrast, fume generation for FSW was below detectable limits. The results of this investigation are shown in Table 6.2.

## 6.6 Summary

Friction stir welding of materials with high softening temperatures has been demonstrated to be technically feasible for a wide range of alloys. Pin tool lengths of up to 7.5 mm (0.3 in.) have been reported in the literature, which should allow single-sided welds of up to 8 mm (0.315 in.). Single-sided welds in thicknesses up to 6.4 mm (0.25 in.) have been successfully achieved. Double-sided welds of up to 13 mm (0.5 in.) have been demonstrated.

Two major classes of tool materials have been used for FSW of high-temperature materials. Refractory metal alloys, primarily W-25%Re, have been used to successfully weld carbon steels, austenitic stainless steels, and titanium. Initially, tool wear was severe, but recent improvements in processing of the tool material have led to decreased tool wear and increased tool life. Tool life as long as 4 m (13 ft) per tool has been reported. Initial problems with tool material contamination of the weld appear to have been greatly reduced.

Superabrasive tools, primarily PCBN, have been used to successfully weld ferritic steels, ferritic stainless steels, austenitic stainless steels, nickel-base superalloys, Invar, and Narloy-Z. Attempts to weld titanium with PCBN tools have been inconclusive. Tool life of 80 m (260 ft) has been demonstrated in FSW of 1018 steel, and very low tool wear has been reported on all other alloys. The primary concern in tool life continues to be fracture, and developments in PCBN grades continue to improve the fracture toughness of the FSW tools. The PCBN tools provide an extremely smooth finish when used for FSW or FSP.

Properties of friction stir welds in all of the alloys tested appear to be excellent. In some cases, they exceed the properties of the base metal. In virtually all cases, they exceed the properties of alternative fusion welding processes. Further, FSW has been demonstrated to produce lower distortion than GMAW and SAW in the welding of 13 mm thick HSLA-65 steel.

**Table 6.2 Airborne emissions associated with welding 304 stainless steel comparing gas tungsten arc welding to friction stir welding**

| | Element emission, mg/m$^3$ | | | |
|---|---|---|---|---|
| Welding process | Chromium | Copper | Manganese | Hexavalent chromium |
| Tungsten inert gas welding | 0.25 | 0.11 | 1.88 | 0.02 |
| Friction stir welding | <0.03 | <0.03 | <0.02 | <0.01 |

Courtesy of M.W. Mahoney, Rockwell Scientific

### REFERENCES

1. W.M. Thomas, E.D. Nicholas, J.C. Needham, M.G. Murch, P. Templesmith, and C.J. Dawes, International Patent Application PCT/GB92/02203 and GB Patent Application 9125978.8, 1991
2. P. Konkol, Characterization of Friction Stir Weldments in 500 Brinell Hardness Quenched and Tempered Steel, *Proceedings of the Fourth International Symposium on Friction Stir Welding,* May 14–16, 2003 (Park City, UT), TWI, paper on CD
3. W.M. Thomas, P.L. Threadgill, and E.D. Nicholas, Feasibility of Friction Stir Welding Steel, *Sci. Technol. Weld. Join.,* Vol 4 (No. 6), 1999, p 365–372
4. T.J. Lienert, and J.E. Gould, Friction Stir Welding of Mild Steel, *Proceedings of the First International Symposium on Friction Stir Welding,* June 14–16, 1999 (Thousand Oaks, CA), TWI, paper on CD

5. C.D. Sorensen, T.W. Nelson, and S.M. Packer, Tool Material Testing for FSW of High-Temperature Alloys, *Proceedings of the Third International Symposium on Friction Stir Welding,* Sept 2001 (Kobe, Japan), TWI, paper on CD
6. M. Collier, R. Steel, T. Nelson, C. Sorensen, and S. Packer, Grade Development of Polycrystalline Cubic Boron Nitride for Friction Stir Processing of Ferrous Alloys, *Mater. Sci. Forum,* Vol 426–432 (No. 4), 2003, p 3011–3016
7. S.M. Packer, T.W. Nelson, C.D. Sorensen, R. Steel, and M. Matsunaga, Tool and Equipment Requirements for Friction Stir Welding of Ferrous and Other High Melting Temperature Alloys, *Proceedings of the Fourth International Symposium on Friction Stir Welding,* May 14–16, 2003 (Park City, UT), TWI, paper on CD
8. M. Posada, J. DeLoach, A.P. Reynolds, M. Skinner, and J.P. Halpin, Friction Stir Weld Evaluation of DH-36 and Stainless Steel Weldments, *Friction Stir Welding and Processing,* TMS, 2001, p 159–171
9. T.J. Leinert, W.L. Stellwag, Jr., B.B. Grimmett, and R.W. Warke, Friction Stir Welding Studies on Mild Steel, *Weld. J.,* Jan 2003, p 1-s to 9-s
10. R. Johnson, J. dos Santos, and M. Magnasco, Mechanical Properties of Friction Stir Welded S355 C-Mn Steel Plates, *Proceedings of the Fourth International Symposium on Friction Stir Welding,* May 14–16, 2003 (Park City, UT), TWI, paper on CD
11. T.J. Lienert, W. Tang, J.A. Hogeboom, and L.G. Kvidahl, Friction Stir Welding of DH-36 Steel, *Proceedings of the Fourth International Symposium on Friction Stir Welding,* May 14–16, 2003 (Park City, UT), TWI, paper on CD
12. P.J. Konkol, J.A. Mathers, R. Johnson, and J.R. Pickens, Friction Stir Welding of HSLA-65 Steel for Shipbuilding, *J. Ship Prod.,* Vol 19 (No. 3), Aug 2003, p 159–164
13. M. Posada, J. DeLoach, A.P. Reynolds, R. Fonda, and J. Halpin, Evaluation of Friction Stir Welded HSLA-65, *Proceedings of the Fourth International Symposium on Friction Stir Welding,* May 14–16, 2003 (Park City, UT), TWI, paper on CD
14. A.P. Reynolds, W. Tang, T. Gnaupel-Herold, and H. Prask, Structure, Properties, and Residual Stress of 304L Stainless Steel Friction Stir Welds, *Scr. Mater.,* Vol 48 (No. 9), May 2003, p 1289–1294
15. A.P. Reynolds, M. Posada, J. DeLoach, M.J. Skinner, and T.J. Lienert, FSW of Austenitic Stainless Steels, *Proceedings of the Third International Symposium on Friction Stir Welding,* Sept 2001 (Kobe, Japan), TWI, paper on CD
16. M. Posada, J. DeLoach, A.P. Reynolds, and J.P. Halpin, Mechanical Property and Microstructural Evaluation of Friction Stir Welded AL-6XN, *Trends in Welding Research, Proceedings of the Sixth International Conference,* April 15–19, 2002 (Pine Mountain, GA), ASM International, p 307–311
17. S.H.C. Park, Y.S. Sato, H. Kokawa, K. Okamoto, S. Hirano, and M. Inagaki, Rapid Formation of the Sigma Phase in 304 Stainless Steel during Friction Stir Welding, *Scr. Mater.,* Vol 49 (No. 12), Dec 2003, p 1175–1180
18. C.D. Sorensen and T.W. Nelson, Sigma Phase Formation in Friction Stirring of Iron-Nickel-Chromium Alloys, *Trends in Welding Research, Proceedings of the Seventh International Conference,* ASM International, 2005
19. C.B. Owen, "Two-Dimensional Friction Stir Welding Model with Experimental Validation," M.S. thesis, Brigham Young University, Provo, UT, 2006
20. T.D. Clark, "An Analysis of Microstructure and Corrosion Resistance in Underwater Friction Stir Welded 304L Stainless Steel," M.S. thesis, Brigham Young University, Provo, UT
21. C.J. Sterling, T.W. Nelson, C.D. Sorensen, R.J. Steel, and S.M. Packer, Friction Stir Welding of Quenched and Tempered C-Mn Steel, *Friction Stir Welding and Processing II,* TMS, 2003, p 165–171
22. Z. Feng, "Friction Stir Welding of API Grade X-65 Steel Pipe," Annual AWS Conference (Dallas, TX), American Welding Society, 2005
23. A. Ozekcin, H. Jin, J.Y. Koo, N.V. Bangaru, R. Ayer, and S. Packer, "A

Microstructural Study of Friction Stir Welded Joints of Carbon Steels," ISOPE 2004, May 23–28, 2004 (Toulon, France), International Society of Offshore and Polar Engineers

24. K. Okamoto, S. Hirano, M. Inagaki, S.H.C. Park, Y.S. Sato, H. Kokawa, T.W. Nelson, and C.D. Sorensen, Metallurgical and Mechanical Properties of Friction Stir Welded Stainless Steels, *Proceedings of the Fourth International Symposium on Friction Stir Welding,* May 14–16, 2003 (Park City, UT), TWI, paper on CD
25. R.J. Steel, C.O. Pettersson, C.D. Sorensen, Y. Sato, C.J. Sterling, and S.M. Packer, Friction Stir Welding of SAF 2507 (UNS S32750) Super Duplex Stainless Steel, Paper PO346, *Proceedings of Stainless Steel World 2003,* KCI Publishing
26. W.A. Palko, R.S. Fiedler, and P.F. Young, Investigation of the Use of Friction Stir Processing to Repair and Locally Enhance the Properties of Large Ni-Al Bronze Propellers, *Mater. Sci. Forum,* Vol 426–432 (No. 4), 2003, p 2909–2914
27. M.W. Mahoney, Friction Stir Welding and Processing: A Sprinter's Start, A Marathoner's Finish, *Trends in Welding Research, Proceedings of the Seventh International Conference,* May 16–20, 2005 (Pine Mountain, GA), ASM International

# CHAPTER 7

# Microstructure and Mechanical Properties of Friction Stir Welded Titanium Alloys

T.J. Lienert, MST Division, Los Alamos National Laboratory

FRICTION STIR WELDING (FSW) of titanium alloys is currently a research area of considerable interest. The objective of this chapter is to summarize the current understanding of FSW of titanium alloys by reporting on their microstructures, microstructural evolution, and mechanical properties. The chapter is organized as follows. It begins with a review of the metallurgy of titanium alloys and a brief discussion on tooling and equipment considerations for FSW of titanium alloys. Subsequently, several studies of FSW of titanium alloys by various researchers are reviewed. Finally, general trends on the subject are summarized, and future needs are discussed.

## 7.1 Titanium Alloys Overview

**General Metallurgy.** Titanium and its alloys possess a unique combination of properties. They are lightweight and can also be processed to give a variety of useful combinations of mechanical properties. Many titanium alloys are found in high-performance applications such as aerospace structures, where their high strength-to-weight ratio provides considerable advantage. Additionally, they exhibit good corrosion resistance in many environments, facilitating their use in chemical-processing, power-generation, and medical prosthesis applications.

Pure titanium experiences an allotropic transformation from the hexagonal close-packed (hcp) alpha (α) phase to the body-centered cubic (bcc) beta (β) phase as its temperature is increased through 882.5 °C (1620.5 °F). Alloying of titanium can be performed to produce a wide variety of microstructures and properties that can be tailored for specific applications. Addition of alloying elements to pure titanium can affect the phase balance (Ref 1). Alloying elements such as aluminum and oxygen tend to promote the α phase and are termed α stabilizers. Other elements, for example, molybdenum, vanadium, and chromium, are called β stabilizers, because they promote the β phase.

Many different titanium alloys have been developed for a large variety of applications. Titanium alloys are generally classified according to the equilibrium phases present in their microstructure at room temperature (Ref 1). They can be classified as commercially pure (CP) alloys and alpha alloys that mainly contain the hcp phase, alpha-beta alloys that contain both phases, and metastable beta alloys and beta alloys that consist largely of the bcc phase.

The schematic pseudobinary phase diagram shown in Fig. 7.1 can be used to understand the classification of titanium alloys. The diagram depicts the different phase fields on a plot of temperature versus the percent of β stabilizers added to a titanium alloy already containing some amount of α stabilizer. The upper solid curve is called the β-transus curve, while the lower solid curve is the α transus. The two dashed curves indicate the locus of the martensite start ($M_s$) and martensite finish ($M_f$) temperatures as a function of composition.

Alloys with compositions less than the point where the α transus meets the composition axis are termed α alloys. The CP alloy discussed in this chapter is an example of an α alloy. Those with compositions greater than the point where the β transus meets the axis are termed β alloys. Those with compositions in between have a microstructure of α and β phases at ambient temperature under equilibrium conditions. Two types of these alloys can be identified. One type has composition limits between the α transus and the $M_s$ curve and can be described by the term α-β alloy. The Ti-6Al-4V alloy discussed later in this chapter is a common α-β alloy. The second type is given the name metastable β alloy. Composition limits for metastable β alloys fall between the $M_s$ and the β transus. Metastable beta alloys can best be described as alpha-beta alloys that contain an appreciable level of beta stabilizers. The low diffusivity of the beta stabilizers promotes complete retention of beta phase to room temperature at moderate cooling rates. The Ti-15V-3Cr-3Al-3Sn and Beta 21-S alloys are common metastable beta alloys.

**Welding of Titanium Alloys.** Welding is an effective manufacturing method for joining components to produce structures. However, welding of titanium alloys is complicated by problems associated with their high reactivity. Titanium alloys rapidly dissolve oxygen, nitrogen, and hydrogen at temperatures above 500 °C (930 °F), resulting in subsequent embrittlement (Ref 2, 3). Dissolution of these gases is especially rapid in the liquid phase. As a result, welding processes must be carried out in inert or vacuum environments to avoid embrittlement. Moreover, parts to be welded and filler metals must be solvent cleaned to remove hydrocarbon-base oils and moisture to prevent embrittlement. Heating of material in the heat-affected zone (HAZ) above the beta-transus temperature can also result in grain growth and produce coarse columnar grains in the fusion zone, resulting in a loss of ductility. Finally, welding can also place certain regions of the welded structure in a state of residual tensile stress, creating concern over subsequent fatigue performance. As a consequence, weldments of titanium alloys are often given a stress-relief heat treatment prior to service. Welding processes used to join titanium alloys include gas tungsten arc welding, plasma arc welding, and electron beam welding. Brazing and solid-state welding processes such as friction welding have also been used to join titanium alloys (Ref. 2, 3).

## 7.2 Overview of Alloys Successfully Welded with FSW

Of the many titanium alloys available, only a few have been studied for FSW. These alloys studied include CP alloys, Ti-6Al-4V, Ti-15V-

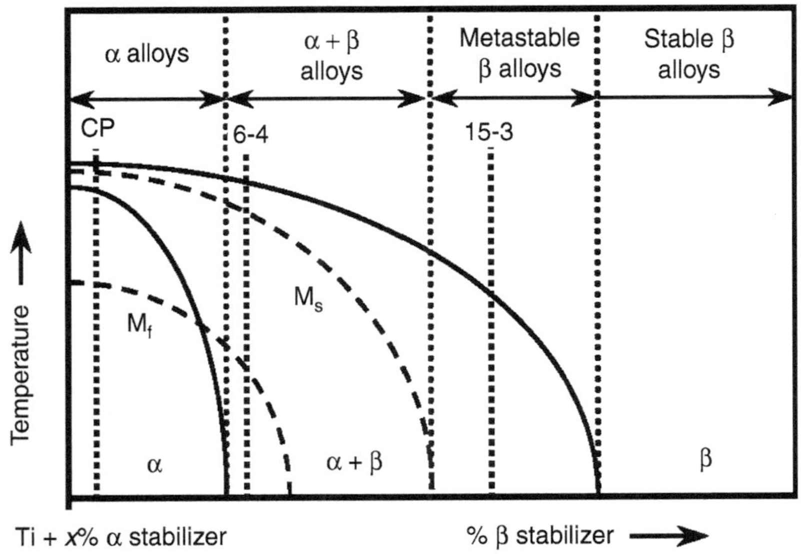


**Fig. 7.1** Schematic pseudobinary phase diagram for titanium alloys. CP, commercially pure; $M_s$, martensite start temperature; $M_f$, martensite finish temperature

3Cr-3Al-3Sn, and Beta 21-S. The following sections provide further background on the compositions, metallurgical information, and uses of each alloy.

**Commercially pure titanium-alloys** are available in four grades that are distinguished according to the amount of impurities, such as carbon, hydrogen, nitrogen, oxygen, and iron, present (Ref 1). These alloys typically have greater than 1000 ppm of total impurities, primarily oxygen. The mechanical properties of CP titanium alloys are strongly affected by even small variations in the impurity content. Consequently, the CP titanium grades are not classified by composition but rather by mechanical properties.

These alloys have an hcp crystal structure known as alpha phase. The beta-transus temperature of CP titanium alloys is ~910 ± 15 °C (1670 ± 27 °F), depending on the oxygen content (Ref 1). These alloys are not strengthened by heat treatment, like some other titanium alloys. They also have excellent corrosion resistance in seawater and marine environments.

The most common grade of CP titanium is grade 2, also known as R50400 in the UNS system (Ref 1). The compositions of grade 2 titanium alloys are summarized in Tables 7.1 and 7.2, while the minimum room-temperature tensile properties are provided in Table 7.3 (Ref 1). Grade 2 titanium has a minimum yield strength of 276 MPa (40 ksi). The greater iron and oxygen contents of grade 2 versus grade 1 impart increased yield and tensile strength with slightly lower ductility. Typical uses for grade 2 titanium include chemical and marine applications, desalination equipment, and airframe skin as well as pump parts and piping systems.

**Ti-6Al-4V Alloys.** Ti-6Al-4V is the "workhorse" titanium alloy, because it is the most widely used of all titanium alloys. It is available in several formulations, including the commercial-impurity level and extra-low interstitial grades (Ref 1). Ti-6Al-4V products can be produced in wrought, cast, and powder metallurgy forms.

Ti-6Al-4V is an alpha-beta alloy that can be modified extensively by both thermal and thermomechanical processing to produce a large variety of microstructures and hence a wide spectrum of mechanical properties. The beta-transus temperature is approximately 1000 °C (1830 °F) and is a function of interstitial content (Ref 1). Samples of Ti-6Al-4V cooled at relatively slow rates from elevated temperatures contain mainly the alpha and beta phases as a result of diffusional transformations, while those cooled rapidly may also contain martensitic phases such as the $\alpha'$ (hcp structure) or the $\alpha''$ (orthorhombic structure) phases.

The composition and minimum room-temperature tensile properties of Ti-6Al-4V alloys are summarized in Tables 7.1 to 7.3 (Ref 1). The alloy is most commonly produced in the mill-annealed condition, where it displays a useful combination of strength, toughness, ductility, and fatigue properties. It is also found in the

**Table 7.1 Nominal compositions of common titanium alloys**

| | Composition, wt% | | | | |
|---|---|---|---|---|---|
| Alloy | Al | Mo | Sn | V | Other |
| CP-Ti (grade 2) | . . . | . . . | . . . | . . . | . . . |
| Ti-6Al-4V | 6 | . . . | . . . | 4 | . . . |
| Ti-15-3 | 3 | . . . | 3 | 15 | 3Cr |
| Beta 21-S | 3 | 15 | . . . | . . . | 3Nb, 0.2Si |

**Table 7.2 Impurity limits for common titanium alloys**

| | Composition, wt% | | | | |
|---|---|---|---|---|---|
| Alloys | N | C | H | Fe | O |
| CP-Ti (grade 2) | 0.03 | 0.08 | 0.015 | 0.30 | 0.25 |
| Ti-6Al-4V | 0.05 | 0.10 | 0.0125 | 0.30 | 0.20 |
| Ti-6Al-4V (ELI)(a) | 0.05 | 0.08 | 0.0125 | 0.35 | 0.13 |
| Ti-15-3 | 0.05 | 0.05 | 0.015 | 0.25 | 0.13 |
| Beta 21-S | 0.05 | 0.05 | 0.015 | 0.40 | 0.15 |

(a) ELI, extra-low interstitial

**Table 7.3 Minimum room-temperature properties of common titanium alloys**

| | | Ultimate tensile strength | | 0.2% yield strength | | |
|---|---|---|---|---|---|---|
| Alloy | Condition | MPa | ksi | MPa | ksi | Elongation, % |
| CP-Ti (grade 2) | . . . | 345 | 50 | 276 | 40 | 20 |
| Ti-6Al-4V | Mill annealed | 896 | 130 | 827 | 120 | 14 |
| Ti-6Al-4V | Solution treated and aged | 1172 | 170 | 1103 | 160 | 10 |
| Ti-6Al-4V (ELI)(a) | Mill annealed | 827 | 120 | 758 | 110 | 15 |
| Ti-15-3 | Solution treated and aged | 786 | 114 | 772 | 112 | 20 |
| Ti-15-3 | Aged (540 °C, or 1000 °F) | 1089 | 158 | 952 | 138 | 10 |
| Beta 21-S | Aged (540 °C, or 1000 °F) | 1413 | 205 | 1345 | 195 | 6.5 |

(a) ELI, extra-low interstitial

beta-annealed condition (annealed above the beta transus) and the solution-treated, quenched, and aged condition.

Ti-6Al-4V was developed for applications requiring high strength and low-to-moderate temperatures. The alloy has a high strength-to-weight ratio and good corrosion resistance in many environments. Ti-6Al-4V finds use in aerospace, automotive, and marine applications as well as for orthopedic implants.

**Ti-15V-3Cr-3Al-3Sn Alloys.** The metastable beta alloy Ti-15V-3Cr-3Al-3Sn-(hereafter referred to as Ti-15-3) is a solute-rich alpha-beta alloy that was developed to lower fabrication costs (Ref 1). Composition ranges and room- temperature tensile properties for Ti-15-3 are summarized in Tables 7.1 to 7.3. It is produced in sheet form and has excellent forming characteristics at ambient temperature. It can be aged after processing to a spectrum of strength levels. Ti-15-3 has a beta-transus temperature of approximately 760 °C (1400 °F) (Ref 1). It is solution annealed at 780 °C (1435 °F) and can be aged between 480 and 540 °C (895 and 1000 °F) to precipitate alpha phase. The Ti-15-3 alloy has lower production costs than the Ti-6Al-4V alloy and finds use in airframe structures. It is normally found in sheet form, owing to the need to achieve cooling rates fast enough to prevent precipitation of the alpha phase.

**Beta 21-S.** The Beta 21-S alloy is a relatively new metastable beta alloy (Ref 1). It was designed to have good formability, similar to Ti-15-3, but also has improved oxidation resistance, creep resistance, and high-temperature strength relative to Ti-15-3. Composition ranges and room-temperature tensile properties for Beta 21-S are listed in Tables 7.1 to 7.3. The alloy contains approximately 15% Mo, 3% Al, and 2.8% Nb, with additions of silicon (Ref 1). It is normally provided in the beta solution-treated condition. Beta 21-S has an elastic modulus close to that of bone and finds use in prosthetic application. It has excellent high-temperature stability and can be used at temperatures up to 290 °C (550 °F).

## 7.3 Tooling and Equipment Considerations

Friction stir welding of titanium alloys differs from FSW of aluminum alloys with regard to the demands placed on the tools and FSW machine. Friction stir welding can be conveniently viewed as a hot working process that is used to join metals. Hot working can be described as deformation processing at temperatures above 50 to 60% of the absolute melting temperature of the metal. The much higher hot working temperatures of titanium alloys relative to Al alloys limit the choice of tool materials to refractory metals such as tungsten (including tungsten-rhenium) and molybdenum alloys or robust cermets such as WC/Co. Tool life is a clear concern for these materials. Hot titanium is an excellent solvent for many of the components of these tools. Strategies to minimize wear and deformation of the tool, especially the pin, must be developed. Successful FSWs have been produced on titanium alloys using CP tungsten, W-25%Re tungsten-rhenium with HfC, and sintered TiC tools.

The reactivity of the titanium alloys as well as the refractory metals tools is another concern. Elimination of atmospheric contamination is required to limit pickup of nitrogen, oxygen, and hydrogen from the atmosphere by both workpiece and tools in order to avoid embrittlement. Hence, the use of inert gas shielding is required during FSW of titanium alloys. Use of an inert gas chamber that can be backfilled with inert gas prior to each weld is preferred.

Finally, considerable heat energy is lost to the tool and then to the tool holder and machine spindle during FSW of titanium alloys. Use of a cooled tool holder, similar to that employed by Lienert and coworkers, is recommended to prevent damage to the FSW machine.

## 7.4 FSW of Mill-Annealed Ti-6Al-4V Plate

A comprehensive study of FSW of mill-annealed Ti-6Al-4V has been completed by Lienert and coworkers (Ref 4). The procedures, results, and discussion provided as follows are excerpted from that reference. Friction stir welds were produced on plates of a Ti-6Al-4V alloy in the mill-annealed condition. The composition of the specific alloy was 6.4% Al, 3.85% V, 0.22% Fe, 0.18% O, and 0.013% H (all weight percent), with the balance titanium. The pin was 0.64 cm (0.25 in.) in length and 0.79 cm (0.31 in.) in diameter. The tool was machined from CP tungsten, and the pin did not feature any threads or other profiling. Welds were made at travel speeds up to 0.17 cm/s (0.067 in./s) using a tool with a 1.9 cm (0.75 in.) diameter shoulder. The tool was rotated at 275 rpm for all of the welds made in this study. The tool and workpiece were protected from surface oxidation by welding in an inert gas chamber. For each weld, thermocouples were firmly attached to the circumference of the tool at two different distances above the shoulder. Thermocouples were also attached at several locations on the top and bottom surfaces of the workpiece. Microstructures of the different weld regions were characterized using light optical microscopy and scanning electron microscopy. Mechanical properties were assessed with microhardness and room-temperature tensile testing.

### 7.4.1 Tool and Workpiece Temperatures

The tool and the flashing surrounding the tool glowed a reddish-orange color during welding, suggesting peak temperatures of at least 1100 °C (2010 °F). Tool thermal cycles from two separate experiments are presented in Fig. 7.2, and data from tool thermocouples from four experiments are summarized in Table 7.4. Plunge time is given for the left side of Fig. 7.2, and tool travel is depicted in the right half of Fig. 7.2. When the tool reached steady state, the thermal gradient along the length of the tool became linear, and the gradient ($\Delta T/\Delta z$, where $z$ is the distance from the shoulder along the height of the tool) was estimated from the two temperatures. Subsequently, the temperature at the shoulder of the tool ($z = 0$) can be estimated by extrapolation. Estimated gradients and shoulder temperatures are also given in Table 7.4.

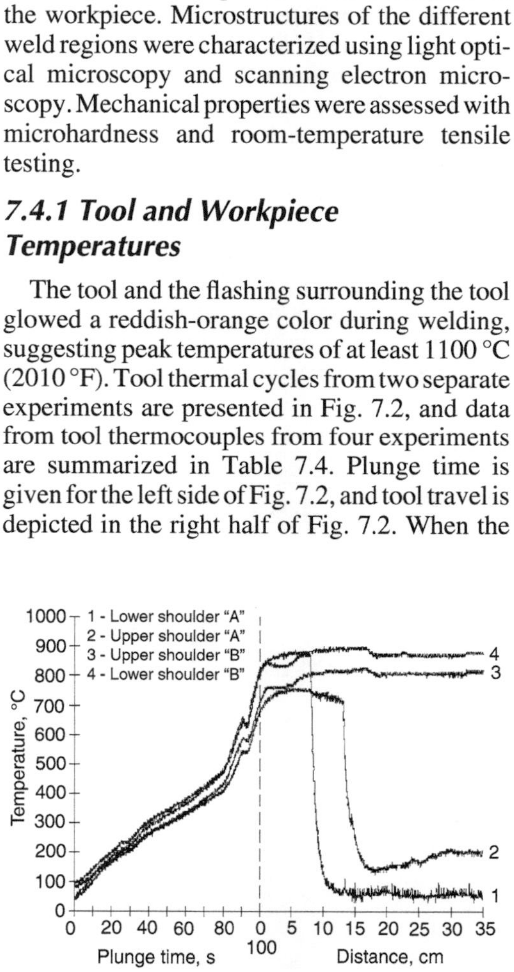


**Fig. 7.2** Tool thermal cycles from two separate friction stir welds on Ti-6Al-4V. Thermocouples were attached at two vertical locations on the tool periphery.

As shown in Table 7.4, the peak temperatures at a distance of 0.64 cm (0.25 in.) from the end of the tool shoulder ranged from 875 to 930 °C (1605 to 1705 °F,) while the peak temperatures at a distance of 0.95 cm (0.37 in.) from the end of the tool shoulder varied from 750 to 830 °C (1380 to 1525 °F). The temperature gradients along the tool, assuming one-dimensional heat flow, ranged from 240 to 500 °C/cm (465 to 930 °F/in.), and the temperatures at the end of the shoulder ($z = 0$), determined by extrapolation from the lower thermocouple (at a known position), varied from 1045 to 1150 °C (1915 to 2100 °F). The average of the four shoulder temperatures from Table 7.4 is 1115 °C (2040 °F). The range of values reported here may stem from inaccuracies in position of the thermocouple placement (±0.3 mm, or 0.012 in.).

Plots of the thermal cycles recorded from thermocouples on the workpiece are shown in Fig. 7.3, and a summary of thermocouple data for the workpiece is presented in Table 7.5. Thermocouples placed to fall within the HAZ at a position ~0.05 cm (0.02 in.) from the end of the pin and ~0.32 cm (0.13 in.) from the weld centerline recorded peak temperatures in the range of 850 to 890 °C (1560 to 1635 °F). Cooling rates through the $M_s$ temperature (~800 °C, or 1470 °F) (Ref 1) were in the range of 40 °C/s (70 °F/)s. Assuming cooling rates of the same order of magnitude throughout the rest of the weld region, as suggested from calculations for arc welds by Adams (Ref 5), the cooling rate through the $M_s$ for these welds was approxi-

**Table 7.4 Summary of tool temperatures ($T$) as a function of distance above the tool shoulder ($z$), temperature gradients along the tool, and extrapolated shoulder temperatures for friction stir welds of Ti-6Al-4V (from four tests)**

| | Position 1 | | Peak temperature ($T$) | | Position 2 | | Peak $T$ | | $\Delta T/\Delta z$ | | $T$ at $z = 0$ | |
|---|---|---|---|---|---|---|---|---|---|---|---|---|
| Test | cm | in. | °C | °F | cm | in. | °C | °F | °C/cm | °F/in. | °C | °F |
| 1 | 0.64 | 0.25 | 895 | 1645 | 0.95 | 0.37 | 820 | 1510 | 240 | 465 | 1045 | 1915 |
| 2 | 0.64 | 0.25 | 875 | 1605 | 0.95 | 0.37 | 750 | 1380 | 390 | 735 | 1125 | 2055 |
| 3 | 0.64 | 0.25 | 930 | 1705 | 0.95 | 0.37 | 825 | 1515 | 330 | 625 | 1140 | 2085 |
| 4 | 0.32 | 0.13 | 990 | 1815 | 0.95 | 0.37 | 830 | 1525 | 500 | 930 | 1150 | 2100 |

mately 40 °C/s across the entire weld region, including the stir zone.

### 7.4.2 Microstructural Characterization

The base metal of the Ti-6Al-4V alloy used in this study was comprised of relatively equiaxed grains of α with smaller amounts of grain-boundary β phase (Figure 7.4a, b). Figure 7.4(a) is an optical micrograph of the base metal. The dark etching phase indicated by the arrows is the grain-boundary β phase. Figure 7.4(b) is a scanning electron microscope (SEM) micrograph of the base metal taken using backscattered electron BE imaging mode. Note the reversal of contrast between the optical and SEM/BE images in Fig. 7.4 and subsequent figures. The average grain diameter was determined by a linear intercept method at approximately 18 μm.

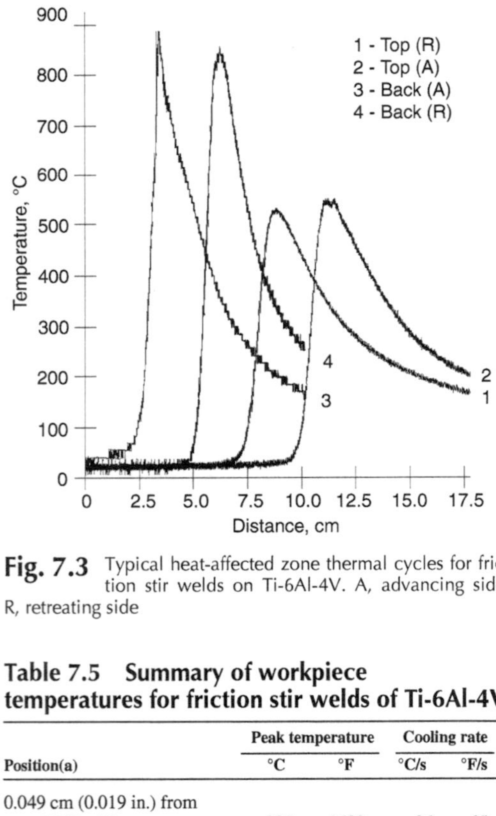


**Fig. 7.3** Typical heat-affected zone thermal cycles for friction stir welds on Ti-6Al-4V. A, advancing side; R, retreating side

**Table 7.5 Summary of workpiece temperatures for friction stir welds of Ti-6Al-4V**

| Position(a) | Peak temperature °C | Peak temperature °F | Cooling rate °C/s | Cooling rate °F/s |
|---|---|---|---|---|
| 0.049 cm (0.019 in.) from weld CL AB | 887 | 1628 | 36 | 65 |
| 0.049 cm (0.019 in.) from weld CL RB | 856 | 1573 | 42 | 76 |
| 0.041 cm (0.016 in.) from SZ/HAZ boundary RT | 525 | 977 | N/A | |
| 0.036 cm (0.014 in.) from SZ/HAZ boundary AT | 540 | 1000 | N/A | |

(a) CL, centerline; A, advancing side; B, bottom of plate; R, retreating side; T, top of plate; SZ, stir zone; HAZ, heat-affected zone

Figure 7.5 is an optical micrograph of the various weld regions from a section taken transverse to the welding direction. Several microstructurally distinct weld regions with different etching can be observed in the figure, including the stir zone or nugget and the HAZ response. Figure 7.6 is an optical micrograph of the stir zone, thermomechanically affected zone (TMAZ), and the HAZ, also from a transverse section. The boundaries between different regions are indicated by dotted lines. Grains of the stir zone and the TMAZ are elongated in a direction parallel to the boundary, indicating evidence of deformation during FSW.

Temperature measurements and microstructures observed in the stir zone suggest that peak temperatures surpassed the β transus. The concentric ring patterns found in the stir zones of friction stir welds on aluminum alloys were not seen in welds on the titanium alloy, owing to the

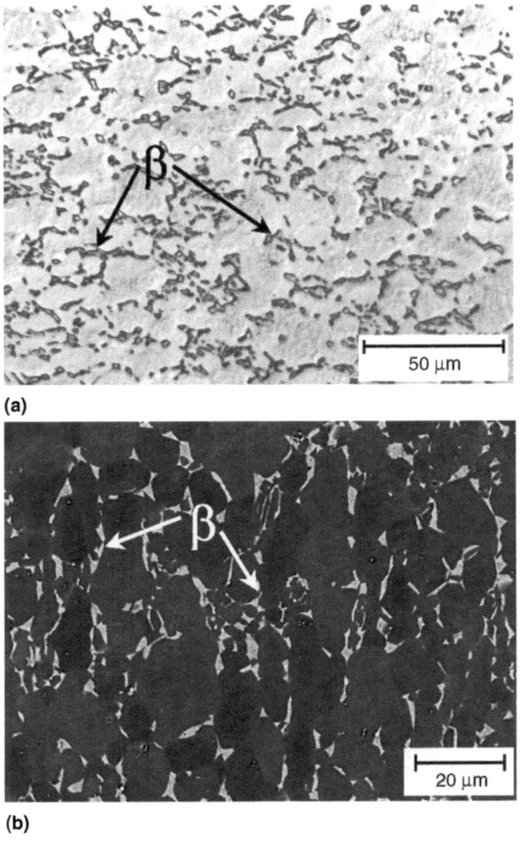


**Fig. 7.4** Micrographs of the Ti-6Al-4V base metal. (a) Optical micrograph. (b) Scanning electron microscope/backscattered electron micrograph. Grains are nearly equiaxed. Microstructure is primarily a phase, with β phase located at grain boundaries. Arrows indicate grain-boundary β phase.

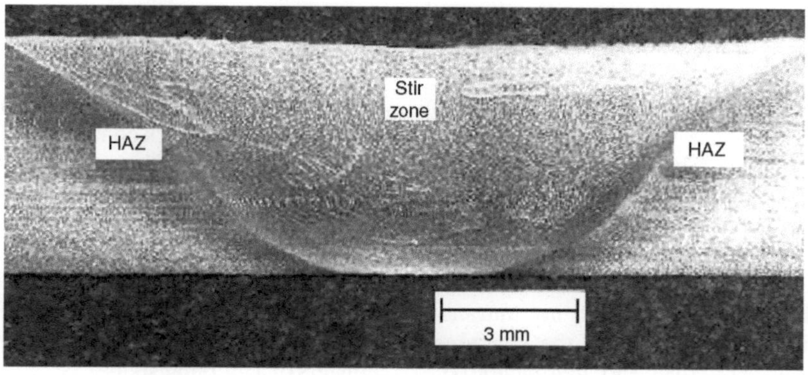


**Fig. 7.5** Optical macrograph of the various weld regions for friction stir welding on Ti-6Al-4V. HAZ, heat-affected zone

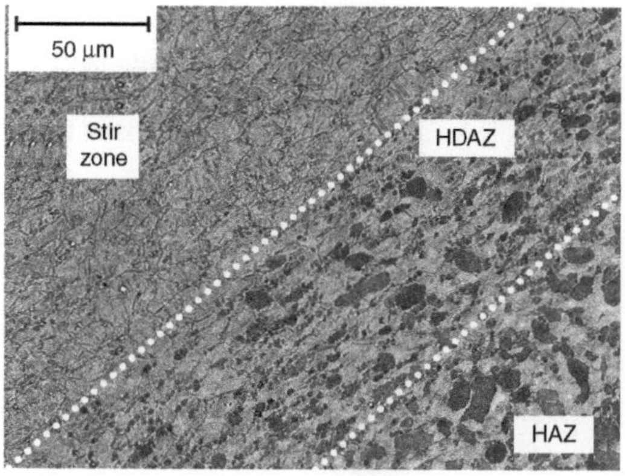


**Fig. 7.6** Optical micrograph of the stir zone heat-and-deformation-affected zone (HDAZ), and heat-affected zone (HAZ) for friction stir welding on Ti-6Al-4V. The dotted lines indicate the boundaries between different regions. Prior-β grains in the stir zone and grains in the HDAZ are elongated parallel to the boundary.

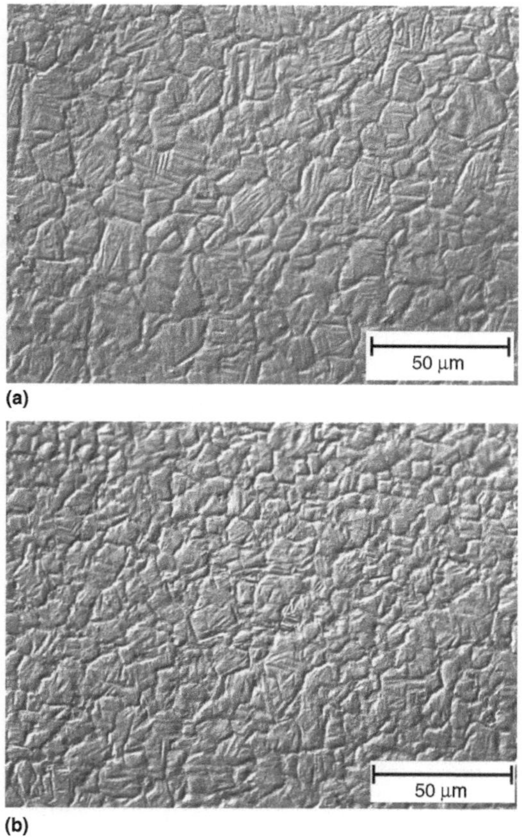


**Fig. 7.7** Optical micrographs of the stir zone for friction stir welding on Ti-6Al-4V. (a) Center of stir zone showing equiaxed grains with grain-boundary α phase and fine acicular α phase in a β matrix. (b) Near top surface, showing finer prior-β grain size than (a)

microstructural modification associated with the solid-state β-to-α transformation on cooling and the lack of threads on the pin used here. The center of the stir zone contained grains with grain-boundary α phase and fine acicular α phase emanating from the grain-boundary α phase into the prior-β grains as a result of the relatively rapid cooling rate (Fig. 7.7a). The grains were not perfectly equiaxed and exhibited some elongation along an axis running from lower left to upper right in Fig. 7.7(a). Average grain diameters were 19.55 ± 5.9 μm along the long axis and 11.45 ± 0.56 μm along the shorter axis. A mean grain diameter of 15.5 μm is obtained by averaging the two diameters. Smaller grain sizes were observed just adjacent to the top surface of the stir zone, presumably resulting from the greater amounts of strain experienced locally due to continued and direct interaction with the tool shoulder (Fig. 7.7b). A gradient in grain size from the bottom of Fig. 7.7(b) to the top was apparent.

These observations were corroborated by SEM/BE images of the stir zone at higher magnifications. Figure 7.8 is an SEM/BE image of the stir zone adjacent to the TMAZ. Note that no untransformed α phase was evident in the stir zone microstructures. The microstructure of the stir zone was characterized by continuous α phase along prior-β boundaries and fine acicular α that grew into prior-β grains that were elongated slightly in a direction parallel to the stir zone/HAZ boundary. Note that no untransformed α phase was evident in the stir zone microstructures. Moreover, no martensitic α′ phase was observed in the stir zone.

Microstructural evidence from optical micrographs revealed that regions corresponding to the HAZ experienced peak temperatures well below the β-transus temperature, resulting in some transformation of the α phase to β during heating (Fig. 7.9a). Examinations also indicated that regions corresponding to the TMAZ underwent peak temperatures just below the β-transus temperature, resulting in considerable transformation of the β phase during heating (Fig. 7.9b). Note that two forms of the α phase were observed in the HAZ. Regions that appear as grain-boundary and acicular α-phase regions were present as β phase at the peak temperature of the thermal cycle imposed by welding. These regions transformed from the prior-β phase during cooling. Consequently, they are referred to as transformed β products. Other regions of a phase never transformed to β during heating and are referred to as untransformed α. Regions of untransformed α are evident in the micrograph shown in Fig. 7.9(b).

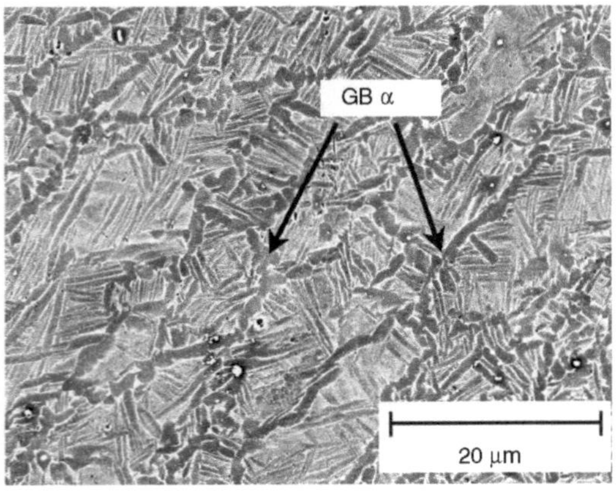


**Fig. 7.8** Scanning electron microscope/backscattered electron micrograph of the stir zone for friction stir welding on Ti-6Al-4V. Arrows indicate grain-boundary (GB) α phase.

These observations were corroborated by SEM/BE images of the same regions at higher magnifications. Figure 7.10 is an SEM/BE image of the near HAZ. As in Fig. 7.9(b), pockets of untransformed α (dark contrast) are apparent. As a result of decomposition of prior-β phase during cooling, fine grain-boundary α phase is also seen along prior-β boundaries, and fine acicular α phase is found throughout the prior-β grains. Note that unlike the continuous grain-boundary α phase seen in the stir zone (Fig. 7.8), the grain-boundary α phase in the near HAZ appears as clusters of a globular shape.

### 7.4.3 Microhardness and Tensile Results

A plot of typical microhardness data is given in Fig. 7.11. Results revealed an increase in hardness from approximately 340 Vickers hard-

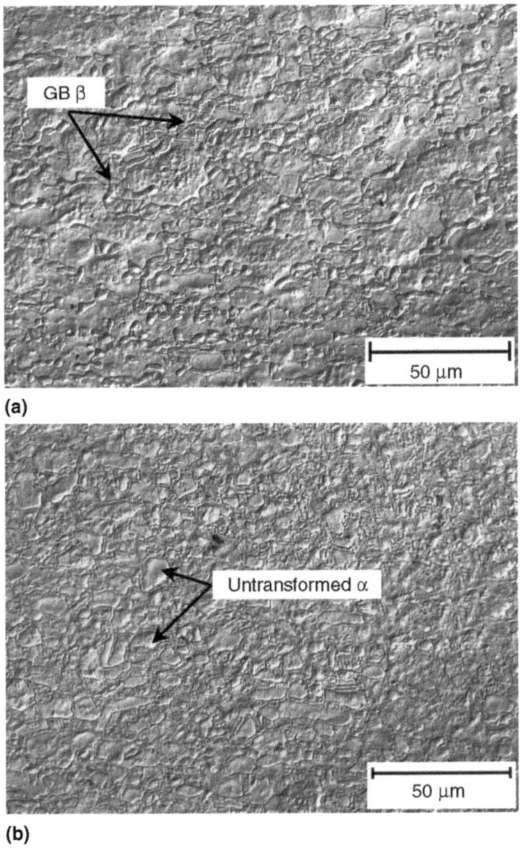


**Fig. 7.9** Optical micrographs of the heat-affected zone (HAZ) for friction stir welding on Ti-6Al-4V, showing different volume fractions of α and β phases. (a) Far HAZ. GB, grain boundary. (b) Near HAZ

ness number (VHN) in the base metal and stir zone to 370 VHN in the HAZ. Tensile data for the base metal and welds are presented in Table 7.6. Reported values for tensile data represent the average of three tests. Both the average and standard deviation are given in the table for each value. The welds exhibited 100% joint efficiency with respect to both yield and tensile strength, where joint efficiency is defined as the strength value (yield or tensile) for the weld sample divided by that of the base metal. The percent elongation for the base metal and weld samples was identical. Failure of the weld tensile samples occurred in the gage section in regions corresponding to the base metal. The lack of oxygen pickup after welding indicated the efficacy of the acrylic glass inert gas chamber and suggests that hardness and tensile properties of the welds were not influenced by oxygen content.

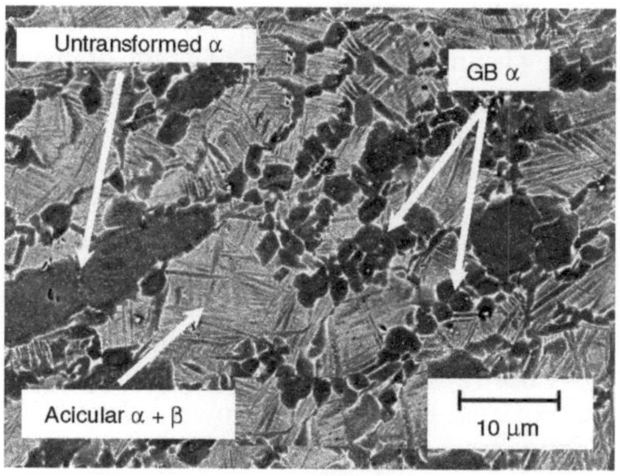


**Fig. 7.10** Scanning electron microscope/backscattered electron micrograph of the near heat-affected zone for friction stir welding on Ti-6Al-4V. GB, grain boundary

## 7.4.4 Discussion

**Stir Zone Temperatures.** Extrapolation of temperature gradients determined from the thermocouple measurements on the tools suggests that temperatures at the tool shoulder exceeded 1115 °C (2040 °F), the average temperature found from the four experiments summarized in Table 7.4, and may have been as high as the maximum temperature measured, 1150 °C (2100 °F). Because the tool was spinning and was held against the workpiece under considerable pressure, asperity contact was eliminated, and the tool was in intimate contact with the workpiece across the entire tool/workpiece interface. Hence, to a first-order approximation,

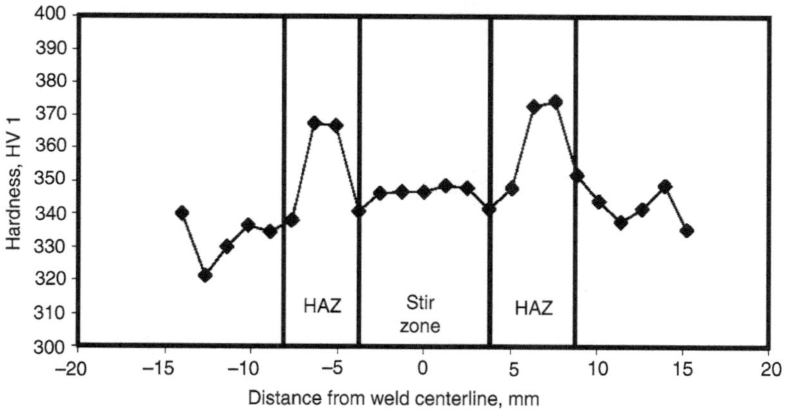


**Fig. 7.11** Microhardness data for a traverse across the entire weld region for friction stir welding on Ti-6Al-4V. Note the increase in hardness of the heat-affected zone (HAZ)

**Table 7.6 Tensile test results for friction stir welds (FSWs) of Ti-6Al-4V**

| | Yield strength | | Tensile strength | | | |
|---|---|---|---|---|---|---|
| Component | MPa | ksi | MPa | ksi | Elongation, % | Failure location |
| Base metal | 897 ± 0.7 | 130.1 ± 0.1 | 957.7 ± 3.4 | 138.9 ± 0.5 | 12.7 ± 0.5 | N/A |
| FSW | 912.9 ± 8.3 | 132.4 ± 1.2 | 1013.5 ± 8.3 | 147.0 ± 1.2 | 12.7 ± 0.9 | Base |

there was no discontinuity in temperature across the interface, and the workpiece material and tool shoulder were at the same temperature during steady state.

Consequently, the material at the top of the stir zone experienced temperatures above 1115 °C and possibly as high as 1150 °C. The β-transus temperature for the Ti-6Al-4V alloy used here is approximately 1010 °C (1850 °F), based on its oxygen content (Ref 1). Note that these temperatures are well above the β-transus temperature for the Ti-6Al-4V alloy and are consistent with microstructural observations described earlier.

**Cooling Rates.** Formation of the α′ martensite phase in Ti-6Al-4V alloys typically requires fairly fast cooling rates, such as those experienced during water quenching (Ref 1). A continuous cooling transformation (CCT) diagram for Ti-6Al-4V, reported by Ahmed and Rack (Ref 6), indicates that a cooling rate in excess of 410 °C/s (740 °F/s) is required to produce a fully martensitic structure, and that cooling rates lower than 20 °C/s (35 °F/s) result entirely in diffusional transformations. Tanner (Ref 7) has also published a partial isothermal time-temperature transformation diagram for Ti-6Al-4V alloys that suggests that cooling rates in excess of 120 °C/s (215 °F/s) are required for any α′ martensite to form.

As discussed earlier, thermocouples placed near the bottom of the stir zone recorded cooling rates of approximately 40 °C/s (70 °F/s) through the $M_s$ temperature of the Ti-6Al-4V alloy. Adams (Ref 5) showed for arc welds on steel that the cooling rates through the $M_s$ (for steels) were of the same order of magnitude throughout the entire weld region, including the fusion zone and the HAZ. This result suggests that the cooling rates through the $M_s$ for the Ti-6Al-4V welds were on the order of 40 °C/s across the entire FSW weld region, including the stir zone. In light of the transformation diagrams described previously (Ref 6, 7) and in accord with the microstructural observations, this cooling rate was too slow to result in formation of α′ martensite.

**Strain-Rate Estimates in the Stir Zone.** Strain rates during FSW have not been measured experimentally. However, several modeling techniques have been used to estimate the strain rates during FSW of aluminum alloys, including a kinematic approach (Ref 8), CTH or hydrocode (Sandia National Laboratories) (Ref 9), computational fluid dynamics models (Ref 10, 11), and solid-mechanics models (Ref 12–15) as well as a formalism using the Zener-Holloman parameter (Ref 16, 17). Plastic strain rates ranging from $10^1$ to $10^3$ $s^{-1}$ have been reported, with the consensus of estimates between $10^2$ and $10^3$ $s^{-1}$.

Using an analogy with metal cutting, Nunes et al. (Ref 8) developed an expression for the mean shear strain rate over the flow path, based on kinematic considerations:

$$d\gamma/dt \approx r\Omega^2 N \quad \text{(Eq 7.1)}$$

where $t$ is time, $r$ is the radius of a plug of material rotating with the pin (taken to be the radius of the pin) and shearing against a single slip surface, $\Omega$ is the angular velocity of rotation of the pin, and $V$ is the forward velocity of the tool. This expression represents a lower bound for the maximum strain rate. Using the pin radius and revolutions per minute (rpm) for this work, the expression gives an approximate strain rate of ~$2.0 \times 10^3$ $s^{-1}$ at the slip surface. Strain rates are expected to decrease to lower values with distance from the slip surface.

Strain rates during friction welding (Ref 16) and FSW (Ref 17) of aluminum alloys have also been estimated with an approach that uses the subgrain size along with the Zener-Holloman ($Z$) parameter. The Zener-Holloman parameter is essentially a temperature-compensated strain rate and is defined as (Ref 18):

$$Z = A\sigma^n = \dot{\varepsilon} \exp(+Q/RT) \quad \text{(Eq 7.2)}$$

where $A$ is a frequency factor, $\sigma$ is the flow stress (true stress), $n$ is the stress exponent, $\dot{\varepsilon}$ is the true strain rate, $Q$ is the apparent activation energy for the controlling process, $R$ is the gas constant, and $T$ is the absolute temperature. An additional relationship between $Z$ and the grain or subgrain size can be determined from experiment. Using published data on the relationship between the subgrain size and $Z$, Frigaard et al. (Ref 17) reported calculated maximum strain rates on the order of $10^1$ for FSW of aluminum alloys.

Using a similar approach, strain rates for FSW of Ti-6Al-4V may be estimated using peak temperatures determined here along with published information on the activation energy and $Z$. Seshacharyulu and coworkers (Ref 19, 20) have developed the following relationship between the prior-β grain size ($d_{p\beta}$) and $Z$ for Ti-6Al-4V processed in the β regime at strain rates below 1 $s^{-1}$:

$$d_{p\beta} = 1954.3 \times Z^{-0.172} \ (\mu m) \quad \text{(Eq 7.3)}$$

Assuming a mean grain size of 15.5 μm, as determined from linear intercept measurements, yields a $Z$ of $1.634 \times 10^{12}$ $s^{-1}$. Li et al. (Ref 21) determined an activation energy of 246 kJ/mole for a Ti-6Al-4V alloy tested in the β regime at strain rates up to 15 $s^{-1}$. Using this value of activation energy and the $Z$ determined previously, strain rates ranging from $1.5 \times 10^3$ to $2.9 \times 10^2$ $s^{-1}$ were determined over the span of temperatures from 1150 to 1045 °C (2100 to 1915 °F), in excellent agreement with those found with the kinematic approach and modeling data outlined previously. However, note that the strain rates found with this method are outside the range used to determine the relationship between the prior-β grain size and $Z$, and some doubt may exist regarding the validity of the results despite the close agreement with other methods.

**Microstructural Evolution in the Stir Zone.** Final microstructural features of interest include grain size and morphology as well as the type of phases, phase fractions, and distribution. First, efforts are made to rationalize the refined grain structure of the stir zone by comparison of FSW conditions with those of published hot working diagrams. Subsequently, the types of phases and their distribution are addressed by relating peak temperatures and cooling rates of FSW with published CCT diagrams for Ti-6Al-4V.

To recapitulate, microstructural observations and extrapolation of tool temperature profiles to the tool/stir zone interface suggest that the stir zone experiences peak temperatures above the beta transus. Peak temperatures in the stir zone may have reached as high as 1150 °C (2100 °F). Moreover, the absence of any retained α phase in the stir zone indicates that time above the β transus was sufficient to allow complete transformation of the stir zone to β phase. Finally, strain rates in the range of $10^2$ and $10^3$ $s^{-1}$ have been estimated.

To facilitate an understanding of grain size evolution in the stir zone, FSW is best viewed as a hot working process used for joining. Important parameters that influence grain size during hot working include the temperature, strain, and strain-rate histories (Ref 18, 22). More specifically, the evolution of grain structures is controlled by peak temperature in addition to any restorative process (i.e., recovery or recrystallization). The operative restorative mechanism is dependent on the total strain, strain rate, temperature, and the stacking fault energy (SFE). For the current discussion, it is assumed that the evolution of grain size in the stir zone is dominated by the peak values of strain rate and temperature.

Comparison of the peak temperature and strain-rate estimates with the pertinent hot-deformation processing maps (Ref 19, 20, 23, 24) suggests that FSW of Ti-6Al-4V involves adiabatic shear banding (β instability) above the β-transus temperature. The instability in this temperature/strain-rate regime is also manifested by the broad oscillations observed on the stress-strain curves at these conditions reported by several researchers for this type of alloy (Ref 19–21, 25).

Shear bands are discrete regions that experience very localized deformation and are frequently found to develop in metals and alloys after large plastic deformation. More specifically, adiabatic shear bands (ASBs) are a type of shear band that can develop during deformation at high strain rates (Ref 26, 27). During deformation involving ASB formation, a large fraction of the plastic work is converted to heat. Especially in alloys with low thermal conductivity (such as Ti-6Al-4V), the heating rate at high strain rates can dominate over the rate of heat loss by conduction, resulting in a local temperature increase and development of a near-adiabatic condition. Subsequently, ASBs may form if the loss of strength due to thermal softening is sufficient to overcome strengthening by strain and/or strain-rate hardening. Refinement of grains within the ASB may occur by dynamic recrystallization or dynamic recovery, depending on the alloy and deformation conditions (Ref 28). Moreover, the dynamic recrystallization can occur in either a continuous or discontinuous fashion (Ref 29).

The ASBs in deformed specimens normally appear as bands with altered microstructure running along directions of maximum resolved shear stress. These bands are seen as distinct, because they are surrounded by larger regions of unaltered microstructure. Note that no such bands of altered microstructure were observed in the stir zones of the welds examined here. The lack of clear evidence for ASBs in the stir zone during FSW of Ti-6Al-4V may be explained by one or more of the following:

- The temperature and strain-rate estimates presented here are wrong.
- The stress state in FSW differs significantly from those used to construct the processing diagrams.

- Evidence for ASBs is obscured by the β-to-α transformation on cooling.

Alternately, this observation may indicate that the entire stir zone may have formed due to continuous and incremental ASB as material is sheared and carried around the pin. The notion of ASB formation during FSW of Ti-6Al-4V may not be as far-fetched as it may first seem. Several related processes involve ASB formation for processing of this alloy. Those familiar with inertia friction welding, a process very similar to FSW, know that a condition of adiabatic shearing (Ref 30) must be achieved to "focus" the mechanical energy on the weld interface to produce proper welds. Moreover, adiabatic shear banding as a process in chip formation during orthogonal machining (also involving high temperatures and shear strain rates) of Ti-6Al-4V (Ref 31, 32) has been reported. Further work is clearly needed to unambiguously determine how the grain structure evolves in these welds.

Assuming that the latter explanation is correct, refinement of grains within the ASBs during FSW of Ti-6Al-4V is likely due to dynamic recovery, owing to the high SFE (Ref 33, 34) and rapid diffusion rates (Ref 35) of the bcc structure present at peak temperatures. The high SFE of the bcc structure prevents dissociation of dislocations into partial dislocations, thereby limiting dislocation tangling and the resulting large increases in dislocation density required for discontinuous dynamic recrystallization. The large diffusion coefficient of the bcc phase aids in recovery, which requires diffusion of atoms to dislocation cores to permit climb of the dislocations into lower-energy configurations.

Regardless of the exact details of the restorative mechanism, the grain size was reduced from approximately 18 μm in the starting material to at least 15.5 μm in the stir zone as a result of FSW. Note that greater reduction of grain size was likely during FSW. However, the relatively long thermal cycle probably allowed for considerable postdeformation grain growth in the stir zone. Experiments involving interrupted welds followed immediately by quenching to room temperature are required to accurately determine the refined grain size.

Figures 7.12 and 7.13 can be used as aids in the discussion of microstructural evolution. Figure 7.12 contains schematics of the tool position versus time, the thermal cycle with superimposed CCT curve, and the pseudobinary phase diagram for Ti-6Al-4V. Actual phase diagrams (Ref 36, 37) and CCT diagrams (Ref 6, 7) are reported in open literature. The positions "a" through "f" on each schematic of Fig. 7.12 correspond to schematics of the microstructure at different points in time depicted in Fig. 7.13(a) through (f), respectively. Microstructural evolution can be followed as a function of time as the tool moves along the plate relative to the point of interest.

As the tool approaches a point of interest in the workpiece (indicated by the circled "X" in Fig. 7.12a), the local temperature begins to rise, and the original microstructure (Fig. 7.13a) begins to evolve. At position "b," the temperature (Fig. 7.12b) has risen into the two-phase α + β region of the phase diagram (Fig. 7.12), and the β phase originally along α grain boundaries grows to consume some of the α (Fig.

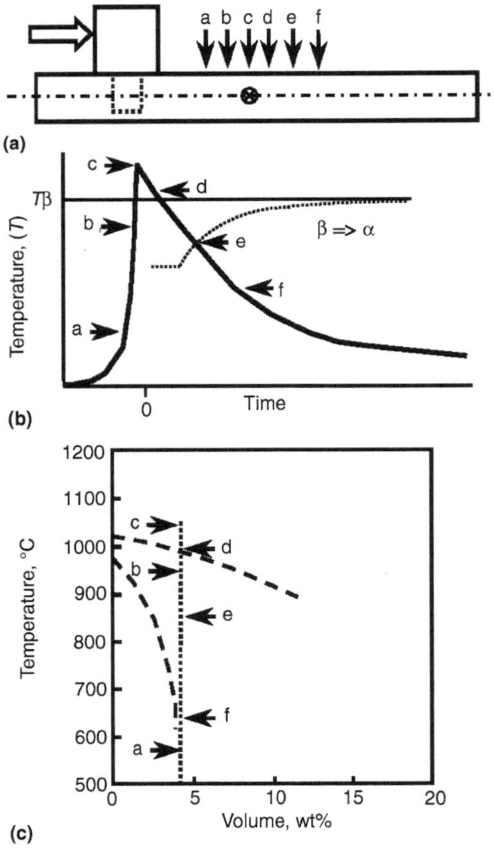


**Fig. 7.12** Schematics of (a) tool position vs. time, (b) thermal cycle with superimposed continuous cooling transformation curve, and (c) pseudobinary phase diagram. Positions "a" through "f" on the diagrams correspond to Fig. 7.13(a) through (f) and are used to describe microstructural evolution in the stir zone for friction stir welding on Ti-6Al-4V.

7.13b). At position "c," the workpiece material is interacting with the pin and is being deformed. The material is near the peak temperature (Fig. 7.12b and c) and undergoes deformation with shear and compressive components (Fig. 7.13c). Dynamic restorative mechanisms may ensue concurrent with the deformation.

Dynamic restoration for Ti-6Al-4V normally involves dynamic recovery (Fig. 7.13d). Once the tool passes the location, deformation and the attendant adiabatic heating cease, and cooling ensues locally. Static recovery and grain coarsening may occur on cooling above the β-transus temperature (Fig. 7.13d). Upon further cooling, subsequent microstructural development is then largely dictated by the phase diagram and continuous cooling diagram (Fig. 7.12b and c). However, note that growth or coarsening of the prior-β grains can continue during cooling until α phase nucleates at prior-β boundaries to limit boundary migration, provided there exists a driving force derived from a difference in local grain sizes. More specifically, larger grains, with a greater number of concave sides, can grow to consume smaller grains with fewer sides.

In accord with microstructural observations, the estimated cooling rates (~40 °C/s, or 70 °F/s) indicate that the phase transformation

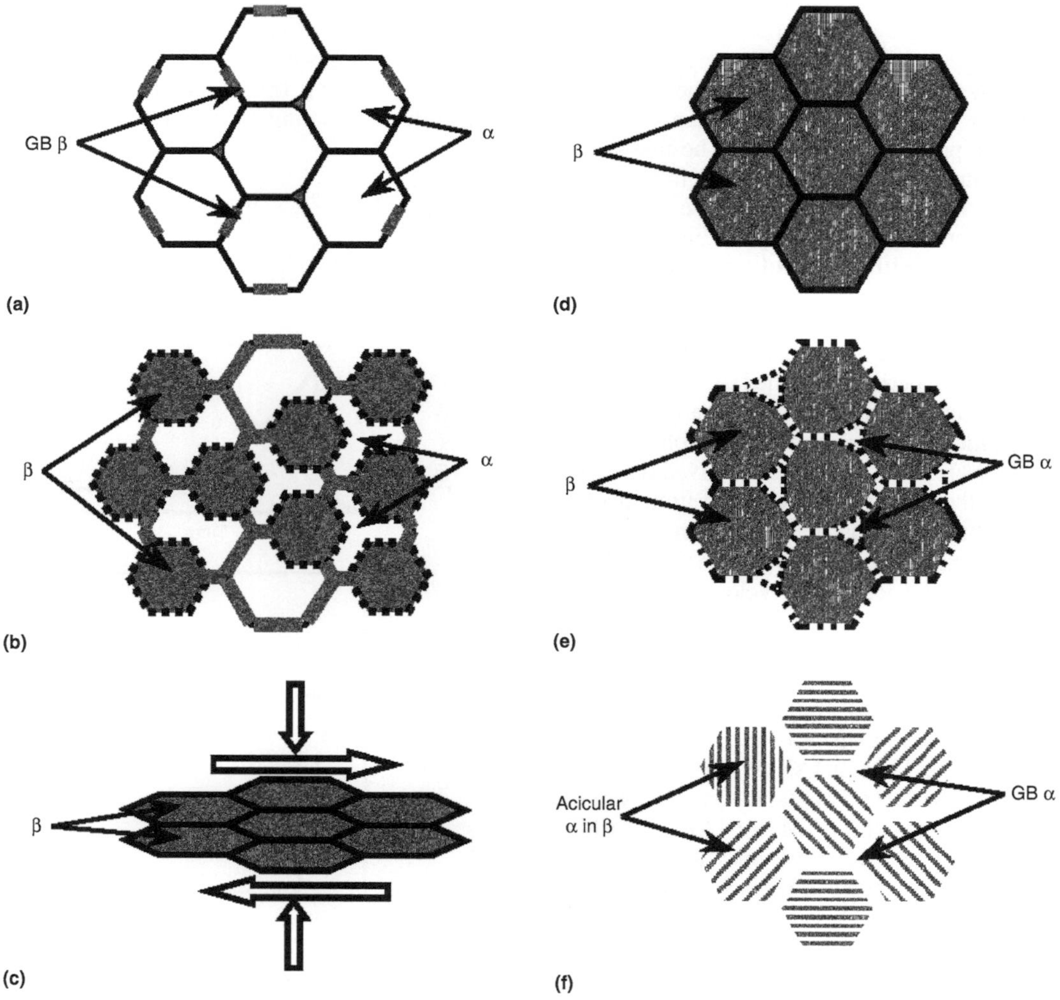


**Fig. 7.13** Schematic of evolution of stir zone microstructures for friction stir welding on Ti-6Al-4V. Schematics (a) through (f) correspond to positions "a" to "f" in Fig. 7.12. (a) Initial base microstructure. α and grain-boundary (GB) β. (b) During heating, GB β grows to consume α. (c) At peak temperature, all β, which undergoes shear and compressive deformation. (d) During/after deformation, β likely undergoes dynamic recovery to static recovery to coarsening. (e) On cooling, α nucleates at triple points to GBs (with Burgers orientation relation). (f) Final microstructure: GB a with fine acicular α in β

occurred via a diffusional process. Cooling rates of ~500 °C/s (900 °F/s) are required to produce α′ martensite in Ti-6Al-4V alloys (Ref 6). Nucleation and growth of the α phase during cooling likely occurred by a well-established mechanism. After sufficient undercooling to the temperature corresponding to point "e" (Fig. 7.12b and c), the α phase nucleated at triple junctions and boundaries of the prior-β grains with a low-energy orientation relation (OR) with one of the prior-β grains (Fig. 7.13e). Subsequently, a series of continuous films of α phase grew to cover the prior-β grain boundaries. The continuous nature of the grain-boundary α phase suggests that the transformation occurred after deformation ceased. With further cooling, acicular α grew from the grain-boundary α into the neighboring prior-β grains, with the interface again defined by the Burgers OR. The α grew as parallel lamellae or colonies with up to twelve different variants possible within a prior-β grain (Ref 1). The fine size of the α lamellae seen here resulted from a rapid cooling rate relative to the coarse lamellar α seen in furnace-cooled samples (Fig. 7.13f).

**Microstructural Evolution in the TMAZ and HAZ.** Because strains in the TMAZ appear too small to cause grain refinement by dynamic restoration processes, and because no strain was experienced in the HAZ of FSWs, the microstructural evolution was mainly dependent on the local thermal history. Published CCT diagrams for Ti-6Al-4V were derived for supratransus thermal cycles and are of limited value for subtransus thermal treatments. Consequently, microstructural evolution in the TMAZ and HAZ of friction stir welds on Ti-6Al-4V alloys can be rationalized with the aid of the appropriate phase diagrams along with knowledge of the local thermal cycles. Figure 7.14 contains schematics similar to those of Fig. 7.12 that can be used to discuss microstructural evolution in the TMAZ/HAZ of the FSWs on Ti-6Al-4V. Figure 7.15 contains schematics of the microstructures of the TMAZ/HAZ in a fashion similar to Fig. 7.13. Positions "a" and "b" in Fig. 7.14 correspond to the schematics in Fig. 7.15(a) and (b).

Data on the thermal cycles experienced in the TMAZ/HAZ were presented previously. Recall that peak temperatures in the TMAZ/HAZ fell below the β-transus temperature. Consequently, complete transformation to β did not occur. The phase balance in the TMAZ/HAZ apparently varied with distance from the weld centerline in accord with the local thermal cycle experienced (Fig. 7.14b). Consistent with the schematic phase diagram for the Ti-Al-V ternary system (Fig. 7.14c), the volume fraction of β in the TMAZ/HAZ increased with decreasing distance from the edge of the stir zone (i.e., increasing peak temperature). Conversely, the fraction of retained (untransformed) α increased with distance from the boundary.

Thermocouple measurements and microstructural evidence revealed that regions corresponding to the HAZ experienced peak temperatures well below the β-transus temperature, resulting in some growth of the β phase (at the expense of the α phase) during heating (Fig. 7.15a). Thermocouple measurements and microstructural observations also indicated that regions corresponding to the TMAZ underwent peak temperatures just below the β-transus temperature, resulting in further growth of the β phase during heating (Fig. 7.15a). Pockets of

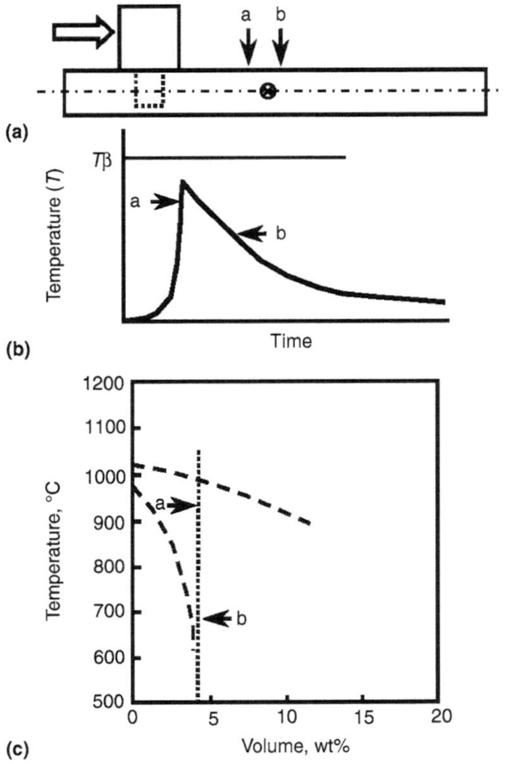


**Fig. 7.14** Schematics of (a) tool position vs. time, (b) thermal cycle with superimposed continuous cooling transformation curve, and (c) pseudobinary phase diagram. Positions "a" and "b" on the diagrams correspond to Fig. 7.15(a) and (b) and are used to describe microstructural evolution in the heat-and-deformation-affected zone/heat-affected zone for friction stir welding on Ti-6Al-4V.

untransformed α phase remained in the microstructure in both the TMAZ and HAZ, and the β phase present at peak temperatures subsequently transformed to α along prior-β grain boundaries, with acicular α phase growing into prior-β grains on cooling in a manner similar to that described previously (Fig. 7.15b). Unlike the continuous grain-boundary α phase seen in the stir zone, the grain-boundary α in the TMAZ took on a globular or blocky form. This observation suggests that the transformation to grain-boundary α may have occurred concurrent with deformation, or that the blocky α phase formed by sympathetic nucleation (Ref 38).

**Hardness and Tensile Properties.** Recall that an increased hardness was observed in the HAZ. This trend was opposite to that found in precipitation-hardened and/or cold-worked aluminum alloys, which exhibit a large drop in hardness in the HAZ due to overaging or recrystallization, respectively. The exact reason for the increase in HAZ hardness here is not known; however, it may have resulted from cold working of the HAZ during FSW and/or from straining during cooling due to coefficient of thermal expansion differences between the α and β phases (Ref 39). Another possibility may involve precipitation of secondary α phase within the β in the TMAZ (Ref 40). Further work is needed to discern the cause.

Note that while the average hardness of the stir zone was nearly identical to that of the base metal, the point-to-point variation in hardness was much smaller for the stir zone region than for the base metal. The smaller variations of the stir zone apparently derived from the greater local uniformity of the microstructure relative to the base metal. For example, indents in the base metal may have encountered different amounts of α and β phase depending on location (to give different hardness values), while indents in the stir zone always sampled the same amounts of each phase.

Consistent with the increased HAZ hardness, welded samples did not fail in the HAZ as a result of tensile testing, as occurs in FSW of aluminum alloys. Rather, welded samples were found to fail in regions of the reduced section corresponding to the base metal. Interestingly, results of tensile testing of the weld samples suggested apparent yield and tensile strength joint efficiencies in excess of 100%. Joint efficiencies greater than 100% result from strain localization, owing to the different microstructures (and thus tensile properties) across the gage length, and are misleading. More specifically, certain regions of the gage length may begin to deform, while other regions with greater hardness and yield strength (for example, the stir zone and TMAZ) do not. Consequently, the deforming regions must be pulled to greater stress values to achieve a given offset strain, thereby giving artificially high yield stress values. In this case, the base-metal region apparently was the weak link among the various regions. Finally, note that properties of the welded regions of the samples were not directly measured here, because failure occurred in the base metal. Testing of miniscale samples taken completely from a given weld region would be needed to directly determine the properties of each region.

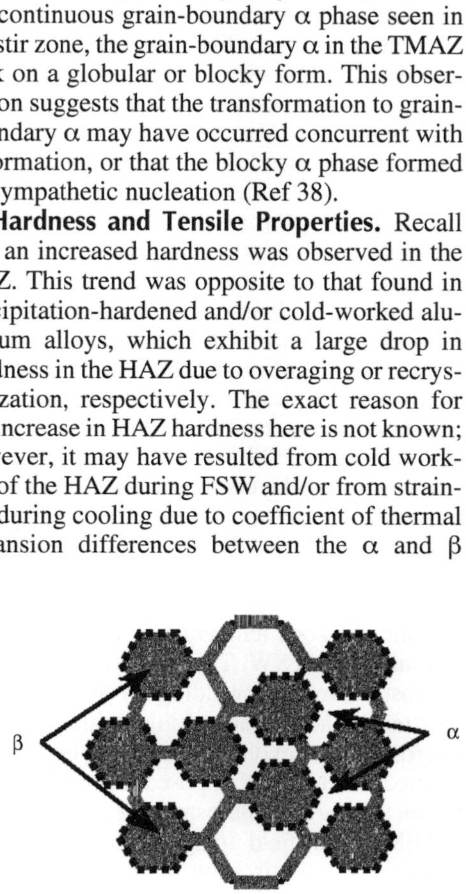


(a)

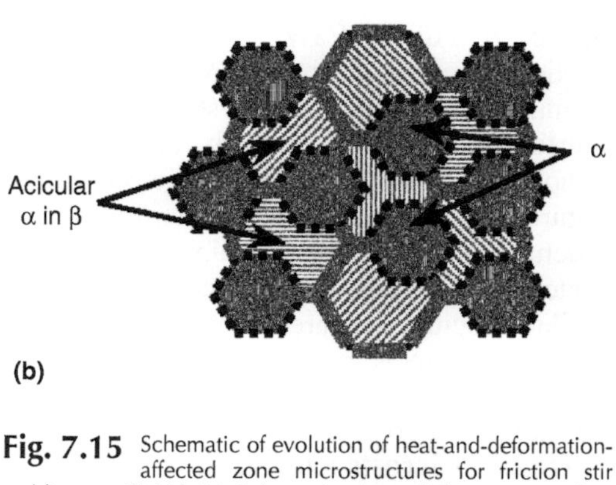


(b)

**Fig. 7.15** Schematic of evolution of heat-and-deformation-affected zone microstructures for friction stir welding on Ti-6Al-4V. Schematics (a) and (b) correspond to positions "a" and "b" in Fig. 7.14. (a) During heating, grain-boundary β grows to consume α. (b) Final microstructure: remnant α with fine acicular α in β

### *7.4.5 Summary and Conclusions*

Extrapolations of temperature measurements from the tool indicated that the temperatures at the tool shoulder were at least 1045 °C (1910 °F) and may have exceeded 1150 °C (2100 °F). To a first-order approximation, the top of the stir zone and tool shoulder were at the same temperature during steady state. Cooling rates of the workpiece were estimated at ~40 °C/s (70 °F/s).

Several microstructurally distinct regions were observed in sections of the FSWs, including the stir zone, TMAZ, and HAZ. Temperature estimates and microstructural observations suggested that peak temperatures experienced in the stir zone exceeded the β-transus temperature. The stir zone or nugget contained α phase outlining the prior-β grains, with fine acicular α phase in a β matrix. The β decomposition in the stir zone occurred by a nucleation-and-growth mechanism involving a diffusional transformation. Consistent with CCT diagrams and the slow cooling rate, no martensitic α′ phase was observed in the stir zone. Temperature measurements and microstructural observations indicated that peak temperatures experienced in the TMAZ and HAZ did not exceed the β-transus temperature. The volume fraction of β phase in the TMAZ and HAZ increased with decreasing distance from the edge of the stir zone. Grains in the TMAZ were elongated in a direction parallel to the stir zone/TMAZ boundary. In accord with the local subtransus thermal cycle, the microstructure of the TMAZ and HAZ contained remnant α phase along with α phase outlining the prior-β grains and fine acicular α phase in a β matrix. The β decomposition in the TMAZ and HAZ also occurred by a nucleation-and-growth mechanism.

The microhardness traverse revealed an increase in hardness from approximately 340 VHN in the base metal and stir zone to 370 VHN in the HAZ. The welds exhibited 100% joint efficiency with respect to both yield and tensile strength, and the average elongation to failure for the weld samples was identical to that for the base metal. Failure of the weld tensile samples occurred in the gage section in regions corresponding to the base metal.

Comparison of temperature and strain-rate estimates with published hot working diagrams suggests that FSW of Ti-6Al-4V may involve ASB formation in the stir zone. Further work is needed to unambiguously determine how the grain structure evolves in the stir zone of these welds. Initial results support the feasibility of FSW for Ti-6Al-4V.

## 7.5 Characterization of FSW Plates of Mill-Annealed and Beta-Annealed Ti-6Al-4V

A characterization study of FSW of mill-annealed and beta-annealed Ti-6Al-4V has been reported by Ramirez and Juhas (Ref 40). Results from that study are summarized in this section. The FSWs were made on 6 mm (0.24 in.) thick plates of Ti-6Al-4V, using parameters identical to those described in the previous section. Plates with two starting heat treat conditions were examined: mill annealed and β annealed. After welding, microstructures of the samples were characterized using light optical microscopy (LOM), scanning electron microscopy (SEM), and transmission electron microscopy (TEM).

**Base-Metal Microstructures.** The mill-annealed base-metal sample exhibited a bimodal microstructure, with bands of α grains and colonies of transformed β. On the other hand, the β-annealed microstructure was composed of large prior-β grains decorated with grain-boundary α. The grain interiors were characterized by a lamellar structure of α + β colonies.

**Stir Zone Microstructures.** The microstructure of the bulk of the stir zone for welds with both starting heat treat conditions was very similar. They exhibited small (~10 μm) prior-β grains with thin layers of grain-boundary α and fine α + β colonies in the grain interior. Grain growth was reportedly limited by the severe deformation and short dwell time near peak temperatures. These microstructures suggested that the stir zone temperatures exceeded the β transus during FSW. The similarity in stir zone microstructure after welding for the two different starting heat treat conditions indicated that microstructural evolution depended on the thermomechanical cycle imposed during FSW and not on the starting microstructure. Figure 7.16 is a TEM bright-field image of an equiaxed α grain from the stir zone of the mill-annealed material. The low dislocation density of this grain suggested that dynamic recrystallization had occurred during FSW.

Microstructures near the TMAZ were also examined. A region near the stir zone/TMAZ interface, called the near-stir zone by the authors, exhibited a distinctive microstructural feature in welds made on both starting microstructures. Small, equiaxed grains of α, approximately 1 μm in size, were reported. Again, their similar structure appeared to indicate that formation of the local microstructure was dependent on the thermomechanical cycle and not on the starting microstructure.

**TMAZ Microstructures.** The microstructures of the TMAZ were somewhat different for the two different starting materials, although the TMAZ for both materials contained fine, equiaxed grains of α phase. In the TMAZ of the mill-annealed weld, the regions of remnant α were thought to have undergone recrystallization, leading to the small size of the grains as a result of the thermomechanical treatment. For the β-annealed sample, the α lamellae within the α + β colonies were reported to have dynamically recrystallized as a result of globularization. For both starting materials, the peak temperature did not exceed the β-transus temperature for the TMAZ region.

Evidence for the formation of acicular particles within larger β grains of the TMAZ of the weld on the β-annealed sample was also reported. Figure 7.17 includes a TEM bright-field image (Fig. 7.17a), a dark-field image (Fig. 7.17b), and a selected area diffraction pattern (Fig. 7.17c) for the grain containing the secondary phase. The acicular precipitates formed an orientation relationship (Burgers OR) with the matrix and were identified as either hexagonal α′ martensite or secondary α. Differentiation between the two phases was not possible.

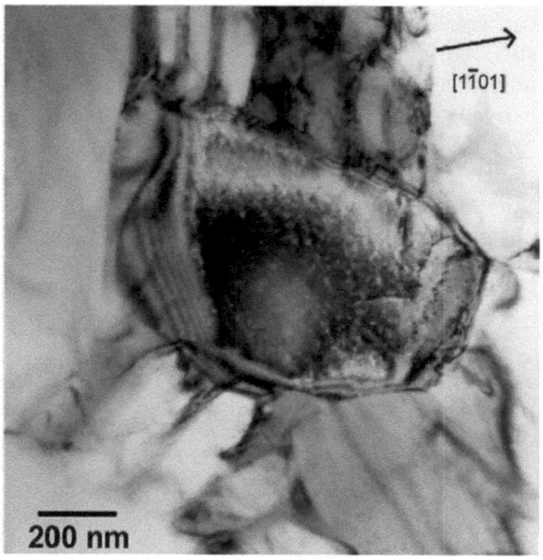


**Fig. 7.16** Transmission electron microscopy bright-field image of an equiaxed alpha particle in the stir zone of the mill-annealed material

## 7.6 FSW of Ti-15V-3Cr-3Al-3Sn Sheet

A study of FSW on sheets of Ti-15-3 has been completed by Lienert (Ref 41). Details of the work are reported in this section. Results of this work are also discussed in a later section involving a comparative study of FSW of different titanium sheets. The composition of the as-received alloy is given in Table 7.7. The material was hot rolled and subsequently cold rolled to a final thickness of ~2 mm (0.08 in.). Following cold rolling, the alloy was annealed to produce a recrystallized microstructure.

Friction stir welds were produced using tools machined from a W-25%Re alloy. A tool with a shoulder diameter of 14 mm ($^{9}/_{16}$ in.) and a pin length of ~/.9 mm (0.075 in.) was used to produce all of the welds discussed here. No threads or other profiles were used on the pin. Tool plunging was completed under displacement control. The tool was maintained at a forward tilt angle of 1° for welding, and the welds were run under load control of the axial ($z$) force. The welds were produced with a tool rotation rate of 200 rpm and a travel rate of 100 mm/min (4.0 in./min), using an axial load of either 9.8 or 10.7

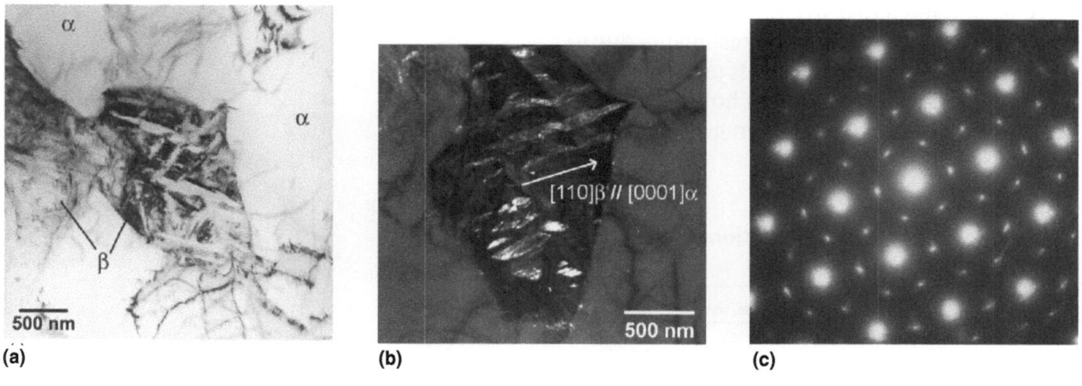


**Fig. 7.17** Transmission electron microscopy (a) bright-field image and (b) dark-field image indicating acicular secondary alpha within large beta grains in the thermomechanically affected zone of the mill-annealed material. (c) Selected area diffraction pattern indicating the Burgers orientation relation between the beta and alpha phases

kN (2200 or 2400 lbf). Tool torque and loads were recorded during welding. To protect the Ti-15-3 alloy workpiece and the tungsten alloy tool, a clear acrylic glass inert gas box with a sliding top that traveled with the tool was fabricated and placed over the entire work area for welding. Tool wear and deformation were monitored before and after each weld by measurements using an optical comparator.

### *7.6.1 Compositions and Microstructural Characterization*

Figure 7.18 is an optical micrograph of the Ti-15-3 base metal. It displays equiaxed grains of beta phase, with an average size ranging from 50 to 100 μm. No evidence of alpha phase can be discerned with optical microscopy. The composition of the base metal is given in Table 7.7 in weight percent. The composition is close to the nominal 15% V, 3% Al, 3% Sn, and 3% Cr composition. The alloy also contains small amounts of iron and oxygen.

Figure 7.19 is a photograph of the top surface of a weld. The surface was clean and relatively free from oxide, indicating that the inert gas box was successful in protecting the workpiece from atmospheric contamination. This statement is corroborated by examining the weld composition data in Table 7.7. No difference in oxygen content was found in the stir zone relative to the base metal, indicating no significant pickup of contaminants from the atmosphere.

Figure 7.20 is an optical macrograph of a transverse section from one of the welds. The weld showed full penetration and no defects. Optical micrographs of the interface between the stir zone and the TMAZ are shown in Fig. 7.21(a) and (b). The stir zone is at the upper left in both micrographs. The grains of the stir zone have been refined to a size of approximately 10 to 20 μm as a result of FSW, presumably due to dynamic recovery (bcc materials tend to recover rather than recrystallize during hot deformation due to their high SFE). The average grain size in the TMAZ was slightly larger than that of the base metal, indicating limited coarsening during welding. Grains of the TMAZ were elongated parallel to the stir zone/TMAZ boundary as a result of material flow. No alpha phase was observed in the stir zone using optical microscopy, and no tungsten was found in the stir zone using energy-dispersive spectroscopy.

Additional optical micrographs of the stir zone region are shown in Fig. 7.22(a) and (b). The microstructure of the flashing at the top edge of the stir zone is shown in the optical micrograph of Fig. 7.22(a). Note the very fine grain size that apparently results from the large strains and high strain rates experienced with the material in direct contact with the shoulder surface. The microstructure of the bottom of the stir zone is shown in Fig. 7.22(b). A region of unrefined grains approximately 30 μm across can be seen at the bottom surface.

After welding, selected samples were aged for 8 h at 635 °C (1175 °F) under vacuum. Microstructures of the base metal and stir zone of aged samples are shown in Fig. 7.23(a) and

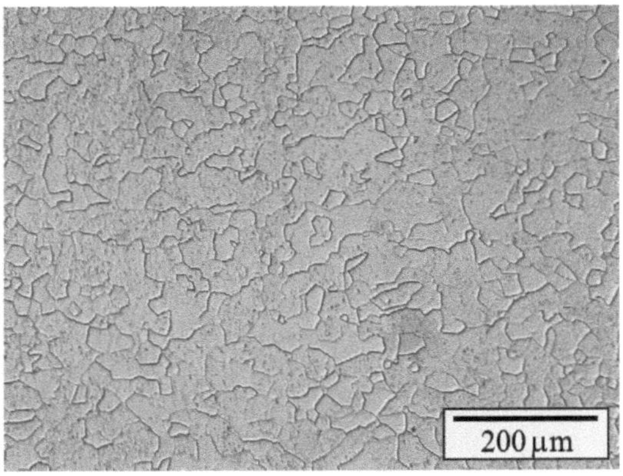


**Fig. 7.18** Optical micrograph of the Ti-15-3 base metal

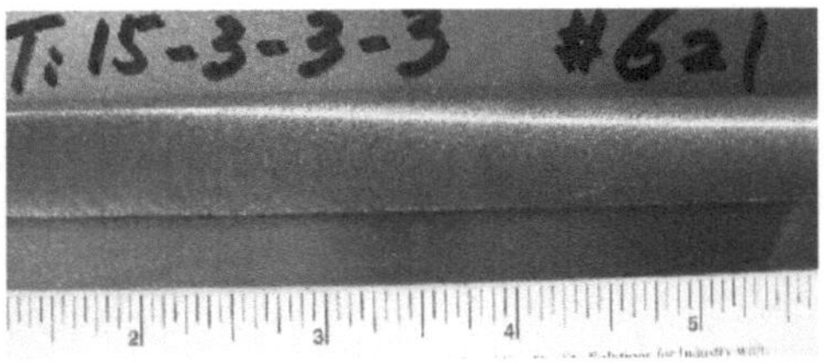


**Fig. 7.19** Photograph of the top surface of a friction stir weld on Ti-15-3. Note the absence of surface oxidation.

**Table 7.7 Base-metal and friction stir weld (FSW) compositions for Ti-15-3**

| | Composition, wt% | | | | | |
|---|---|---|---|---|---|---|
| Component | V | Al | Sn | Cr | Fe | O |
| Base metal | 15.5 | 3.19 | 2.99 | 2.97 | 0.10 | 0.12 |
| FSW | 15.6 | 3.14 | 3.00 | 3.02 | 0.74 | 0.12 |

(b). Aging in this temperature range results in precipitation of alpha phase on grain boundaries and throughout grain interiors (Ref 42–44). A nearly continuous film of alpha phase can be seen along grain boundaries of the base metal in Fig. 7.23(a). The alpha phase nucleated along the boundaries with an orientation relation with respect to one of the grains forming the boundary. Widmanstätten alpha phase can also be seen in some grain interiors. Aging also creates a continuous grain-boundary film of alpha phase along beta-phase grain boundaries of the stir zone, as seen in Fig. 7.23(b).

### *7.6.2 Microhardness and Tensile Properties*

Microhardness profiles of a weld are plotted in Fig. 7.24. The spatial limits of the stir zone/TMAZ boundary and the TMAZ/base-metal boundary are indicated by vertical lines. The nominal hardness of the base metal was

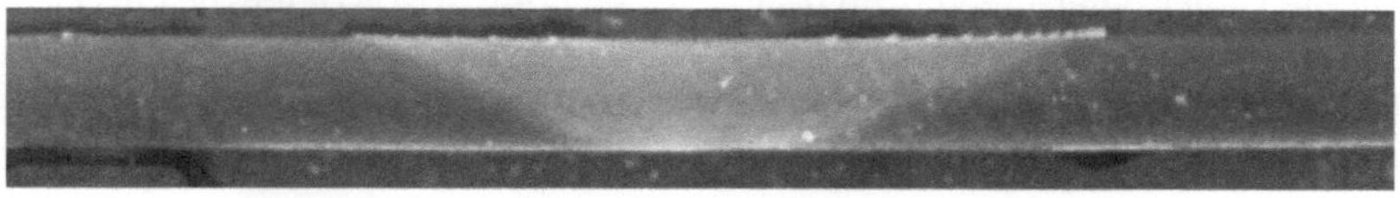

**Fig. 7.20** Optical macrograph of a transverse section of a friction stir weld on Ti-15-3

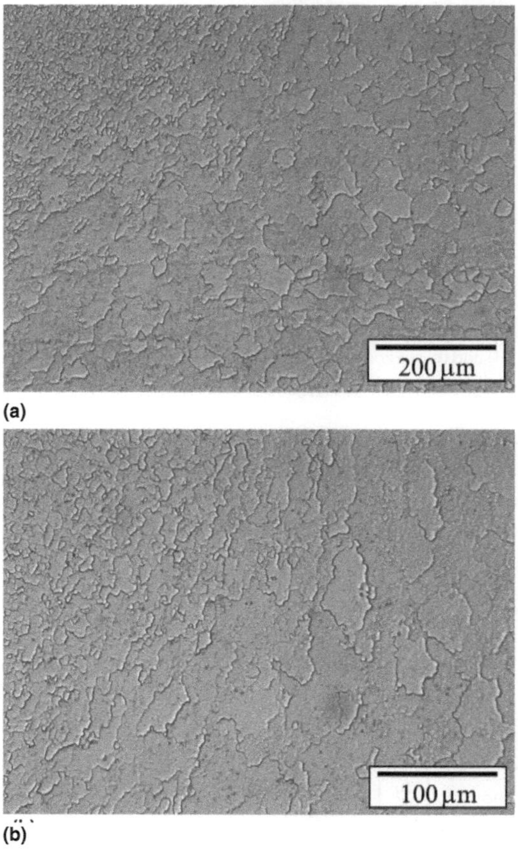


**Fig. 7.21** Optical micrographs of the stir zone/heat-and-deformation-affected zone boundary of a friction stir weld on Ti-15-3

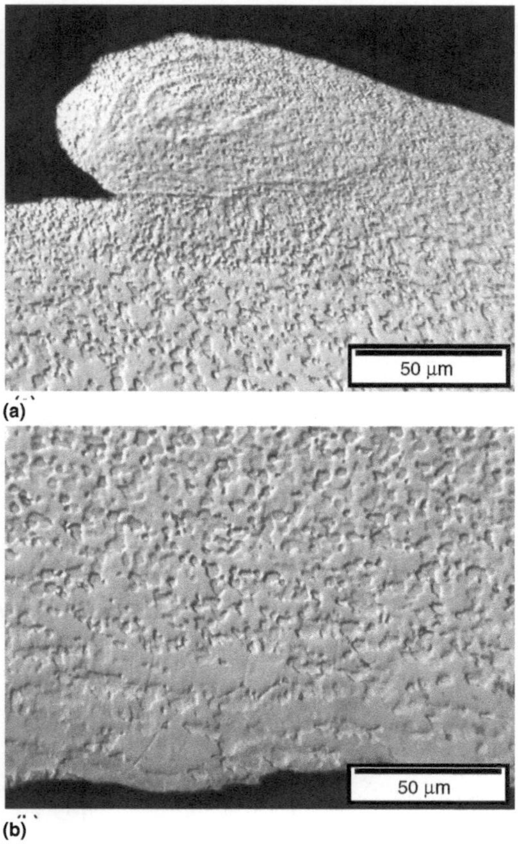


**Fig. 7.22** Optical micrographs of the stir zone of a friction stir weld on Ti-15-3. (a) Top surface. Note the fine grain size along the top surface. (b) Bottom surface. Note the lack of grain refinement near the bottom surface.

~260 VHN. Hardness through the weld region varied between approximately 240 and 280 VHN, with an average hardness of approximately 260 VHN. A slight increase in hardness in the stir zone may be inferred from the results. The increased hardness in the stir zone may have stemmed from the locally refined grain size through the Hall-Petch relation.

Figure 7.25 is a comparison of the tensile results between the base metal and a representative weld sample. These results are also summarized in Table 7.8. The welds displayed higher yield and tensile strength relative to the base metal. The average values determined from four tests were 817 MPa (118.5 ksi) yield strength, 822 MPa (119.2 ksi) tensile strength, and 6.4% elongation. All of the welds failed in locations outside the weld region, indicating the absence of weld defects, as shown in Fig. 7.26.

Microhardness results for welds that were aged after welding (postweld heat treated) are shown in Fig. 7.27. The spatial limits of the stir zone/TMAZ boundary and the TMAZ/base-metal boundary are indicated by vertical lines. The aging treatment produced little change in the hardness response in the base-metal and weld regions relative to the nonaged samples. A slight increase in hardness can be seen in the stir zone region. The average hardness of the base metal remained at ~260 VHN.

A comparison of tensile curves of as-received base metal, aged base metal, and aged weld samples is shown in the plot in Fig. 7.28. Average tensile results for each of these conditions are sum-

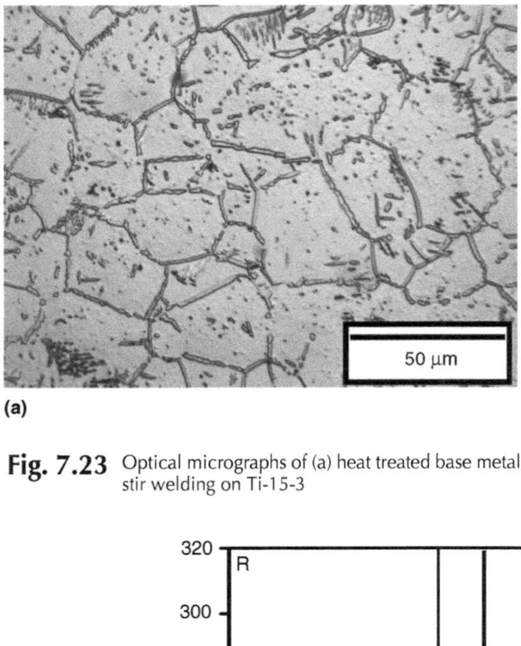


(a)

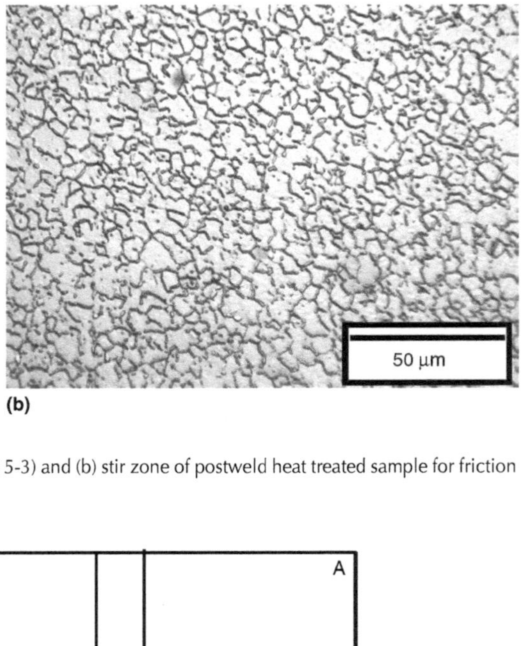


(b)

**Fig. 7.23** Optical micrographs of (a) heat treated base metal (Ti-15-3) and (b) stir zone of postweld heat treated sample for friction stir welding on Ti-15-3

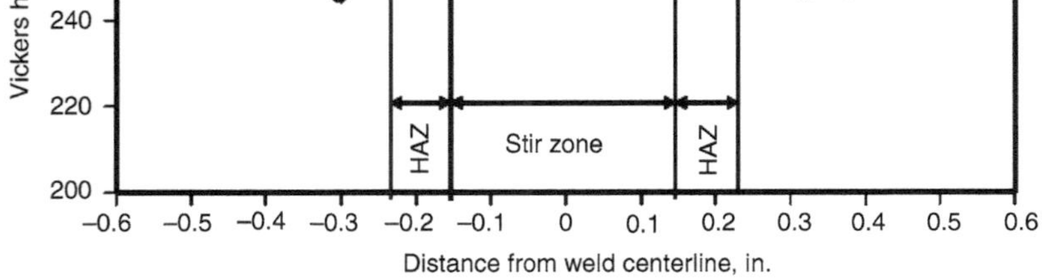


**Fig. 7.24** Microhardness results for as-welded friction stir weld on Ti-15-3. HAZ, heat-affected zone; R, retreating; A, advancing

marized in Table 7.8. The aged base metal evidenced greater yield and tensile strength relative to the as-received base metal but lower strain to failure. Average values for the aged base metal were 807 MPa (117.0 ksi) yield strength, 811.5 MPa (117.7 ksi) tensile strength, and 8.2% elongation. The aged weld sample shown here had a very similar stress-strain response to the aged base metal, with slightly greater percent elongation. Average values for the aged weld specimens were 817 MPa (118.5 ksi) yield strength, 822 MPa (119.2 ksi) tensile strength, and 6.4% elongation. Three of the four weld tensile samples failed outside of the weld region, as shown in Fig. 7.29.

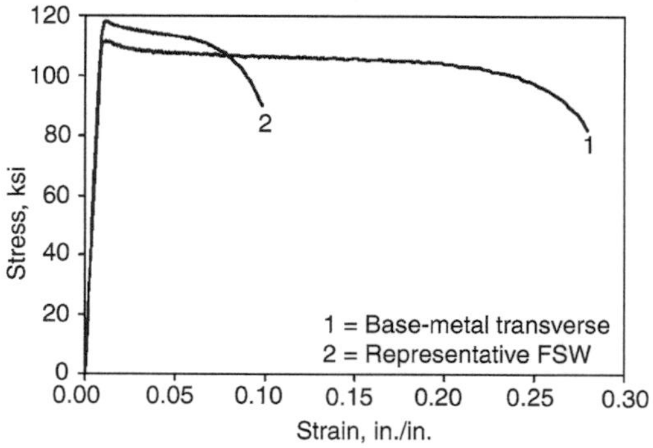


**Fig. 7.25** Stress vs. strain plots for as-received base metal and as-welded friction stir weld (FSW) samples of Ti-15-3

### 7.6.3 Discussion

**Microstructural Evolution in the Stir Zone.** Important aspects of the microstructure of the stir zone include the grain size as well as the volume fraction and distribution of phases. Microstructural evolution in the stir zone is dictated by the thermomechanical cycle imposed during FSW. More specifically, the microstructure develops in accord with the local strain/strain-rate/temperature path. Little is known

**Table 7.8 Tensile properties for base metal and friction stir welds (FSWs) of Ti-15-3**

| Specimen type | 0.2% offset yield strength | | Tensile strength | | Elongation, % |
|---|---|---|---|---|---|
| | MPa | ksi | MPa | ksi | |
| Annealed base metal: longitudinal | 810.2 | 117.5 | 813.6 | 118.0 | 31 |
| Annealed base metal: transverse | 765.3 | 111.0 | 768.8 | 111.5 | 28 |
| Heat treated base metal: transverse | 807 | 117.0 | 811.5 | 117.7 | 8.2 |
| 19 mm (0.75 in.) FSW (average of three) | 728.8 | 105.7 | 768.1 | 111.4 | 5.5 |
| 14.3 mm (0.563 in.) FSW (average of four) | 817 | 118.5 | 822 | 119.2 | 6.4 |
| 14.3 mm (0.563 in.) FSW postweld heat treat (average of three) | 815.7 | 118.3 | 825.3 | 119.7 | 6.0 |

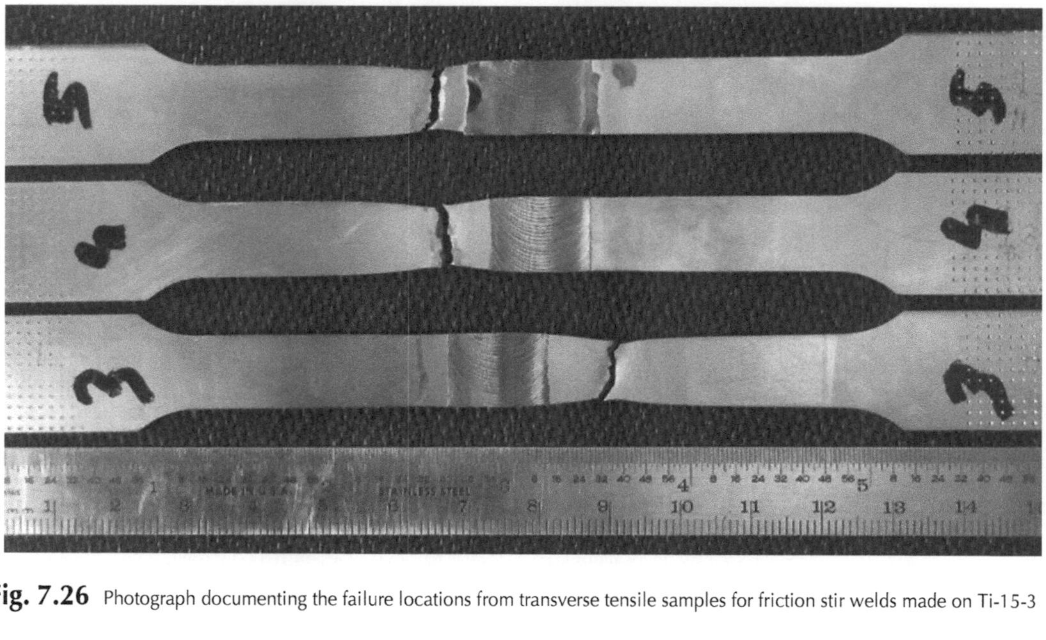

**Fig. 7.26** Photograph documenting the failure locations from transverse tensile samples for friction stir welds made on Ti-15-3

about the temperature cycle experienced in the stir zone for the work reported here, beyond the fact that the tungsten-rhenium tool glowed red during FSW, suggesting temperatures above 1000 °C (1830 °F). Microstructural evidence and heat flow analysis reported for FSW of Ti-6Al-4V indicated that peak temperatures in the stir zone exceeded the beta transus of that alloy (~1000 ° C) (Ref 4). Given that the beta transus of the Ti-15-3 alloy investigated here is ~760 °C (1400 °F), it is not unreasonable to assume that peak temperatures experienced in the stir zone were in excess of the transus temperature.

Refinement of the grain size in the stir zone of the welds produced here occurred as a result of some restorative process. The restoration processes include recovery or recrystallization, and they may occur either statically (during heating after cold deformation), dynamically (during hot deformation), or metadynamically

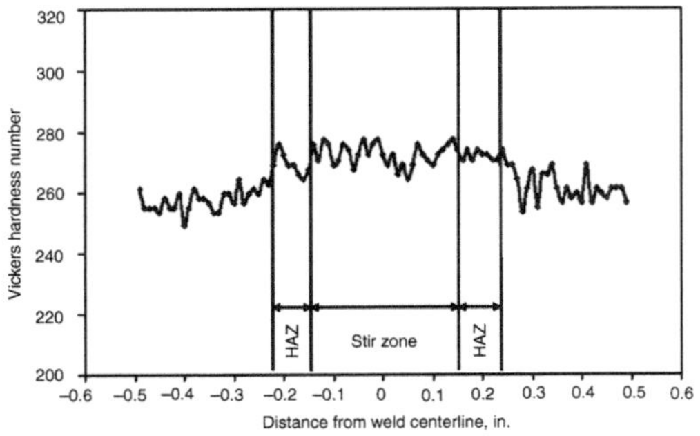


**Fig. 7.27** Microhardness results for postweld heat treated friction stir weld sample on Ti-15-3. HAZ, heat-affected zone

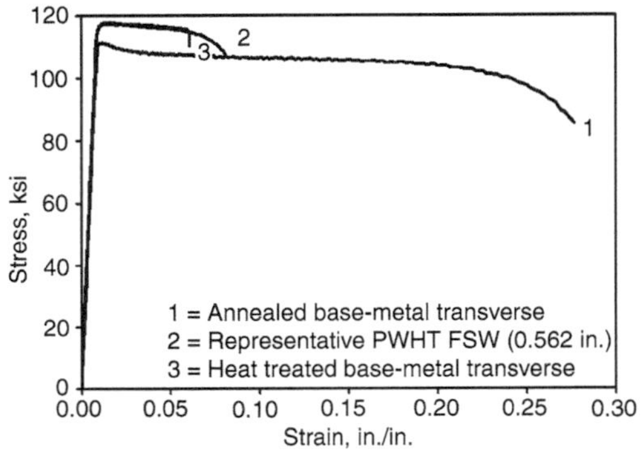


**Fig. 7.28** Stress vs. strain plots for annealed base metal, heat treated base metal, and postweld heat treated (PWHT) friction stir weld (FSW) samples on Ti-15-3

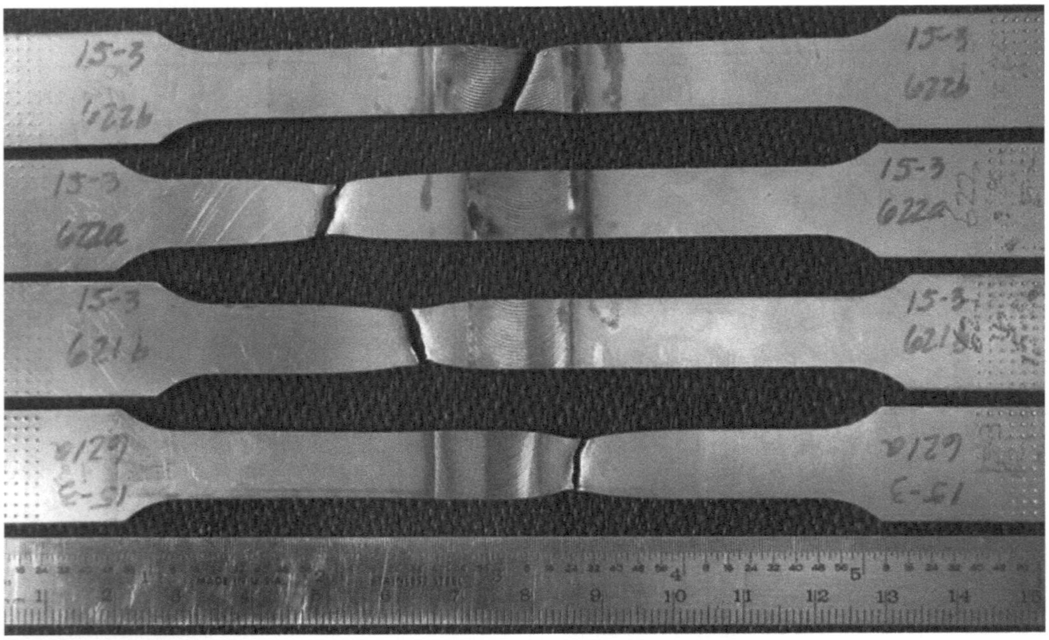

**Fig. 7.29** Photograph documenting the failure locations from transverse tensile samples on Ti-15-3 (postweld heat treated)

(after hot working) (Ref 18). As mentioned previously, hot deformation in the beta-phase field of the Ti-15-3 alloy normally involves dynamic recovery due to the low SFE of the alloy. The apparent activation energy for this process is reported to be very close to that for self-diffusion (Ref 45). However, metadynamic recrystallization may ensue at heavily deformed regions, such as grain boundaries, after deformation has ceased and during subsequent cooling (Ref 45). Hence, the most likely scenario for grain refinement in the stir zone during FSW of the Ti-15-3 alloy involves dynamic recovery during deformation, followed by metadynamic recrystallization.

After FSW, no evidence of alpha phase was observed in the stir zone or TMAZ, using LOM and SEM characterization. However, very fine alpha precipitates too small to resolve using LOM or SEM may exist in the microstructure. Metastable beta alloys, in general, and the Ti-15-3 alloy, in particular, are essentially alpha-beta alloys that are rich in beta stabilizers. They are designed to have sluggish beta decomposition to alpha during cooling from above the beta transus in order to retain a 100% beta microstructure during rapid cooling (Ref 1, 44, 46).

An approximate time-temperature transformation (TTT) diagram for alpha precipitation during reheating of Ti-15-3 has been reported (Ref 44). Assuming that this TTT diagram may be used to estimate cooling transformations, a bound or limit may be determined for cooling rates that would promote alpha precipitation. Cooling rates slower than 0.5 to 1 °C/s (1 to 2 °F/s) through the temperature range of 700 to 500 °C (1300 to 930 °F) would be required to promote formation of any alpha phase. Even slower cooling rates would be required to allow appreciable alpha formation.

**Microstructural Evolution during Postweld Aging.** Aging of the Ti-15-3 alloy can be used to increase strength after processing of the beta-phase microstructure. Aging in the temperature range of 480 to 540 °C (900 to 1000 °F) promotes precipitation of alpha phase. Higher-temperature aging treatments tend to result in precipitation of grain-boundary alpha, while lower-temperature treatments give homogeneous distributions of alpha (at grain boundaries and throughout grain interiors), with better toughness. Two-step heat treats that involve lower-temperature treatments to develop homogeneous nucleation followed by higher-temperature aging to hasten growth rates can also be employed. Aging at too high a temperature allows overaging and a loss of strength and hardness.

In this work, an 8 h aging treatment at 635 °C (1175 °F) was employed after welding to mimic the arc welding study of Becker and Baeslack (Ref 42). This aging treatment resulted in the formation of continuous films of alpha along beta grain boundaries of both the base metal and stir zone. This treatment produced a small increase in yield and tensile strength of the base metal and virtually no increase in strength for the weld samples. Moreover, no discernible change in microhardness was observed in the aged weld samples relative to the as-welded samples. Consequently, the aging treatment used here appears to have been performed at too high a temperature to give any real strength or hardness improvements over the annealed base-metal samples and as-welded samples. The heat treatment used here apparently resulted in overaging. Lower-temperature aging treatments are suggested for better strengths and hardness.

**Mechanical Properties.** One peculiar feature of the stress-strain curves for the samples tested here is the lack of work hardening. In fact, the samples exhibit a slight work-softening effect. The features noted here are consistent with those seen by other researchers (Ref 1, 46) and result from the balance of beta-stabilizing elements (Ref 46). Beta isomorphous stabilizer additions to titanium alloys, such as vanadium, promote low solid-solution strengthening rates but do not form embrittling compounds. On the other hand, beta eutectoid stabilizers, such as chromium, provide greater solid-solution stabilizing rates but tend to promote formation of embrittling eutectoid compounds. The largest alloy addition to the Ti-15-3 alloy is vanadium (15%), a beta isomorphous stabilizer. The lack of work hardening in Ti-15-3 is believed to stem from co-planar slip in bands that widen as strain is increased (Ref 46).

Note that the percent elongation of the weld samples was much lower than that found for the base-metal samples. The percent elongation of defect-free welds normally appears low relative to the base metal, owing to nonuniform elongation throughout the gage length stemming from microstructural gradients created by the welding process. For welded samples, the strain is usually carried by narrow regions of lower strength, such as the TMAZ, and assumptions made concerning uniform elongation throughout the gage length are not valid. Hence, interpretation of the lower elongation experienced by samples produced in this study is complicated by nonuniform elongation.

### *7.6.4 Summary*

Friction stir welds were successfully produced on 2 mm (0.08 in.) thick sheets of the Ti-15-3 alloy without gross defects. No measurable tool wear/deformation or pickup of material from the W-25%Re tool during welding was found. The W-25%Re tool material is suitable for FSW of this alloy. Moreover, no evidence of appreciable atmospheric contamination was observed in the weld area, proving the efficacy of the inert gas box.

The FSW resulted in considerable grain refinement in the stir zone. No evidence for the presence of alpha phase was observed in the TMAZ or stir zones of welds made on the as-received base metal, using LOM and SEM techniques. Aging of the base metal and weld samples resulted in the formation of a nearly continuous film of alpha phase along grain boundaries of the base metal and of the weld stir zone. A slight increase in hardness was observed in the stir zone of welds on as-received material and aged material.

Welds exhibited high tensile joint efficiencies with acceptable ductility. Defect-free welds failed in regions corresponding to the base metal. The FSW of Ti-15-3 is feasible, but more work is needed for a more complete understanding.

## 7.7 Texture of FSWs in Beta 21-S

Microstructures and crystallographic texture of FSWs on Beta 21-S have been reported by Reynolds et al. (Ref 47). The FSWs were produced on 1.6 mm (1/16 in.) thick sheets of Beta 21-S using a tool made from a tungsten alloy. All welds were made in an inert gas box backfilled with argon. Welds were produced at 200 rpm at travel speeds ranging from 0.85 to 5.08 mm/s (0.0335 to 0.2 in./s). After welding, microstructures of the welds were examined using LOM, and textures were examined using orientation imaging microscopy (OIM). The OIM results were given as pole figures. Stir zone pole figures were rotated to align the welding direction with the vertical direction of the figure and the tangent to the pin at the trailing edge parallel to the horizontal direction. This arrangement placed the rotation axis of the tool nearly parallel to the normal direction of the figure.

Defect-free welds were produced with a refined grain size in the stir zone. Average grain sizes in the stir zone decreased with increasing weld travel speed. The OIM was used to determine crystallographic texture at the centerline of the stir zone at the sheet midplane. Base metal {111} and {110} pole figures (Fig. 7.30) as well as stir zone {110} pole figures (Fig. 7.31) were reported.

Stir zone pole figures for the four different welds are presented in Fig. 7.31. All of the stir zone pole figures from the sheet midplane showed similar textures, with {110} pole closely aligned with the normal direction. The pole figures for each of the stir zones could be brought into coincidence with small rotations about the normal directions. Pole figures from planes near the top and bottom of the plate thickness were also similar to that for the midplane for the weld made at the slowest travel speed. This observation suggested that the texture was relatively homogeneous through the sheet thickness. Welds made at the slower travel speeds had a stronger texture than those made at faster travel speeds. Textures for the stir zones of the welds made in this study were shown to closely match those reported for torsion of another bcc metal, tantalum (Ref 48). The torsion axis was closely aligned with the tool rotation axis.

## 7.8 FSW of CP Titanium

Lee and coworkers (Ref 49) have reported on a study of FSW on pure titanium. Although no details were given on the composition of material, it was similar to some type of CP titanium alloy. Plates of CP titanium, 5.6 mm (0.22 in.) in thickness, were joined by FSW using a sintered TiC tool with a water-cooling system. Welds were produced at 1100 rpm at a welding speed of 500 mm/min (20 in./min).

The base metal had equiaxed grains with an average diameter of ~25 μm. Welds were produced with no apparent defects. Optical microscopy revealed that the microstructure of the stir zone was characterized by a high density of deformation twins within the grains. The twin density varied with position relative to the position of the tool shoulder, with denser twins found near the upper part of the weld. The stir zone was reported to have undergone recrystallization.

The TEM results indicated that the base metal had an equiaxed grain structure with a low twin content. The TEM characterization of the stir zone showed a large amount of twin embedded structure, with many grains having a high dislocation density within a network structure. The observation of dislocation walls suggested that recovery was incomplete or continuous in nature. The high dislocation density and presence of dislocation walls indicated that the initial stage of

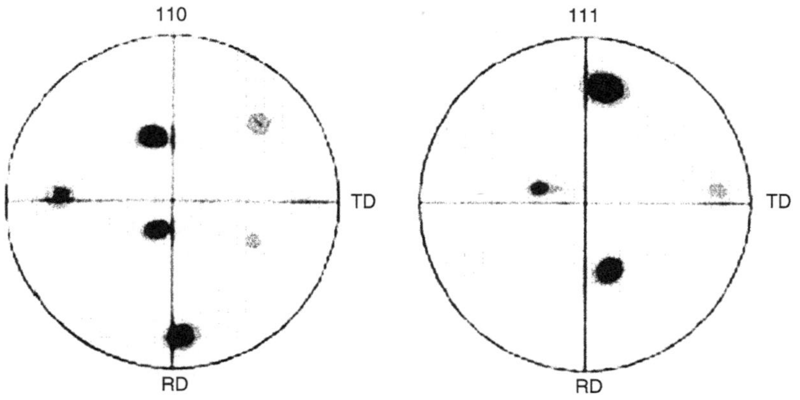


**Fig. 7.30** Base-metal {110} and {111} pole figures from Beta 21-S

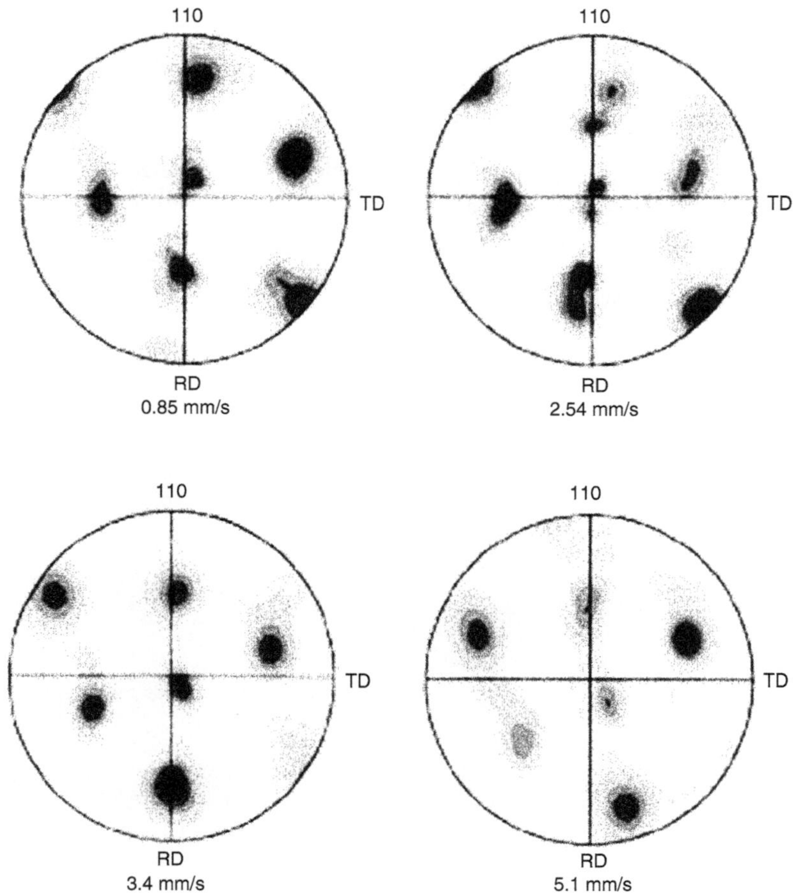


**Fig. 7.31** {110} pole figures from friction stir welds on Beta 21-S

deformation during FSW occurred by slip. However, the observation of twinning suggested that slip subsequently ceased, and further deformation was accommodated by twinning.

The microhardness trace across the weld showed scattered results, with the average hardness similar to that of the base metal. Slight softening of the HAZ due to annealing was reported. Peaks in the hardness data in the stir zone were shown to correspond to regions of higher twin content. The increase in hardness in densely twinned areas was attributed to the Basinski effect. The average tensile strength of the weld samples was 430 MPa (62 ksi) compared with 440 MPa (64 ksi) of the base metal. Elongations of the weld samples averaged 20% versus 25% for the base-metal samples. Fractures were reported to have occurred in the HAZ on the retreating side of the weld.

## 7.9 Comparative FSW Study of Titanium Sheet Alloys

A comparative study of FSW of three different titanium sheet alloys was reported by Lienert (Ref 41). The three alloys were CP titanium (grade 2), Ti-6Al-4V, and Ti-15V-3Cr-3Al-3Sn. All materials were in sheet form and were in the range of 2.1 to 2.3 mm (0.084 to 0.090 in.) thick. Results for the Ti-15-3 alloy have been presented in a previous section. Results for the CP titanium alloy and the Ti-6Al-4V are discussed in this section.

The FSWs were produced using the same parameters and methods described previously in the study on Ti-15-3. Tool plunging was completed under displacement control. The CP titanium welds were run under load control of the axial ($z$) force, while the Ti-6Al-4V welds were run under displacement control. The welds were produced with a tool rotation rate of 200 rpm and a travel rate of 100 mm/min (4.0 in./min).

**Process Results.** Successful welds could not be produced on the Ti-6Al-4V material using load control. The loads required for the CP titanium welds (in load control) were much greater than for the Ti-15-3 welds discussed earlier, despite the same approximate sample thickness. Forward loads for the CP titanium and Ti-6Al-4V were also much greater than for the Ti-15-3 welds. No measurable wear or deformation of the single tool used to run all of the welds was found after total weld lengths of over 9 m (30 ft).

**Compositions and Microstructures.** The compositions of CP titanium alloy and the Ti-6Al-4V alloy are given in Table 7.9. The main alloying elements in the CP alloy were iron and oxygen. A photograph of the top surface of the CP weld is shown in Fig. 7.32. This alloy was very difficult to weld and exhibited a flaky sur-

**Table 7.9 Base-metal compositions for Ti-6Al-4V and commercially pure (CP) titanium**

| | Composition, wt% | | | | | |
|---|---|---|---|---|---|---|
| Component | V | Al | Sn | Cr | Fe | O |
| 6–4 base metal | 6.28 | 3.73 | ... | ... | 0.17 | 0.15 |
| CP base metal | ... | ... | ... | ... | 0.30 | 0.25 |

Fig. 7.32 Photograph of the top surface of a friction stir weld on commercially pure titanium. Note the absence of surface oxidation.

face appearance. An optical micrograph of a transverse weld section is given in Fig. 7.33. The weld surface showed considerable flash and sheet thinning. The base-metal microstructure of the CP titanium alloy is presented in Fig. 7.34. The α-phase grains had intercept lengths between 10 and 40 μm in size and often contained twins. The grains in the stir zone of the CP titanium welds were refined to a size of less than ~5 μm, presumably by dynamic recrystallization (Fig. 7.35). The density of twins found in the stir zone with LOM was very low.

A photograph of the top surface of the Ti-6Al-4V weld is shown in Fig. 7.36. This alloy was very difficult to weld but exhibited a smooth surface. An optical micrograph of a transverse weld section of the Ti-6Al-4V weld is given in Fig. 7.37. A slight lack of penetration is apparent. The base-metal microstructure of the Ti-6Al-4V alloy is presented in Fig. 7.38. The base-metal microstructure was characterized by fine grains of α phase (10 to 30 μm in size) that were slightly flattened, with a nearly continuous distribution of β phase along grain boundaries. An optical micrograph of the stir zone/TMAZ boundary is shown in Fig. 7.39. The grains of the stir zone for this weld were refined to a size of less than ~5 μm, again presumably by dynamic recovery. Although not shown, the TMAZ exhibited regions of remnant

**Fig. 7.33** Optical macrograph of a transverse section of a friction stir weld on commercially pure titanium

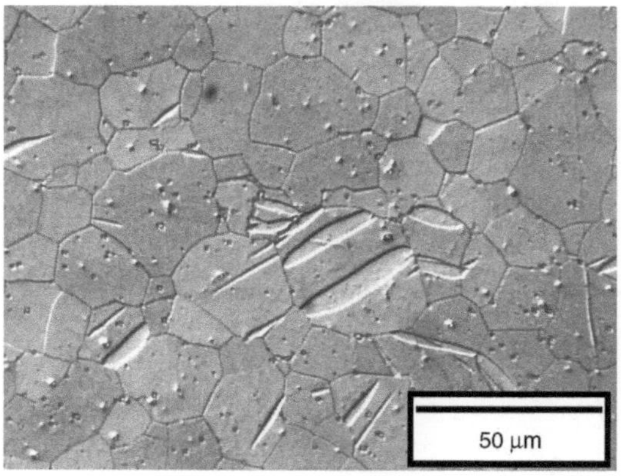


**Fig. 7.34** Optical micrograph of the commercially pure titanium base metal

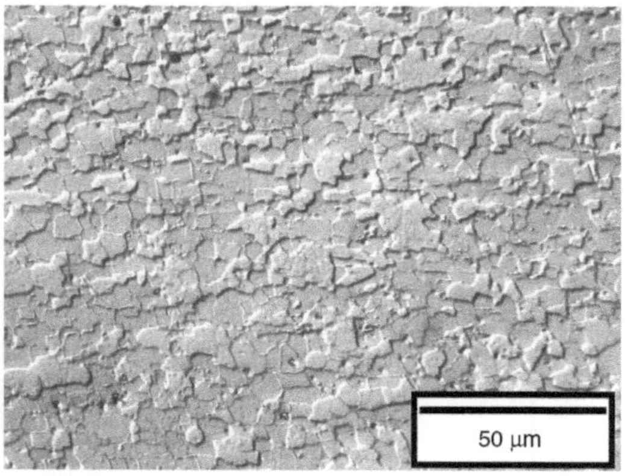


**Fig. 7.35** Optical micrograph of the stir zone of a friction stir weld on commercially pure titanium

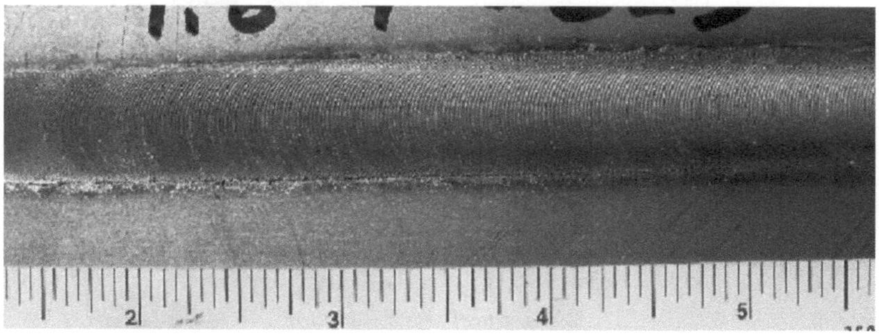

**Fig. 7.36** Photograph of the top surface of a friction stir weld on Ti-6Al-4V. Note the absence of surface oxidation.

α phase with grains of α and grain-boundary β phase.

**Hardness and Tensile Results.** Results of the microhardness traverse across the weld region in the CP titanium weld are shown in Fig. 7.40. There was considerable scatter in the data. The stir zone exhibited increased hardness relative to the average base-metal hardness, most likely due to the refined grain size. A slight loss in hardness in the HAZ can be seen from the

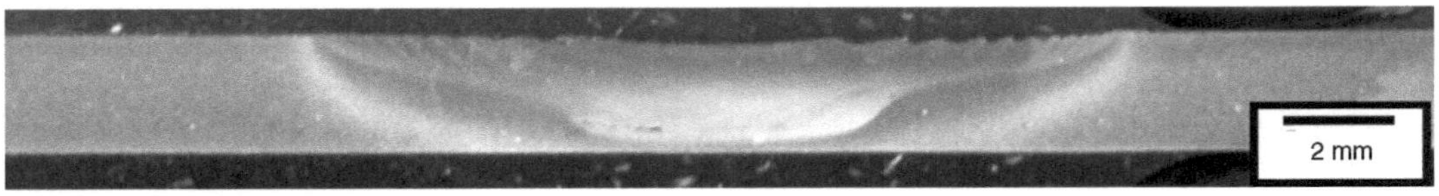


**Fig. 7.37** Optical macrograph of a transverse section of a friction stir weld on Ti-6Al-4V. Note the lack of full penetration.

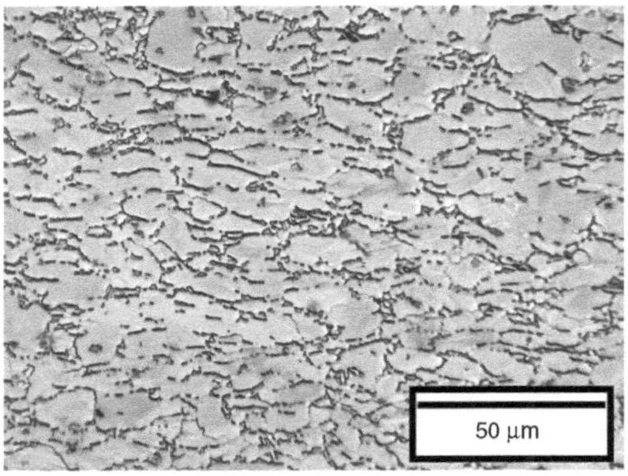


**Fig. 7.38** Optical micrograph of the Ti-6Al-4V base metal

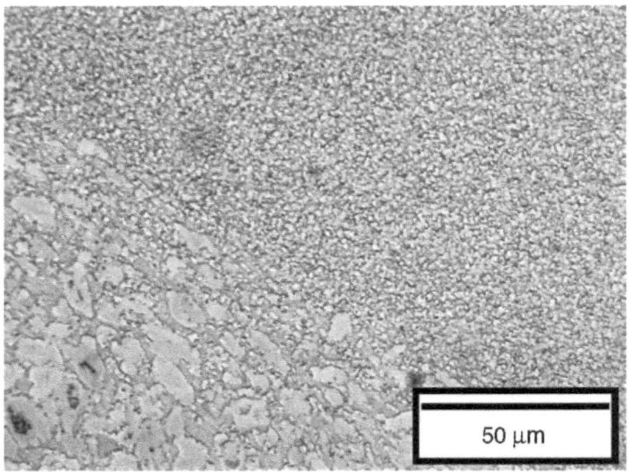


**Fig. 7.39** Optical micrograph of the stir zone/heat-and-deformation-affected zone boundary of a friction stir weld on Ti-6Al-4V

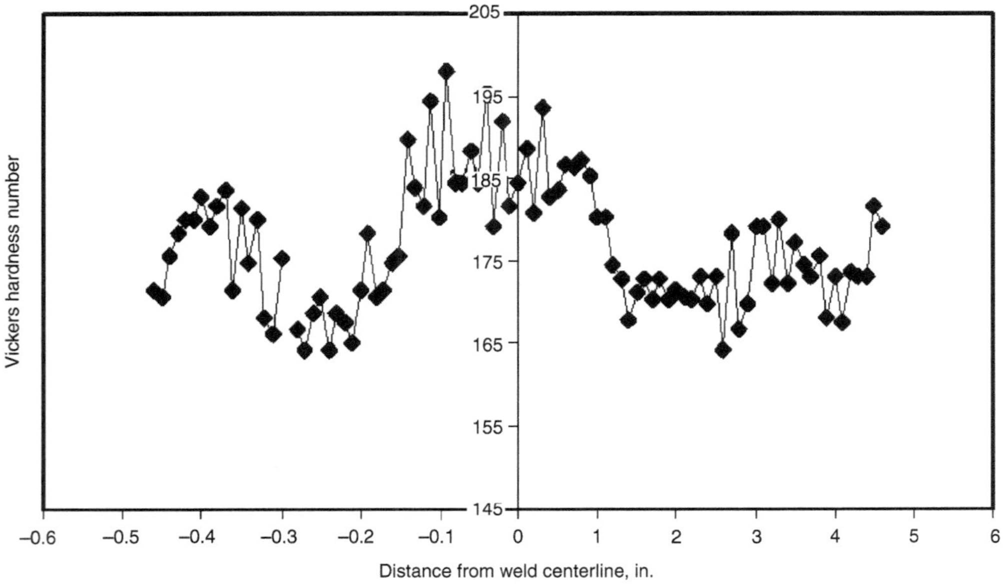


**Fig. 7.40** Microhardness results for as-welded friction stir welds on commercially pure titanium

data. A summary of the tensile results for the CP titanium base metal and welds is given in Table 7.10. The welds had a joint efficiency of ~85% for yield and tensile strength. However, the elongation for the weld samples was much lower than for the base metal. Failures occurred through the stir zone due to excessive sheet thinning.

Results of the microhardness traverse across the weld region for the Ti-6Al-4V weld are shown in Fig. 7.41. Again, there was considerable scatter in the data. The boundaries of the stir zone are indicated by the vertical lines. A large increase in hardness is evident in the stir zone relative to the average base-metal hardness, possibly due to the refined grain size. A summary of the tensile results for the Ti-6Al-4V base metal and welds is given in Table 7.11. The welds had a joint efficiency in excess of 95% for yield and tensile strength. Elongations for the weld samples averaged 4.5%, much lower than the base metal. Three of the four samples tested failed in locations corresponding to the stir zone as a result of the lack of penetration defect.

**Comparison of the Weldability of α, α + β, and β Alloys.** In the comparative study discussed here, three titanium alloys with different compositions and phase balance (α, α + β, β) have been FSWed under nominally identical conditions. Their response to FSW varied considerably. The Ti-15-3 alloy was the easiest to weld, with the largest process window and lowest axial and forward loads. Production of welds without defects was easiest with the Ti-15-3 alloy. In contrast, welds on the Ti-6Al-4V alloy could not be produced using load control and exhibited large forward loads. The CP titanium alloy also produced large forward loads and large axial loads. The ease of welding ranking, from easiest to hardest, was Ti-15-3, Ti-6Al-4V, and CP titanium. However, the reasons

**Table 7.10 Tensile properties of base metal and friction stir welds (FSWs) on commercially pure titanium**

| | 0.2% offset yield strength | | Tensile strength | | |
|---|---|---|---|---|---|
| Specimen type | MPa | ksi | MPa | ksi | Elongation, % |
| Base-metal transverse | 376.5 | 54.6 | 453.0 | 65.7 | 22 |
| FSW as-welded (avg of four) | 319.2 | 46.3 | 393.0 | 57.0 | 2.6 |

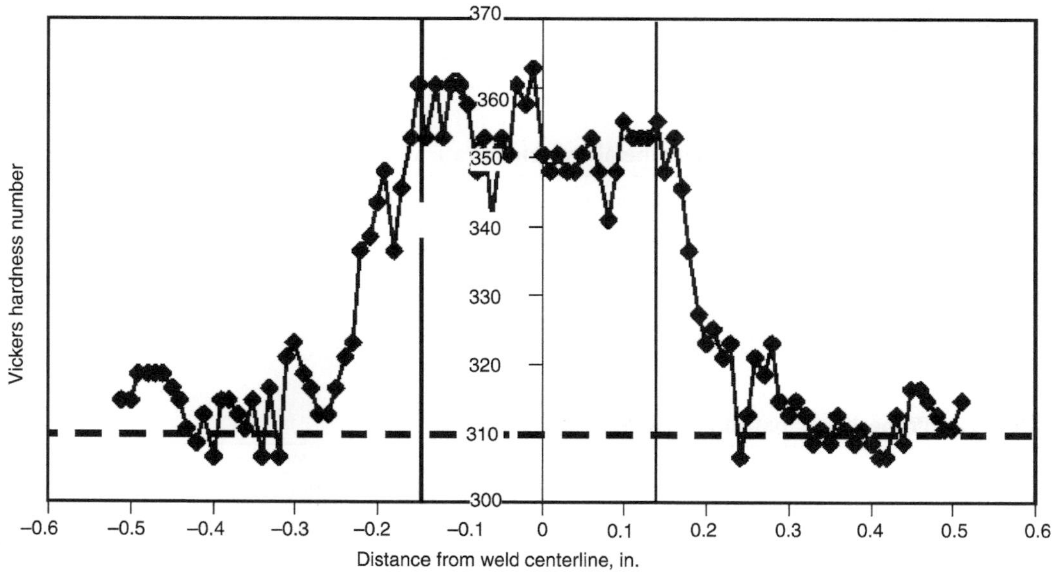


**Fig. 7.41** Microhardness results for as-welded friction stir welds on Ti-6Al-4V

for the differences in ease of welding are not clear.

A comparison of thermophysical and thermomechanical properties for all three alloys is given in Table 7.12. Values for thermal conductivity, heat capacity, density, thermal diffusivity, flow stress (800 °C, or 1470 °F, and strain rate =10 $s^{-1}$) and beta-transus temperature are given. Also recall that the CP titanium alloy has an hcp crystal structure, the Ti-15-3 alloy has a bcc structure, and the Ti-6Al-4V alloy has a dual hcp/bcc structure. Furthermore, it is important to note that the total alloy content increases from CP titanium to Ti-6Al-4V to Ti-15-3. Study of the various properties suggests that ease of welding may be dependent on crystal structure, thermal conductivity, and beta-transus temperature. However, much more work is needed to understand differences in the FSW response of the three alloys.

One key factor was not investigated here. Several researchers have reported in presentations that there may be an interaction between the tungsten-rhenium tool and some titanium alloys during FSW that makes welding difficult. At present, no quantitative explanation has been offered for this interaction. Nonetheless, the choice of tool material in the current study may have had unintended consequences on the results.

## 7.10 Summary

To summarize, FSW of titanium alloys appears feasible and promising despite the limited work to date, and further studies are warranted. Initial results indicate that acceptable tensile properties can be achieved. Microstructures of the various weld regions evolve in accord with the local thermomechanical cycle, and phase diagrams, CCT curves, and hot working data are useful in rationalizing microstructural evolution.

Despite the successes to date, more work is needed for a complete understanding of FSW of titanium alloys. Development of new tool materials/designs is needed to increase tool life to a point where FSW of titanium alloys is cost-competitive with other joining processes. Moreover, an explanation for tool/workpiece material interactions is required. Better designs for FSW machines purpose-built for processing of titanium and higher-flow-stress/higher-temperature materials that can accommodate the heat lost to the tool holder are probably necessary.

Only a handful of titanium alloys have currently been FSWed. Considerable scope for investigation of other titanium alloys exists. Finally, property databases for first-tier (tensile) and second-tier (fatigue, fracture) mechanical properties are mandatory if designers are to use FSW of titanium alloys in future designs. Corrosion databases are also required for designers.

### ACKNOWLEDGMENTS

The author wishes to thank Los Alamos National Laboratory for support during the preparation of this manuscript. Appreciation is

**Table 7.11 Tensile properties of base metal and friction stir welds (FSWs) on Ti-6Al-4V**

| | 0.2% offset yield strength | | Tensile strength | | |
|---|---|---|---|---|---|
| Specimen type | MPa | ksi | MPa | ksi | Elongation, % |
| Base-metal transverse | 1010.1 | 146.5 | 1054.9 | 153.0 | 18 |
| FSW as-welded (avg of four) | 951.5 | 138.0 | 1028.7 | 149.2 | 4.5 |

**Table 7.12 Property comparisons for sheet titanium alloys**

| | | | | | Flow stress(a) | | Beta-transus temperature | |
|---|---|---|---|---|---|---|---|---|
| Alloy | Thermal conductivity, W/mK | Heat capacity, J/kgK | Density, gm/cm³ | Thermal diffusivity, m²/s | MPa | ksi | °C | °F |
| Commercially pure | 21.8 | 523 | 4.51 | $6.78 \times 10^{-6}$ | 180 | 26 | 915 | 1680 |
| 6–4 | 6.6 | 580 | 4.43 | $2.57 \times 10^{-6}$ | 350 | 51 | 995 | 1825 |
| 15–3 | 8.1 | 508 | 4.76 | $3.34 \times 10^{-6}$ | 375 | 54.3 | 770 | 1420 |

(a) Flow stress at 800 °C (1470 °F) and strain rate = 10 $s^{-1}$

also extended to Dr. M.C. Juhas of The Ohio State University and Professor A.P. Reynolds of the University of South Carolina for helpful discussions and the use of figures. This chapter is dedicated to my daughter, Marisa, and my wife, Kellie.

## REFERENCES


1. R. Boyer, G. Welsch, and E.W. Collings, Ed., *Materials Properties Handbook: Titanium Alloys,* ASM International, 1994, p 5–11, 165–259, 483–636, 899–919, 921–929, 1051–1060
2. D.L. Olson, T.A. Siewert, S. Liu, and G.R. Edwards, Ed., *Welding, Brazing, and Soldering,* Vol 6, *ASM Handbook,* ASM International, 1993, p 507–523
3. W.R. Oates and A.M. Saitta, Ed., *Materials and Applications,* Vol 4, *Welding Handbook,* 8th ed., American Welding Society, 1998, p 488–540
4. T.J. Lienert, R. Wheeler, W.L. Stellwag, Jr., and K.V. Jata, Microstructures and Tensile Properties of Friction Stir Welds on Ti-6Al-4V, *Metall. Mater. Trans. A,* Dec 2005, accepted for publication
5. C.M. Adams, *Weld. J.,* Vol 37 (No. 5), 1958, p 210s-215s
6. T. Ahmed and H.J. Rack, *Mater. Sci. Eng. A,* Vol 243, 1998, p 206–211
7. G.F. VanderVoort, Ed., *Atlas of Time-Temperature Diagrams for Nonferrous Alloys,* ASM International, 1991, p 342
8. A.C. Nunes, Jr., E.L. Bernstein, and J.C. McClure, *Abstracts of Papers from 2000 AWS Convention,* 2000, p 46
9. A. Askari, S. Silling, B. London, and M.W. Mahoney, *Friction Stir Welding and Processing,* Proceedings from TMS Fall Meeting, Nov 4–8, 2001 (Indianapolis, IN), K.V. Jata, M.W. Mahoney, R.S. Mishra, S.L. Semiatin, and D.P. Field, Eds., TMS, p 43–53
10. T.U. Seidel, Ph.D dissertation, University of South Carolina, 2001, p 155–200
11. T.U. Seidel and A.P. Reynolds, *Sci. Technol. Weld. Join.,* Vol 8 (No. 3), 2003, p 175–183
12. S. Xu, X. Deng, A.P. Reynolds, and T.U. Seidel, *Sci. Technol. Weld. Join.,* Vol 6 (No. 3), 2001, p 191–193
13. X. Deng and S. Xu, *Trans. NAMRI/SME,* Vol XXIX, 2001, p 631–638
14. S. Xu and X. Deng, Session 10B, Paper A, *Proceedings from the Fourth International Friction Stir Welding Symposium,* May 14–16, 2003 (Park City, UT), P. Threadgill, Ed., TWI, Granta Park, U.K.
15. R.L. Goetz and K.V. Jata, *Friction Stir Welding and Processing,* Proceedings from TMS Fall Meeting, Nov 4–8, 2001 (Indianapolis, IN), K.V. Jata, M.W. Mahoney, R.S. Mishra, S.L. Semiatin, and D.P. Field, Ed., TMS, p 35–42
16. O.T. Midling and Ø. Grong, *Acta Metall. Mater.,* Vol 42 (No. 5), 1994, p 1595–1609
17. Ø. Frigaard, Ø. Grong, J. Hjelen, S. Gulbrandsen-Dahl, and O.T. Midling, Session 11, Paper B, *Proceedings from the First International Friction Stir Welding Symposium,* June 14–16, 1999 (Thousand Oaks, CA), P. Threadgill, Ed., TWI, Granta Park, U.K.
18. J.J. Jonas, C.M. Sellars, and W.J. McG. Tegart, *Metall. Rev.,* Vol 14 (No. 130), 1969, p 1–24
19. T. Seshacharyulu, S.C. Medeiros, W.G. Frazier, and Y.V.R.K. Prasad, *Mater. Sci. Eng. A,* Vol 284, 2000, p 184–194
20. Y.V.R.K. Prasad, T. Seshacharyulu, S.C. Medeiros, and W.G. Frazier, *J. Eng. Mater. Technol.,* Vol 123, 2001, P 355–360
21. L.X. Li, K.P. Rao, Y. Lou, and D.S. Peng, *Z. Metall.,* Vol 94 (No. 9), 2003, p 1006–1011
22. T.H. Courtney, *Mechanical Behavior of Materials,* 2nd ed., McGraw-Hill Book Co., 2000, p 340–345
23. Y.V.R.K. Prasad, T. Seshacharyulu, S.C. Medeiros, W.G. Frazier, J.T. Morgan, and J.C. Malas, *Adv. Mater. Process.,* Vol 158 (No. 2), 2000, p 85–89
24. Y.V.R.K. Prasad, T. Seshacharyulu, S.C. Medeiros, and W.G. Frazier, *J. Mater. Process. Technol.,* Vol 10, 2001, p 320–327
25. R. Ding, Z.X. Guo, and A. Wilson, *Mater. Sci. Eng. A,* Vol 327, 2002, p 233–245
26. S.P. Timothy, *Acta Metall.,* Vol 35 (No. 2), 1987, p 301–306
27. S.L. Semiatin, M.R. Staker, and J.J. Jonas, *Acta Metall.,* Vol 32 (No. 9), 1984, p 1347–1354
28. J.A. Hines and K.S. Vecchio, *Acta Mater.,* Vol 45 (No. 2), 1997, p 635–649
29. F.J. Humphreys and M. Hatherly, *Recrys-*

*tallization and Related Annealing Phenomenon,* 1st ed., Elsevier Science Ltd., Oxford, U.K., 1995, p 363–392
30. P.C. Johnson, B.A. Stein, and R.S. Davis, "Inertia Welding," Report prepared for the Caterpillar Tractor Company, Peoria, IL, Nov 1966
31. A.E. Bayoumi and J.Q. Xie, *Mater. Sci. Eng. A,* Vol 190, 1995, p 173–180
32. J.Q. Xie, A.E. Bayoumi, and H.M. Zbib, *Int. J. Mach. Tools Manuf.,* Vol 36 (No. 7), 1996, p 835–847
33. T.H. Courtney, *Mechanical Behavior of Materials,* 1st ed., McGraw-Hill, Inc., 1990, p 309–317
34. J.J. Jonas and H.J. McQueen, Plastic Deformation of Materials, *Treatise on Materials Science and Technology,* Vol 6, Academic Press, p 394–490
35. A.M. Brown and M.F. Ashby, *Acta Metall.,* Vol 28 (No. 8), 1980, p 1085–1101
36. J.J. Rausch, F.A. Crossley, and H.D. Kessler, *J. Met. (Trans. AIME),* Vol 206, 1956, p 211–214
37. T. Tsujimoto, *Trans. JIM,* Vol 10, 1969, p 281–286
38. H.I. Aaronson, G. Spanos, R.A. Masamura, R.G. Vardiman, D.W. Moon, E.S.K. Menon, and M.G. Hall, *Mater. Sci. Eng. B,* Vol 32, 1995, p 107–123
39. J.W. Elmer, T.A. Palmer, S.S. Babu, and E.D. Specht, *Mater. Sci. Eng. A,* Vol 391, 2005, p 104–113
40. A.J. Ramirez and M.C. Juhas, *Mater. Sci. Forum,* Vol 426–432, 2003, p 2999–3004
41. T.J. Lienert, Unpublished research at the University of South Carolina, 2002
42. D.W. Becker and W.A. Baeslack III, Property-Microstructure Relationships in Metastable Titanium Alloy Weldments, *Weld. J. Res. Suppl.,* Vol 59 (No. 3), 1980, p 85-s to 92-s
43. J. Ma and Q. Wang, Aging Characterization and Application of Ti-15-3 Alloy, *Mater. Sci. Eng. A,* Vol 243, 1998, p 150–154
44. O.M. Ivasishin, P.E. Markovsky, Y.V. Matviychuk, and S.L. Semiatin, Precipitation and Recrystallization Behavior of Beta Titanium Alloys During Continuous Heat Treatment, *Metall. Mater. Trans. A,* Vol 34, 2003, p 147–158
45. I. Weiss and S.L. Semiatin, Thermomechanical Processing of Beta Titanium Alloys—An Overview, *Mater. Sci. Eng. A,* Vol 243, 1998, p 46–65
46. H.W. Rosenberg, Ti-15-3: A New Cold-Formable Sheet Titanium Alloy, *J. Met.,* Vol 35 (No. 11), 1983, p 30–34
47. A.P. Reynolds, E. Hood, and W. Tang, Texture in Friction Stir Welds of Timetal 21S, *Scr. Mater.,* Vol 52, 2005, p 491–494
48. A.D. Rollett and S.I. Wright, *Texture and Anisotropy: Preferred Orientations in Polycrystals and Their Effect on Materials Properties,* Cambridge University Press, 1998, p 178–238
49. W.-B. Lee, C.-Y. Lee, W.-S. Chang, Y.-M. Yeon, and S.-B. Jung, Microstructural Investigation of Friction Stir Welded Pure Titanium, *Mater. Lett.,* Vol 52, 2005, p 3315–3318

CHAPTER 8

# Microstructures and Properties of Copper Alloys after Friction Stir Welding/Processing

Terry R. McNelley, Keiichiro Oh-Ishi, and Alexander P. Zhilyaev
Naval Postgraduate School, Department of Mechanical and Astronautical Engineering

FRICTION STIR WELDING (FSW) and friction stir processing (FSP) of higher-melting metals and alloys, including copper, steels, stainless steels, nickel, and titanium, are emerging from the laboratory and moving into industrial use. Many potential applications of friction stir technology to copper and copper-base alloys have been identified. However, few of these applications have been evaluated, and corresponding microstructure-property data are limited in scope. The current understanding of FSW/FSP of copper and its alloys, with particular concern for microstructure evolution and microstructure-property relationships, is summarized in this section.

## 8.1 Physical Metallurgy Considerations

Copper and copper-base alloys offer unique combinations of conductivity (both thermal and electrical), strength, formability, and corrosion resistance and are used in a wide range of engineering applications. Additional valuable attributes of these materials include color, resistance to sparking, and nonmagnetic behavior.

The thermal and electrical conductivities of copper are highest for the pure metal and decrease significantly with alloying. Unlike iron and titanium, pure copper does not undergo phase changes after solidification and remains as a face-centered cubic $\alpha$ phase in the solid state. Several elements exhibit extensive solid solubility in copper, and so, the corresponding alloys are strengthened by the solutes and by cold work. The solubility of zinc in copper exceeds 30 wt% at 25 °C (77 °F), and brasses exhibit excellent strength-toughness combinations over wide composition and temperature ranges; they are also readily formed and strengthen by cold deformation and annealing treatments.

With sufficient alloying additions, several copper-base alloys become heat treatable and respond to quenching and tempering treatments that are analogous to those employed with steels. Aluminum bronzes containing ~10 wt% Al transform to the body-centered cubic $\beta$ phase upon heating to temperatures >850 °C (1560 °F). The microstructures of such alloys reflect the decomposition of the $\beta$ phase during subsequent cooling; the various decomposition products of the $\beta$ phase depend sensitively on the details of the alloy composition and the heat treatment. Microstructure/mechanical property relationships in these alloys are complex, and the hardening response due to quenching is not as pronounced as that in carbon steels.

Finally, precipitation hardening is attainable with the addition of 1.5 to 2.0 wt% Be to copper. Such alloys are typically solution heat treated, quenched, and then aged to develop refined dispersions of the $\gamma$ (CuBe) phase. Strength and hardness values comparable to those of quenched and tempered steels are readily attained in copper-beryllium alloys.

## 8.2 Conventional Joining of Copper and Copper Alloys

Joining of copper and copper-base alloys in manufacturing is routinely carried out by various welding, brazing, and soldering processes. Arc welding can be accomplished by shielded metal arc, gas tungsten arc, gas metal arc, and submerged arc welding as well as by many variants of these processes. The main factors determining weldability are the thermal conductivity, the solidification range of the materials being joined, and the presence of low-melting constituents.

The high thermal conductivity of pure copper dictates that high heat source intensity must be employed in order to achieve localized melting in the pure metal and dilute alloys. Several alloying additions, including zinc and tin, reduce weldability by increasing susceptibility to cracking. Adherent oxides of aluminum, nickel, and beryllium may inhibit welding and often must be removed to ensure sound welds. Various elements that may be present in copper alloys (e.g., zinc) are both volatile and toxic, and this dictates control of ventilation and facilities to contain fumes and dust in order to protect welders and the surrounding environment.

Brazing and soldering of copper and its alloys are also well-developed techniques for joining during manufacture and repair and are sometimes preferred in order to avoid problems that may be associated with fusion welding processes. Almost all copper-base alloys can be joined by conventional brazing techniques; these include torch, furnace, dip, induction, and resistance brazing. Likewise, most copper-base alloys exhibit good solderability, although joint strength is typically lower than materials being joined and lower than the joint strengths attainable by welding or brazing processes.

## 8.3 Temperature Considerations in FSW/FSP of Copper and Its Alloys

Pure copper melts at 1083 °C (1981 °F), which is the lowest melting temperature among the higher melting metals discussed in this book. However, peak temperatures approaching 1000 °C (1830 °F), which is $0.94T_{Melt}$ for copper, have been reported for FSP of cast NiAl bronze (Cu-9.4Al-5Ni-4Fe; compositions are in weight percent) (Ref 1). Thus, temperatures as well as forces developed in FSW/FSP of copper and its alloys will impose limits on the choice of tool materials. Similarly, peak temperatures have been estimated to be in the range of 0.7 to $0.95T_{Melt}$ in FSW of oxygen-free copper, phosphorus-deoxidized copper, an aluminum bronze (Cu-Al-5Zn-5Sn), and copper-nickel (Cu-25Ni) (Ref 2). Conventional hot work die steels, such as H-13, and pure tungsten performed well with the nominally pure copper materials but poorly with the alloys. This apparently reflected the higher flow stresses of the alloys for the processing conditions chosen. Various sintered carbide tools performed poorly due to brittleness, while polycrystalline cubic boron nitride tools performed well with all of these alloys when care was exercised in tool and process design. In the development of tooling for FSP of the cast NiAl bronze material, excessive tool wear was encountered with tools prepared from MP159 (25Ni-36Co-19Cr-9Fe-7Mo-3Ti), while tools fabricated using Densimet 176 (92.5W-Fe,Ni; a sintered powder metallurgy material) have performed consistently well (Ref 3).

## 8.4 FSW of Oxygen-Free Copper

Following a decade of development, FSW has emerged as the preferred process in the sealing of copper canisters for encapsulation of nuclear waste material (Ref 4–6). The canisters are to be fabricated from seamless copper tubes that are nominally 4.8 m (16 ft) in length, 1 m (3⅓ ft)in diameter, and 50 mm (2 in.) in wall thickness. Top and bottom caps must be joined to the tube to complete the encapsulation of the waste material. To meet the requirements of this application, oxygen-free copper was chosen for the tubes and end closures. The high heat source intensity of electron beam welding and careful joint preparation were required in order to achieve high weld quality with adequate control of melting during fusion welding of the end closures to the tube cylinders. For FSW, numerous tool designs and tool materials, including Nimonic 105 and Densimet, were evaluated, and high-quality welds were obtained routinely. This solid-state process appears to produce thick-section welds in pure copper reliably and reproducibly. By tailoring FSW and tool parameters, welds were produced that exhibited uniform, equiaxed grain structures throughout the weld nugget, with a grain size matching that of the base metal and 100% efficiency in the resulting weld joints (Ref 4). In contrast, in a more limited study, softening and a nonuniform stir zone (SZ) grain size were reported in FSW of 4 mm (0.16 in.) thick strain-hardened and annealed pure copper sheet (Ref 7).

## 8.5 Microstructure Evolution during FSW of Oxygen-Free Copper and Selected Copper-Base Alloys

Dynamic recrystallization has been cited as the predominant mechanism of microstructure evolution during FSW of oxygen-free copper. This mechanism also appears to explain weld nugget microstructures in phosphorus-deoxidized copper as well as in an aluminum bronze (Cu-Al-5Zn-5Sn) and a copper-nickel (Cu-25Ni) alloy (Ref 2). Typical orientation imaging microscopy data in support of this conclusion are summarized in Fig. 8.1. The grain maps in the images of Fig. 8.1(a) and (b) are for the oxygen-free copper-base material, which is apparently in a cold-worked condition. Boundaries surrounding elongated grains are indicated in black, and twin boundaries are light lines in Fig. 8.1(a); the twins account for 2.1% of the boundaries in the base metal. The different shades indicate different lattice orientations from grain to grain in this representation. These same data are classified according to the state of

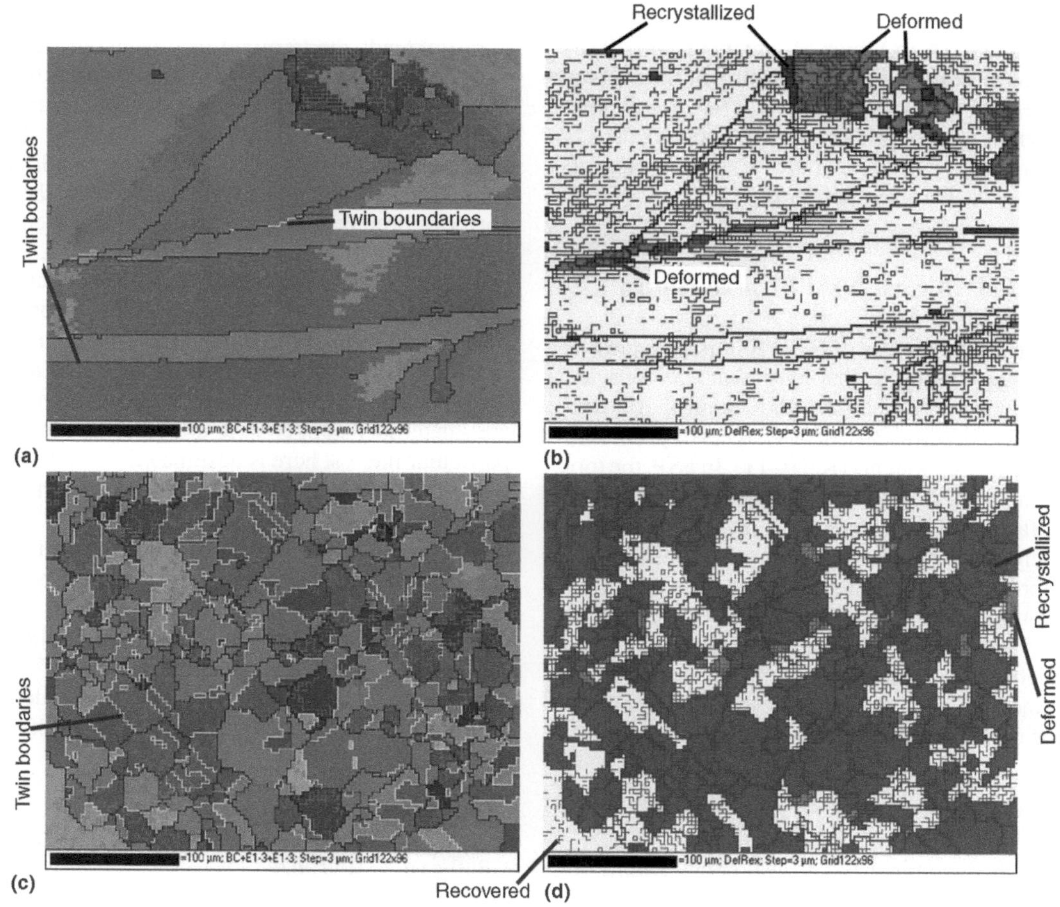


**Fig. 8.1** Orientation imaging microscopy data for oxygen-free copper-base material are represented as (a) a grain map and (b) according to the state of strain as follows: deformed (medium gray), recovered (light gray), and recrystallized (dark gray). For the weld nugget, the grain map in (c) shows an equiaxed grain structure with a large fraction of twin boundaries and in (d) that most of these grains are recrystallized (dark gray). Courtesy of T. Saukkonen and K. Savolainen, Helsinki University of Technology, Espoo, Finland

strain as deformed (medium gray), recovered substructure (light gray) or recrystallized (dark gray) in Fig. 8.1(b). The predominance of subgrains suggests that the cold-worked copper-base metal had been given a recovery anneal prior to FSW (Ref 2).

Representative microstructure data for the weld nugget are shown in Fig. 8.1(c) and (d). The data of Fig. 8.1(c) show that the weld nugget microstructure comprises refined, equiaxed grains that contain annealing twins. The fraction of twin boundaries is ~35% in the weld nugget. Figure 8.1(d) indicates that most of the weld nugget grains are free of substructure and therefore are recrystallized. Altogether, these data indicate that restoration in the weld nugget and surrounding thermomechanically affected zone (TMAZ) takes place by recrystallization during FSW of oxygen-free copper (Ref 2). Highly refined SZ grains 0.8 to 1.5 μm in size were produced from ~22.5 μm base-metal grains by FSW of 2 mm (0.08 in.) thick Muntz metal (60-40Zn) sheets (Ref 8). Distinct hardening of the SZ was observed. The mechanism of grain refinement was not reported, although such a composition is fully β above ~800 °C (1470 °F), and it is an α/β alloy at ordinary temperatures.

## 8.6 FSP of Cast NiAl Bronze Alloys

An allied process of FSW/FSP is emerging as a metalworking technology that can provide localized modification and control of microstructures in near-surface layers of processed metallic components (Ref 9–11). In FSP, the tool is traversed in a predetermined pattern over the surface of a single workpiece in order to achieve microstructure modification and corresponding improvement of properties in selected regions of wrought or cast metals and alloys. Severe plastic deformation and restoration during the thermomechanical cycle of FSP may create highly refined SZ microstructures, especially in alloys. For cast metals, FSP also results in closure of casting porosity as well as homogenization refinement of the as-cast microstructure and converts the as-cast microstructure to a wrought condition in the absence of macroscopic shape change (Ref 1).

Cast NiAl bronze alloys are used for components in a wide range of marine systems due to good combinations of corrosion resistance, strength, toughness, friction coefficients, and nonsparking behavior (Ref 12). Many cast components produced in NiAl bronze involve thick sections, and the resulting slow cooling rates contribute to coarse microstructures and reduced physical and mechanical properties (Ref 13). In such applications, NiAl bronze materials may not be readily heat treatable, and so, FSP represents an alternative means of selectively strengthening the surfaces of such components (Ref 14, 15).

**Physical Metallurgy of NiAl Bronze.** The addition of nickel and iron to copper-aluminum alloys extends the terminal face-centered cubic (fcc) α-phase field and suppresses γ-phase formation that occurs in binary copper-aluminum alloys (Ref 16–18). The γ phase forms by the eutectoid reaction $\beta \rightarrow \alpha + \gamma$ in binary alloys containing more than 9.5 wt% Al (Ref 19); the γ corrodes preferentially in marine environments due to its high aluminum content, and so, its presence is deleterious (Ref 16, 20). The nickel and iron additions increase NiAl bronze mechanical properties through the precipitation of complex κ phases that form in both the α and the β phases (Ref 18). Altogether, NiAl bronzes are quaternary copper-base alloys; the alloy of particular interest here is designated UNS95800 (Ref 21), and composition data are given in Table 8.1.

The constitution and transformation characteristics of NiAl bronze materials have been described in detail elsewhere (Ref 16–18, 20, 22–30). An as-cast Cu-9Al-5Ni-4Fe alloy solidifies as a single-phase β solid solution. The sequence of transformations during subsequent equilibrium cooling is summarized in Fig. 8.2(a),

**Table 8.1 Nominal and typical compositions of UNS95800 cast NiAl bronze**

| Composition | Element, wt% | | | | | | |
|---|---|---|---|---|---|---|---|
| | Cu | Al | Ni | Fe | Mn | Si | Pb |
| Min/max | 79.0 min | 8.5–9.5 | 4.0–5.0 | 3.5–4.5 | 0.8–1.5 | 0.10 max | 0.03 max |
| Nominal | 81 | 9 | 5 | 4 | 1 | ... | ... |
| Typical | 81.2 | 9.39 | 4.29 | 3.67 | 1.20 | 0.05 | <0.005 |

while the micrograph in Fig. 8.2(b) was obtained from a cast NiAl bronze component that required 10 days to cool to room temperature. The corresponding cooling rate is ~$10^{-3}$ · °C · $s^{-1}$. The as-cast alloy remains fully β until cooled to approximately 1030 °C (1890 °F). At this temperature, the primary α phase begins to form with a Widmanstätten morphology. Meanwhile, nucleation of globular κ, which is nominally $Fe_3Al$, takes place in the β, beginning at 930 °C (1705 °F). The globular morphology is apparent in the micrograph of Fig. 8.2(b) and is usually termed $\kappa_{ii}$ (in Cu-Al-Ni-Fe alloys containing >5 wt% Fe, an $Fe_3Al$ phase forms with a dendritic morphology and is termed $\kappa_i$) (Ref 25, 27). At approximately 860° C (1580 °F), the solubility of iron is exceeded in the α, and fine κ precipitates begin to form; these fine precipitates are also nominally $Fe_3Al$ and are usually termed $\kappa_{iv}$. The remaining β decomposes by a eutectoid reaction at approximately 800 °C (1470 °F), which results in the formation of a nickel-rich κ phase, $\kappa_{iii}$, that has a lamellar morphology. Proeutectoid $\kappa_{iii}$ may exhibit a globular morphology and may form by epitaxy on the $\kappa_{ii}$.

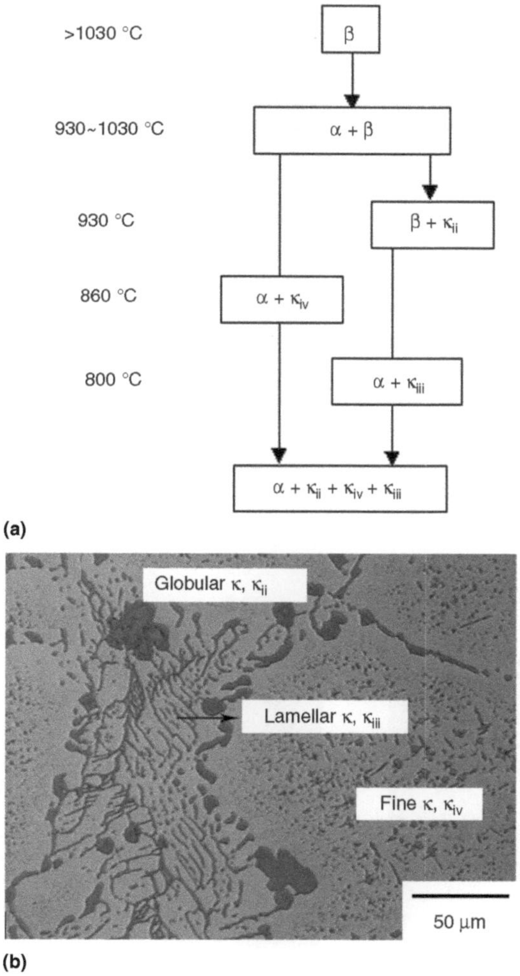


**Fig. 8.2** (a) Sequence of transformations during equilibrium cooling of a Cu-9Al-5Ni-4Fe alloy. (b) Typical microstructure of slowly cooled (rate ~$10^{-3}$ · °C · $s^{-1}$) material

The α phase is an fcc terminal solid solution having a lattice parameter $a_0$ = 0.364 nm (Ref 25). The $Fe_3Al$ phases ($\kappa_{ii}$ and $\kappa_{iv}$) have a $DO_3$ structure; the lattice parameter of the $\kappa_{ii}$ is 0.571 nm, while that of $\kappa_{iv}$ is 0.577 nm (Ref 25, 27, 29). The NiAl ($\kappa_{iii}$) phase has a B2 structure with a lattice parameter of 0.288 nm (Ref 25, 27, 28). Fully ordered $Fe_3Al$ ($\kappa_{ii}$ and $\kappa_{iv}$) and NiAl ($\kappa_{iii}$) will have interatom spacing that differs by less than 1% and is therefore difficult to distinguish by diffraction methods alone.

**Microstructure Evolution in NiAl Bronze due to FSP.** Montages of micrographs from transverse and longitudinal sections through the SZ of a representative example of a single FSP pass on an NiAl bronze material are shown in Fig. 8.3. In this instance, the FSP was accomplished with a tool fabricated from MP159 (25Ni-36Co-19Cr-9Fe-7Mo-3Ti). The tool shoulder diameter was 23.8 mm, while the pin was 7.95 mm in diameter, 6.95 mm in length, and machined with a spiral groove. The tool rotation rate was 1000 rpm, and the traversing rate was 20.3 cm · $m^{-1}$ (Ref 31). Both montages include base metal as well as the SZ. In the transverse section shown in Fig. 8.3(a), the boundary between the SZ and surrounding material is distinct on the advancing side and beneath the tool but is indistinct on the retreating side. The longitudinal section shown in Fig. 8.3(b) was obtained along the centerline of the SZ, denoted A-A′ in Fig. 8.3(a), and the distinct character of the SZ boundary is apparent in this image as well. Base-metal grains are distorted in the TMAZ, although the extent and direction of shearing varies with location along the SZ-TMAZ boundary. The dark-etching features in the TMAZ and nearby base metal reflect local reversion of the lamellar α + $\kappa_{iii}$ to form β due to the heating associated with the process, followed by rapid cooling and transformation of the β to various nonequilibrium transformation products.

Comparison of the SZ and the as-cast NiAl bronze base metal shows that the microstructure is much finer in the SZ, but that it also varies with depth. Detailed analyses of regions such as those indicated in Fig. 8.3 (1 to 4 in the transverse plane and 1′ to 4′ in the longitudinal plane) have shown that the observed variation in microstructure can be correlated with peak temperature attained during the FSP thermomechanical cycle (Ref 1, 3). The microstructure data are summarized in Fig. 8.4 to 8.6. In locations nearest the surface in contact with the tool shoulder, that is, region 1 in Fig. 8.4 and region 1′ in Fig. 8.5, the microstructure reflects full transformation to β. During subsequent cooling after passage of the tool, the β begins to decompose by the formation of α with a Widmanstätten morphology and then by the formation of dark-etching constituents during further cooling. In regions 2 and 2′, elongated band- or blocklike clusters of equiaxed primary α grains that contain annealing twins are interspersed with elongated regions that comprised fine β-transformation products. Elongation of the primary α in region 2 is more notable in the transverse plane (Fig. 8.4); in region 2′ in the longitudinal plane, the clusters of primary α grains are more irregular in shape, but the grains within these clusters are still equiaxed. The prior-β regions exhibit fine Widmanstätten α and fine, unresolved β-transformation products in the dark-etching regions. Transmission electron microscopy investigations have shown that these products include bainitic and martensitic constituents formed by decomposition of the β. The central regions of the SZ exhibit distinct "onion ring" flow patterns. The ringlike character of these patterns is seen most clearly in Fig. 8.3(a); these features appear as alternating layerlike structures that are inclined away from the direction of tool travel in the longitudinal section in Fig. 8.3(b). At higher magnification (region 3 in Fig. 8.4; region 3′ in Fig. 8.5), these layers appear to consist of bands of α having a Widmanstätten morphology interspersed with elongated bands of primary α containing fine, equiaxed grains. The apparent horizontal spacing of these bands is ~230 μm in region 3′, while the tool advance per revolution is ~203 μm · rev$^{-1}$ (Fig. 8.5). Thus, it is likely that these features constitute bands that have experienced different thermomechanical histories and are then brought into proximity behind the tool on successive revolutions. Finally, a highly refined but unresolved structure is apparent in regions 4 and 4′ (Fig. 8.4 and 8.5, respectively) at the bottom of the SZ in both transverse and longitudinal planes.

Transmission electron microscopy in Fig. 8.6(a) illustrates the highly refined structure of α grains that are 1 to 2 μm in size at the bottom of the SZ (region 4) for this same tool and material after processing at 800 rpm and a traversing rate of 15.2 cm · m$^{-1}$. The absence of β-transformation products at the bottom of the SZ likely reflects heating only to the vicinity of the eutectoid temperature during FSP. In this location, microstructure evolution appears to have occurred mainly by deformation and recrystallization of the primary α. Convergent beam electron diffraction methods were employed to obtain grain-specific orientation data in this region, and the corresponding grain-to-grain disorientations are indicated by the line width in

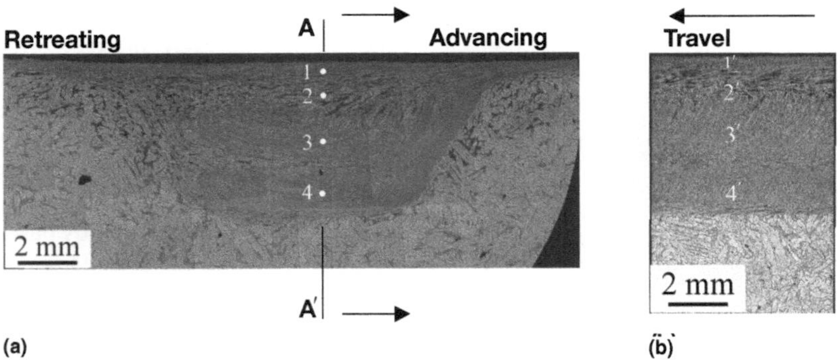


**Fig. 8.3** Montages of micrographs for a single friction stir processing pass in as-cast NiAl bronze, showing the stir zone from the tool shoulder region downward into base metal. (a) Transverse section. (b) Longitudinal section along A-A′

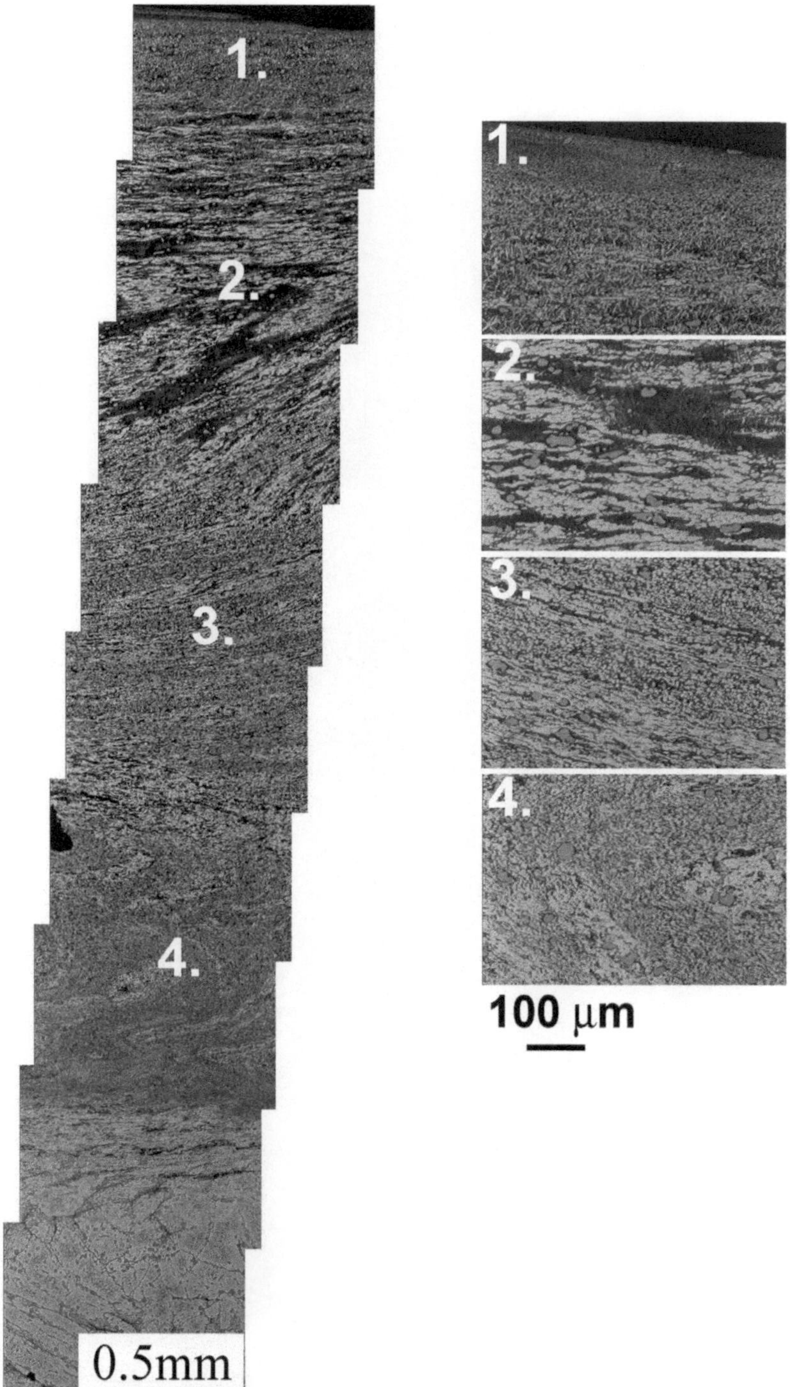


**Fig. 8.4** Variation in microstructure in the transverse plane. In region 1, transformation of β with a Widmanstätten morphology is evident. In region 2, a mixture of deformed primary α and β-transformation products has formed. Bands from the "onion rings" are shown in region 3, and a grain-refined region is seen in region 4 near the bottom of the stir zone.

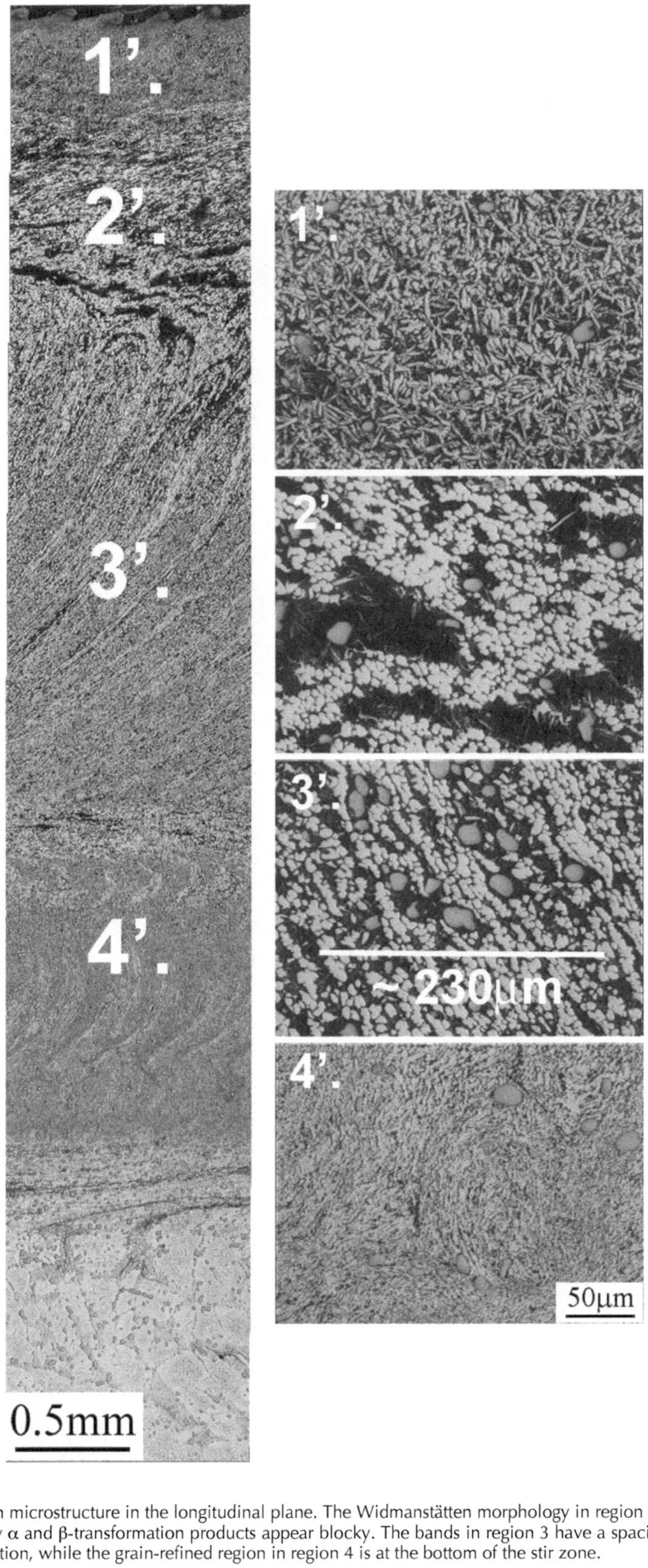


**Fig. 8.5** Variation in microstructure in the longitudinal plane. The Widmanstätten morphology in region 1 is distinct. In region 2, the primary α and β-transformation products appear blocky. The bands in region 3 have a spacing corresponding to the tool advance per revolution, while the grain-refined region in region 4 is at the bottom of the stir zone.

the tracing of the boundary structure shown in Fig. 8.6(b). (The term *disorientation* refers to minimum angle among all crystallographically equivalent rotations that bring adjacent lattices into coincidence.) Many of the straight boundaries have disorientations corresponding to twins, and so, these features reflect the formation of annealing twins following recrystallization during the FSP thermomechanical cycle.

Various annealing and hot rolling experiments were conducted in order to establish a basis for estimation of local SZ peak temperatures during FSP. Typical results of these experiments are illustrated in Fig. 8.7(a); these data were obtained by annealing small coupons for 1 h at the indicated temperatures, followed by cooling of the coupons in laboratory air to give cooling rates of $\sim 10^0$ °C·s$^{-1}$. Additional experiments involving shorter anneals or concurrent hot rolling (with reheating between passes) were also conducted. Upon heating into the range of the eutectoid reaction, the lamellar $\alpha + \kappa_{iii}$ eutectoid constituent redissolves to form β. Then, during subsequent cooling, various β transformation form, while the primary α remains unaffected. The reaction apparently has not yet begun upon heating at 770 °C (1420 °F) but has clearly taken place during heating at ≥820 °C (1510 °F); full transformation to β is only apparent upon heating to 1000 °C (1830 °F). Altogether, the annealing data demonstrated that equilibrium fractions of β were attained within 6 min of heating at temperature. However, the globular $\kappa_{ii}$ apparently dissolved more slowly. With concurrent hot rolling at this same temperature, the α and β phases apparently deform compatibly; upon cooling after hot rolling, the deformed and recrystallized primary α remains, while various transformation products form from the β.

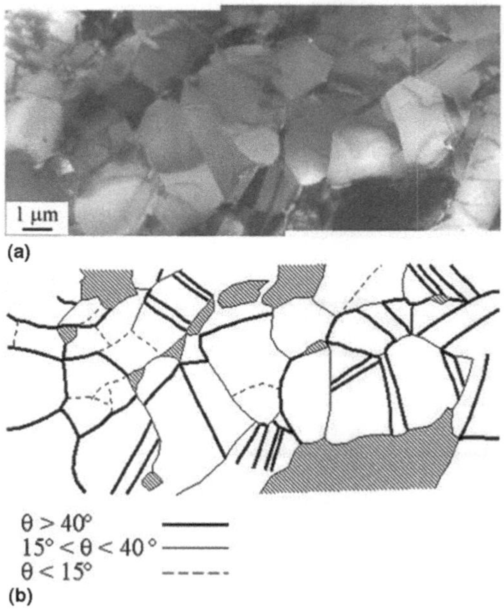


**Fig. 8.6** (a) Transmission electron microscopy images of refined grain structure. (b) Results of convergent beam electron diffraction analysis from the lower stir zone corresponding to region 4 in Fig. 8.3. Grain boundaries are delineated by various lines, depending on grain-to-grain disorientation angle: thick for θ > 40°, thin for 15° < θ < 40°, and dotted for θ < 15°, respectively.

The microstructure data in Fig. 8.7(a) show that the volume fraction of β-transformation products increases as the annealing temperature increases and that the β-transformation products include α with a Widmanstätten morphology. Figure 8.7(b) is a plot of the volume fraction of β-transformation products as a function of the heating temperature in this annealing experiment; identical results were obtained from hot-rolled samples. The volume fraction of the globular $\kappa_{ii}$ was measured as well; this phase dissolved more slowly because of its morphology and the low iron diffusion rate, but it had disappeared upon heating above 950 °C (1740 °F). Estimates of the local peak temperature were made by measuring the corresponding volume fraction of β-transformation products in SZ microstructures. Concurrent deformation has been shown to result in order of magnitude increases in spheroidization rates during warm working of high-carbon steels (Ref 32, 33). On this basis, the dissolution of the lamellar $\alpha + \kappa_{iii}$ constituent will be accelerated by the severe concurrent deformation, and near-equilibrium microstructures should develop in the SZ during FSP. The schematic in Fig. 8.8 illustrates an SZ peak temperature distribution for the material processed at 800 rpm and 15.2 cm · m$^{-1}$ that corresponds to the microstructures of Fig. 8.3 to 8.6. This distribution assumes that the reversion of the as-cast microstructure occurs during deformation and heating to the local peak temperature and that the reversion reactions are greatly accelerated by the concurrent deformation. The average volume fraction of β-transformation products was determined in regions exhibiting distinct onion-ring formation.

**Distribution of SZ Mechanical Properties after FSP.** Because microstructures vary significantly with location, a miniature sheet-type tension test coupon design was developed to evaluate the distributions of strength and ductility throughout SZs. In the current study, FSP was conducted using a 12 mm (0.5 in.) step-spiral Densimet 176 tool, as illustrated in Fig. 8.9(a). Processing was conducted at 800 rpm, with a traversing rate of 10.2 cm·m$^{-1}$. Test coupons according to the design in Fig. 8.9(b) were obtained by wire electric discharge machining, and care was taken to maintain sample location relative to the SZ centerline and surface. This is suggested in the schematic of Fig. 8.9(c) for tensile axes that are aligned with the direction of tool travel. The registry with SZ microstructure is illustrated in Fig. 8.9(d) by superposition of the gage cross sections of the tensile coupons on a montage of micrographs from a transverse section of the corresponding SZ.

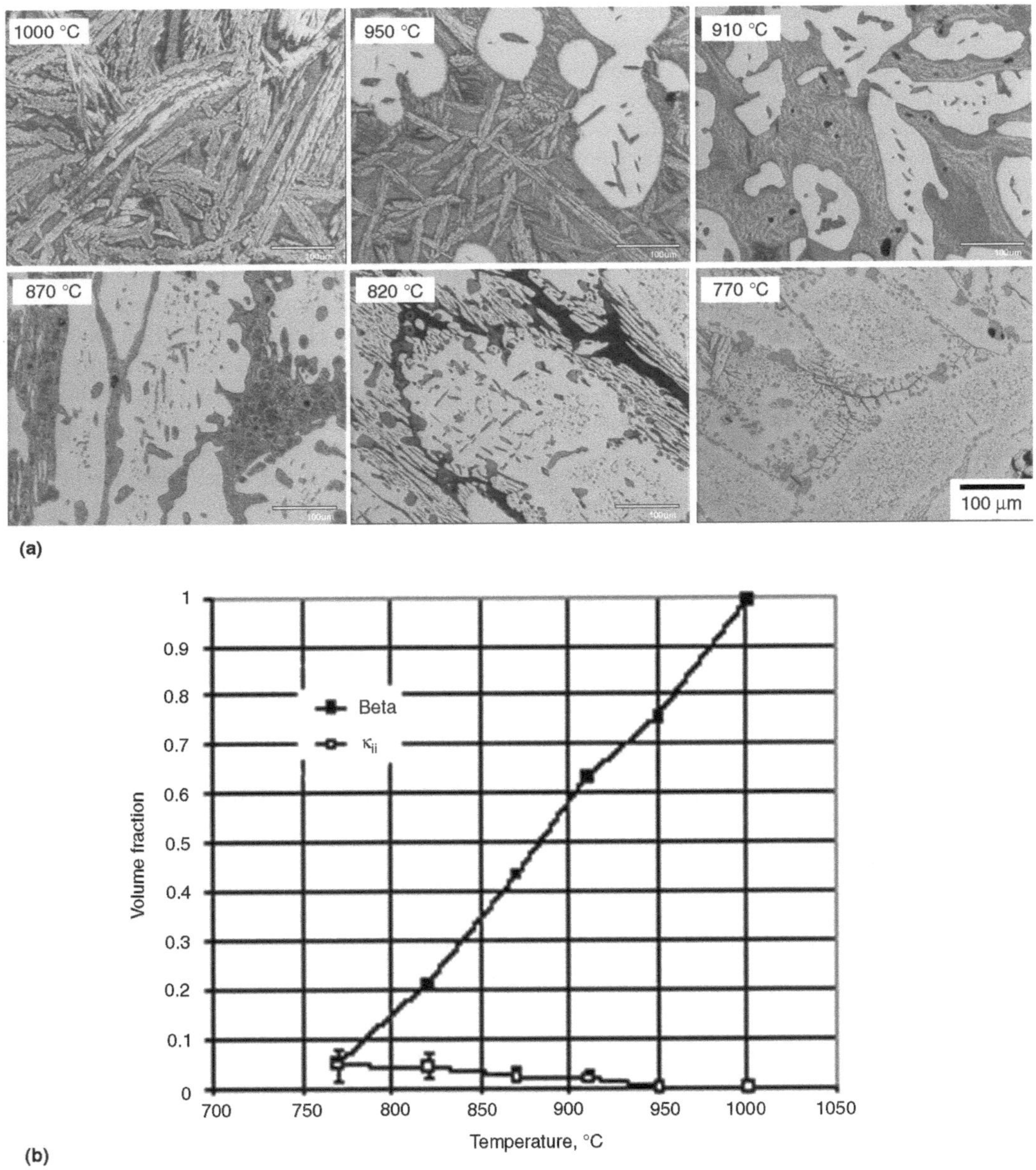


**Fig. 8.7** (a) Influence of annealing temperature on microstructure of a cast NiAl bronze. (b) Plot of corresponding dependence of the volume fraction of β-transformation products, determined by quantitative metallography, on temperature

Tension testing involved standard procedures. Corresponding nominal base-metal mechanical properties were 220 MPa (32 ksi) yield strength, 450 MPa (65 ksi) tensile strength, and 10% elongation to fracture. Typical results are shown in Fig. 8.10. Altogether, these data show that FSP may result in increased elongations as well as in increased yield and tensile strengths. The distributions of yield and tensile strengths are shown in Fig. 8.10(a) and (b), while the corresponding ductility data are in Fig. 8.10(c). Of particular note is that yield and tensile strength values near the plate surface in contact with the tool shoulder are highest, where yield strength has been raised to approximately 500 MPa (73 ksi) and tensile strength to 760 MPa (110 ksi); ductility was approximately 20% elongation to failure at this location. This reflects the high local SZ temperature and the predominance of Widmanstätten α as a transformation product of a fully β microstructure. The yield and tensile strengths as well as the ductilities all appear to remain above base-metal values throughout the SZ. However, there appear to be locations of low ductility (~5% elongation) in the TMAZ underneath the tool shoulder. Similar sites of low ductility appear in the heat-affected zones of fusion welds and appear to be associated with martensitic transformation products of β formed during rapid cooling after heating to ~800 °C (1470 °F). At this temperature, the lamellar $\alpha + \kappa_{iii}$ reverts to form β of relatively high aluminum content, which would decrease upon heating to higher temperatures.

A similar investigation was performed following multipass FSP using a raster pattern involving overlapping of adjacent passes. The FSP was again conducted using a 12 mm (½ in.) step-spiral Densimet 176 tool operated at 800 rpm with a traversing rate of 10.2 cm·m$^{-1}$. A montage of the microstructure and location of tensile coupons is illustrated in Fig. 8.11(a). The micrographs in Fig. 8.11(b) show microstructures at two SZ locations and illustrate highly refined microstructures and a predominance of the Widmanstätten morphology in the SZ for this material and processing condition. The corresponding distributions of the tensile properties are shown in Fig. 8.12. The distributions of yield and ultimate strength are summarized in Fig. 8.12(a) and (b), and the elongation data are in Fig. 8.12(c). These data span a region corresponding to two overlapping passes and show that the region of low ductility under the tool shoulder has been eliminated by this multipass process. Yield strengths at the plate surface now approach 550 MPa (80 ksi), tensile strengths are 800 MPa (115 ksi), and the ductility is also consistently high at 30% elongation to failure. This emphasizes that FSP of cast NiAl bronze materials results in distinct surface hardening of the material. Nevertheless, in locations below the SZ, there still appears to be a region of low ductility corresponding to the TMAZ, and presumably, there would be such a region at the outer edge of the region processed by such a raster procedure.

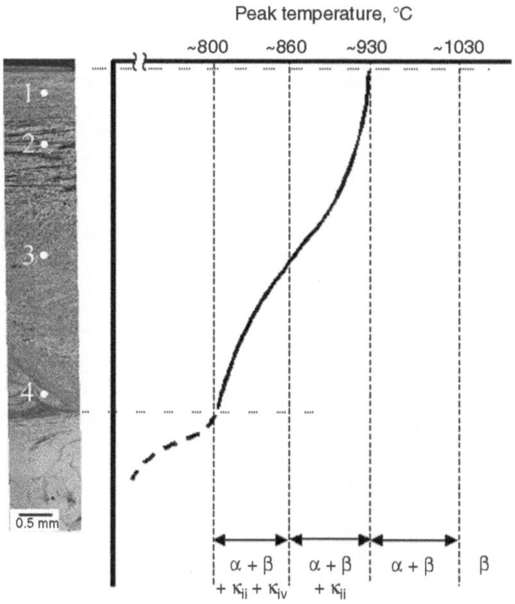


**Fig. 8.8** Variation in peak temperature with depth in the stir zone. The local peak temperature was estimated from the local apparent fraction of β-transformation products and the data of Fig. 8.7.

**Monotonic and Cyclic Mechanical Properties Following FSP.** In developing FSP for application to large marine castings, 38.1 mm (1.5 in.) thick NiAl bronze plates were processed by the Densimet 176 tool (Fig. 8.9). The FSP parameters are included in Table 8.2. The processing involved either a linear raster pattern or a rectangular spiral raster pattern in order to process large areas of the as-cast plate material (Ref 34). Macrographs of sections that are transverse to the long axis of the linear or rectangular rasters are shown in Fig. 8.13(a) or (b), respectively. Mechanical properties were then evaluated by means of cylindrical test samples that were machined approximately from the mid-depth of the SZ, so that the gage section contained only FSPed material (Ref 35). Test samples were always machined having their long axes perpendicular to the local direction of tool travel. Conventional tension testing was conducted using tensile bars of 4 mm (0.16 in.) gage diameter. Fatigue testing was accomplished using samples of 6 mm (0.24 in.) gage diameter. The fatigue testing involved either fully reversed

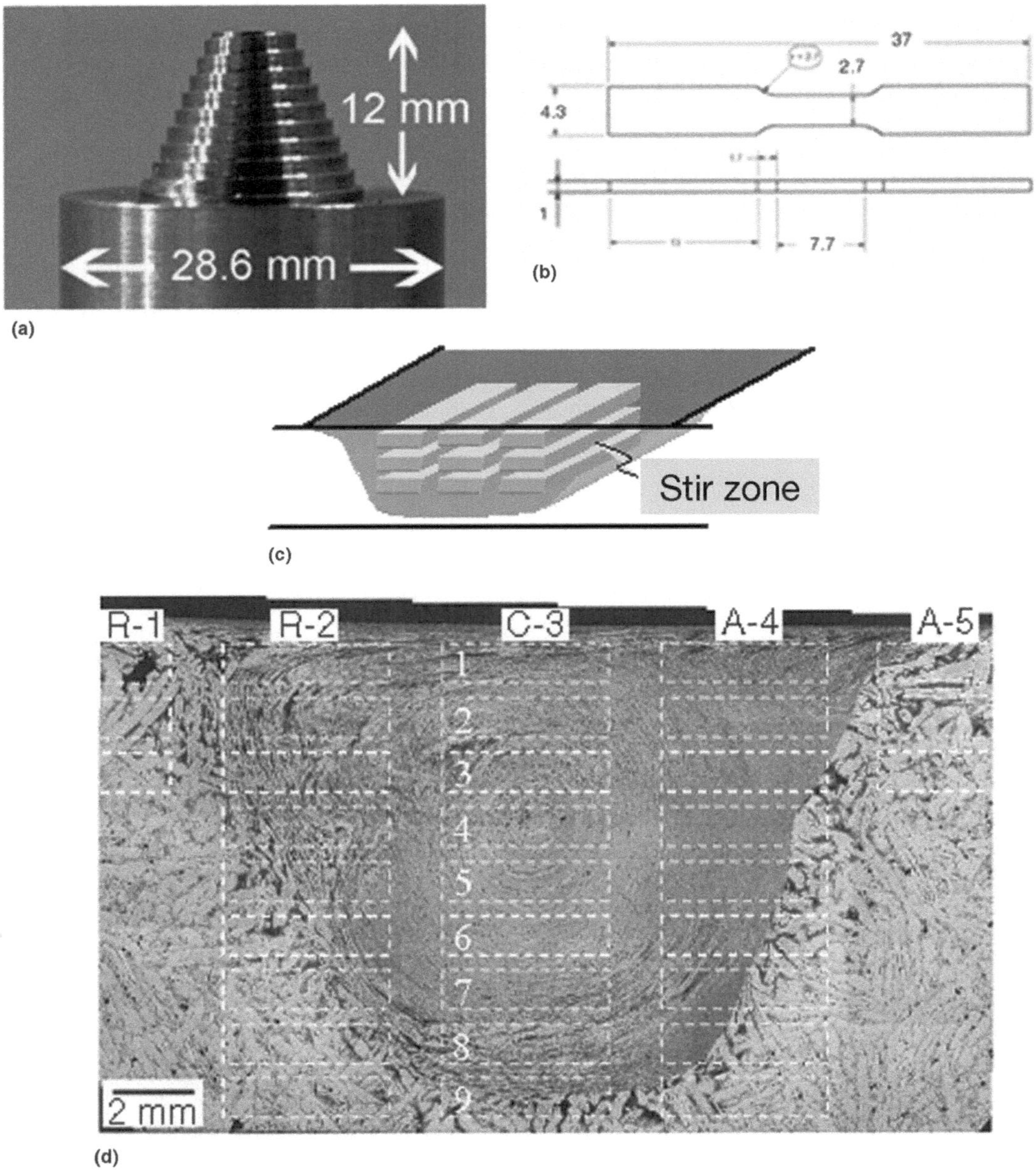


**Fig. 8.9** (a) Densimet 176 tool used to process as-cast NiAl bronze. (b) Miniature tensile sample design (dimensions in millimeters (c) Schematic representation of the distribution of tensile coupons in a stir zone (SZ). (d) Transverse section of the SZ for processing at 800 rpm/10.2 cm · $m^{-1}$. (d) The cross sections of the tensile coupons are indicated by the rectangles.

rotating-bending fatigue ($R \equiv \sigma_{min}/\sigma_{max} = -1$) or tension-tension fatigue ($R = 0.1$).

Table 8.2 provides a summary of the influence of FSP parameters on the conventional monotonic tensile properties of an as-cast NiAl bronze material.

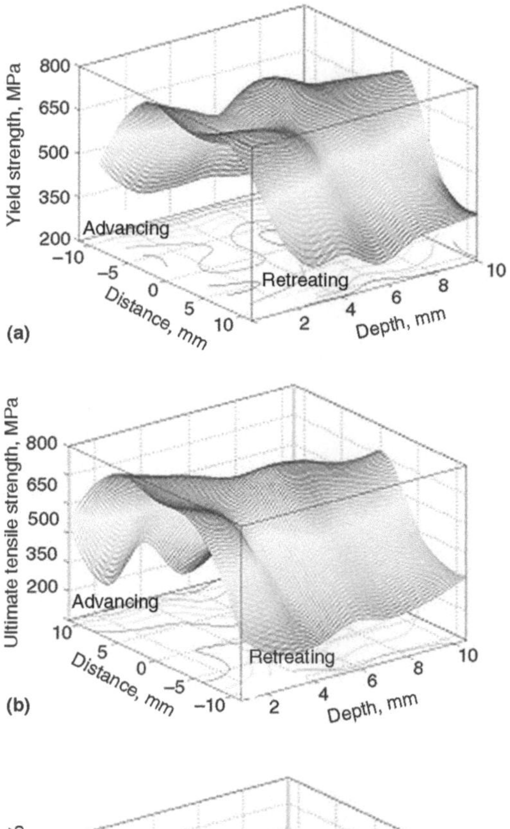


**Fig. 8.10** The mechanical properties for the stir zone in Fig. 8.9, showing the distribution of (a) yield strength, (b) tensile strength, and (c) ductility in for tensile test coupons aligned with the longitudinal axis. Both strength and ductility are raised relative to the as-cast material, although regions of low ductility are apparent in the thermomechanically affected zone under the tool shoulder.

The tensile data are presented as a function of raster pattern and tool parameters for the SZ tests. The yield and tensile strength data for as-cast material agree well with results of testing with the miniature samples. However, the as-cast ductility here is twice as high, and this may reflect the effect of sample size relative to the size and distribution of casting defects in the material as well as variations in the as-cast materials. The yield and tensile strengths in the linear raster here also compare well with the results from the miniature coupons, despite the different orientations of the tensile axes relative to the local direction of the tool traverse. These larger samples may have an averaging effect on the SZ ductility values in that the miniature coupon data indicated a gradient in ductility from the plate surface downward through the SZ.

Nevertheless, from these data, FSP modification of the as-cast material produces a 140 to 172% increase in yield strength and a 40 to 57% increase in tensile strength. The raster patterns give slight yield and tensile strength differences as a function of tool parameters, but no consistent trend is evident. These increases in strength may be attributed to the elimination of casting defects and refinement of the microstructure. In particular, FSP produces microstructures (e.g., fine grained, Widmanstätten, and lamellar) that generally exhibit greater yield and tensile strength values compared to the as-cast material. In these data, the FSP produced either an 18 to 41% decrease (linear raster) or a 12 to 38% increase (rectangular spiral raster) in percent elongation. However, all of the FSP elongation values are above 10%, which is the minimum specified for as-cast NiAl bronze, as well as uniformly above results from testing with the miniature coupons. The observed differences in percent elongation may reflect differences in grain flow patterns between the linear and rectangular raster patterns. Profile views of the crack path in samples from the linear and rectangular raster patterns are shown in Fig. 8.14. Tensile samples from linear raster patterns exhibit strain localization at uplifted grains, as shown in Fig. 8.14(a), while spiral raster patterns tend to give increased uniformity of microstructure, as seen in Fig. 8.14(b), through the sample depth and therefore reduced strain localization.

The results of rotating-bending fatigue tests are provided in Fig. 8.15(a) for as-cast NiAl bronze and for this material after FSP using the linear raster pattern. Corresponding data for uniaxial fatigue tests conducted with $R = 0.1$ are included in Fig. 8.15(b). For both loading modes, the processed material has significantly higher fatigue resistance when compared to the as-cast condition. This is not surprising, insofar as fatigue strength would be expected to scale with static tensile strength. However, these data also show that the fatigue resistance of the processed material is dependent on processing parameters as well as loading mode. In Fig. 8.15(a), the material processed using 1000 rpm/7.6 cm·m$^{-1}$ has the highest fatigue resistance, the material

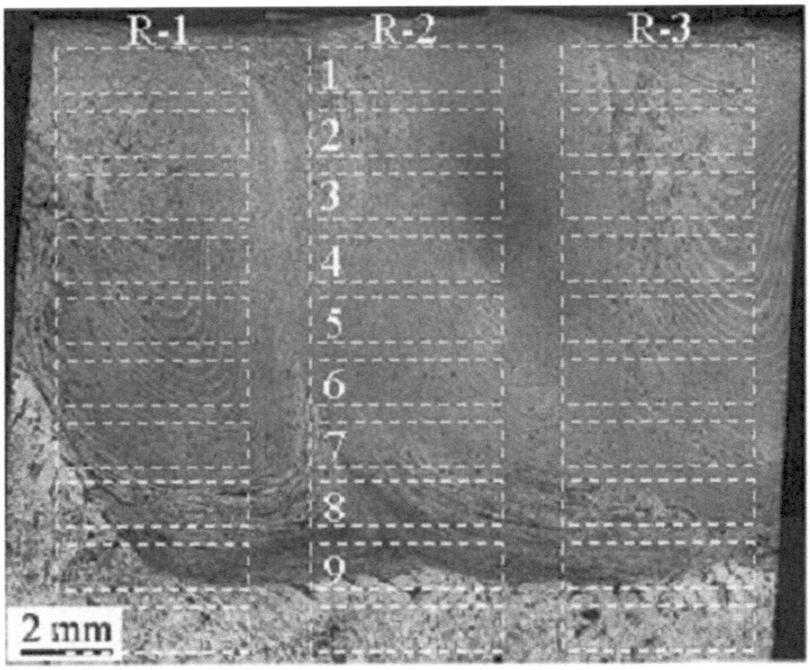


(a)

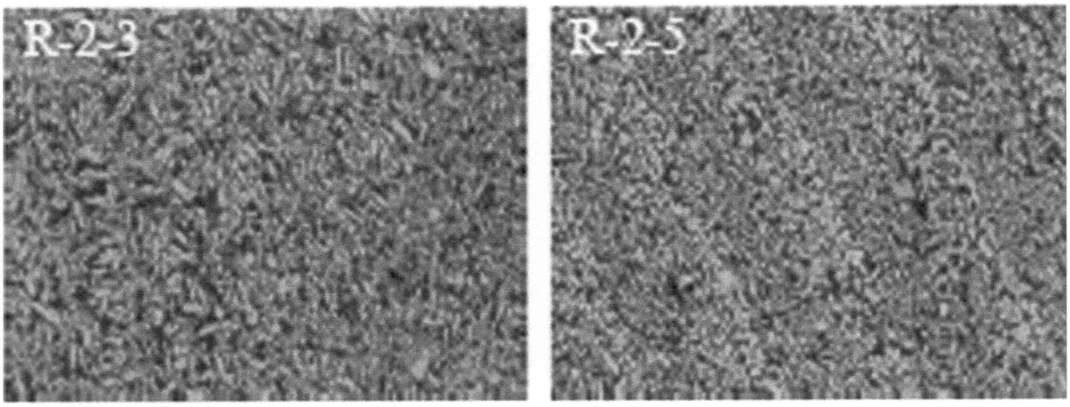


(b)

**Fig. 8.11** (a) Transverse section of the stir zone for a multipass raster process involving friction stir processing at 800 rpm/10.2 cm · m$^{-1}$, with the distribution of tensile coupon cross sections also highlighted. (b) At higher magnification, the microstructure consists of refined Widmanstätten α; also, the bands in the "onion ring" structures are less distinct for multipass processes.

processed using 800 rpm/10.2 cm·m$^{-1}$ has intermediate fatigue resistance, and the material processed at 1200 rpm/5.1 cm·m$^{-1}$ has the lowest fatigue resistance among the processed materials. These data show that the material with the highest tensile strength exhibits the lowest fatigue resistance. Models to describe the influence of processing parameters on the thermomechanical history or on microstructure evolution during FSW/FSP remain to be developed. In the current study, neither metallography nor fractography has revealed the microstructural basis for the influence of processing parameters on the fatigue behavior of the processed materials.

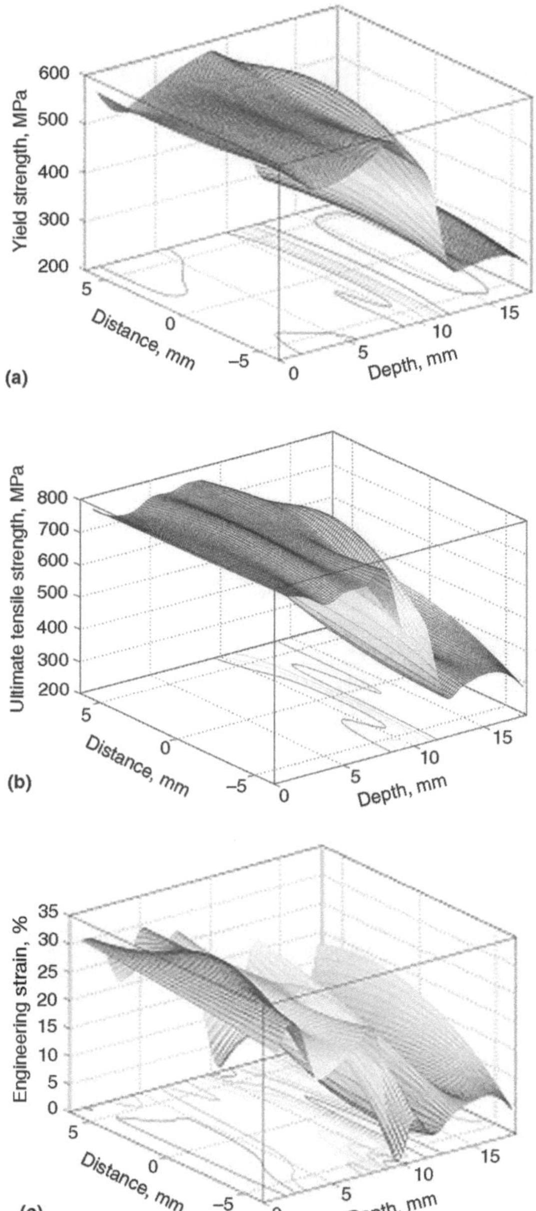


**Fig. 8.12** The mechanical properties for the stir zone in Fig. 8.11, showing the distribution of (a) yield strength, (b) tensile strength, and (c) ductility for tensile test coupons aligned with the local longitudinal axis for the multipass raster pattern. Exceptional strength/ductility combinations are achieved near the plate surface, although low ductility is apparent in the thermomechanically affected zone under the stir zone.

Under uniaxial conditions, the FSP material again exhibits a significant improvement in fatigue resistance in comparison to the as-cast material. A lesser dependence on processing parameters is evident in the data for the FSP material in Fig. 8.15(b), but the same trend observed for the rotating-bending data is evident. This may reflect the more aggressive nature of the loading under rotating-bending conditions in this study.

## 8.7 Summary

Applications of friction stir technologies to welding and processing of copper and several copper-base alloys have been described. As an alternative to fusion welding, joining of oxygen-free copper and dilute solid-solution copper alloys may be readily and reliably accomplished by FSW, and resulting joints may exhibit uniform microstructures and offer 100% joint efficiency. In cast NiAl bronzes, FSP may enable localized modification and improvement of properties by closure of porosity and refinement of microstructures in near-surface regions of cast components. In combination with FSP, transformations in NiAl bronze materials may also enable selective strengthening of surfaces and improved resistance to fatigue.

## ACKNOWLEDGMENTS

The authors acknowledge the provision of friction stir processed materials and data on monotonic and cyclic behavior of the NiAl bronze by M.W. Mahoney and C.B. Fuller, Rockwell Scientific Corporation. The Naval Surface Warfare Center (Carderock, MD) supplied the NiAl bronze materials, and the Defense Advanced Research Projects Agency (DARPA), with Dr. L. Christodolou as program sponsor, provided the funding for this work.

**Fig. 8.13** Cross-sectional views of microstructures produced by (a) linear raster and (b) rectangular spiral raster patterns. Processing used the tool shown in Fig. 8.9. The distinct uplift pattern in (a) reflects the switching between advancing and retreating sides as the processing takes place (the tool is alternately moving into or out of the plane of the micrograph). Courtesy of C.B. Fuller and M.W. Mahoney, Rockwell Scientific Corporation, Thousand Oaks, CA

## Table 8.2 Monotonic tensile properties of NiAl bronze

| Friction stir processing parameters, rpm/cm·m$^{-1}$ | Linear raster(a) | | | | | Rectangular spiral raster(b) | | | | |
|---|---|---|---|---|---|---|---|---|---|---|
| | Yield strength | | Tensile strength | | | Yield strength | | Tensile strength | | |
| | MPa | ksi | MPa | ksi | Elongation, % | MPa | ksi | MPa | ksi | Elongation, % |
| 800/10.2 | 504 | 73.1 | 765 | 111 | 12.8 | 460 | 66.8 | 743 | 108 | 30.0 |
| 1000/7.6 | 522 | 75.7 | 769 | 112 | 14.3 | 507 | 73.6 | 761 | 110 | 25.8 |
| 1200/5.1 | 518 | 75.1 | 805 | 117 | 17.4 | 472 | 68.4 | 743 | 108 | 24.4 |
| As-cast | 192 | 27.9 | 530 | 77 | 21.8 | | | | | |

(a) Average of four samples. (b) Average of six samples

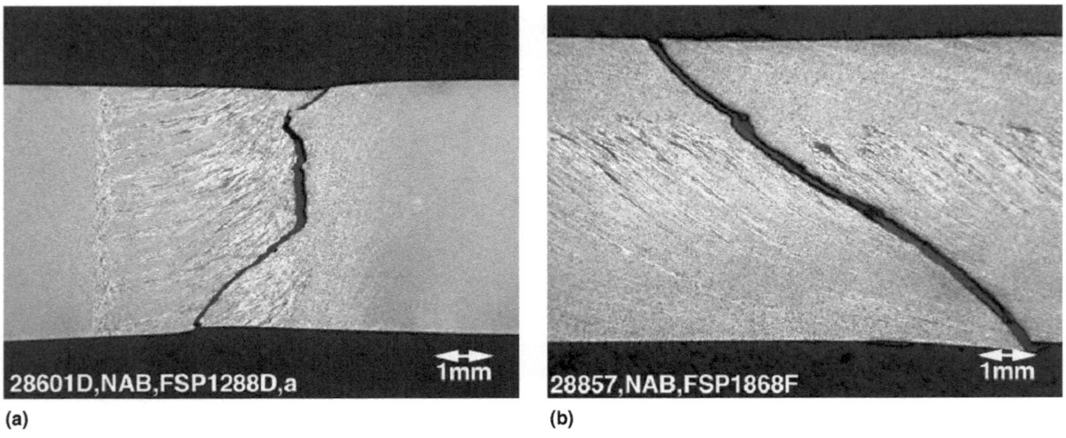


**Fig. 8.14** Macrographs of fractured tensile samples having tensile axes perpendicular to the local longitudinal direction of the raster for either (a) linear or (b) rectangular spiral raster patterns. Courtesy of C.B. Fuller and M.W. Mahoney, Rockwell Scientific Corporation, Thousand Oaks, CA

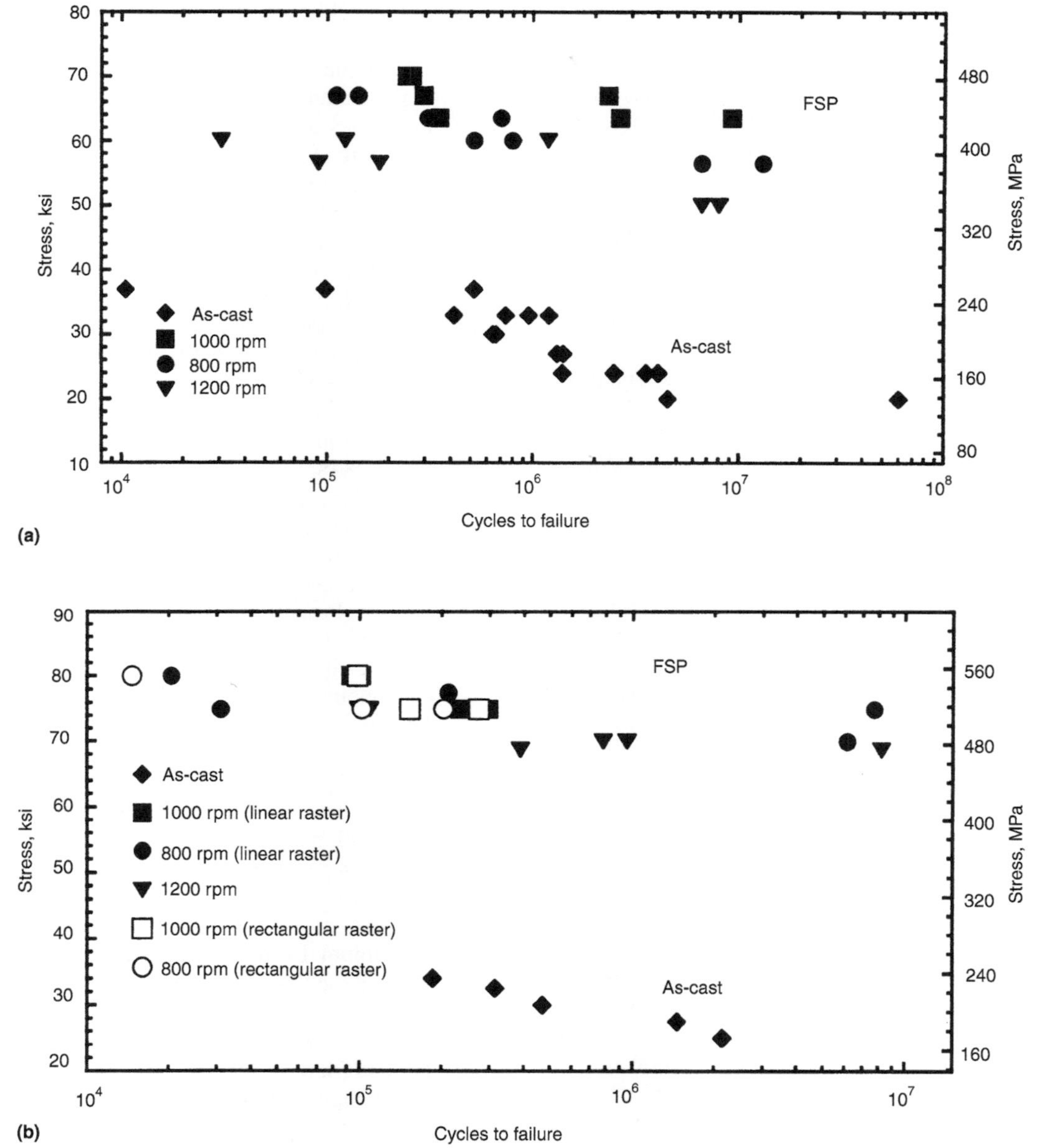


**Fig. 8.15** Plots of maximum applied stress versus cycles to failure for (a) rotating-bending fatigue ($R = -1$) and for (b) uniaxial fatigue ($R = 0.1$). (a) Data for base metal and the linear raster. (b) Data for the rectangular spiral raster are also included. Courtesy of C.B. Fuller and M.W. Mahoney, Rockwell Scientific Corporation, Thousand Oaks, CA

## REFERENCES


1. K. Oh-Ishi and T.R. McNelley, Microstructural Modification of As-Cast NiAl Bronze by Friction Stir Processing, *Metall. Mater. Trans. A,* Vol 35, 2004, p 2951–2961
2. K. Savolainen, J. Mononen, T. Saukkonen, H. Hänninen, and J. Koivula, Friction Stir Weldability of Copper Alloys, *Fifth International Conference on Friction Stir Welding, Proceedings* (Metz, France), TWI, 2004
3. K. Oh-Ishi and T.R. McNelley, The Influence of Friction Stir Processing Parameters on Microstructure of As-Cast NiAl Bronze, *Metall. Mater. Trans. A,* Vol 36, 2005, p 1575–1585
4. C. Andersson, "Development of Fabrication Technology for Copper Canisters with Cast Inserts," Technical Report TR-02-07, Svensk Kärnbränslehantering AB, April 2002, p 7–66
5. L. Cederqvist, FSW to Seal 50 mm Thick Copper Canisters—A Weld that Lasts for 100,000 Years, *Fifth International Conference on Friction Stir Welding, Proceedings* (Metz, France), TWI, 2004
6. TWI World Center for Materials Joining Technology, Decision Made—It's Official, *Connect,* No. 138, Sept–Oct 2005, p 1, http://www.twi.co.uk/j32k/ (accessed Aug 11, 2006)
7. W.-B. Lee and S.-B. Jung, The Joint Properties of Copper by Friction Stir Welding, *Mater. Lett.,* Vol 58, 2004, p 1041–1046
8. H.-S. Park, T. Kimura, T. Murakami, Y. Nagano, K. Nakata, and M. Ushio, Microstructures and Mechanical Properties of Friction Stir Welds of 60%Cu-40%Zn Copper Alloy, *Mater. Sci. Eng. A,* Vol 371, 2004, p 160–169
9. R.S. Mishra, Friction Stir Processing Technologies, *Adv. Mater. Process.,* Vol 161 (No. 10), 2003, p 43–46
10. R.S. Mishra, Z.Y. Ma, and I. Charit, Friction Stir Processing: A Novel Technique for Fabrication of Surface Composite, *Mater. Sci. Eng. A,* Vol 341, 2003, p 307–310
11. Z.Y. Ma, R.S. Mishra, and M.W. Mahoney, Friction Stir Processing for Microstructural Modification of an Aluminum Alloy Casting, *Friction Stir Welding and Processing II, Proceedings,* March 2003 (San Diego, CA), TMS, 2003, p 221–230
12. J.A. Duma, Heat Treatments for Optimizing Mechanical and Corrosion-Resisting Properties of Nickel-Aluminum Bronzes, *Nav. Eng. J.,* Vol 87, 1975, p 45–64
13. R.J. Ferrara and T.E. Caton, Review of Dealloying of Cast Aluminum Bronze and Nickel-Aluminum Bronze Alloys in Sea Water Service, *Mater. Perform.,* Vol 21, 1982, p 30–34
14. M.W. Mahoney, W.H. Bingel, S. Sharma, and R. Mishra, Microstructural Modification and Resultant Properties of Friction Stir Processed Cast NiAl Bronze, *Mater. Sci. Forum,* Vol 426–432, 2003, p 2843–2848
15. W. Palko, R. Fielder, and P. Young, Investigation of the Use of Friction Stir Processing to Repair and Locally Enhance the Properties of Large NiAl Bronze Propellers, *Mater. Sci. Forum,* Vol 426–432, 2003, p 2909–2914
16. E.A. Culpan and G. Rose, Corrosion Behaviour of Cast Nickel Aluminum Bronze in Sea Water, *Br. Corros. J.,* Vol 14, 1979, p 160–166
17. G.M. Weston, "Survey of Nickel-Aluminum Bronze Casting Alloys on Marine Applications," DSTO MRL-R807, Australia Department of Defense Report, Melbourne, April 1981, p 1–21
18. P. Brezina, Heat Treatment of Complex Aluminum Bronzes, *Int. Met. Rev.,* Vol 27, 1982, p 77–120
19. M. Hansen and K. Anderenko, *Constitution of Binary Alloys,* 2nd ed., McGraw-Hill, 1958, p 84–89
20. E.A. Culpan and G. Rose, Microstructural Characterization of Cast Nickel Aluminium Bronze, *J. Mater. Sci.,* Vol 13, 1978, p 1647–1657
21. A. Cohen, Ed., *Properties and Selection: Nonferrous Metals and Special-Purpose Materials,* Vol 2, *Metals Handbook,* 10th ed., ASM International, 1990, p 386–387
22. G.W. Lorimer, F. Hasan, J. Iqbal, and N. Ridley, Observation of Microstructure and Corrosion Behaviour of Some Aluminum Bronzes, *Br. Corros. J.,* Vol 21, 1986, p 244–248
23. D.M. Lloyd, G.W. Lorimer, and N. Ridley, Characterization of Phases in a Nickel-Aluminium Bronze, *Met. Technol.,* Vol 7, 1980, p 114–119
24. F. Hasan, G.W. Lorimer, and N. Ridley,

Precipitation during the Tempering of a Nickel-Aluminum Bronze, *Int. Conf. on Solid-to-Solid Phase Transformations, Proceedings* (Pittsburgh, PA), TMS, 1982, p 745–749
25. F. Hasan, A. Jahanafrooz, G.W. Lorimer, and N. Ridley, The Morphology, Crystallography, and Chemistry of Phases in As-Cast Nickel-Aluminum Bronze, *Metall. Trans. A,* Vol 13, 1982, p 1337–1345
26. F. Hasan, G.W. Lorimer, and N. Ridley, Crystallography of Martensite in a Cu-10Al-5Ni-5Fe Alloy, *J. Phys.,* Vol 43, 1982, p C4 653–658
27. A. Jahanafrooz, F. Hasan, G.W. Lorimer, and N. Ridley, Microstructure Development in a Complex Nickel-Aluminum Bronze, *Metall. Trans. A,* Vol 14, 1983, p 1951–1956
28. F. Hasaan, G.W. Lorimer, and N. Ridley, Tempering of Cast Nickel-Aluminium Bronze, *Met. Sci.,* Vol 17, 1983, p 289–295
29. F. Hasan, J. Iqbal, and N. Ridley, Microstructure of As-Cast Aluminium Bronze Containing Iron, *Mater. Sci. Technol.,* Vol 1, 1985, p 312–315
30. P. Weill-Couly and D. Arnaud, Influence de la Composition et de la Structure des Cupro-Aluminiums sur leur Comportment en Service, *Fonderie,* No. 322, 1973, p 123–135
31. M.W. Mahoney, Rockwell Scientific Company, Thousand Oaks, CA, private communication, Dec 2002
32. O.D. Sherby, B. Walser, C.M. Young, and E.M. Cady, Superplastic Ultra-High Carbon Steels, *Scr. Metall.,* Vol 9, 1975, p 569–574
33. B. Walser and O.D. Sherby, Mechanical Behavior of Superplastic Ultrahigh Carbon Steels at Elevated Temperature, *Metall. Trans. A,* Vol 10, 1979, p 1461–1471
34. M.W. Mahoney, C. Fuller, W.H. Bingel, and M. Calabrese, Friction Stir Processing of Cast NiAl Bronze, *Mater. Sci. Forum,* in press
35. C.B. Fuller, M.W. Mahoney, W.H. Bingel, M. Calabrese, and B. London, Tensile and Fatigue Properties of Friction Stir Processed NiAl Bronze, *Mater. Sci. Forum,* in press

## SELECTED REFERENCES

- C.R. Brooks, *Heat Treatment, Structure and Properties of Nonferrous Alloys*, American Society for Metals, 1982, p 275–327
- E.F. Nippes, Ed., *Welding Soldering and Brazing*, Vol 6, *Metals Handbook*, 9th ed., American Society for Metals, 1983, p 400–427, 1033–1048

# CHAPTER 9

# The Corrosion Properties of Friction Stir Welded Aluminum Alloys

Jesse Lumsden, Rockwell Scientific Company

THE MOST COMMON application of friction stir welding is joining aluminum alloys. Of particular interest is that the technology has been shown to be an effective method of joining the previously considered unweldable precipitation-hardened high-strength 2*xxx* and 7*xxx* aluminum alloys. The strength of friction stir welds in these alloys is 30 to 50% greater than that of conventional fusion welds, while the fatigue life is comparable to that of riveted components (Ref 1). Although the welds have excellent mechanical properties, investigations show that corrosion in the weld zones of some friction stir welded (FSW) aluminum alloys is a serious issue and may limit their application. Relatively few investigations have been performed on the corrosion properties of other FSW alloy systems. Thus, the scope of this chapter is limited to FSW aluminum alloys.

It is well known that corrosion properties of aluminum alloys depend critically on microstructure (Ref 2). Because the friction stir welding process produces unique microstructures (Ref 3, 4), the weld zone can have corrosion properties different from those of the parent material. An example of the type of microstructure produced is shown in the optical cross-sectional view of a weld zone in Fig. 9.1. This figure is a transverse view in the weld travel direction. The weld advancing side, that is, the side where the tangential rotating tool velocity opposes the travel direction, is characterized by the extended flow at the top of the nugget. The weld nugget, or stir zone, has a fine-grain, fully recrystallized microstructure. Adjacent to the nugget there is a partially recrystallized zone, due to the coexistence of both deformed grains and a necklace of recrystallized fine grains in the grain boundaries of the deformed grains. This zone is referred to by some authors as the thermomechanically affected zone (TMAZ). There is a small transition zone with parent metal grains deformed upward (Fig. 9.1) toward the top surface, with fine grains located within the grain boundaries. The amount and direction of deformation depends on tool design. For example, a scroll shoulder tool has recently been developed that produces downward-directed grains. The TMAZ is a transition-zone microstructure between the fully recrystallized nugget and the conventional pancake-grain parent metal.

In general, friction stir welding creates a sharp temperature gradient from the weld nugget over a short distance into the parent metal. Typically, in the TMAZ immediately adjacent to the weld nugget, the peak temperature approaches 500 °C (930 °F), the time at peak temperature is short (measured in seconds), and cooling to room temperature is relatively rapid (~1 min). Figure 9.2 shows typical temperature transients measured with embedded thermocouples at the distances shown from the weld centerline. As far as 18 mm (0.7 in.) from the centerline, the temperature peaks at approximately 70% of the melting point and stays above the T7 and T6 tempering temperature for AA7075 for almost a minute. These temperatures are sufficiently high and times at temperature are long enough to cause dissolution, nucleation, and/or coarsening of the strengthening precipitates in precipitation-hardened aluminum alloys. The change in the temperature-time profile with distance from the nugget causes a heat-affected zone (HAZ) with a gradient in microstructure and precipitate morphology different from the parent material (Ref 3, 4). Chapters 3, 4, and 10 give a detailed description of thermal and microstructural aspects.

## 9.1 Sensitization

Some of the initial investigations (Ref 5, 6) of the corrosion properties of FSW 2*xxx* and 7*xxx* aluminum alloys showed that areas of the weld zone are sensitized. The weld zone is defined as that area where the temperature increased during friction stir welding. The thermal transients experienced during friction stir welding alter the grain-boundary chemistry and the chemistry in the grain near the boundary to selectively sensitize grain-boundary regions to varying degrees. This observation is in accordance with an early investigation of sensitization of aluminum alloys, which showed that there are well-defined time-temperature conditions where a transformation from pitting to intergranular corrosion occurs (Ref 7). It was shown that AA2024 followed a time-temperature-corrosion C-curve behavior. The samples used to generate this curve were quenched rapidly from the solution treatment temperature and held for varying periods at a series of intermediate temperatures. It has been reported (Ref 8) that time-temperature-corrosion curves developed for AA7150-W and AA7055-W are qualitatively similar to those for AA2024-T4, which suggests that time-temperature-corrosion C-curve behavior exists for all 2*xxx* and 7*xxx* aluminum alloys. The peak temperature, time at temperature, and cooling rate during friction stir welding determine the final chemical imbalance in the grain-boundary regions. Tool travel speed influences the time at peak temperature and, in all cases, times above critical temperatures are measured in seconds. Figure 9.3 shows a time-temperature-corrosion C-curve for AA2024 (Ref 7), modified by superimposing the time-temperature band simulating the cooling behavior in the most susceptible FSW corrosion region. Clearly, the peak temperature during friction stir welding is above the

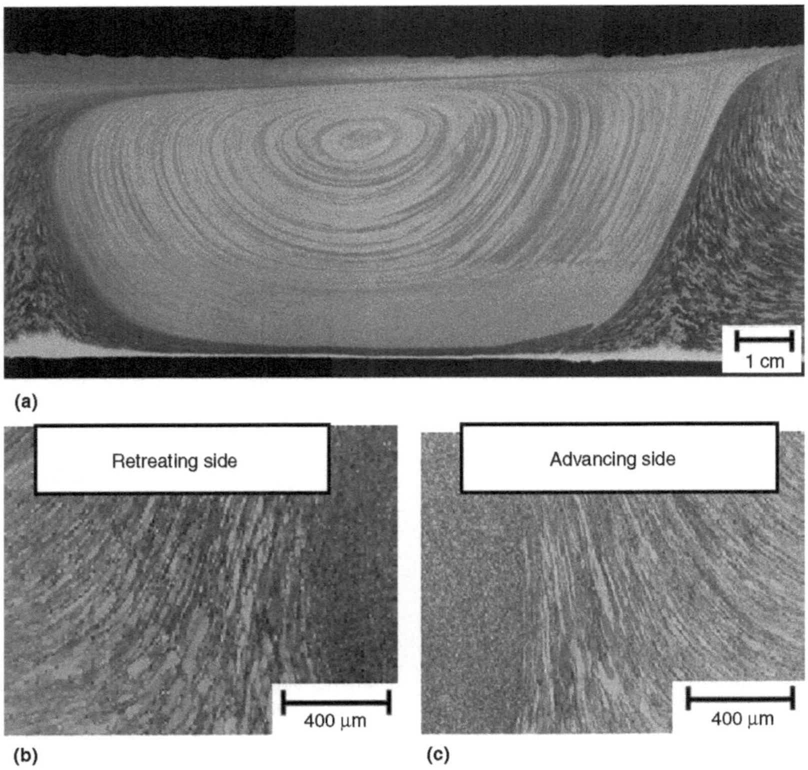


**Fig. 9.1** Macrograph of the weld zone in friction stir welded AA7050-T7651, and micrographs illustrating uplifted grains on the advancing and retreating sides

knee of the transformation curve, and the cooling rate is sufficiently slow to traverse through the intergranular corrosion zone, creating grain boundaries susceptible to attack. The time-temperature band is from results using a conventional tool (flat shoulder and threaded pin) at a rotation rate of 250 rpm and a travel speed of 2 cm/min. These will vary for different tool designs and processing parameters.

A simple immersion test (Fig. 9.4) shows the location and variation in the degree of sensitization relative to the stir zone in FSW AA7075-T651 (Ref 9). After the initial 24 h of exposure, the attack is limited to the HAZ and the outer

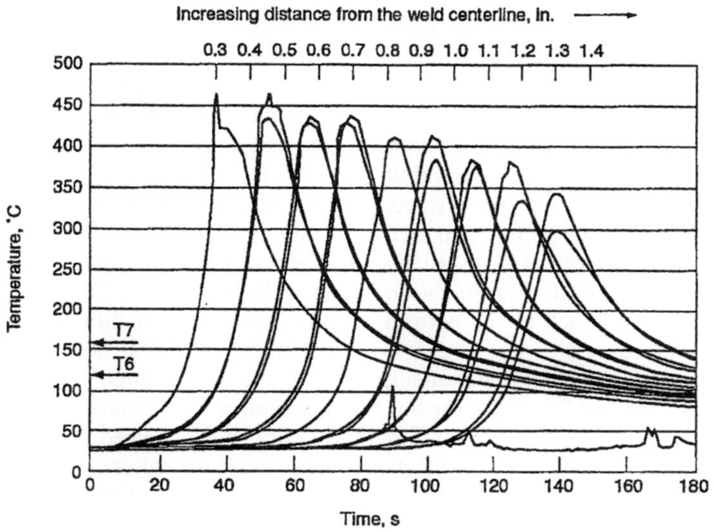


**Fig. 9.2** Temperature-time transients during friction stir welding at various distances from the weld centerline. Also shown are age temperatures for the T6 and T7 tempers

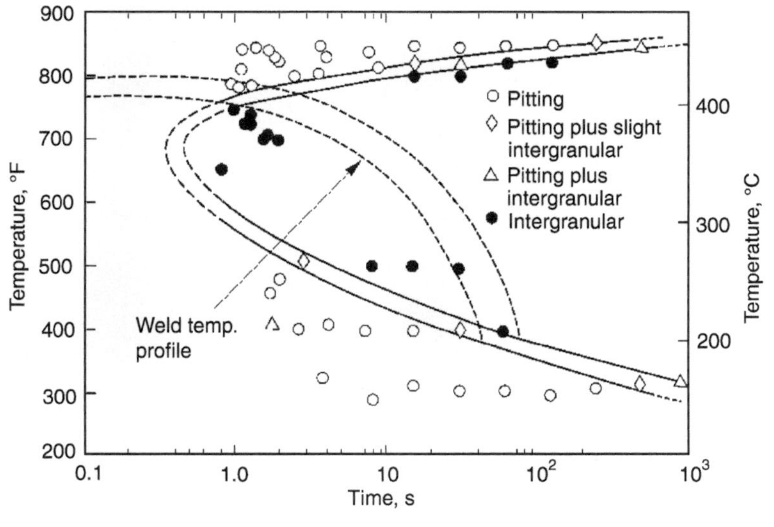


**Fig. 9.3** Effect of temperature and time in interrupted-quenching experiments on type of corrosion attack developed in 2024-T4 aluminum sheet by an accelerated corrosion test. Time-temperature band added to illustrate the cooling path associated with the most sensitized region of the heat-affected zone following friction stir welding. Source: Ref 7

edges of the TMAZ, indicating that this is the most heavily sensitized region. For this exposure time, neither the stir zone nor the parent material is attacked. After approximately 72 h of exposure, the intergranular attack in the HAZ has become more severe and now includes all regions of the previously unattacked TMAZ. In the TMAZ, the attack is unevenly distributed, even after 120 h, indicating an inhomogeneous microstructure and/or an inhomogeneous precipitate morphology following friction stir welding deformation. For all observations, the corrosion attack in the weld region was intergranular, whereas no intergranular corrosion was detected in the parent metal. The shape of the initial corrosion zone in Fig. 9.4 is indicative of the thermal behavior associated with friction stir welding. That is, the shoulder of the rotating tool acts as a heat source on the top surface, whereas support tooling on the bottom acts as a heat sink. Thus, as heat penetrates into the plate, it diffuses, creating a larger area of sensitization and subsequent corrosive attack as the bottom of the plate is approached.

It is widely believed that intergranular attack in aluminum-copper 2*xxx* and 7*xxx* alloys is due to the precipitation of copper-rich intermetallics in the grain boundaries. This leads to copper-depleted zones adjacent to the grain boundaries. Indeed, energy-dispersive spectroscopy, performed in conjunction with transmission electron spectroscopy, has detected a lower copper concentration in the sensitized grain boundaries of FSW AA7075 relative to that in the grain boundaries of the parent material (Ref 9, 10). Two mechanisms, which are not mutually exclusive, have been proposed for how precipitation of copper-rich intermetallics increases susceptibility to intergranular corrosion. Intermetallics have been shown (Ref 11, 12) to be cathodically active by catalyzing the oxygen reduction reaction. It is hypothesized (Ref 12) that the copper-rich intermetallics in the grain boundary establish local galvanic cells with the adjacent regions, accelerating their corrosion. Another hypothesized mechanism links intergranular corrosion in the weld zones with increased pitting susceptibility. Galvele et. al (Ref 13) have shown, using aluminum-copper binaries, that pitting resistance decreases (pitting potential decreases) as the copper content decreases. They have correlated this with intergranular corrosion. Thus, intergranular corrosion occurs only if chloride ions are present and the electrochemical potential is above the pitting potential, which is lower in the copper-depleted grain-boundary region than the grain interior. It was indeed found that FSW AA7075 had severe intergranular corrosion after exposure to a 4*N* NaCl-0.5*N* $KNO_3$-0.1*N* $HNO_3$ solution diluted to 10%, but there was no intergranular corrosion when 4*N* $Na_2SO_4$ was substi-

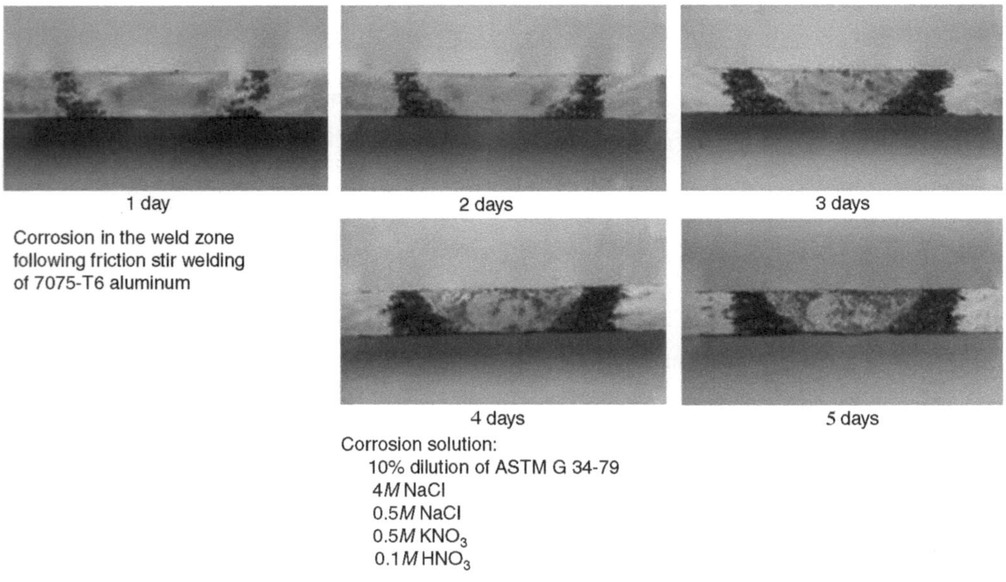


**Fig. 9.4** Corrosion attack of AA7075-T651 aluminum following extended exposure to a solution of 4*N* NaCl-0.5N $KNO_3$-0.1*N* $HNO_3$ diluted to 10%.

tuted for 4*N* NaCl, although the pH was the same (Ref 9).

## 9.2 Test Methods

Immersion tests provide rapid assessment of susceptibility to exfoliation corrosion and the location and severity of sensitization (Ref 14, 15). Whereas the HAZ and TMAZ have microstructures that can potentially be susceptible to exfoliation corrosion, the microstructure of the stir zone makes it resistant to exfoliation corrosion. This type of corrosion occurs in aluminum alloys because of the combination of microstructural effects and susceptibility to intergranular corrosion. The insoluble corrosion products resulting from integranular attack occupy a volume greater than the dissolved metal. These corrosion products produce a wedging stress on the highly elongated, pancakelike grain structure, thereby imparting the characteristic layered appearance of exfoliation corrosion. The equiaxed 5 to 15 μm grains in the stir zone are much less susceptible to exfoliation corrosion than the conventionally processed, approximately 200 μm flat-grain material. The highly aggressive solutions used in immersion tests sometimes severely attack the parent material as well as the weld zone, not giving the selectivity to identify the sensitized area and/or the degree of sensitization. A diluted form (10% by volume) of the ASTM solution (Ref 14) used in the exfoliation test, which is 4*N* NaCl-0.5*N* $KNO_3$-0.1*N* $HNO_3$, has been found to give satisfactory results (Ref 5, 9). Figure 9.5 shows examples of the type of results obtained after exposing FSW samples to the diluted 4*N* NaCl-0.5*N* $KNO_3$-0.1*N* $HNO_3$. These low-magnification micrographs were taken after exposing FSW AA7075-T651 for 24 h (Ref 9). Others (Ref 12) have immersed specimens in NaCl or used the ASTM procedure (Ref 15) for testing for susceptibility to intergranular corrosion, which is immersion in a 1 *M* NaCl-0.1 *M* $H_2O_2$ solution.

A useful technique (Ref 16) allows the in situ visual evaluation of the corrosion active region. Figure 9.6 shows the concepts involved. An agar gel containing NaCl and a wide-range pH indicator is placed on the specimen. The gel is cast into a 1 to 2 mm (0.04 to 0.08 in.) thick plate. Pieces are cut from the plate and placed onto the specimen. When aluminum ions from an active corrosion site enter the moisture-laden gel, electrons flow from the active corrosion site to other areas on the surface, where charge is conserved by oxygen reduction. The aluminum ions in the gel undergo hydrolysis, generating hydrogen ions. This creates a localized acidic region. Oxygen reduction produces OH ions and a basic region on the surface. The colors produced by the wide-range pH indicator locate the acidic and basic regions. The chemical reactions are as follows:

$Al_{solid} \rightarrow Al^{3+}{}_{aq} + 3e^-$ Corrosion

$Al^{3+}{}_{aq} + H_2O \rightleftharpoons AlOH^{2+} + H^+{}_{aq}$ Hydrolysis

$O_2 + 2H_2O + 4e^- \rightarrow 4OH^-{}_{aq}$ Reduction of oxygen

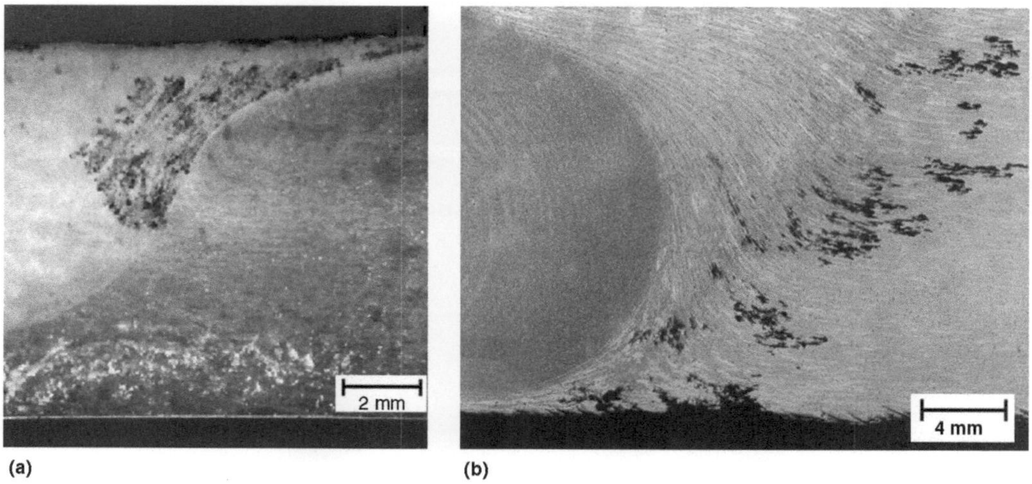


(a) (b)

**Fig. 9.5** Intergranular attack in the most heavily sensitized region of the weld zones in (a) friction stir welded (FSW) AA7050-T651 and (b) FSW AA7075-T651 after a 2-h immersion in a 10% dilution of 4*N* NaCl-0.5*N* $KNO_3$-0.1 *N* $HNO_3$

An example (Ref 12) of the application of the gel technique to FSW AA2024-T351 is shown in Fig. 9.7. This figure shows how varying the welding parameters changes the location of the sensitized region. At low rotation rates, the pH indicator shows that the whole weld zone is acidic, indicating corrosion. The adjacent region is basic and thus cathodicaly protected

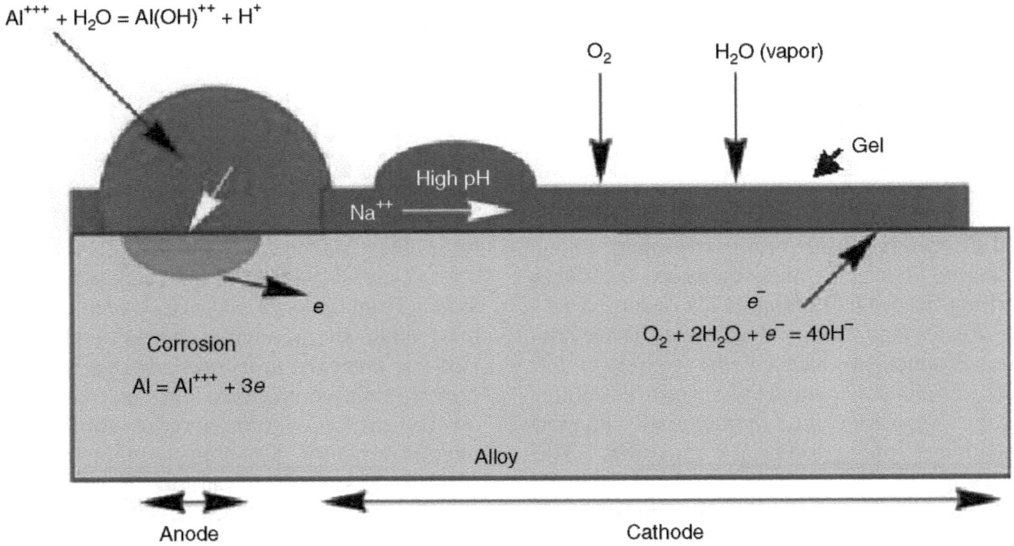


**Fig. 9.6** Schematic indicating how an agar gel containing sodium chloride and a wide-range pH indicator shows the corrosion and cathodic sites on a freely corroding nickel-aluminum-bronze material

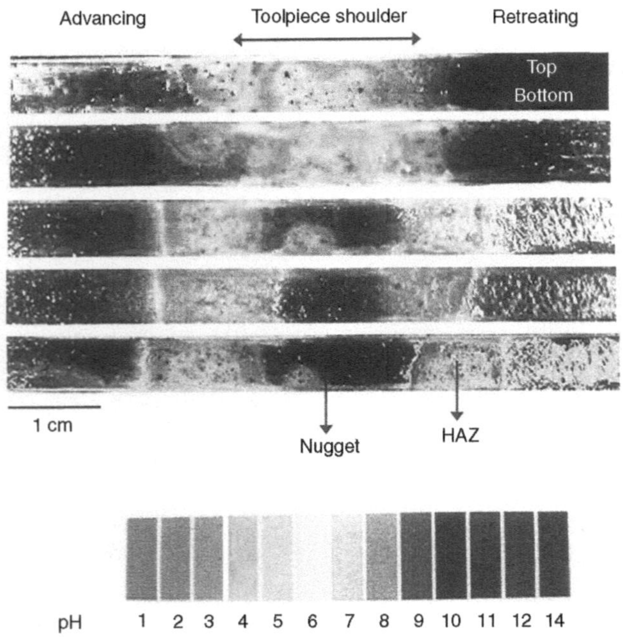


| Sample | rpm | mm/min |
|---|---|---|
| FSW *SF* | 215 | 154 |
| FSW *SS* | 215 | 75 |
| FSW *MM* | 350 | 95 |
| FSW *FF* | 468 | 154 |
| FSW *FS* | 468 | 75 |

**Fig. 9.7** An example of results using the gel visualization technique showing how changing the tool rotation rate and travel speed changes the location of the cathodic and anodic (corroding) regions of the weld zone in friction stir welded (FSW) AA2024-T351. HAZ, heat-affected zone. Source: Ref 12

against corrosion. At higher rotation rates, the HAZ is corroding (acidic), and the stir zone is cathodically protected (basic).

Pitting is a major corrosion mechanism of aluminum alloys. The sensitized regions of a weld zone will be more susceptible to pitting than the parent material. Electrochemical techniques provide a means of quantifying the pitting susceptibility. Potentiodynamic scans of the anodic polarization kinetics are the most frequently used technique. The sample is placed in a deaerated chloride solution and scanned in the anodic direction. When the passive film breaks down by pitting and loses its protective properties, there is an abrupt order of magnitude or larger increase in the anodic current density. The potential at which this occurs is called the breakdown or pitting potential. The lower this potential is relative to that of the parent material, the more susceptible to pitting corrosion. Experimentally, it is difficult to mask a sample so that only a small area of the weld zone is exposed for making potentiodynamic polarization. One technique (Ref 9) is to make cuts through the sample so that the surface of interest is a uniform microstructure, either all in the stir zone or, as a first approximation, a plane in the TMAZ or HAZ with all points the same distance from the stir zone. Recently, a microelectrode technique (Ref 12) has been used that has a spatial resolution of approximately 1 mm (0.04 in.). This technique allows polarization curves to be taken at multiple positions across the weld zone.

The test that has been most frequently used to evaluate the stress-corrosion cracking (SCC) susceptibility of FSW alloys has been the slow-strain rate (SSR) test, described in ASTM G 129 (Ref 17). The procedure involves the application of very slow strain rates ($10^{-7} \cdot$ to $10^{-4} \cdot s^{-1}$), achieved by a constant extension rate on a tension specimen while monitoring load and extension of the specimen. The test solution is typically 3.5% NaCl, which is the approximate concentration of NaCl in seawater. The solution is sometimes made more aggressive by adding $H_2O_2$. The sample geometry is such that the gage length is sufficiently long to include the stir zone and both HAZs. The grips are electrically isolated from the test apparatus, preventing galvanic couples. The degree of susceptibility to SCC is assessed through measurement of differences between the responses in tension tests to samples in a potentially environmentally aggressive test environment compared to tests conducted in a controlled environment. Changes in time to failure, specimen elongation or reduction in area, and/or metallographic evidence of selective corrosion attack are used to determine susceptibility to SCC. The SSR test has been chosen for several reasons. First, the smooth specimen configuration allows initiation of SCC at any sensitized grain boundary that intersects the surface. Second, the exact location of the sensitized region relative to the FSW fusion zone is not well defined and changes location during natural and artificial aging. The SSR test evaluates the entire FSW zone. Third, the test is rapid and gives easily quantifiable results in terms of elongation to failure, reduction in cross-sectional area, or percentage of intergranular fracture. Fourth, the SSR technique accelerates the processes believed to be key elements of SCC, that is, crack tip strain rate and rupture of the passive film.

The deformation across a friction stir weld is mainly concentrated in the HAZs, resulting in necking in both regions (Ref 4). When testing transverse specimens, the inhomogeneous distribution of stress across the weld must be considered in interpreting the results, especially when comparing results with the parent metal.

The alternate immersion test method provides a second procedure for accelerated stress-corrosion testing of high-strength aluminum alloys. Conventional dog-bone tensile specimens are spring loaded in tension (in accordance with ASTM G 49, Ref 18) and exposed to alternate immersion in 3.5% NaCl (in accordance with ASTM G 44, Ref 19) for a defined test duration or to failure.

Other SCC test procedures exist. However, these involve bending the specimens or applying local stresses. It is always a concern with such techniques that the load is applied in the most sensitized region.

## 9.3 Alloy Systems

**2*xxx* Aluminum Alloys.** Test results have been reported on FSW AA2014 for the resistance to exfoliation corrosion, intergranular corrosion, and SCC (Ref 20). Tests were run on 6 mm (¼ in.) plates. The details of the welding procedure were not given. Intergranular corrosion resistance was assessed using ASTM G 110-92. The depth and severity of intergranular corrosion was the same for stir zone, TMAZ, and parent material. A portion of the HAZ was not susceptible to intergranular corrosion. Exfoliation corrosion tests were conducted in accordance with ASTM G 34-97.

The most extensively studied FSW alloy in the 2*xxx* system is FSW AA2024. One of the earliest papers was by Biallas et al. (Ref 20). They joined 1.6 and 4 mm (0.06 and 0.15 in.) thick plates of AA2024-T3 using a conventional milling machine and the procedure described in Ref 21. The ASTM G 34 test procedure produced severe exfoliation corrosion in the TMAZ, with pitting and blistering in the stir zone. There was no dependence of the exfoliation corrosion in the 1.6 mm specimens when the rotation speed was varied from 800 to 2400 rpm and the travel speed from 80 to 240 mm/min. Intergranular corrosion and pitting occurred in the stir zone when specimens were immersed in the ASTM G 110 solution for 6 h. The pit depth in the stir zone, 150 μm, was less than that found in the region adjacent to the weld and the parent material, which was 200 mm. Bent-beam specimens, fabricated from the 4 mm FSW plates, did not fail during 40 days of alternate immersion testing. Metallographic examination of the specimens after testing showed pitting and intergranular corrosion in all areas of the weld zone as well as the parent material.

A wide-ranging investigation of the effects of processing parameters (rotation rate and travel speed) on the electrochemical and corrosion properties of FSW AA2024-T351 has been carried out by other investigators (Ref 12, 22, 23). The gel visualization technique and immersion tests in 0.1 *M* NaCl for 24 h show that the location of the sensitized region is dependent on the rotation rate and that the width of the sensitized region depends on the rotation rate and travel speed (Fig. 9.7). Rotation rates of 215 to 468 rpm and travel speeds of 75 to 154 mm/min (3 to 6 in./mm) were used. At low rotation rates, the corrosion occurred in the stir zone and spread into the HAZ. At a given rotation rate, the width of the corrosion zone decreased with increasing travel speed. At a given travel speed, the width of the corrosion zone increased with increasing rotation rate. Microscopy showed that the corrosion was intergranular attack (IGA). The rate of IGA in the stir zone was greater at slow rotation rates than at medium and fast rotation rates. The rate of IGA was greater in the HAZ at higher rotation rates. Pitting potentials were measured by potentiodynamic polarization curves. The pitting potentials in the stir zone and the HAZ were similar and significantly lower than in the parent material. Microscopy after the polarization curves were obtained showed pits around constituent particles, whereas IGA occurred in the stir zone and HAZ. This shows that the composition in the grain boundaries was changed in such a way that the pitting potential in the boundaries was lowered. The continuous attack occurred as the pits linked up after they initiated.

There is little information available in the literature about FSW AA2*xxx* alloys other than FSW AA2024. Some SCC results for FSW AA2219 (Ref 24, 25) and corrosion fatigue results for FSW AA2519 (Ref 26) have been reported. No SCC was observed in FSW AA2219 in the four-point bend test under alternate immersion conditions in a 3.5% NaCl solution for 90 days. Specimens were loaded at 50, 65, and 85% of the ultimate tensile strength (UTS). There was also no loss in the UTS in specimens tested by the SSR technique in air after pre-exposure to alternate immersion conditions for 90 days. A strain rate of $3.3 \times 10^{-5} \cdot s^{-1}$ was used. Some SSR results indicate that the weld zone in FSW AA2219 is more resistant to SCC than the parent material (Ref 27). Corrosion fatigue tests on FSW AA2519 in 3.5% NaCl had a similar fatigue life and fatigue threshold as the parent material.

**5*xxx* Aluminum Alloys.** Evaluations of the corrosion properties of FSW AA5454 (Ref 27, 28) and FSW AA5456 (Ref 29) have been reported. The pitting potentials of FSW AA5454 in the O and H34 tempers were determined from potentiodynamic polarization curves and compared with those from the parent material and gas tungsten arc welded (GTAW) material. The pitting potentials of the FSW material were slightly lower than those of the parent material, but significantly higher than the GTAW material. This suggests that the FSW material has approximately the same pitting resistance as the parent material, but substantially more pitting resistance than the GTAW material (Ref 27). ASTM International tests for susceptibility to exfoliation corrosion and intergranular corrosion susceptibility showed no difference in FSW AA5454-O and the parent material (Ref 28).

The SCC resistance of FSW AA5454 in the O and H34 tempers has been evaluated by the SSR technique and by U-bend specimens. Both tests were performed with specimens immersed in a deaerated 0.5 *M* NaCl solution while polarized at 60 mV above the open-circuit potential (Ref 28) and at the open-circuit potential (Ref 29). This potential was chosen because test material exposed to an air-saturated 0.5 *M* NaCl solution has an electrochemical potential approximately 60 mV above open circuit. The results are inconclusive. The SSR results indicate that both tempers are susceptible to SCC, where failure occurred in the HAZ close to the stir zone. The U-bend tests indicate no susceptibility.

The pitting resistance of FSW AA5456-H116 was measured by determining the pitting potential in 0.1 *M* NaCl using the potentiodynamic polarization technique. A microelectrochemical method was used. This enabled polarization scans from several points in the HAZ, TMAZ, and stir zone. There was no significant difference between the pitting potentials in the HAZ, TMAZ, and parent material. The pitting potentials in the stir zone were significantly less. When the material was immersed in phosphoric acid, attack was primarily confined to the stir zone. Phosphoric acid has been shown to preferentially attack regions of beta-phase ($Mg_2Al_3$) precipitation. This suggests that the reduced pitting resistance of the stir zone is caused by beta-phase intermetallics (Ref 12).

**6*xxx* Aluminum Alloys.** A limited amount of corrosion test data exists for FSW AA6056-T6, 6013-T6, and 6061-T4. The ASTM tests on FSW AA6056-T6 and FSW AA6013-T6 for intergranular corrosion susceptibility, SCC, and exfoliation corrosion susceptibility indicate that the FSW material has the same susceptibility as the parent material (Ref 30). The same tests on FSW AA6061-T4 give the same results as those on the other two 6*xxx* alloys, except for localized corrosion at the weld interface.

**7*xxx* Aluminum Alloys.** Many of the investigations (Ref 5, 9, 10, 31–36) of the corrosion properties of FSW alloys have been on AA7050 and AA7075. Both of these alloys undergo severe sensitization when they are friction stir welded. Potentiodynamic polarization curves give a 50 to 75 mV decrease in the pitting potential in the weld zones relative to the parent material. The SSR technique, using 3.5% NaCl, has shown that the weld zone in FSW AA7050-T7651 and both tempers, T7651 and T651, of FSW AA7075 are susceptible to SCC. Although these alloys have very similar compositions, the location of the most heavily sensitized region in the weld zone is different for each alloy. As shown in Fig. 9.5, intergranular corrosion initially occurs at the stir zone/TMAZ boundary in FSW 7050 and at the TMAZ/HAZ boundary in FSW AA7075.

The SCC fracture zone in these alloys, produced by the SSR technique, is in accordance with the immersion test results. When the welds in these two alloys are tested for SCC susceptibility in the transverse direction, the FSW AA7050 fractures on the stir zone side of the stir zone/TMAZ boundary (Fig. 9.8), and FSW AA7075 fractures on the HAZ side of the TMAZ/HAZ interface (Fig. 9.9). An analysis of the strain-rate effects on the elongation to failure and the fractography suggests that FSW AA7050 is much less resistant to SCC than FSW AA7075, although the situation is reversed for the parent material. Results for FSW AA7050 at a strain rate of $3 \times 10^{-5} \cdot s^{-1}$ produced a fracture that was totally intergranular through the fine-grain microstructure of the stir zone (Fig. 9.8), whereas the fracture mode of FSW AA7075 at this strain rate was transgranu-

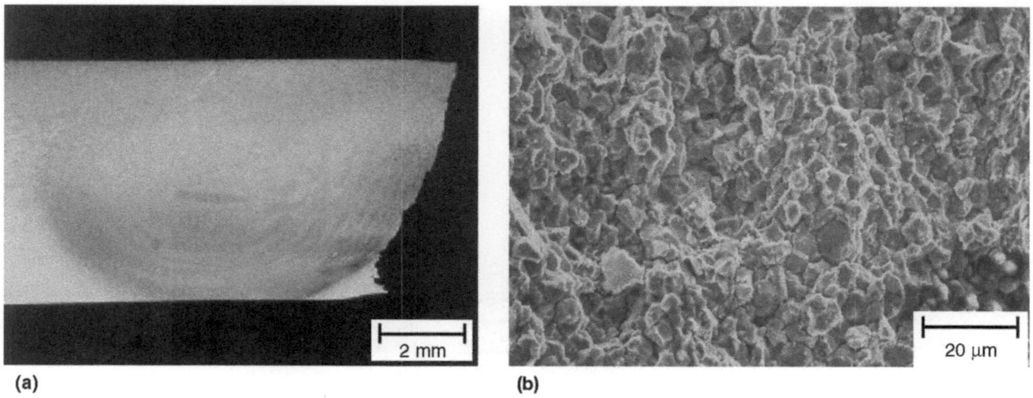

(a) (b)

**Fig. 9.8** Stress-corrosion cracking fracture in friction stir welded AA7050-T7651. (a) Optical micrograph showing cross-sectional view. (b) Scanning electron microscopy micrograph of the fracture face lar and ductile. When FSW AA7075 is tested at a strain rate of $3 \times 10^{-6} \cdot s^{-1}$, the fracture is approximately 20% intergranular (Fig. 9.9). The fracture surface on the FSW AA7075 specimen has deep grooves on the edges where the elongated grains pulled out when decohesion occurred. The tensile axis is in the long transverse direction.

Other 7*xxx* alloys investigated include FSW AA7476-T7351 (Ref 35, 36), FSW AA7349-T6 (Ref 37), and FSW AA7010-T7 (Ref 38). The corrosion properties of all are similar to FSW AA7050 and FSW AA7075, although the location of the most heavily sensitized zone varies. This shows that relatively small changes in composition allow a different grain-boundary microstructure and composition.

The metallurgical transformations in the 7*xxx* aluminum alloys produced during friction stir welding are very sensitive to small changes in the temperature-time transients that occur during the joining procedure. Even changes in tool design can affect these transients and improve or worsen the resistance to SCC (Ref 39.) The microstructures of the 7*xxx* aluminum alloys are unstable in the as-welded conditions. They naturally age and respond to artificial aging (Ref 40, 41), which has been shown to change the susceptibility to SCC and the location of the fracture zone.

## REFERENCES


1. P.F. Mende and T.A. Eager, *Adv. Mater. Process.* Vol 159, 2001, p 39
2. D.O. Sprowls and R.H. Brown, Stress-Corrosion Mechanisms for Aluminum Alloys, *Fundamental Aspects of Stress-Corrosion Cracking*, R.W. Staehle, Ed., NACE, 1969
3. C.G. Rhodes, M.W. Mahoney, W.H. Bingel, R.A. Spurling, and C.C. Bampton, Effects of Friction Stir Welding on Microstructure of 7075 Aluminum, *Scr. Metall.*, Vol 36, 1997, p 69–75
4. M.W. Mahoney, C.G. Rhodes, J.G. Flintoff, R.A. Spurling, and W.H. Bingel, Properties of Friction Stir Welded 7075 T651 *Aluminum, Metall. Mater. Trans. A*, Vol 29, July 1998, p 1955–1964
5. J. Lumsden, M. Mahoney, G. Pollock, D. Waldron, and A. Guinasso, "Stress-Corrosion Susceptibility in 7050-T7541 Following Friction Stir Welding," First International Symposium on Friction Stir Welding (Thousand Oaks, CA), TWI, 1999
6. G. Biallas, R. Braun, C.D. Donne, G. Staniek, and W.A. Kaysser, "Mechanical Properties and Corrosion Behavior of Friction Stir Welded 2024-T3," First


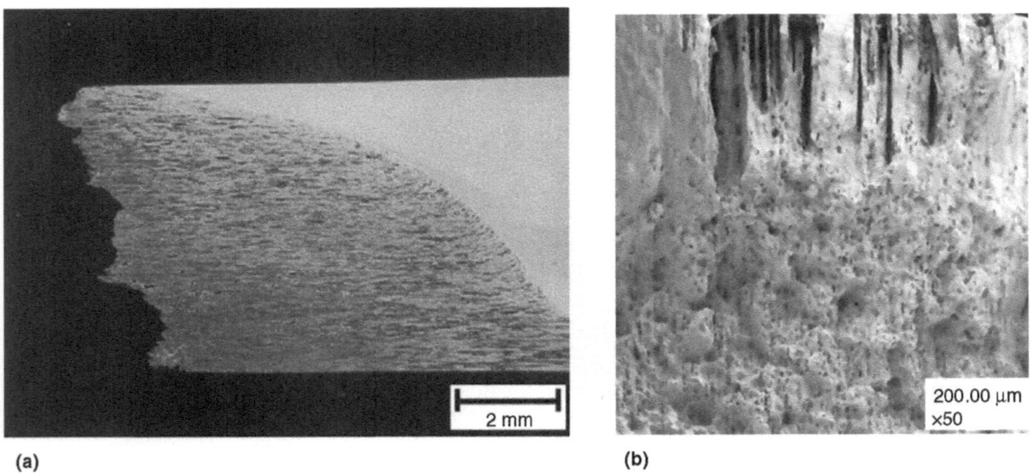


**Fig. 9.9** Stress-corrosion cracking fracture in friction stir welded AA7075-T7651. (a) Optical micrograph showing cross-sectional view. (b) Scanning electron microscopy micrograph of the fracture face

International Symposium on Friction Stir Welding (Thousand Oaks, CA), TWI, 1999
7. K.R. Van Horn, Ed., *Aluminum*, Vol 1, *Properties, Physical Metallurgy and Phase Diagrams*, American Society for Metals, 1967, p 140
8. J.T. Staley, S.C. Byrne, E.L. Colvin, and K.P. Palmer, *Mater. Sci. Forum*, Vol 217, 1996, p 1587
9. J.B. Lumsden, M.W. Mahoney, and C.G. Rhodes, Intergranular Corrosion Following Friction Stir Welding of 7075-T6 Aluminum, *Corrosion*, Vol 55, 1999, p 69–74
10. R.G. Buchheit and C.S. Paglia, Localized Corrosion and Stress-Corrosion Cracking of Friction Stir Welded 7075 and 7050, *Corrosion and Protection of Light Metal Alloys*, The Electrochemical Society, 2004, p 94
11. P. Schmutz and G.S. Frankel, Characterization of AA2024-T3 by Scanning Kelvin Probe Force Microscopy, *J. Electrochem. Soc.*, Vol 145, 1998, p 2285
12. B.J. Connolly, A. Davenport, S. Williams, R. Ambat, D. Price, M. Jariyaboon, A. Wescott, and C.J. Goodfellow, "Localized Corrosion of Friction Stir Welds in Aluminum Alloys," Fifth International Symposium on Friction Stir Welding (Metz, France), TWI, 2004
13. J.R. Galvele and S.M. De Micheli, *Corros. Sci.*, Vol 10, 1970, p 179
14. "Standard Test Method for Exfoliation Corrosion Susceptibility in 2*xxx*- and 7*xxx*-Series Aluminum Alloys (EXCO Test)," G 34-97, ASTM
15. "Standard Practice for Evaluating Intergranular Corrosion Resistance for Heat Treatable Aluminum Alloys by Immersing in Sodium Chloride + Hydrogen Peroxide Solution," G 110-92, ASTM
16. H.S. Isaacs, G. Adzic, and C.S. Jeffcoate, Visualizing Corrosion, *Corrosion*, Vol 56, 2000, p 971
17. "Standard Practice for Slow Strain Rate Testing to Evaluate the Susceptibility of Metallic Materials to Environmentally Assisted Cracking," G 129, ASTM
18. "Standard Practice for Preparation and Use of Direct Tension Stress-Corrosion Test Specimens," G 49-85, ASTM
19. "Standard Practice for Evaluating Stress-Corrosion Cracking Resistance of Metals and Alloys by Alternate Immersion in 3.5% Sodium Chloride Solution," G 44, ASTM
20. G. Biallas, R. Braun, C.D. Donne, G. Staniek, and W.A. Kaysser, "Mechanical Properties and Corrosion Behavior of Friction Stir Welded 2024-T3," First International Symposium on Friction Stir Welding (Thousand Oaks, CA), TWI, 1999
21. W.M. Thomas, E.D. Nicholas, J.C. Needham, M.G. Murch, P. Templesmith, and C.J. Dawes, Improvements Relating to Friction Welding, European Patent EP 0 615 480 B1, 1992
22. A. Ambat, M. Jariyaboon, A.J. Davenport, S.W. Williams, D. Price, and A. Wescott, "Micro-Electrochemical Investigation of Friction Stir Welds in Aluminum Aerospace Alloy 2024," 15th International Corrosion Congress (Granada, Spain), 2002
23. M. Jariyaboon, A.J. Davenport, R. Ambat, B.J. Connolly, S.W. Williams, and D.A. Price, The Effect of Welding Parameters on the Corrosion Behavior of Friction Stir Welded AA2024-T351, *Corros. Sci.*, in press
24. E.I. Meletis, P. Gupta, and F. Nave, Stress-Corrosion Cracking Behavior of Friction Stir Welded High-Strength Aluminum Alloys, *Friction Stir Welding and Processing II*, TMS, 2003, p 107
25. Z.X. Li, W.J. Arbergast, P.J. Hartley, and E.I. Meletis, Microstructure Characterization and Stress-Corrosion Evaluation of Friction Stir Welded Al 2195 and Al 2219 Alloys, *Proceedings of the Fifth International Conference on Trends in Welding Research*, ASM International and American Welding Society, 1998, p 568
26. P.S. Pao, E. Lee, C.R. Feng, H.N. Jones, and D.W. Moon, Corrosion Fatigue in FSW Welded Al 2519, *Friction Stir Welding and Processing II*, TMS, 2003, p 113
27. C.S. Paglia and R.G. Buchheit, Microstructure, Microchemistry, and Environmental Cracking Susceptibility of Friction Stir Welded 2219-T87, *Mater. Sci. Eng.*, in press
28. G.S. Frankel and Z. Xia, Localized Corrosion and Stress-Corrosion Cracking Resistance of Friction Stir Welded Aluminum Alloy 5454, *Corrosion,* Vol 55, 1999, p 139

29. A.P. Reynolds, Mechanical and Corrosion Performance of TGA and Friction Stir Welded Aluminum for Tailor Welded Blanks: Alloys 5454 and 6061, *Proceedings of the Fifth International Conference on Trends in Welding Research*, ASM International and American Welding Society, 1998, p 563
30. D. Lohwasser, Friction Stir Welding of Aerospace Alloys, *Proceedings of the Fifth International Conference on Trends on Welding Research*, 2003
31. J.B. Lumsden, M.W. Mahoney, C.G. Rhodes, and G.A. Pollock, Corrosion Behavior of FSW 7050-T7651, *Corrosion*, Vol 59, 2003, p 212
32. C.S. Paglia, M.C. Carroll, A.P. Reynolds, and R.G. Buchheit, Correlation between Localized Corrosion, Hardness and Microstructure in a 7075-T651 Friction Stir Weld, *Proceedings of Critical Factors in Localized Corrosion IV*, 2002–24, The Electrochemical Society, 2003, p 347–358.
33. R. Buchheit and C. Paglia, Localized Corrosion and Stress-Corrosion Cracking of Friction Stir Welded 7075 and 7050, *Proceedings of the Electrochemical Society*, PV2003-23, The Electrochemical Society, 2004, p 94
34. A.J. Leonard, "Corrosion Resistance of Friction Stir Welds in Aluminum Alloys 2014-T651 and 7075-T651," Third International Symposium on Friction Stir Welding (Kobe, Japan), TWI, 2004
35. A. Merati, K. Sarda, D. Raizenne, and C. Dalle Donne, Improving Corrosion Properties of Friction Stir Welded Aluminum Alloys by Localized Heat Treatment, *Friction Stir Welding and Processing II*, The Metals Society, 2003, p 77
36. M. Kumagai, S. Tanaka, H. Hatta, and H. Yoshida, "Integral Wing Panel for Airplanes Produced by Friction Stir Welded Extrusion," Third International Symposium on Friction Stir Welding (Kobe, Japan), TWI, 2001
37. D. Allehaux, G. Petit, M.H. Campagnac, G. Lapasset, and A. Denquin, "Microstructure and Properties of Friction Stir Welded 7349-T6 Aluminum Alloy," Fourth International Symposium on Friction Stir Welding (Park City, UT), TWI, 2005
38. F. Hannour, A.J. Davenport, and M. Strangwood, "Corrosion of Friction Stir Welds in High-Strength Aluminum Alloys," Second International Symposium on Friction Stir Welding (Gothenburg, Sweden), TWI, 2000
39. J. Lumsden, G. Pollock, and M. Mahoney, Effect of Tool Design on Stress-Corrosion Resistance of FSW AA7050-T7451, *Friction Stir Welding and Processing III*, The Metals Society, 2005, p 19
40. J. Lumsden, G. Pollock, and M. Mahoney, Effect of Post Weld Heat Treatments on the Corrosion Properties of FSW AA7050, *Friction Stir Welding and Processing II*, The Metals Society, 2003, p 99
41. J. Lumsden, G. Pollock, and M. Mahoney, The Effect of Thermal Treatments on the Corrosion Behavior of Friction Stir Welded 7050 and 7075 Aluminum Alloys, *Mater. Sci. Forum*, 2003, p 2867

# CHAPTER 10

# Process Modeling

Hugh R. Shercliff and Paul A. Colegrove
Department of Engineering, University of Cambridge

PROCESS INNOVATIONS invariably evolve empirically, using accumulated experience from laboratory trials. Friction stir welding (FSW) is no exception, with development initially based on aluminum alloys and later for other engineering alloys. The modern power of process modeling should, however, be exploited to support and guide experimental development work, in particular to accelerate takeup in industrial applications and reduce development costs. Modeling based on scientific understanding of the mechanisms and physical phenomena of FSW has still lagged behind but has great potential for guiding tool design, predicting likely operating conditions in new materials or joint geometries, and then optimizing process conditions for maximum process speed. Process modelers also seek to address the performance of FSW joints, for example, to predict the occurrence of voids and defects, the extent of microstructural and property changes in the deformed and heat-affected regions, and the development of residual stress.

This chapter discusses the status of modeling of FSW, based on reviews (Ref 1) by the authors to which readers are referred for more detail. Friction stir welding presents a multiphysics modeling challenge, because it combines closely coupled heat flow, plastic deformation at high temperature, and microstructure and property evolution. All three contribute to the processability of a material by FSW and to the subsequent properties of the weld. Figure 10.1 illustrates the key physical interactions involved in linking process and material parameters to the outputs needed by designers. The core process model governs the heat generation by plastic dissipation and friction between tool and workpiece and the subsequent thermal history imposed on the material. The metal flow needed to produce sound joints is determined by tool design and control of process parameters, such as the downforce, but is intimately linked to heat generation. The thermal history is central to the process—controlling material softening to enable efficient stirring and rapid traverse speeds, microstructure and property changes, and final residual stress and distortion.

In all process modeling, it is essential to keep the goals of the model in view and to adopt an appropriate level of complexity. Analytical and numerical methods each have a role to play, although numerical methods dominate due to the power and ease of use of modern workstations and software. Numerical modeling is based on discretized representations of specific welds, using finite element, finite difference, or finite volume techniques. These methods can capture much of the complexity in material constitutive behavior, boundary conditions, and geometry, but the computational penalty means that, in practice, a limited range of conditions tends to be studied in depth. Therefore, it is good modeling practice to explore simplifications to the problem that give useful insight across a wider domain, for example, making valid two-dimensional (2-D) approximations to inherently three-dimensional (3-D) behavior. It is also essential to deliver a model that is properly validated and whose sensitivity is known to uncertainty in the input material and process data—ideals that are rarely carried through in practice.

## 10.1 Analytical Estimates of Heat Generation

As discussed in Chapters 1 to 3, the tool shoulder provides heating and constrains the deformation zone, while the probe shapes the deformation path that seals the joint and also generates a proportion of the heat, depending on the tool dimensions. The tool rotates at high speeds, such that the peripheral speed of the shoulder and probe is very much greater than the translational speed. Friction stir welding primarily uses viscous dissipation in the workpiece material, driven by high shear stresses at the tool/workpiece interface. However, the boundary conditions in FSW are complex. Material at the interface may either stick to the tool, in which case it has the same local velocity as the tool, or it may slip, in which case the velocity may be lower and not in the same direction. The temperature and normal contact stresses vary widely over the tool, so it is unlikely that a single contact condition will be valid. Contact may be partially slipping and partially sticking, and if local melting occurs, there may be oscillating stick-slip behavior. The effect (or even existence) of local melting in FSW is a heavily debated topic. Peak temperatures close to the solidus temperature have certainly been measured, but heating above this temperature is physically limited; local melting of second-phase particles or eutectic microstructures will rapidly reduce the shear stress effectively to zero, leading to a steep drop in local heat input and temperature. Hence, the heat generation is self-stabilizing at near-solidus temperatures, and any melt volume must remain very small—too small to be evident in the final microstructure or (critically) to lead to problems associated with melting, such as liquation cracking.

Modeling the heat generation therefore requires some representation of the interface contact behavior together with the viscous dissipation behavior of the material. Simple analytical approaches are discussed in this section; numerical methods, in which heat generation is calculated directly from coupled models of the temperature field and metal deformation, are discussed in section 10.3, "Metal Flow."

The simplest estimates of average heat-generation rate (Ref 1) consider a purely rotating tool shoulder (neglecting the translation velocity and the probe) by analogy with conventional rotary friction welding. For a slipping contact, the power $q$ is given by:

$$q = \frac{2\pi}{3} \mu\, p\, \omega\, R_s^3 \qquad \text{(Eq 10.1)}$$

where $\mu$ is the coefficient of friction, $p$ is the normal pressure, $\omega$ is the angular velocity (radians/s), and $R_s$ is the tool shoulder radius (neglecting the central area occupied by the probe). For the limit of sticking friction, $\mu p$ is replaced by a constant shear yield stress $k$, so the average power generation is:

$$q = \frac{2\pi}{3} k\, \omega\, R_s^3 \qquad \text{(Eq 10.2)}$$

Both approaches require calibration. The average pressure below the tool may be estimated from the downforce (if measured), but the coefficient of friction is a parameter that can be adjusted within the physically meaningful range (≈0.2 to 0.5). Similarly, the shear yield stress will be of the order of half the tensile yield stress at temperatures approaching the solidus but will again be adjusted empirically.

Probe heating can also be estimated using sticking friction (normal pressure not being straightforward in this case). For a cylindrical probe of radius $R_p$ and length $L_p$, rotating at angular velocity $\omega$, the heat-generation rate is:

$$q = 2\pi\, k\, \omega\, L_p R_p^2 \qquad \text{(Eq 10.3)}$$

For given tool dimensions and assuming the same shear yield stress on probe and shoulder, the relative contribution from each may be estimated from Eq. 10.1 and 10.2. This shows that heat generation from the probe is negligible in thin plate but is typically 10% or more for thick plate. Finally, it should be noted that a fraction of the heat is also lost by conduction into the tool itself (typically on the order of 10% or less). This may either be estimated from a simple heat flow model for the tool or introduced as an adjustable efficiency factor in the net heat input.

Given the need to calibrate these estimates of heat input, they are best regarded as simple checks on experimental data. Modern FSW equipment routinely outputs torque, $T$, as well as angular velocity, so the total heat input from the machine may be directly estimated from the product $T\omega$ (neglecting the heat generation from translation that is approximately 1% of this value). Thermocouple measurements are then used for further refinement of the net heat input. A model based on heating at the tool interface must also describe the spatial distribution in heat input over the tool. This is considered further in the analysis of heat conduction in the subsequent section.

## 10.2 Heat Conduction

Thermal modeling to predict the temperature field in FSW is central to the problem (Fig. 10.1). It has been applied to optimize welding conditions and as input to the prediction of microstructure, properties, distortion, and residual stress. Thermal modeling is also closely coupled to the metal flow (section 10.3, "Metal Flow").

The thermal analysis of FSW means solving the partial differential equations for heat flow, subject to the imposed boundary conditions, to find the temperature field as a function of position and time. Modeling the heat flow in FSW must consider the following factors:

- Distribution and intensity of the heat input
- Heat losses, particularly to the tooling and backing plate
- Influence of the initial stationary dwell
- Transients along the weld induced by finite plate effects (e.g., heatup of the workpiece and backing plate may mean that steady-state conditions are not obtained)

### 10.2.1 Analytical Methods

The classical starting points for heat flow analysis, originally for arc welding, are the point and line source solutions for a moving heat source, due to Rosenthal (Ref 2, 3) and elegantly reassessed and extended by Myhr and Grong (Ref 4, 5). These solutions approximate the plate as being infinite in extent in two or

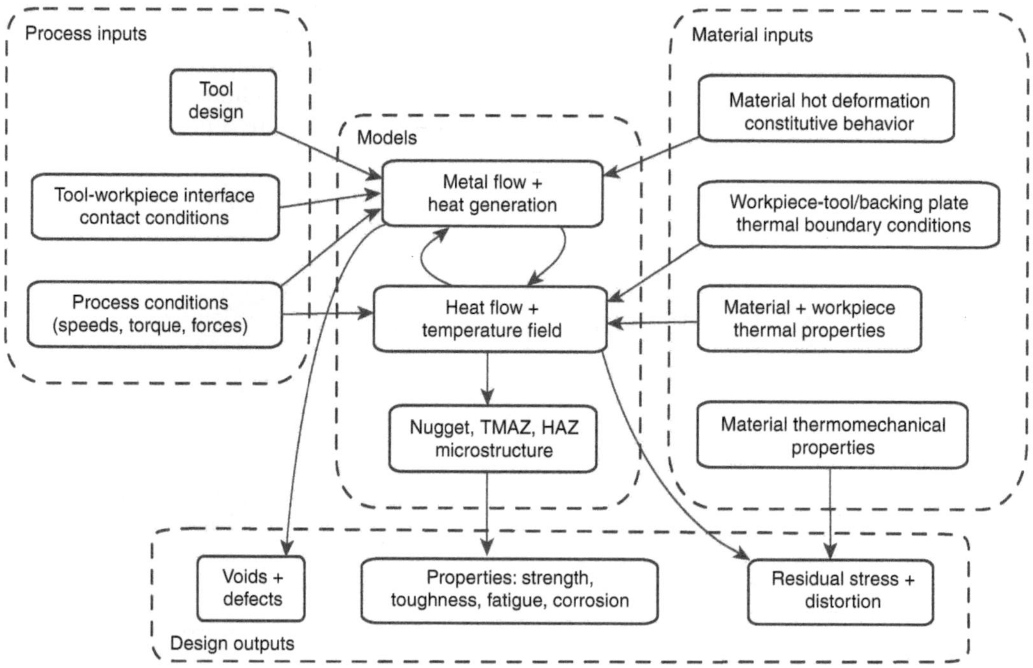


**Fig. 10.1** Summary of the key physical interactions in friction stir welding and the models linking process and material input parameters to the outputs needed by designers. TMAZ, thermomechanically affected zone; HAZ, heat-affected zone

three dimensions and use constant thermal properties. They provide a valuable reference point before embarking on more complex heat flow analyses. Two significant differences between arc and FSW are:

- The heat input is distributed around and beyond the interface of a solid tool in FSW, whereas an arc can be better approximated by a concentrated source at the surface of the melt pool.
- The plunge in FSW generates an initial thermal field around the tool, which greatly shortens the transient-to-steady-state conditions compared to a source moving over an initially cold workpiece.

Early work lumped all of the heat input to the workpiece into a point source and used the Rosenthal solutions directly, demonstrating via instrumented welds that this was reasonably accurate for the far field (i.e., for distances greater than the shoulder radius) (Ref 6). Given the distributed nature of the heat source, it is also common to use the Rosenthal equations for FSW by integrating infinitesimal heat inputs distributed over the tool area, for example, assuming a constant power density or a power intensity that varies with the radius from the tool axis (Ref 7–9). If this level of detail is applied to the heat source, then other issues become significant, for example, the thermal boundary conditions between the workpiece and the backing plate and the temperature dependence of the thermal properties. This all incurs a significant programming penalty and offers little that cannot be achieved using techniques such as finite element analysis, for which commercial codes are routinely available. Analytical methods have therefore largely been superseded by thermal analysis using finite element or finite volume techniques.

### 10.2.2 Numerical Methods

Numerical methods offer many advantages over analytical methods, being better suited to finite plate geometries, temperature-dependent thermal properties, and complex boundary conditions (such as heat losses to the backing plate, distributed heat sources, or frictional heating). The tool and backing plate can also be explicitly included as conducting solids in the thermal analysis. Finite element analysis is the most common numerical tool used for this problem (Ref 10–24), although finite difference methods have also been used (Ref 25). Finite volume solvers are equally suitable but tend to be used only when a simultaneous metal flow solution is sought (see section 10.3, "Metal Flow"). Figure 10.2(a) summarizes the ingredients required for a numerical solution of the thermal field: the heat input, material thermal properties, and thermal boundary conditions, discussed in turn as follows. All require a degree of calibration

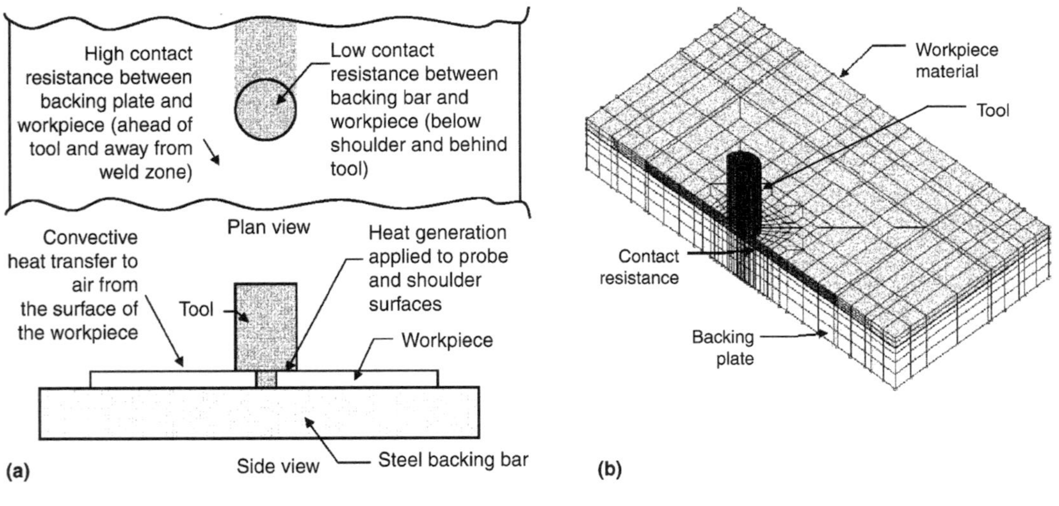


**Fig. 10.2** (a) Inputs and boundary conditions required for numerical thermal analysis of friction stir welding. (b) Typical finite element mesh for analysis of the temperature field in both workpiece and backing plate

using experimentally determined temperatures (and heat inputs). It is most important to appreciate the sensitivity of the results to uncertainty in all of the inputs. Repeat analyses with parameters set at upper and lower realistic values are essential in order to appreciate the uncertainty in the results and to identify which parameters have the greatest influence for the problem of interest. Recognizing that some parameters are secondary may allow simplifications to be made, giving gains in computational efficiency. Unfortunately, there is a strong tendency with modern software to add complexity, because it is possible and gives the illusion of greater precision, and to interrogate the models fewer times rather than more. This is inherently bad practice in modeling and should be resisted.

Numerical methods are able to capture full transient heat flow behavior, that is, in which the temperature at a given position with respect to the heat source evolves with time. Transient heat flow is important in the early stages of welding or as an effect of finite plate geometry. The initial plunge provides a degree of preheat, so steady-state conditions, in which the thermal field with respect to the source is unchanging, are reached more rapidly in FSW than, for example, in arc welding. Long friction stir welds, as used in shipbuilding, for example, are in any case predominantly in the steady-state regime. Steady-state thermal analysis is much faster than a full transient solution and is a good starting point in building a model, even if transient analysis is required to capture later detail. The key difference in implementation is the nature of the mesh used to discretize the volume. In a Eulerian formulation, the mesh is fixed, and the material is allowed to flow through the mesh, which is suitable for steady-state analysis; the material flows through a stationary thermal field "attached" to the tool. The Lagrangian formulation is more general and used for transient analysis, with the mesh being fixed to the material, and temperature evolving everywhere as the heat source moves with respect to the mesh.

There are several other important issues in the choice of discretization of the material volume in the software implementation, that is, the choice of mesh and element size and element type (linear, quadratic, 3-D, shell, etc.). Numerical solvers approximate the temperature field over the volume, constrained by both boundary conditions and type of meshed representation used. Figure 10.2(b) shows a typical finite element mesh for solving the thermal field in both workpiece and backing plate. A fine mesh is used near the heat source, where the temperature gradients are steep. These effects are not discussed in detail here, but it should be recognized that they can have as much of an influence on the results as the material and process input data. It is important for the modeler to have a proper awareness of the effects of the mesh size and type on both results and computation time. Sensitivity analysis has an equal role to play in validating a numerical model.

**Heat Input.** Most thermal analyses use a heat source distributed over the tool surface, with a heat flux per unit area (in $W/m^2$) of constant intensity or prescribed spatial variation (e.g., with radius from the tool axis). This recognizes that the heat is generated at the interface by friction or by viscous dissipation in a layer that is sufficiently thin to consider it to be at the interface. Alternatively, for the probe, its share of the heat input may be distributed over the probe volume. In heat flow problems, such approximations rapidly lose significance as the distance from the heat source increases. The second assumption is usually to ignore tool tilt and the effect of metal flow on the distribution of heat generation and to treat the source as axisymmetric. The high rotational speed and consequent convective heat flow by motion of the material act to smooth out the circumferential distribution of heat. As an example, the heat input intensity has been represented as follows (Ref 17):

$$\text{For the shoulder } (W/m^2)\text{: } q_s(r) = \frac{3}{2\pi} \times \frac{Q_s \times r}{R_s^3 - R_p^3} \quad \text{(Eq. 10.4)}$$

that is, the intensity increases with radius $r$ for $R_p \leq r \leq R_s$, where $R_p$ and $R_s$ are the radius of the probe and shoulder, respectively, and $Q_s$ is the power input from the shoulder (at the workpiece surface).

$$\text{For the probe } (W/m^3)\text{: } q_p(r) = \frac{Q_p}{\pi R_p^2 L_p} \quad \text{(Eq. 10.5)}$$

that is, a uniform volumetric heat source occupying the space filled by the probe, of length $L_p$, supplying a power $Q_p$. The parameters $Q_s$ and $Q_p$ are adjusted empirically, usually informed by experimental measurements of the total power input, as discussed in section 10.1, "Analytical Estimates of Heat Generation." The radial increase in shoulder heat input is one way to recognize the local tool velocity over the workpiece (neglecting the translational velocity). Other distributions are, of course, possible and indeed likely; the nature of the contact will be temperature dependent, with the heat generation limited as the temperature approaches melting. The radial source will tend to overestimate the peripheral temperature in consequence. It is difficult to be more physically realistic without fully solving the coupled heat-generation and conduction problem (discussed in section 10.3, "Metal Flow"), although this introduces further assumptions and approximations. It is important to recognize that the total heat input is the key parameter, and there are limits to the refinements of the spatial distribution of the heat that are meaningful in fitting experimental temperature data.

**Material Thermal Properties.** The partial differential equation for heat flow depends on three material properties: density, $\rho$; specific heat (per unit mass), $C_p$; and thermal conductivity, $\lambda$. Specific heat and conductivity, in particular, are temperature-dependent properties, and numerical methods are able to incorporate this in solving for the temperature field. There are practical difficulties, however, in obtaining reliable data. The situation is particularly complicated in heat treatable aluminum alloys in which the microstructure evolves significantly during welding. The thermal conductivity is influenced primarily by the amount of solid solution and the presence of fine-scale hardening precipitates. The initial temper therefore has an influence on the room-temperature properties, and the properties will evolve as precipitates dissolve and reform during the thermal cycle, on top of the normal temperature dependence of a stable alloy microstructure (caused by phonon scattering). Published data tend to be for material after some fixed (long) hold time, which may not be representative of the state in a weld for which thermal cycles last tens of seconds. Figure 10.3 shows typical data for aluminum alloy 2024 (Ref 26, 27). A degree of pragmatism is again required in using such data. It is common to take published temperature-dependent values that neglect microstructure evolution, as in Fig. 10.3, or simply to take average constant values at a midrange (or room) temperature. Because heat input (and boundary conditions) requires calibration, uncertainty in thermal properties is, to a large degree, masked by comparable uncertainties elsewhere in the model.

**Thermal Boundary Conditions.** Heat is lost to the tool, to the surrounding atmosphere, and to the backing plate (and any clamps applied to the plate). The effect of the tool was discussed previously, because allowance is made for this in the net heat input. Convection to the atmosphere is modeled via a heat-transfer coefficient. The value for air convection is low; little heat is lost this way, and the temperature in the plate is insensitive to the value of the heat-transfer coefficient to air. The most important heat loss is to the backing plate, usually modeled with a thermal conductance between the two solids (workpiece and backing plate). Because a significant downforce is applied to the tool, and the metal under the tool is hot, the thermal contact is intimate under the tool. Furthermore, the good contact is retained beneath the weld after the tool moves on. A constant conductance does not capture this behavior well, but the conductance should preferably evolve depending on the tool position, which makes the numerical analysis more complex. One approach is to have a high conductance under the tool itself and a lower value every place where there is contact with the backing plate. An alternative is to calibrate a temperature-dependent conductance, with increasing thermal coupling as the temperature rises. This has the numerical advantage that the boundary condition depends on the local temperature alone, independent of the tool position, which is numerically similar to using a heat-transfer coefficient. Faster solutions still can be achieved by replacing the backing plate altogether with an equivalent convective fluid, cali-

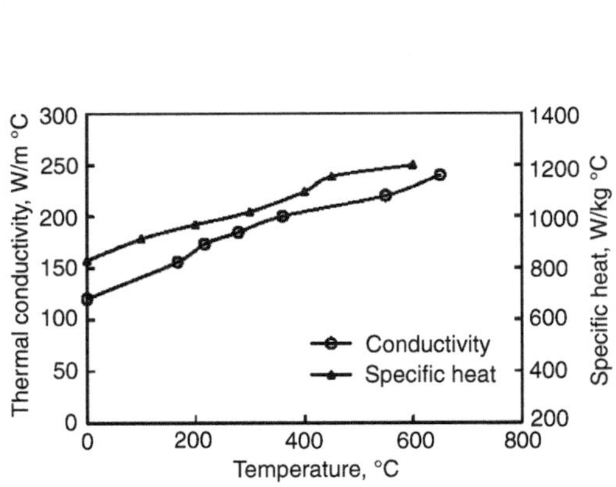


**Fig. 10.3** Example of thermal property data for aluminum alloy 2024. Adapted from Ref 26, 27

brating an appropriate notional heat-transfer coefficient. Figure 10.4(a) shows a typical thermal field predicted using a finite volume solver, with a high conductance under the tool. The difference in workpiece and backing plate temperatures away from the tool can be seen.

### *10.2.3 Experimental Validation*

The experimental measurement of the temperature field was introduced in Chapter 3, "Temperature Distribution and Resulting Metal Flow." Most evaluation of the thermal field uses thermocouples, although thermal cameras and pyrometers have been used to indicate the surface temperatures around the shoulder. The difficulties with using standard K-type thermocouples are:

- Their finite size, such that an average temperature is indicated over a finite range of temperature in a thermal gradient
- Their response time, which may not keep up with the temperature when this varies rapidly, although FSW is a relatively slow welding process
- Accurate location of the thermocouple at the desired position in the depth and transverse directions
- Ensuring good contact between the thermocouple and the root of the associated hole

With care, an accuracy on the order of 10 °C (18 °F) can be achieved, and this sets a limit to the degree of calibration in heat input and boundary conditions that can be justified.

Figure 10.4(b) shows a typical set of thermal cycles measured by thermocouple, together with the predicted curves (in this case, calculated using a finite volume solver). Note that the experimental history for the hottest thermocouple is truncated, because the weld deformation ran into the thermocouple and it was destroyed. This has the advantage of giving direct evidence of peak temperatures in the weld nugget, although the cooling history is lost.

The predicted thermal histories are most sensitive to the heat input and to the heat loss to the backing plate. These affect the thermal history differently in each location, so the more thermocouples the better. There is a tendency for the two parameters to compensate for one another to some extent, particularly in thin plate where the through-thickness temperature gradient is small. Increasing the heat input raises the peak temperature everywhere and changes the cooling rate, while increasing the thermal conductance primarily modifies the cooling part of the curve. It is beneficial in decoupling these two effects to have an instrumented backing plate, measuring its temperature in at least one location. Figure 10.4(a) shows a typical predicted 3-D temperature field, with the backing plate. Efficient and physically realistic adjustment of heat-transfer conditions is not easy and improves with experience of thermal modeling in many different situations. As always, greater insight into the problem is obtained by running multiple analyses, showing the sensitivity of the thermal histories to systematic variations in key parameters, such as conductance to the backing plate. Note also that different interpretations of the quality of fit are obtained, depending on the way the predicted temperature field is evaluated against the thermocouple data. For example, different conclusions may be drawn on the model quality by comparing the full thermal cycles $T(t)$, as opposed to a plot of the peak temperature $T_p$ as a function of transverse position.

## 10.3 Metal Flow

Modeling the metal flow in FSW is a challenging problem but is fundamental to understanding and developing the process. As introduced in Fig. 10.1, flow models seek to simultaneously capture the thermal and mechanical aspects of the problem in enough detail to address a subset of the following issues:

- Flow visualization and insight into the mechanisms by which the joint line is broken down and forged in a sound metal-metal bond, including the flow of dissimilar metals
- Improved evaluation of the heat generation and the related heat flow that governs the temperature field (and hence microstructure, properties, loads, and residual stress)
- Virtual tool design, to optimize tool profiling for different materials and thicknesses
- Accelerate the optimization of process conditions (minimize force and maximize speed), particularly for new alloys
- Susceptibility to formation of defects (e.g., voids) and sensitivity to process variability, such as fluctuations in the plunge depth or initial plate fit-up

Some conceptual descriptions of the flow behavior were introduced in Chapter 3, "Temperature Distribution and Resulting Metal Flow," and further experimental evidence is dis-

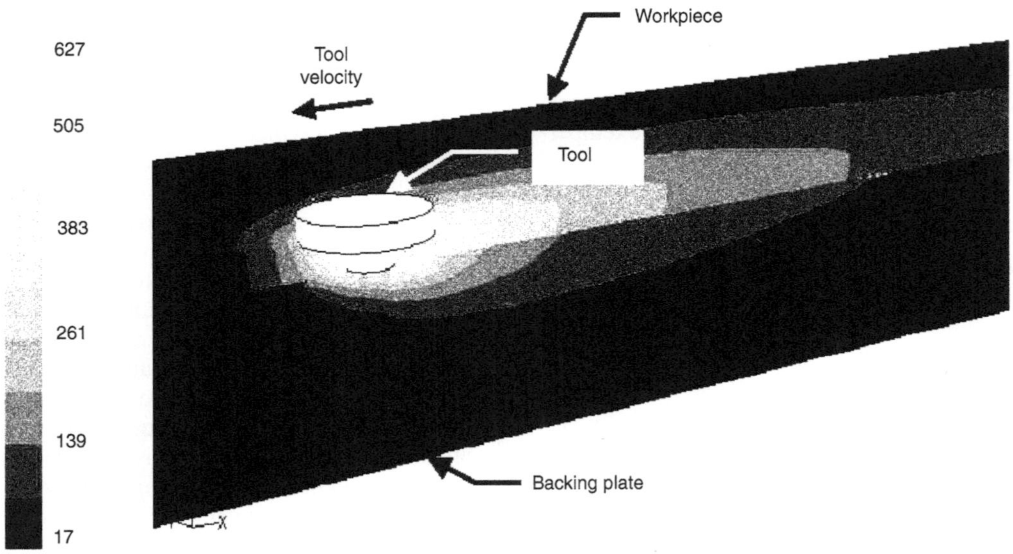


(a)

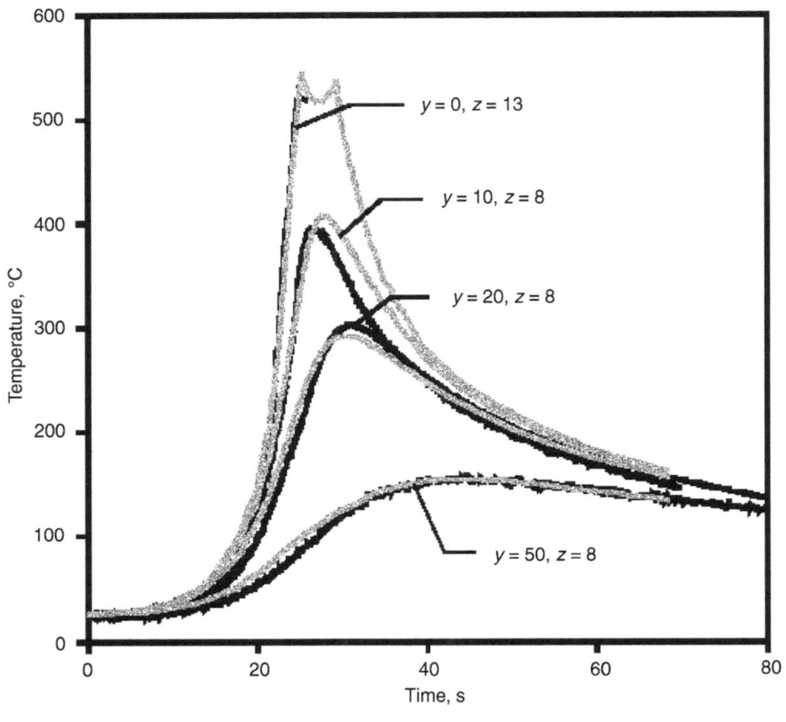


(b)

**Fig. 10.4** (a) Typical predicted global three-dimensional temperature field for moving friction stir welding (FSW) heat source, including heat transfer to backing plate. Source: Ref 28. (b) Measured and predicted thermal cycles for a typical FSW in aluminum alloy 7075, using a finite volume solver for the numerical analysis. Source: Ref 29 cussed in section 10.3.7, "Experimental Flow Validation." These experiments both guide the development of flow models and provide some direct validation of predicted flow patterns. A number of key points are worth emphasizing at the outset:

- The shoulder, pin, and backing plate provide significant kinematic constraint on the incompressible metal flow; that is, the path is, to a large extent, geometrically determined by the process. Continuity dictates that the flow separates on the advancing side, with the material ahead of the pin being swept around the retreating side in an extrusion-like process, forming a longitudinal friction weld as the metal streams are forced together.
- The circumferential speed of points on the tool interface (both pin and shoulder) is higher than the translational speed of the tool, usually significantly so.
- The material speed at the interface may reach the local tool speed if sticking occurs, but slip will limit the speed to a lower value, and this aspect will be highly sensitive to the local temperature (and hence local viscosity or shear flow stress).
- Heat generation and conduction, combined with the typical softening response of alloys with increasing temperature, leads to a temperature gradient away from the tool and an intense, thin deformation zone close to the interface.
- Tool features on pin and shoulder and the overall kinematics of flow to bypass the tool induce some flow in the through-thickness direction, superimposed on the essentially in-plane flow around the pin.

The flow around the pin is at the heart of the process and is one of the main determinants of success in FSW. Away from the shoulder and backing plate, the kinematics of the process noted previously dictate that the flow is predominantly in the plane of the plate. Hence, various authors have first analyzed the 2-D flow around the pin at midthickness rather than the full 3-D flow, giving significant benefits in computational efficiency. When this has been successfully modeled, the analysis can be extended with more confidence to three dimensions.

### 10.3.1 Analytical Flow Modeling

Some analytical approaches have been tried to model the flow in FSW, assuming a simplified view of the flow pattern and tool/workpiece interface conditions, in order to estimate tool forces, torques, or heat generation. For example, Stewart et al. (Ref 30) compared two alternative theories for material flow: a mixed-zone model, with deformation distributed over the nugget and thermomechanically affected zone (TMAZ), and a single slip surface model, with slip concentrated on a surface. The latter appeared more consistent with experimental observations. An upper-bound analysis was presented by Shercliff and Colegrove (Ref 1) to predict the size of the deformation zone around the pin, based on the inherent kinematic constraint of the in-plane 2-D flow around the pin and continuity of flow. Schmidt and Hattel (Ref 31) extended this to solve for the continuum velocity field analytically, using continuity with a linear velocity profile away from the tool. Heurtier et al. (Ref 32) assumed a velocity field superimposing rotation, translation, and vortex flows (conceptually similar to the qualitative kinematic description, Ref 33, discussed in Chapter 3, "Temperature Distribution and Resulting Metal Flow"). This approach delves considerably deeper into the prediction of deformation and temperature.

Metal flow in FSW is inherently complex, and, as with analytical thermal models, a point is rapidly reached where it is preferable to switch to numerical meshed methods using commercial codes. These are discussed in the next section.

### 10.3.2 Numerical Flow Modeling

The deformation aspect of FSW suggests that the underlying physics has parallels in other thermomechanical processes outside the usual domain of thermal welding research. Numerical FSW flow modeling can therefore draw on analyses and codes used for other processes, such as friction welding, extrusion, machining, forging, rolling and ballistic impact. The FSW flow modeling uses finite element (Ref 1, 12, 34–41), finite volume (Ref 28, 29, 42–51), or shock wave physics (Ref 52) codes. Most of these are used for computational fluid dynamics (CFD) analyses, which treat the problem as one of viscous fluid flow rather than solid plasticity. The validity of this approach stems from the large inelastic strain, with hot metal flow corresponding to viscoplastic behavior at very low Reynolds number. Hence, the Navier-Stokes equations may be solved with the convective and time-dependent terms neglected, using an appropriate temperature and strain-rate-dependent viscosity. Key aspects of all flow analyses are thus:

- Coupling between flow and thermal fields
- Choice of an appropriate constitutive response for the material (and the availability of data)
- Contact conditions at the tool workpiece interface

These issues are discussed in the following sections.

As for heat flow analyses, numerical flow models can use either a Eulerian or Lagrangian formulation for the mesh (or a hybrid of the two, arbitrary Lagrangian-Eulerian, or ALE), but the distinction is more significant than in thermal analysis. This is because in a Lagrangian analysis, the mesh is attached to the material, but the material itself (and thus the mesh) deforms. This limits the strains to only a moderate level before mesh distortion leads to failure of the analysis, so frequent remeshing and a fine mesh in the deforming regions are required. This leads to computationally intensive analyses that can only simulate the initial plunge and dwell or relatively short distances of real welding.

The difficulty of solving the full 3-D metal flow in FSW as an elastic-plastic problem has therefore led most researchers to concentrate on CFD viscoplasticity approaches. One consequence is that some mechanical effects are excluded from the scope of the analysis, for example, studying the effect of varying the downforce. Free surfaces also present difficulties in CFD, because the deforming material must fill the available space between the solid boundaries of the tool, backing plate, and so on. However, experimental evidence exists (discussed later) that for some tools and welding conditions, there is a stable cavity behind the pin or within the tool features, and it is also well-established empirically that defects such as tunnel voids can be left in the wake of the tool. These cavities are assumed to have a second-order effect on the flow as a whole, but it is important not to overinterpret the predicted flow in the wake of the pin if, in practice, some form of separation occurs. It would be of great benefit if the likelihood of voids could be predicted by CFD analysis of the material state in the wake of the pin; for example, a state of high hydrostatic tension may correlate with a tendency to cavitate and generate a stable pore. This has so far proved elusive in CFD modeling, but a few researchers have pursued this via 3-D elastic-plastic finite element analysis of FSW using an ALE formulation (Ref 39, 40). The results provide interesting physical insight but show great sensitivity to the assumed material response and contact conditions. The very long computation times also make it unlikely that these analyses will be used routinely as design tools.

An intermediate style of analysis is illustrated by the application of the CTH code (developed by Sandia National Laboratory) to FSW (Ref 52). This code is primarily used to simulate high-speed impact and penetration phenomena encountered in ballistics. It has the advantage over CFD that it captures the elastic as well as the viscoplastic response of the material and has been shown to produce detailed flow predictions for a wide range of geometries.

The CFD analysis of FSW ranges from 2-D flow around a cylindrical pin to full 3-D analysis of flow around a profiled pin. Some of the numerical issues are the same in all cases, in particular, building an efficient mesh for rapid computation. One difficulty is the steep gradient in flow velocity near the tool. Most analyses divide the mesh into zones, as illustrated in Fig. 10.5(a). Because the flow near the tool is predominantly rotational, the mesh in this region rotates with the tool. In the far field, all material that undergoes some deformation is treated as a fluid, with the mesh stationary and the metal flowing through it at the traverse velocity. The rotating zone is made large enough to contain the entire deformation zone, such that the velocity to be matched across the interface between the two zones is equal to the traverse velocity everywhere. However, the mesh size in the rotating zone is much finer and graded toward the tool, to capture the intense deformation (Fig. 10.5b). A further simplification is to model the workpiece to either side as a translating solid, rather than a fluid (which, of course, in reality it is). This reduces the number of elements for which the fluid solver must operate. Note that 3-D analysis, as illustrated in Fig. 10.5(a), is able to handle some of the process complexities: a concave shoulder, tool tilt, and threaded pin profiles.

A particular complexity of FSW flow modeling is that the rotation of a profiled tool generates a geometry that varies cyclically. The deformation field evolves continuously but repeats on a period determined by the sequential positions at which the tool orientation is identical with respect to the traverse direction. This is once every revolution for the original threaded tools but every one-third revolution for more recent tools with threefold rotational symmetry, such as the Triflute (TWI Ltd.). In principle, a full transient flow solution is required to capture the flow during this period of revolution. Faster solutions of sufficient accuracy can be obtained by finding the steady-state solution for given angular orientations of the tool; each solution is like a snapshot of the flow at a particular instant in time (Ref 28, 45, 46). The full transient analy-

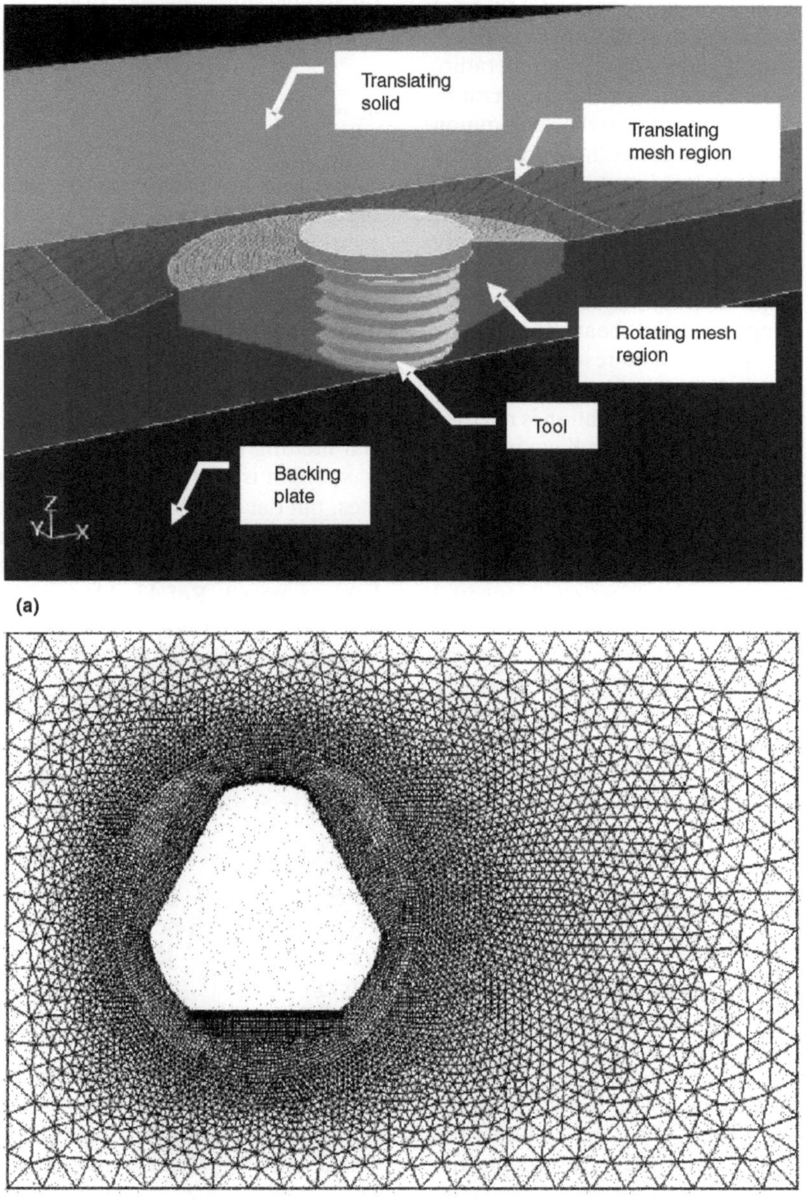


**Fig. 10.5** Mesh definition for computational fluid dynamics analysis of friction stir welding. (a) Subdivision into translating and rotating zones. (b) Example of two-dimensional mesh

sis, with the tool rotating incrementally, is much more laborious but was undertaken periodically as a check on this steady-state simplification.

### 10.3.3 Coupled Thermal and Flow Modeling

The close coupling between heat flow and metal flow in FSW was discussed earlier (Fig. 10.1). Viscous dissipation and friction generate heat at or near the tool interface, and a thermal gradient develops away from the tool by conduction, but the loop is closed by the temperature dependence of the viscosity and friction conditions at the interface. As noted earlier, a common approximation is to recognize that most of the heat generation takes place in a thin layer close to the tool. Hence, if an estimate of the heat input can be measured or estimated independently (section 10.1, "Analytical Estimates of Heat Generation"), the thermal problem may be solved first, imposing the heat flux at the tool/workpiece interface. The resulting thermal field is then imposed on a flow model (in which heat generation by viscous dissipation is not then strictly required). Some analyses have imposed isothermal conditions on the basis that the temperature difference across the deformation zone is small, and heat transfer by convective material flow will also smooth out temperature differences in the circumferential direction. Sequential thermal and flow analyses are much quicker than fully coupled analysis, which needs to converge on temperature and flow fields that are spatially self-consistent with the heat generation and conduction. It is important, however, to compare the approximate analyses with occasional fully coupled solutions to check the influence on the predicted flow field.

### 10.3.4 Material Constitutive Behavior for Flow Modeling in FSW

All flow modeling to date has been on FSW of aluminum alloys, with the exception of Goetz and Jata (Ref 36), who also analyzed titanium alloys. For aluminum alloys, it is reasonable to assume that steady-state hot deformation conditions apply. At typical FSW temperatures, the large strain deformation is perfectly viscoplastic, with work hardening balanced by dynamic recovery or dynamic recrystallization. Hence, the hot forming literature provides relevant information from hot testing in torsion, tension, or compression, although much of the best data may be proprietary. Aluminum extrusion data are most relevant, because there are many similarities in the deformation conditions (high strain rates and near-melt temperatures), while the greatest commercial application of FSW is in welded extrusions of heat treatable aluminum alloys.

The most common approach to modeling steady-state hot flow stress, $\sigma$, is the Sellars-Tegart law (Ref 53), combining the dependence on temperature, $T$, and strain rate, $\dot{\varepsilon}$, via the Zener-Hollomon parameter, $Z$:

$$Z = \dot{\varepsilon} \exp\left(\frac{Q}{RT}\right) = A\,(\sinh \alpha\sigma)^n \qquad \text{(Eq 10.6)}$$

where $Q$ is an effective activation energy, $R$ is the gas constant, and $\alpha$, $A$, and $n$ are material constants. If the data cannot be well fitted to this equation, an alternative option in numerical modeling is to store the data as a look-up table and to interpolate directly.

A difficulty with using hot deformation data for FSW modeling is that the test temperatures rarely extend right up to the solidus, when material melting commences. The near-solidus loss in strength is critical to the way that FSW operates, but data are not yet routinely available for this regime. Recognizing the physical cutoff of the solidus, empirical softening regimes have been proposed (Ref 43, 44, 48, 50, 51). A typical fit of Eq 10.6 to experimental data, with an empirical near-solidus response, is shown in Fig. 10.6(a) for aluminum alloy 7449.

Askari et al. (Ref 52) and Schmidt et al. (Ref 40, 41) have used an alternative constitutive response developed by Johnson and Cook (Ref 54) for modeling ballistic impacts:

$$\sigma = (A + B\varepsilon^n)\,(1 + C \ln \dot{\varepsilon}^*)\,(1 - T^{*m}) \qquad \text{(Eq 10.7)}$$

where, $A$, $B$, $n$, $C$, and $m$ are material constants; $\dot{\varepsilon}^*$ ($= \dot{\varepsilon}/\dot{\varepsilon}_0$) is the dimensionless plastic strain rate for $\dot{\varepsilon}_0 = 1$); and $T^*$ is the homologous temperature, given by $T^* = (T - T_0)/(T_m - T_0)$, with $T_m$ and $T_0$ being the melt and ambient temperatures, respectively.

Figure 10.6(b) shows the form of this constitutive law for aluminum alloy 2024 (neglecting the strain-hardening term for hot deformation of aluminum alloys, i.e., setting $B = 0$). While this law captures the desirable feature that the strength falls to zero at the solidus temperature, it is somewhat at odds physically with the well-established strain-rate dependence exhibited by aluminum alloys in the Zener-Hollomon regime. The general flow pattern predicted is somewhat insensitive to the constitutive law, due to the inherent kinematic constraint of the process. However, the heat generation, temperature, and flow stress near the tool and the loading on the tool will depend closely on the material law. Hence, predictions using the Johnson-Cook law should be treated with caution, and more physically realistic constitutive data should be used.

A further complexity is that the standard tests used to obtain hot deformation data involve holding the specimen at temperature for a period of time before conducting the test. While this has little effect on non-heat-treatable alloys (1000, 3000, and 5000 series), it may be significant for heat treatable alloys (2000, 6000, and 7000 series). This is because commercial tempers such as T3 and T6 have unstable microstructures, which evolve when heated to tem-

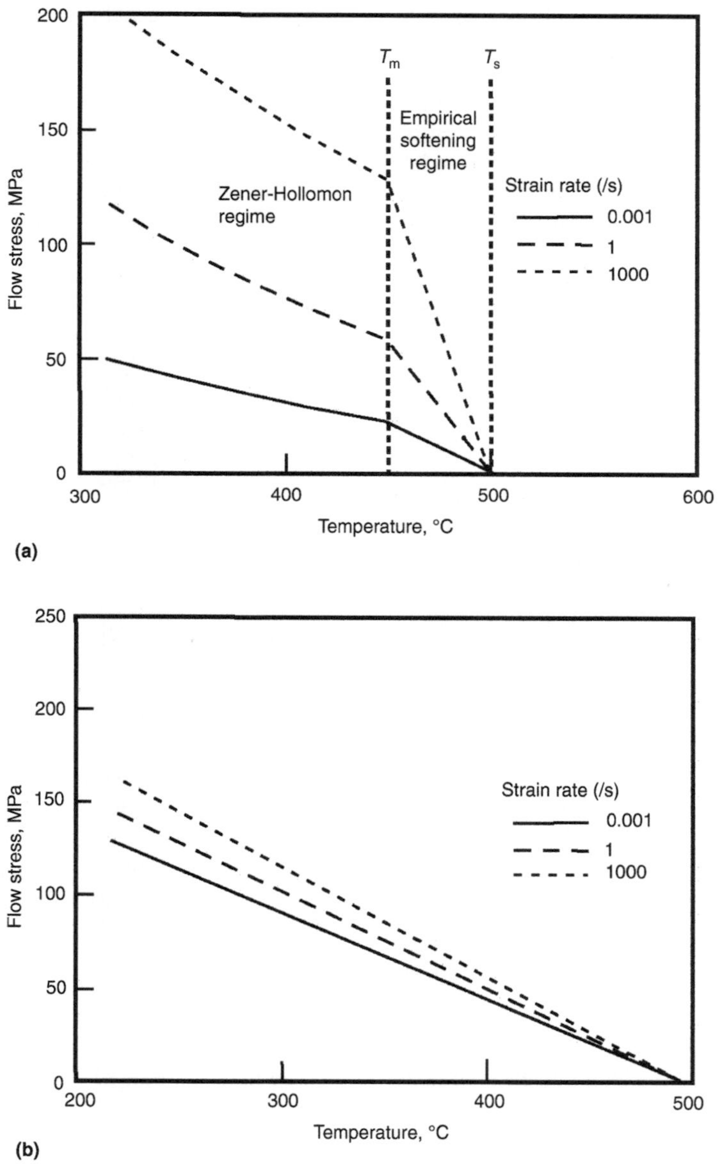


**Fig. 10.6** (a) Constitutive data for hot deformation of aluminum alloy 7449, fitted to the Sellars-Tegart law, with a linear empirical softening regime applied between the limit of the data and the solidus temperature. $T_m$, melt temperature; $T_s$, solidus temperature. Source: Ref 50, 53. (b) Typical form of the Johnson-Cook constitutive law for aluminum alloy 2024, neglecting strain hardening. Source: Ref 47

peratures of 200 °C (390 °F) or more. In the relatively short thermal cycles in FSW, the flow stress of heat treatable alloys will evolve at a rate determined by the dissolution of the hardening precipitates, giving a different strength to that observed using standard test procedures. Furthermore, the deformation itself may accelerate the dissolution of precipitates by dislocations cutting shearable precipitates and short-circuit diffusion along dislocations, so it is not purely an effect of thermal history. For most commercial alloys and tempers, however, dissolution is rapid and does not lead to coarse equilibrium precipitation within the thermal cycles induced by FSW. The important temperature regime for metal flow is 100 to 200 °C (180 to 360 °F) below the solidus, when a solid solution can usually be assumed. Microstructural modeling is also now capable of tracking the complex evolution of precipitation in commercial alloys (section 10.4, "Microstructure and Property Evolution in FSW of Aluminum Alloys"). It is clear, however, that better experimental flow data are needed in the near-solidus regime.

### *10.3.5 Tool Material Interface Conditions*

The contact conditions between the tool and the material are central to the friction stir process. The process sweeps material from the leading to the trailing edge of the tool, around the retreating side of the tool. The shear stresses at the tool/workpiece interface control this behavior, in particular, the dragging of material into the probe wake to generate the seam between the separated flows around the advancing and retreating sides. The nature of the contact is virtually impossible to observe directly, so modelers have used various physical assumptions to capture the problem numerically.

The boundary conditions applied at the tool interface either prescribe the material velocity field at the interface or the interfacial shear-stress distribution:

- Full sticking, with the local material velocity matching that of the tool interface everywhere, or the applied shear stress being equal to the shear yield stress (sticking friction)
- Slipping, with the rotational speed of the material being an arbitrary constant fraction of the tool rotation speed
- Stick/slip conditions, in which the shear stress is limited to an arbitrary constant value
- Coulomb friction, with the shear stress being limited to a maximum value dependent on the local normal pressure. Because Coulomb friction depends on the normal stress, this is only valid for finite element analysis in which the elasticity is included. It is most commonly applied to model the shoulder contact but requires the assumption of a constant coefficient of friction, usually calibrated via the net measured torque or indirectly through the temperature field, which reflects the frictional heat input.

Analyses range widely in complexity from isothermal models with a constant shear stress to represent sticking friction at the interface, to fully coupled thermal and flow models with temperature-dependent stick/slip conditions over the whole interface. The particular problem of the limited knowledge of flow properties close to the solidus was discussed in section 10.3.4, "Material Constitutive Behavior for Flow Modeling in FSW." Coupled analyses that do not include softening tend to predict excessive torques and heat generation. This is overcome empirically by setting a relatively low limiting frictional stress, so that slipping conditions prevail. Alternatively, empirical softening regimes (as in Fig. 10.6a) achieve the same effect by rapidly reducing the shear flow stress near the solidus, with sticking conditions throughout. It is therefore difficult to determine from first principles whether stick or slip occurs. The latter approach is numerically more robust, because the interfacial velocity is prescribed everywhere, and it captures the inherent self-stabilization of the process as the temperature approaches the solidus. For the shoulder, sticking conditions tend to predict excessive heating, even with softening. This probably reflects the lower degree of constraint on material near the periphery of the shoulder, where hot metal can be extruded out of the contact, and also the lower contact pressure (or incomplete contact) on the leading edge of the shoulder due to tool tilt.

The main outputs of flow modeling are flow visualizations, to illustrate the process mechanism and to compare with experimental marker techniques (section 10.3.7, "Experimental Flow Validation"). These include streamlines, particle tracks, velocity maps, and strain-rate contour plots. Flow models can also be validated indirectly, through the coupling to the thermal field (i.e., heat generation) or the net loading on the machine, by integration of the stresses acting on the tool (torque, traverse force, and downforce). The CFD models cannot predict absolute forces, because elasticity is neglected. Nonetheless, the trends in forces, torque, and heat generation can be investigated as the rotation and traverse speeds are varied or as the tool profile is changed (see section 10.3.6, "Influence of Tool Profile and Process Conditions").

**Two-Dimensional Flow Modeling.** Because the intense deformation is limited to a relatively thin layer near the interface, the shoulder and probe can, to some extent, be considered separately. Cross sections of friction stir welds (Chapters 1 to 3) show a characteristic sweep of material across the joint line near the surface, driven by the trailing edge of the shoulder. At midthickness, however, the influence of the shoulder is primarily as a remote heat source contributing to the temperature field along the probe, particularly for thick-section welds. This enables simpler, faster 2-D analyses to be considered as an approximation to the flow around the probe. The 2-D analyses assume the tool is prismatic and reasonably long compared to its radius. The high-speed rotation primarily drives the flow circumferentially, but tool features promote some out-of-plane flow. However, 2-D analyses are suitable to explore the effect of those features that are predominantly longitudinal, such as the deep grooves in a Triflute or machined flats. Threads cannot be modeled in 2-D, because their pitch is small compared to the probe diameter.

A number of 2-D analyses have been presented, initially using finite element analysis but later switching to CFD using the commercial code FLUENT (Fluent, Inc.) (Ref 35, 38, 43–46, 48, 50), evolving from steady-state analyses for a plain cylinder to full transient analyses of profiled tools.

Figure 10.7 shows the basic 2-D flow field for a rotating, translating cylinder (Ref 44). A stagnation point is observed on the advancing side, with the flow separating and all material in the path of the probe being swept around the retreating side, with a friction weld being generated on the advancing side in the tool wake. Note how the streamlines are packed into a thin zone on the retreating side, where the material is accelerated from the traverse velocity to values close to the probe velocity. Note also that there are sharp changes in direction predicted on the advancing side. On the leading edge, the stress field is essentially compressive, and this can be achieved. However, on the trailing edge, the stress field will become tensile, and there will be a strong tendency for separation at or near the interface, with a stable cavity behind the tool (and the potential for generating a longitudinal void).

The effect of changing the interfacial boundary conditions is illustrated in Fig. 10.8. This is for a cylindrical tool with small concave features (discussed further in section 10.3.6, "Influences of Tool Profile and Process Conditions") but shows the effect of stick versus slip. Two flow representations of each analysis are shown: velocity vectors and streamlines. The streamline plots show that under sticking conditions, it is predicted that a layer of material adheres to and rotates with the tool (Fig. 10.8c). With slipping conditions (Fig. 10.8d), the flow past the retreating side is more diffuse, and the stagnation point is much closer to the tool interface on the advancing side. The width of the nugget and TMAZ is therefore influenced by the nature of the interface conditions. Another way to illustrate this is shown in Fig. 10.8(a,b). A boundary is superimposed on the arrow plot, indicating where the effective strain rate exceeds the (arbitrary) value of 2 $s^{-1}$. This boundary is indicative of the deformation region size and is closer to the tool in the slipping case. Finally, further insight into the process mechanism is obtained by tracking material through the deformation zone. A straight line of points across the workpiece, normal to the welding direction, has been tracked for a constant time along the streamlines. The change in shade of the streamlines indicates their final posi-

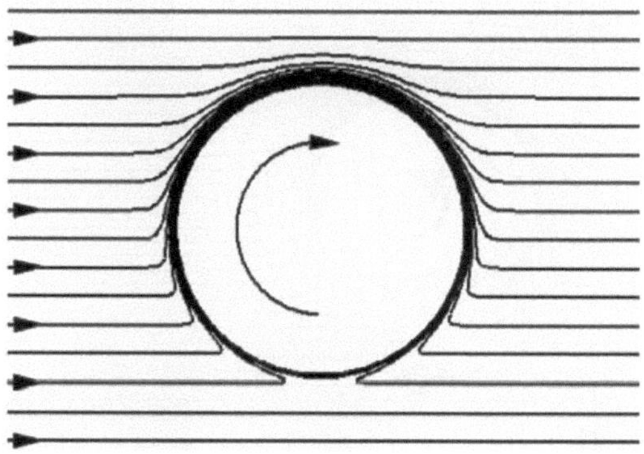

**Fig. 10.7** Typical generic flow path around the probe in friction stir welding, illustrated from a two-dimensional computational fluid dynamics model with a cylindrical tool. Adapted from Ref 44

tion. This indicates a backward bulge in the material behind the probe, with a thin region swept forward on the advancing side, with some differences in profile for stick and for slip. Marker experiments confirm this general pattern of deformation (see section 10.3.7, "Experimental Flow Validation").

### 10.3.6 Influence of Tool Profile and Process Conditions

Friction stir tooling has evolved empirically, based on observation of forces, microstructures, and defects. Early flow models analyzed cylindrical probes (Ref 36–38, 44), moving on to idealized 2-D profiles (usually with threefold symmetry) (Ref 45–47, 50), and complete 3-D models incorporating threads and flutes, including the commercial tools such as the 5651 tool and the MX-Triflute (TWI, Ltd.) (Ref 1, 12, 28, 48, 49, 52). Tool profiles described in computer-aided design can routinely be transferred to CFD and finite element models of the deforming solid, but mesh generation for the flow around 3-D tool features is nontrivial, given the complexity of shape and the need to capture steep velocity gradients without excessive computation times.

There is an almost infinite variety of possible tool shapes, and fabricating tool steel profiles and conducting experimental trials is expensive. Hence, the potential of modeling is particularly great in the area of tool design. Colegrove and Shercliff (Ref 45–47, 50) used the 2-D CFD methods discussed in section 10.3.5, "Tool Material Interface Conditions," to study a series of profiles, as illustrated in Fig. 10.9. The sensitivity of the flow pattern, torque, and traverse force to tool shape was compared for 2024, 7075, and 7449 aluminum alloys. Experimental

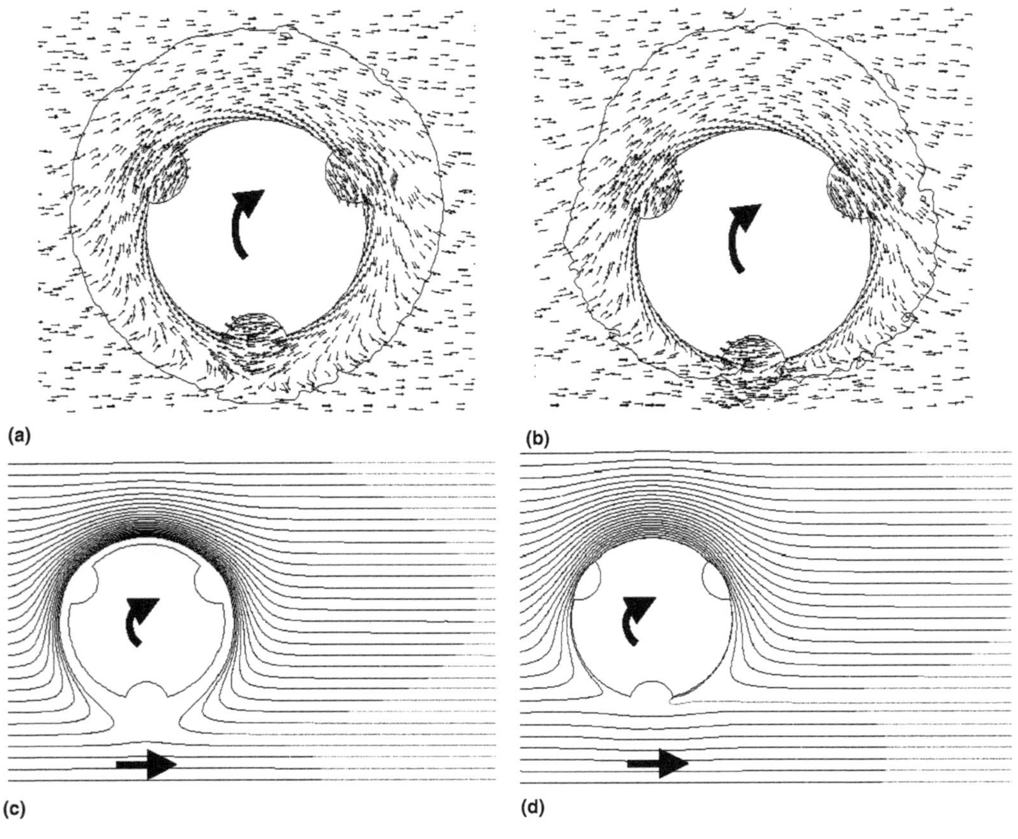


**Fig. 10.8** Effect of interfacial boundary conditions on the predicted flow from a two-dimensional computational fluid dynamics model with a profiled tool. (a, b) Velocity vectors and the boundary at which the effective strain rate is 2 $s^{-1}$. (c, d) Streamline plots, with the change in shade indicating the final position of points that were initially in a line perpendicular to the weld. (a) and (c) use a stick boundary condition; (b) and (d) use a slip model, with a limiting shear stress of 40 MPa (6 ksi). Adapted from Ref 46

trials were conducted on the Trivex tool (TWI Ltd.) (Ref 28), because the model indicated that a lower traverse force was required than with the Triflute for the same rotation and traverse conditions. It was also predicted that the tool would reduce "hooking" in lap welds (Ref 48). Both proved to be the case experimentally, but the Trivex also proved to be prone to generating voids (Ref 50), something that the flow models struggle to predict explicitly, as discussed earlier.

Figure 10.10 shows the flow vectors and streamlines for a Triflat tool. These use the slip version of the model, so they may be compared directly with Fig. 10.8(b, d). Again, the broad pattern of flow around the probe is similar, but the detail is significantly altered. These figures also explore the variation in the flow as the tool orientation with respect to the translation direction changes; the two extreme positions 60° apart are illustrated. The instantaneous flow path around the tool oscillates significantly be-

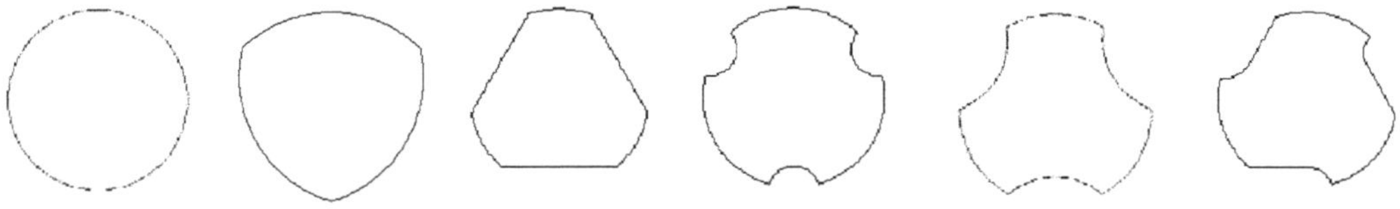

**Fig. 10.9** Example of two-dimensional tool profiles tested by computational fluid dynamics modeling. Source: Ref 45–47, 50

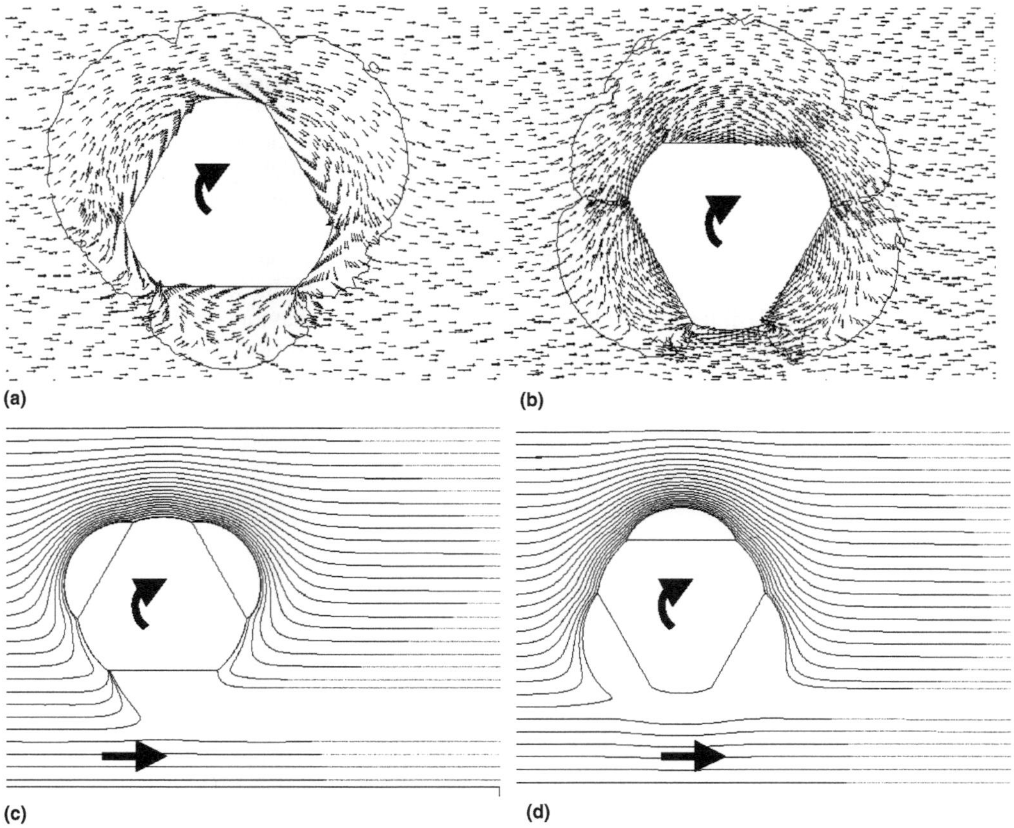


**Fig. 10.10** Effects of tool profile and orientation on the predicted flow from a two-dimensional computational fluid dynamics model using interfacial slip, with a limiting shear stress of 40 MPa (6 ksi). (a, b) Velocity vectors and the boundary at which the effective strain rate is 2 $s^{-1}$. (c, d) Streamline plots, with the change in shade indicating the final position of points that were initially in a line perpendicular to the weld. Adapted from Ref 46

tween these limits, three times per revolution. This adds weight to the hypothesis that the cyclic patterns of the nugget ("onion rings") directly reflect the tool rotation between positions of identical orientation of the profiling. A numerical marker experiment was also conducted to trace the material path throughout the deformation. Figure 10.11 shows a series of stills of material points (initially in a straight line) passing the tool and being deposited downstream in the characteristic curve. The curve breaks up somewhat on the advancing side of the tool center, this being the region where the onion ring pattern is most pronounced. Limits on mesh density and the approximations made in the flow model do not enable the onion ring to be predicted directly, but the results are nonetheless interesting pointers to the process mechanism and the effect of changing the tool profile, particularly when the model output is animated as a video clip.

Three-dimensional flow analysis is much more demanding computationally but is necessary to capture the full detailed characteristics of FSW. Added detail in 3-D includes complete tools with shoulder and probe; tool tilt; probe features such as threads, helical flutes, and taper in diameter; and flow below the pin—important for avoiding root defects. The output of 3-D CFD is essentially the same as in 2-D, but flow paths are more difficult to present graphically, requiring 2-D slices or isometric views. Figure 10.12 shows an example of the output from 3-D flow modeling: streamlines in an incoming horizontal plane passing a Triflute and a Trivex profile. Vertical movement of the material is now apparent. Note that the Triflute shows significantly more material being captured and taken around the tool more than once, whereas the Trivex struggles to fill the space behind the tool on the advancing side. This is therefore consistent with the generation of a void in the wake of a Trivex tool.

Many combinations of tool profile, material properties, and boundary conditions have been investigated by flow modeling. It is common in the literature for great effort to go into modeling an individual weld, partly due to availability of samples and partly due to the computational overhead in complex analyses. From an industrial perspective, it is essential for modeling tools to be fast enough to explore the parameter space, for example, the effect of rotation and traverse speeds on heat generation, torque, and traverse force. As discussed earlier, not all analyses can predict all of these parameters, due to the way the model is constructed. Figure 10.13 shows (in schematic form) the typical trends in recent modeling work (Ref 51) as rotational speed is varied (for a given weld speed in 2024 aluminum alloy). These trends correspond qualitatively with experience. Heat input saturates at a certain rotation speed, and the model suggests that this corresponds to the interface approaching the solidus temperature and stabilizing. The reduced interfacial stress is unable to generate more heat to take the material above the solidus. The minimum in force is of particular interest. Modeling suggests that this is achieved when the material condition around the tool corresponds to temperatures and strain rates close to the onset of near-solidus softening, as discussed in section 10.3.4, "Material Constitutive Behavior for Flow Modeling in FSW" (Fig. 10.6) (Ref 50, 51). The results are preliminary, but it would clearly be of great benefit in reducing experimental trials if near-optimal welding conditions could be predicted

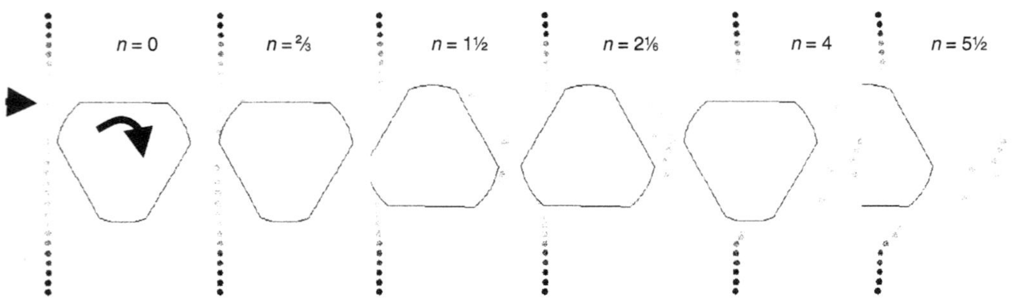


**Fig. 10.11** Predicted particle tracks through multiple revolutions of a Triflat tool from a two-dimensional computational fluid dynamics model. The cumulative number of revolutions ($n$) in each case is indicated. Adapted from Ref 47

directly from a knowledge of the constitutive response of the material.

### *10.3.7 Experimental Flow Validation*

Experimental validation of flow modeling in FSW is particularly challenging, because observing flow paths in real-time is extremely difficult. As noted in Chapter 3, "Temperature Distribution and Resulting Metal Flow," a range of experiments has been devised to study the flow by subsequent sectioning and optical microscopy and by x-ray tomography. Marker experiments use embedded contrast materials to observe the movement of material elements from their initial to final position. A range of materials and marker geometries has been used in aluminum alloy FSW: steel or lead balls (Ref 55, 56), contrasting aluminum alloy pins (Ref 57–59), SiC or copper foil (Ref 57, 60–64), tungsten wire (Ref 65), and titanium powder (Ref 66). Care is needed to ensure that the markers do not influence the deformation behavior. This can be checked by comparing metallographic sections with and without the marker and by logging the machine force and torque as the marker passes the tool (Ref 60). Welding dissimilar materials enables the redistribution of the joint interface and the materials to either side to be seen clearly, using the etching contrast in the alloys (Ref 67–69), and dissimilar alloy welding is, of course, of commercial interest in its own right. Stop-action techniques have been used to "freeze" the complete deformation zone (Ref 55, 56, 60, 63, 64, 70). Careful synchronization of tool withdrawal and rotation can enable the thread to be disengaged, leaving the deforming material behind (Ref 55, 56). A selection of examples of flow validation experiments is shown in Fig. 10.14 and 10.15.

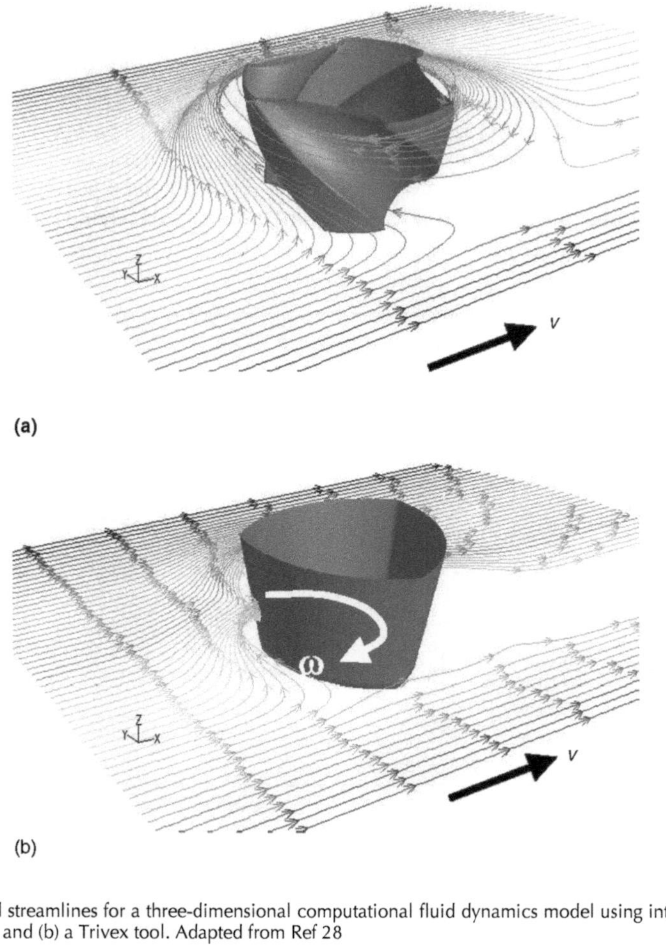


**Fig. 10.12** Predicted streamlines for a three-dimensional computational fluid dynamics model using interfacial slip for (a) a Triflute tool and (b) a Trivex tool. Adapted from Ref 28

Figure 10.14(a) shows an initially straight transverse marker that has been welded through (Ref 58). Note that the material bulges backward in a curve, and a thin zone is swept forward on the advancing side, both characteristics seen in the predicted streamlines and particle tracks of Fig. 10.8, 10.10, and 10.11. Stop-action welds with discrete thin SiC layers (Ref 63, 64) also match the streamline pattern. Figure 10.14(b) shows parallel incoming markers being swept around the retreating side and the flow separation near the advancing side, while Fig. 10.14(c) shows the path of the original joint line around the retreating side and its breakup into layers of the onion ring. Figure 10.14(d) shows Colligan's early stop-action micrograph (Ref 56) of a longitudinal section after tool extraction. In this example, it is apparent that the threads are not full on the rear of the tool, and that the onion ring pattern is generated in the tool wake but compressed into the upper half of the section by the flow of material under the tool root. Experiments such as these offer great detail for model validation, but models currently only have the capability to be tested on the broad pattern of flow and not in the detail.

Optical microscopy is limited to plane sections through welds, but x-ray tomography offers the potential for 3-D visualization. Figure 10.15(a) shows an isometric image of a copper foil placed on the joint line of a stop-action weld in 2024 aluminum alloy (Ref 60). The initial sweeping of the joint line by the shoulder is apparent, and the copper is dispersed over a significant distance from the joint line. However, transverse slices of the x-ray image downstream of the tool (Fig. 10.15b) indicate a concentration of copper particles in a characteristic curve on the retreating side of the nugget, and this is confirmed by optical micrographs in the plane of the workpiece (Fig. 10.15c). This view also reveals several contrasting zones being generated in the deformation zone, seen at higher magnification in Fig. 10.15(d, e). The deforming material closest to the tool etches much darker than the surrounding material extruding past the tool, and the dark etching material itself divides into two layers ("A" and "B" in Fig. 10.15e). Downstream, the paler etching material occupies the retreating side (containing most of the copper marker material), with the darker etching material on the advancing side. The onion ring appears to be made of thin, alternating layers of the two in a remarkably uniform repeating pattern, not a chaotic flow, as has been suggested by some authors. The darker material completely encircles the tool, suggesting that this material makes many revolutions of the tool. The paler etching material extrudes past the retreating side, making less than one revolution, capturing and breaking up the copper and thus the joint line (Fig. 10.15d). Incoming material must steadily transfer into the

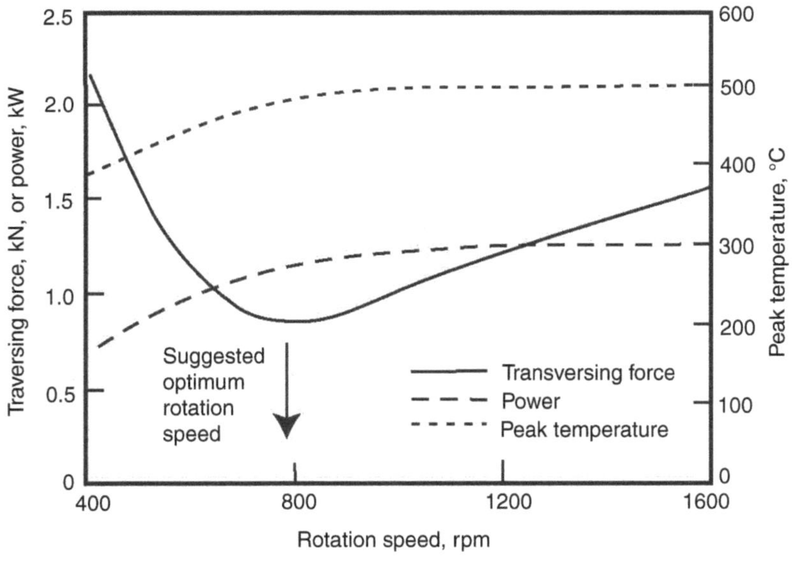


**Fig. 10.13** Schematic of computational fluid dynamics predicted trends with rotation speed in heat generation, peak temperatures, and traversing force for friction stir welding of 2024 aluminum alloy

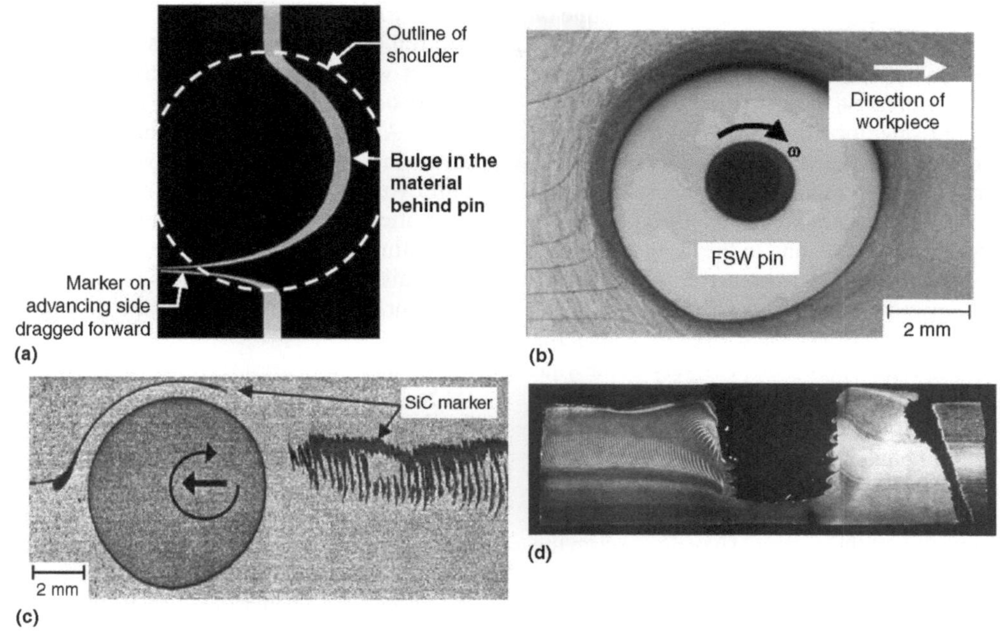


**Fig. 10.14** Metallographic techniques for tracking the flow pattern in friction stir welding (FSW). (a) Transverse copper foil. Adapted from Ref 58. (b, c) Longitudinal SiC markers. Adapted from Ref 63, 64. (d) Longitudinal section of exit hole after synchronized pin retraction. Adapted from Ref 56

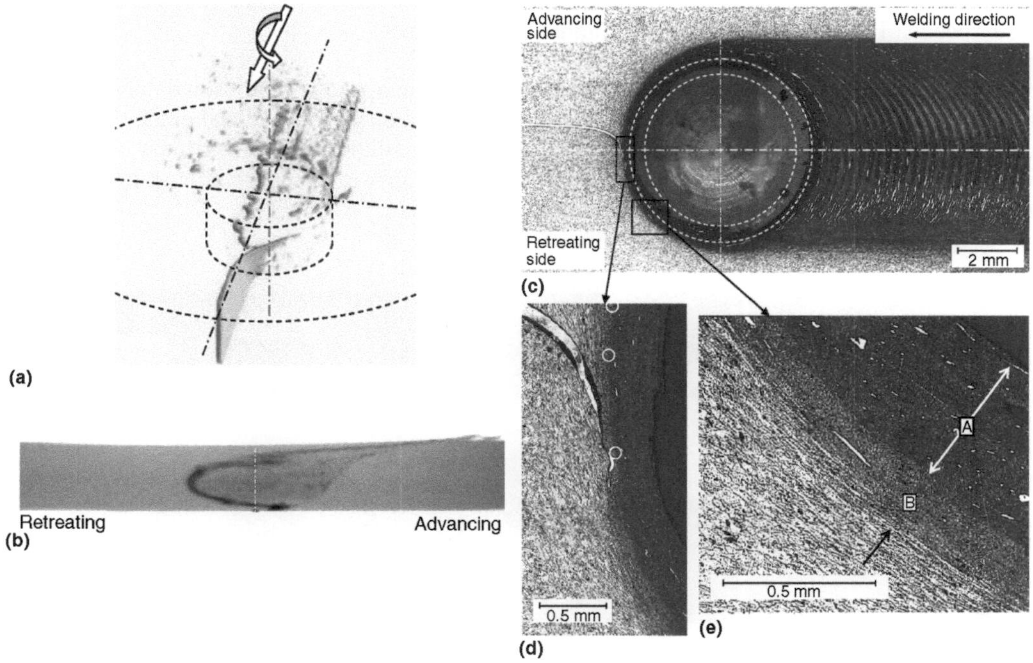


**Fig. 10.15** X-ray tomographic and corresponding metallographic interpretation of friction stir weld flow mechanism. (a) Three-dimensional x-ray tomography showing breakup of copper foil placed on joint line. (b) Longitudinal view of transverse slice through x-ray tomograph. (c–e) Corresponding optical micrographs in plane of workpiece at midthickness. Adapted from Ref 60

intense deformation zone on the leading edge of the tool to maintain the supplies of outgoing "A" and "B" material into the nugget. The transition zone "B" is clear on the retreating side, but on the advancing side, incoming material may be captured directly into zone "A." Some material must also transfer from zone "B" into the dark etching zone "A," because some of the copper is observed in zone "A" and in the nugget material on the advancing side downstream (Fig. 10.15c, e). It is unwise to generalize too much from a single section through one weld, and, of course, the detail of the interpretation will vary somewhat with different tool geometries, process conditions, and alloys. However, the figures illustrate the experimental detail potentially available for validation of flow models and for understanding the mechanism of breakup of the joint line and the formation of the nugget. They do confirm the general interpretation of the mechanism of FSW as being an intense deformation zone near the tool, with a surrounding extrusion zone, as described by Arbegast (Ref 71) and outlined in Chapter 3, "Temperature Distribution and Resulting Metal Flow."

Figure 10.16 shows transverse sections in welds between dissimilar alloys (6082-T6 and 2024-T3) (Ref 68), with the placing of the alloys reversed with respect to advancing and retreating sides but identical rotation and traverse speeds. The etching contrast highlights the separate alloys, and it is clear that the "handedness" of the weld has a significant influence on the formation of the nugget. The dominant contact with the shoulder is with the alloy on the retreating side, because this is swept across the joint line on the trailing edge. Experiments by Palm (Ref 67) showed that, in a dissimilar-material weld, the entire intense plastic zone in contact with the pin could consist of the alloy placed on the advancing side (consistent with zone "A" in Fig. 10.15 being formed from advancing-side material). Modeling the flow with two dissimilar incoming materials has yet to be attempted and presents a major challenge in dealing with the fine-scale layering of the materials and the complexities of handling two different deformation laws for the flow stress.

## 10.4 Microstructure and Property Evolution in FSW of Aluminum Alloys

Chapter 4, "Microstructure Development in Aluminum Alloy Friction Stir Welds," discusses the main microstructural observations in the FSW of wrought aluminum alloys. The evolution of microstructure in welded heat treatable aluminum alloys has been modeled in great detail. The methods were mostly developed for arc welding and have been more recently applied to the thermal cycles in FSW (Ref 4, 72–79). For the heat-affected zone, the problem is purely thermal; for the TMAZ and nugget, there is the potential added complexity of coupling between the deformation microstructure and precipitation.

These microstructure models fall into two categories:

- Semiempirical (with some physical basis), based on isothermal heat treatments and indirect calibration via hardness measurement and able to predict hardness profiles across welds

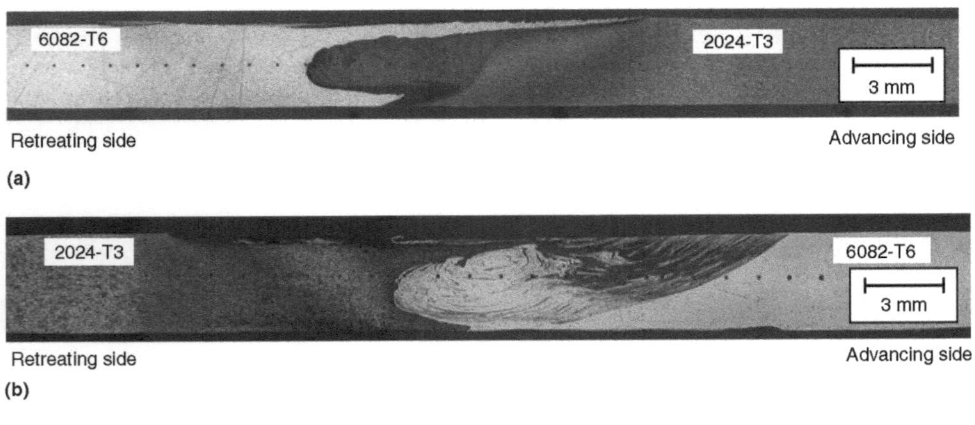


**Fig. 10.16** Metallographic cross sections of dissimilar friction stir welds between 6082-T6 and 2024-T3. (a, b) Welds under identical conditions with the two alloys reversed. Adapted from Ref 68

- Physically based, using detailed thermodynamics and kinetics of phase transformations, calibrated on direct measurement of microstructural features and able to predict hardness and strength, with the potential for extension to ductility, fracture toughness, fatigue, and corrosion properties.

**Semiempirical Microstructure Model.** The semiempirical methodology has been applied to FSW in 2000-, 6000-, and 7000-series alloys (Ref 6, 25, 48, 68, 80–84). It is currently limited to artificially aged tempers (T5, T6, or T7) for reasons discussed subsequently. The procedure is as follows.

Isothermal softening experiments are conducted, typically from 200 °C (390 °F) up to the solidus temperature for times ranging from 1 to $10^5$ s. Figure 10.17(a) shows a typical data set for alloy 2014-T6 (Ref 68). Softening of a peak-aged condition stems from two possible mechanisms: dissolution of hardening precipitates or overaging to a more stable (nonhardening) phase. To distinguish between these, the samples are naturally aged (which may take 3 months or more for 2000- and 7000-series alloys) (Fig. 10.17b). Subtracting the two data sets reveals the change in hardness by natural aging (Fig. 10.17c). It is apparent that the strength recovery is determined primarily by the hold temperature. Maximum recovery after the highest hold temperature corresponds to full dissolution. As the hold temperature falls, the degree of dissolution falls, with the hardening increment scaling directly with the available solute. Below a temperature of 350 °C (660 °F), there is no strength recovery; softening is thus permanent and is due to overaging. For naturally aged tempers (T3 or T4), the data are more complex and beyond the scope of the semiempirical approach. For intermediate temperatures, further artificial aging occurs, the hardness increases, and the response is also sensitive to the heating rate.

The softening data are fitted to a simple model, based on dissolution kinetics. Even though softening also stems from overaging, the underlying mechanism is still governed by the kinetics of precipitate dissolution, so a single model suffices for both. The time $t_1^*$ for full dissolution at a temperature $T$ is given by:

$$t_1^* = t_{r1} \exp\left[\left(\frac{Q_{eff}}{R}\right)\left(\frac{1}{T} - \frac{1}{T_r}\right)\right] \qquad \text{(Eq 10.8)}$$

where $R$ is the universal gas constant, $Q_{eff}$ is an effective activation energy for precipitate dissolution in the particular alloy, and $T_r$ is a reference temperature at which the time for full dissolution is $t_{r1}$. For one-dimensional dissolution (i.e., assuming platelike precipitates), the particle fraction (normalized by the initial value) depends on time at constant temperature as:

$$\frac{f}{f_0} = 1 - \left(\frac{t}{t_1^*}\right)^{1/2} \qquad \text{(Eq 10.9)}$$

The volume fraction of hardening precipitates is inferred from the hardness data as:

$$\frac{f}{f_0} = \left(\frac{\text{HV} - \text{HV}_{min}}{\text{HV}_{max} - \text{HV}_{min}}\right) \qquad \text{(Eq. 10.10)}$$

where HV is the measured hardness, and $\text{HV}_{max}$ and $\text{HV}_{min}$ are the maximum and minimum hardness corresponding to peak precipitation and full dissolution, respectively. Figure 10.17(d) shows the model for 2014-T6, plotted as log $(1 - f/f_0)$ versus log $(t/t_1^*)$. By adjusting $Q_{eff}$, the data converge to a single master curve. From Eq 10.9 a straight line of gradient 0.5 is expected. The early stages of dissolution follow this slope, but the slope steadily decreases in the later stages of dissolution, due to impingement of adjacent diffusion fields. Hence, a pragmatic semiempirical approach, which retains the physical basis of the model, is to use the master curve as a "look-up table."

The second step is to apply the isothermal model to the thermal cycles $T(t)$ predicted from heat flow analysis. Writing the microstructure evolution law (Eq 10.9) in differential form, this may be integrated directly over the cycle, such that Eq 10.9 is replaced by:

$$\frac{f}{f_0} = 1 - \left[\int \frac{dt}{t_1^*}\right]^{1/2} \qquad \text{(Eq 10.11)}$$

The integral in Eq 10.11 represents the kinetic strength of the thermal cycle with respect to precipitate dissolution. Grong and Shercliff (Ref 74) discuss in detail the circumstances in which single internal-state variable models for microstructure evolution can be integrated via a kinetic strength. Essentially, the differential evolution law must be isokinetic and therefore additive (i.e, $df/dt$ is a separable function of $f$ and $T$). The thermal profile is therefore converted into a series of short isothermal steps of duration $\delta t$; $\delta t/t_1^*$ is calculated for each isothermal step and the values summed over the thermal cycle to give the net effective $t/t_1^*$. This is converted into a fraction of precipitates dissolved using the master curve in 10.17(d) and converted to as-welded hardness using Eq 10.10.

The kinetic strength may also be used to estimate an equivalent isothermal hold time at the peak temperature $T_p$ of the thermal cycle. The kinetic strength is set equal to $(t_{eq}/t_1^*)$, with $t_1^*$ evaluated for a temperature equal to $T_p$. Typically, $t_{eq}$ is of the order of 2 to 5 seconds in FSW. Natural aging after welding is accounted for, using the hardness change data (Fig. 10.17c), by finding the hardness change at a temperature

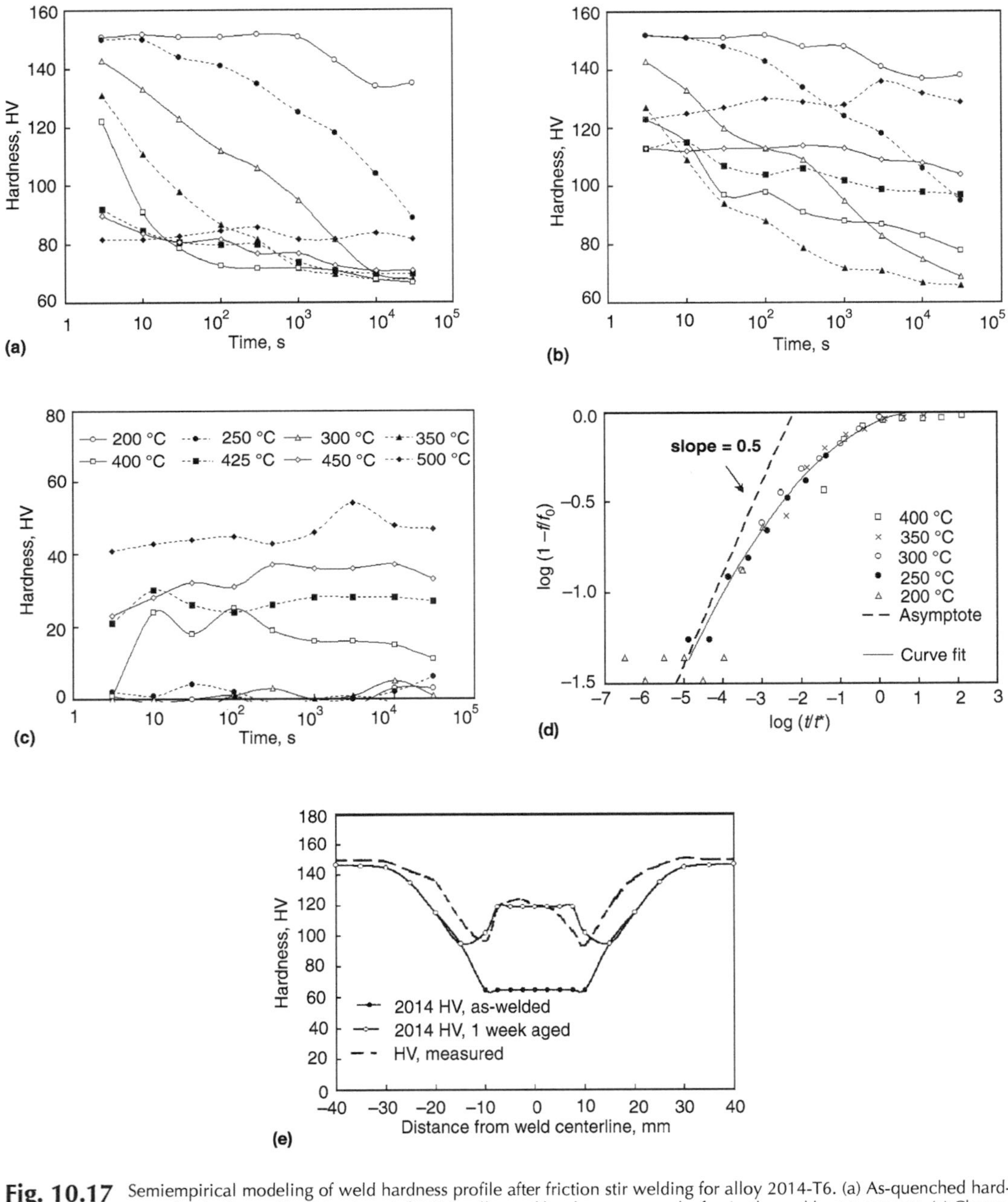


**Fig. 10.17** Semiempirical modeling of weld hardness profile after friction stir welding for alloy 2014-T6. (a) As-quenched hardness after isothermal heat treatment. (b) Naturally aged hardness one week after isothermal heat treatment. (c) Change in hardness by natural aging (data from a and b). (d) Data fit to master curve for softening model based on precipitate dissolution. (e) Measured hardness profile compared with predicted hardness (as-quenched and after natural aging). Adapted from Ref 68

equal to $T_p$ for hold times of the order of $t_{eq}$. Figure 10.17(e) shows the predicted hardness profiles in 2014-T6 using this procedure, both as-welded and naturally aged. The form of the hardness profile is captured reasonably well, in particular, the minimum hardness, which is critical in design. The model is sufficiently accurate to investigate the effect of changing process conditions, for example.

Similar methods have been applied to softening in non-heat-treatable alloys (Ref 82). Alloys in H-conditions (cold worked by rolling or extrusion) soften by recovery and recrystallization. This may also be described in simple kinetic terms using isothermal data and applied to thermal cycles, as mentioned previously.

**Physically Based Microstructure Model.** More sophisticated approaches to the evolution of precipitation in heat treatable aluminum alloys have recently been proposed (Ref 75–79). In these analyses, the evolution of the full size distribution of precipitates is modeled, because this governs the competition between dissolution, coarsening, and transformation from one phase to another. Isothermal and ramp heating and weld thermal cycles have been modeled for previously aged conditions. Extensive use is made of direct measurement of volume fractions and particle radii by small-angle x-ray scattering and electron microscopy (transmission electron microscopy or field emission gun/scanning electron microscopy, or FEG/SEM) for calibration and validation of the model. The models have been applied to ternary extrusion alloys in the 6000 and 7000 series and more recently to the more complex copper-bearing aerospace alloys of the 2000 and 7000 series.

The key ingredients of the physically based methodology are:

- Thermodynamic calculation of phase stability for both metastable and equilibrium precipitates, employing thermodynamic database software
- Classical isothermal nucleation, growth, and coarsening theory, applied to thermal cycles

More than one population of precipitates may be considered simultaneously, with the competition between phases and evolution of each phase determined by the instantaneous microstructural state and temperature. A fundamental concept is the relationship between the size distribution of a given precipitate and the critical radius for stability. The microstructure evolution is tracked continuously in small time-steps and involves complex "bookkeeping" to maintain conservation of solute as the populations of different precipitates change in response to the temperature of each time-step. The models are dependent on several thermodynamic and kinetic parameters, which must often be calibrated for a given alloy. While there is a significant computational penalty and the need for considerable expertise in implementation and validation of such a model, the potential benefits are large. For example, FEG/SEM is able to provide independent data for grain bulk and grain boundaries. This opens up the potential for modeling the effect of dislocation structures on precipitation within the TMAZ and nugget, including quench sensitivity effects (i.e., precipitation of nonhardening phases during the cooling part of the thermal cycle).

The desired output from the models is not microstructure but properties. Strength (and hardness) predictions can be made at a more detailed level than in the semiempirical approach, using the predicted volume fraction and average radius (Ref 85). Detailed validation of strength distributions has become possible by the experimental technique of electronic speckle pattern interferometry (Ref 86). In this technique, the surface of a tensile specimen is analyzed to determine the local stress-strain curve at every point in the weld cross section. In principle, the detailed description of the precipitate state (including distinctions between grain interiors and boundaries) can be used to predict more complex but industrially critical properties, such as ductility, fracture toughness, fatigue, and corrosion. The development of robust microstructure property relationships in this context remains a challenge for future research.

## 10.5 Residual Stress

Residual stress and distortion are important in any welding process. Modeling of these effects in FSW again draws primarily on earlier work on arc welding of heat treatable aluminum alloys, adapted to the thermal histories and mechanical constraint imposed in FSW. Residual stress in welding is primarily caused by the transient thermal cycles in the vicinity of the weld. The intense local heating around the heat source generates nonuniform expansion and contraction. The hot expanding metal close to the weld yields due to its reduced strength and the constraint of the cooler surrounding metal. On cooling, a misfit in strain results between the yielded and unyielded regions. These strains lead to residual tensile stress (predominantly parallel to the weld) in the near-weld region, or distortion, or a combination of the two. Modeling of residual stress in FSW therefore requires a good thermal model for the whole workpiece, coupled to a mechanical model for the elastic-plastic response at temperature.

The relatively few modeling studies to date all use finite element analysis (Ref 10, 11, 14, 17, 22, 24, 84, 87). The thermomechanical aspect of residual stress introduces the need for further input data, in addition to the thermal data discussed in section 10.2, "Heat Conduction." The plastic strains responsible for residual stress are small, due to the constraint of the surrounding workpiece. They are much less than the strains in the flow region due to the FSW process, but it is the strain outside the flow region and in the nugget region as it cools behind the tool that matters. Input data required therefore include the Young's modulus as a function of temperature, the coefficient of thermal expansion, and the temperature dependence of the flow stress. The complexity of the flow response with temperature in heat treatable alloys was discussed in the context of flow modeling earlier (section 10.3, "Metal Flow"). For residual-stress modeling, the low strain-rate data are relevant, but there are similar issues about the influence of hold time in standard tests prior to measurement of the yield stress. However, the plastic strains occur at relatively high temperature, when most hardening precipitates have dissolved (and work hardening may also be neglected). These aspects have been investigated for 2024-T3 arc welds (Ref 88, 89).

The mechanical constraint imposed on the workpiece during any welding process is critical in determining the residual stress and distortion. In contrast to arc welding, the FSW process also applies mechanical loads directly via the tool: downforce, traverse force, and torque. Preliminary residual-stress models have been attempted to investigate this effect (Ref 17). Figure 10.18 shows the predicted longitudinal stress in a 2024-T3 alloy FSW, first with the heat input alone and second with a superimposed downforce and torque under the tool shoulder. Figure 10.18(a) shows the characteristic pattern of parallel bands of tensile residual stress on either side of the joint line, also observed in arc welds (Ref 88, 89). Superposition of the torque (Fig. 10.18b) produces a degree of asymmetry in the residual-stress profile across the weld, typically shifting the peak stress by 10%. These predictions have not yet been properly validated, but modest asymmetry has been observed experimentally. The maximum tensile stress predicted for 2024-T3 is on the order of 200 MPa (29 ksi), comparable to the room-temperature (postwelding) yield strength.

Robust validation data for residual-stress models require experimentally intensive and costly diffraction testing, using neutrons or x-rays. The particular value of synchrotron x-ray techniques has been illustrated for several aluminum welding studies, including FSW applied to dissimilar alloys (Ref 88–90). Bringing together the finite element analysis of residual stress and the extensive synchrotron data is a matter of current research.

## 10.6 Summary

Modeling of FSW has followed the empirical development of the process for the last decade. Numerical methods now dominate, due to their ability to capture essential complexity in the underlying physics. Heat-generation and temperature-prediction techniques are now sufficiently robust and fast to be used routinely as inputs to models that build on the thermal history of the weld. The prediction of metal flow remains challenging, but CFD models in particular have shed considerable light on the fundamental mechanisms of FSW and the influence of changing the tool design and process conditions. Robust prediction of the formation of defects remains elusive, however. Modeling of microstructure evolution and residual stress has mainly concentrated on the commercially dominant heat treatable aluminum alloys. Semiempirical predictions of hardness profiles have been tested for many alloys, but the emerging physically based models hold out the promise of predictive capability for properties such as ductility, fracture, fatigue, and corrosion. This and the validation of numerical models for residual stress are the next major challenges in FSW modeling.

## ACKNOWLEDGMENTS

The authors acknowledge the many contributions to the work presented here, but in particular thank the following for discussions over many years of the problems of modeling FSW: Professor Philip Withers and Dr. Joe Robson (University of Manchester, United Kingdom), Dr. Terry Dickerson (University of Cambridge, United Kingdom), Professor Stewart Williams (University of Cranfield, United Kingdom), Professor Øystein Grong (NTNU Trondheim, Norway), and Dr. Mike Russell (TWI, United Kingdom). Professor Tony Reynolds (University of South Carolina, United States), Dr. B. London (Cal Poly-SLO, United States), and Dr. Kevin Colligan (CTC, United States) are thanked for permission to use their figures in this chapter.

## REFERENCES

1. H.R. Shercliff and P.A. Colegrove, Modeling of Friction Stir Welding, *Mathematical Modelling of Weld Phenomena 6,*

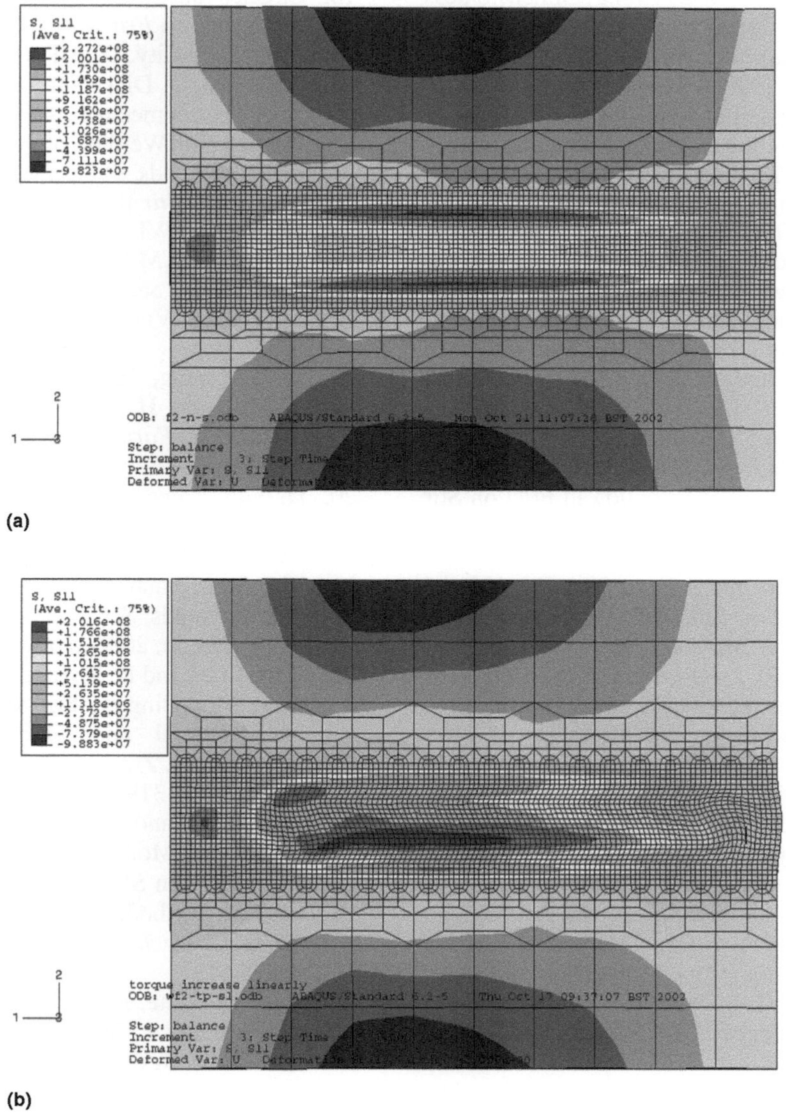


**Fig. 10.18** Longitudinal residual stress in friction stir welds of 2024-T3, predicted by finite element analysis. (a) Thermal input only. (b) Thermal input with mechanical downforce and torque superimposed. Adapted from Ref 17

H. Cerjak and H. Bhadeshia, Ed., Maney Publishing, London, U.K., 2002, p 927–974
2. D. Rosenthal, Mathematical Theory of Heat Distribution during Welding and Cutting, *Wed. J.* Vol 20 (No. 5), 1941, p 220s–234s
3. D. Rosenthal, The Theory of Moving Sources of Heat and Its Application to Metal Treatments, *Trans. ASME,* Vol 68, 1946, p 849–866
4. Ø. Grong, *Metallurgical Modelling of Welding,* 2nd ed., Institute of Materials, London, U.K., 1997
5. O.R. Myhr and Ø. Grong, Dimensionless Maps for Heat Flow Analyses in Fusion Welding, *Acta Metall. Mater.,* Vol 38 (No. 3), 1990, p 449–460
6. M.J. Russell and H.R. Shercliff, Analytical Modeling of Microstructure Development in Friction Stir Welding, *Proc. First Int. Symp. on Friction Stir Welding,* June 1999 (Thousand Oaks, CA), TWI
7. J.C. McClure, Z. Feng, T. Tang, J.E. Gould, and L.E. Murr, A Thermal Model of Friction Stir Welding, *Proc. Fifth Int. Conf. on Trends in Welding Research* (Pine Mountain, GA), ASM International, 1998
8. W. Tang, X. Guo, J.C. McClure, L.E. Murr, and A.C. Nunes, Heat Input and Temperature Distribution in Friction Stir Welding, *J. Mater. Process. Manuf. Sci.,* Vol 7, Oct 1998, p 163–172
9. J.E. Gould and Z. Feng, Heat Flow Model for Friction Stir Welding of Aluminum Alloys, *J. Mater. Process. Manuf. Sci.,* Vol 7 (No. 2), Oct 1998, p 185–194
10. J.C. Yue and Q. Xinhai, Thermal and Thermo-Mechanical Modeling of Friction Stir Welding of Aluminum Alloy 6061-T6, *J. Mater. Process. Manuf. Sci.,* Vol 7 (No. 2), Oct 1998, p 163–172
11. J.C. Yue and Q. Xinhai, Heat Transfer and Thermal-Mechanical Analysis of Friction Stir Joining of AA6061-T6 Plates, *Proc. First Int. Symp. on Friction Stir Welding,* June 1999 (Thousand Oaks, CA), TWI
12. P.A. Colegrove, M. Painter, D. Graham, and T. Miller, 3-Dimensional Flow and Thermal Modeling of the Friction Stir Welding Process, *Proc. Second Int. Symp. on Friction Stir Welding* (Gothenburg, Sweden), TWI, 2000
13. M.Z.H. Khandkar and J.A. Khan, *J. Mater. Process. Manuf. Sci.,* Vol 10 (No. 2), 2001, p 91–105
14. Q.-Y. Shi, T. Dickerson, and H.R. Shercliff, Thermo-Mechanical Analyses of Welding Aluminum Alloy with TIG and Friction Stir Welding, *Proc. Sixth Int. Conf. on Trends in Welding Research,* April 2002 (Phoenix, AZ), ASM International, p 247–252
15. K. Lindner, Z. Khandkar, J. Khan, W. Tang, and A.P. Reynolds, *Proc. Fourth Int. Symp. on Friction Stir Welding,* May 2003 (Park City, UT), TWI
16. C. Chen and R. Kovacevic, *Proc. Fourth Int. Symp. on Friction Stir Welding* May 2003 (Park City, UT), TWI
17. Q.-Y. Shi, T. Dickerson, and H.R. Shercliff, Thermomechanical FE Modeling of Friction Stir Welding of Al-2024 Including Tool Loads, *Proc. Fourth Int. Symp. on Friction Stir Welding,* May 2003 (Park City, UT), TWI
18. P. Heurtier, M.J. Jones, C. Desrayaud, J.H. Driver, and F. Montheillet, *Mater. Sci. Forum,* Vol 426–432, 2003, p 2927–2932
19. W.-S. Chang, H.-S. Bang, S.-B. Jung, Y.-M. Yeon, H.-J. Kim, and W.-B. Lee, *Mater. Sci. Forum,* Vol 426–432, 2003, p 2953–2958
20. D. Lawrjaniec, A. Abisror, C. Decker, M. Koçak, and J. dos Santos, *Mater. Sci. Forum,* Vol 426–432, 2003, p 2993–2998
21. M.Z.H. Khandkar, J.A. Khan, and A.P. Reynolds, *Sci. Technol. Weld. Join.,* Vol 8 (No. 3), 2003, p 66–174
22. C.M. Chen and R. Kovacevic, Finite Element Modelling of Friction Stir Welding—Thermal and Thermomechanical Analysis, *Int. J. Mach. Tools Manuf.,* Vol 43, 2003, p 1319–1326
23. C.M. Chen and R. Kovacevic, Thermomechanical Modelling and Force Analysis of Friction Stir Welding by the Finite Element Method, *Proc. Inst. Mech. Eng., Part C: J. Mech. Eng. Sci.,* Vol 218, 2003, p 509–519
24. R.W. McCune, H. Ou, C.G. Armstrong, and M. Price, Modeling Friction Stir Welding with the FE Method—A Comparative Study, *Proc. Fifth Int. Symp. on Friction Stir Welding,* Sept 2004 (Metz, France), TWI
25. Ø. Frigaard, Ø. Grong, and O.T. Midling,

*Metall. Mater. Trans. A,* Vol 32 (No. 5), 2001, p 1189–1200

26. Y.S. Touloukian, *Specific Heat: Non Metallic Solids,* Plenum, New York, 1970
27. Y.S. Touloukian, *Thermal Conductivity: Non Metallic Solids,* Plenum, New York, 1970
28. P.A. Colegrove and H.R. Shercliff, Development of the Trivex Friction Stir Welding Tool, Part II: 3-Dimensional Flow Modeling, *Sci. Technol. Weld. Join.,* Vol 9 (No. 4), 2004, p 352–361
29. P.A. Colegrove and H.R. Shercliff, Experimental and Numerical Analysis of Aluminum Alloy 7075-T7351 Friction Stir Welds, *Sci. Technol. Weld. Join.,* Vol 8 (No. 5), 2003, p 360–368
30. M.B. Stewart, A.C. Nunes, G.P. Adams, and P. Romine, *Devel. Theor. Appl. Mech.,* Vol XIX, 1998, p 472
31. H. Schmidt and J. Hattel, An Analytical Model for Prescribing the Flow Around the Tool Probe in Friction Stir Welding, *Proc. TMS Spring Meeting, Friction Stir Welding III,* March 2003 (San Diego, CA), TMS
32. P. Heurtier, C. Desrayaud, and F. Montheillet, *Mater. Sci. Forum,* Vol 396–402, 2002, p 1537–1542
33. J.A. Schneider and A.C. Nunes, Characterisation of Plastic Flow and Resulting Microtextures in a Friction Stir Weld, *Metall. Mater. Trans. B,* Vol 35, 2004, p 777–783
34. L.D. Oosterkamp, O.J. Emmerhoff, and O.T. Midling, Friction Stir Welding of Aluminum Extrusions, *Recent Advances in Welding Simulation,* IMechE Seminar Publication 2000–13, Institution of Mechanical Engineers, London, U.K., 2000, p 79–88
35. S. Xu, X. Deng, A.P. Reynolds, and T.U. Seidel, Finite Element Simulation of Material Flow in Friction Stir Welding, *Sci. Technol. Weld. Join.,* Vol 6 (No. 3), 2001, p 191–193
36. R.L. Goetz and K.V. Jata, in *Friction Stir Welding and Processing,* K.V. Jata et al., Ed., TMS, 2001, p 35–42
37. P. Ulysse, *Int. J. Mach. Tools Manuf.,* Vol 42, 2002, p 1549–1557
38. S. Xu and X. Deng, *Proc. Fourth Int. Symp. on Friction Stir Welding,* May 2003 (Park City, UT), TWI
39. L. Fourment, S. Guerdoux, M. Miles, and T. Nelson, *Proc. Fifth Int. Symp. on Friction Stir Welding,* Sept 2004 (Metz, France), TWI
40. H. Schmidt and J. Hattel, A Local Model for the Thermomechanical Conditions in Friction Stir Welding, *Model. Simul. Mater. Sci. Eng.,* Vol 13, 2005, p 77–93
41. H. Schmidt and J. Hattel, CFD Modeling of the Shear Layer Around the Tool Probe in Friction Stir Welding, *Friction Stir Welding III,* March 2005 (San Diego, CA), TMS
42. G.J. Bendzsak, T.H. North, and C.B. Smith, An Experimentally Validated 3-D Model for Friction Stir Welding, *Proc. Second Int. Symp. on Friction Stir Welding* (Gothenburg, Sweden), TWI, 2000
43. T. Long, T.U. Seidel, W. Tang, and A.P. Reynolds, in *Hot Deformation of Aluminum Alloys III,* Z. Jin et al., Ed., TMS, 2003
44. T.U. Seidel and A.P. Reynolds, *Sci. Technol. Weld. Join.,* Vol 8 (No. 3), 2003, p 175–183
45. P.A. Colegrove and H.R. Shercliff, Development of the Trivex Friction Stir Welding Tool, Part I: 2-Dimensional Flow Modeling and Experimental Validation, *Sci. Technol. Weld. Join.,* Vol 9 (No. 4), 2004, p 345–351
46. P.A. Colegrove and H.R. Shercliff, 2-Dimensional CFD Modeling of Flow Round Profiled FSW Tooling, *Sci. Technol. Weld. Join.,* Vol 9 (No. 6), 2004, p 483–492
47. P.A. Colegrove and H.R. Shercliff, Modelling the Friction Stir Welding of Aerospace Alloys, *Proc. Fifth Int. Symp. on Friction Stir Welding,* Sept 2004 (Metz, France), TWI
48. P.A. Colegrove, T. Hyoe, and H.R. Shercliff, Development of the Trivex Friction Stir Welding Tool for Making Lap Welds, *Proc. Fifth Int. Symp. on Friction Stir Welding,* Sept 2004 (Metz, France), TWI
49. P.A. Colegrove and H.R. Shercliff, 3-Dimensional CFD Modeling of Flow Round a Threaded Friction Stir Welding Tool Profile, *J. Mater. Process. Technol.,* Vol 169, 2005, p 320–327
50. P.A. Colegrove and H.R. Shercliff, CFD Modeling of the Friction Stir Welding of Thick Plate 7449 Aluminum Alloy, *Sci. Technol. Weld. Join.,* 11(4), July 2006, p 429–441

51. P.A. Colegrove, H.R. Shercliff, and R. Zettler, A Model for Predicting the Heat Generation in Friction Stir Welding, *Sci. Technol. Weld. Join.*, 2006, in press
52. A. Askari, S. Silling, B. London, and M. Mahoney, in *Friction Stir Welding and Processing,* K.V. Jata et al., Ed., TMS, 2001, p 43–54
53. C.M. Sellars and McG. Tegart, Hot Workability, *Int. Met. Rev.,* Vol 17, 1972, p 1–24
54. G.R. Johnson and W.H. Cook, *Proc. Seventh Int. Symp. on Ballistics* (The Hague, Netherlands), 1983, p 541–548
55. K. Colligan, Dynamic Material Deformation during Friction Stir Welding of Aluminum, *Proc. First Int. Symp. on Friction Stir Welding,* June 1999 (Thousand Oaks, CA), TWI
56. K. Colligan, Material Flow Behavior during Friction Stir Welding of Aluminum, *Weld. J. Suppl.,* Vol 78 (No. 7), 1999, p 229s–237s
57. A.P. Reynolds, T.U. Seidel, and M. Simonsen, Visualisation of Material Flow in an Autogenous Friction Stir Weld, *Proc. First Int. Symp. on Friction Stir Welding,* June 1999 (Thousand Oaks, CA), TWI
58. A.P. Reynolds, Visualisation of Material Flow in Autogenous Friction Stir Welds, *Sci. Technol. Weld. Join.,* Vol 5 (No. 2), 2000, p 120–124
59. Y.-H. Zhao, S.-B. Lin, F.-X. Qu, and L. Wu, Influence of Pin Geometry on Material Flow in FSW Process, *Mater. Sci. Technol.,* Vol 22 (No. 1), 2006, p 45–50
60. T.L. Dickerson, H.R. Shercliff, and H. Schmidt, A Weld Marker Technique for Flow Visualisation in Friction Stir Welding, *Proc. Fourth Int. Symp. on Friction Stir Welding,* May 2003 (Park City, UT), TWI
61. M. Guerra, J.C. McClure, L.E. Murr, and A.C. Nunes, in *Friction Stir Welding and Processing,* K.V. Jata et al., Ed., TMS, 2001, p 25–34
62. M. Guerra, C. Schmidt, J.C. McClure, L.E. Murr, and A.C. Nunes, Flow Patterns during FSW, *Mater. Charact.,* Vol 49, 2003, p 95–101
63. B. London, M. Mahoney, W. Bingel, M. Calabrese, and D. Waldron, *Proc. Third Int. Symp. on Friction Stir Welding,* Sept 2001 (Kobe, Japan), TWI
64. B. London, M. Mahoney, W. Bingel, M. Calabrese, R.H. Bossi, and D. Waldron, in *Friction Stir Welding and Processing II,* K.V. Jata et al., Ed., TMS, 2003, p 3–12
65. J.A. Schneider and A.C. Nunes, in *Friction Stir Welding and Processing II,* K.V. Jata et al., Ed., TMS, 2003, p 43–51
66. R. Zettler, S. Lomolino, J. dos Santos, T. Donath, F. Beckmann, T. Lippmann, and D. Lohwasser, A Study on Material Flow in FSW of AA2024-T351 and AA6056-T4 Alloys, *Proc. Fifth Int. Symp. on Friction Stir Welding,* Sept 2004 (Metz, France), TWI
67. F. Palm, Charakterisierung des Materialflusses beim Reibrührschweissen (Friction Stir Welding) von Al-Werkstoffen, *Materials Week '99,* Oct 1998 (Munich, Germany), (in German)
68. H.R. Shercliff, M.J. Russell, A.D. Taylor, and T.L. Dickerson, Microstructural Modelling in Friction Stir Welding of 2000-Series Aluminium Alloys, *Méc. Ind. (J. Assoc. Fr. Méc.),* Vol 6, 2005, p 25–35
69. M.J. Peel, A. Steuwer, P.J. Withers, H.R. Shercliff, T.L. Dickerson, and Q.-Y. Shi, Dissimilar Friction Stir Welds in AA5083-AA6082, Part I: Process Parameter—Thermal History and Weld Property Effects, *Metall. Mater. Trans. A,* in press
70. K.J. Colligan and S.K. Chopra, Examination of Material Flow in Thick Section FSW of Aluminum Using a Stop-Action Technique, *Proc. Fifth Int. Symp. on Friction Stir Welding,* Sept 2004 (Metz, France), TWI
71. W.J. Arbegast, Modeling Friction Stir Joining as a Metal Working Process, *Hot Deformation of Aluminum Alloys,* Z. Jin, Ed., TMS, 2003
72. O.R. Myhr and Ø. Grong, Process Modeling Applied to 6082-T6 Aluminum Weldments, Part 1: Reaction Kinetics, *Acta Metall. Mater.,* Vol 39 (No. 11), 1991, p 2693–2702, 2703–2711
73. O.R. Myhr, Ø. Grong, S. Klokkehaug, H.G. Fjær, and A.O. Kluken, Process Model for Welding of Al-Mg-Si Extrusions, *Sci. Technol. Weld. Join.,* Vol 2 (No. 6), 1997, p 245–253
74. Ø. Grong and H.R. Shercliff, Microstruc-

tural Modeling in Metals Processing, *Prog. Mater. Sci.,* Vol 47, 2002, p 163–282
75. Ø. Grong and O.R. Myhr, Additivity and Isokinetic Behavior in Relation to Diffusion Controlled Growth, *Acta Mater.,* Vol 48, 2000, p 445–452
76. O.R. Myhr and Ø. Grong, Modelling of Non-Isothermal Transformations in Alloys Containing a Particle Distribution, *Acta Mater.,* Vol 48, 2000, p 1605–1615
77. O.R. Myhr, Ø. Grong, and S.J. Andersen, Modelling of the Age Hardening Behavior of Al-Mg-Si Alloys, *Acta Mater.,* Vol 49, 2001, p 65–75
78. M. Nicolas and A. Deschamps, Characterisation and Modeling of Precipitate Evolution in an Al-Zn-Mg Alloy during Non-Isothermal Heat Treatments, *Acta Mater.,* Vol 51 (No. 20), 2003, p 6077–6094
79. N. Kamp, A. Sullivan, R. Tomasi, and J.D. Robson, Modelling of Heterogeneous Precipitate Distribution Evolution during Friction Stir Welding Process, *Acta Mater.,* Vol 54, 2006, p 2003–2014
80. T. Hyoe, H.R. Shercliff, and P.A. Colegrove, Thermal and Microstructure Modeling in Thick Plate Aluminum Alloy 7075 Friction Stir Welds, *Friction Stir Welding II,* March 2003 (San Diego, CA), TMS
81. J.D. Robson, A. Sullivan, G. McShane, and H.R. Shercliff, Microstructural Evolution during Friction Stir Welding of AA7449, *Proc. Fifth Int. Symp. on Friction Stir Welding,* Sept 2004 (Metz, France), TWI
82. M.J. Peel, A. Steuwer, and P.J. Withers, Dissimilar Friction Stir Welds in AA5083-AA6082, Part II: Process Parameter—Effects on Microstructure, *Metall. Mater. Trans. B,* Vol. 37, July 2006, p 2195–2206
83. C. Gallais, A. Denquin, A. Pic, A. Simar, T. Pardoen, and Y. Brechet, Modelling the Relationship between Process Parameters, Microstructural Evolutions and Mechanical Behavior in a Friction Stir Welded 6*xxx* Aluminum Alloy, *Proc. Fifth Int. Symp. on Friction Stir Welding,* Sept 2004 (Metz, France), TWI
84. Z. Feng, X.-L. Wang, S.A. David, and P. Sklad, Modeling of Residual Stresses and Property Distributions in Friction Stir Welds of Al Alloy 6061-T6, *Proc. Fifth Int. Symp. on Friction Stir Welding,* Sept 2004 (Metz, France), TWI
85. A. Deschamps and Y. Brechet, Influence of Predeformation and Aging of an Al-Zn-Mg Alloy, *Acta Mater.,* Vol 47 (No. 1), 1999, p 281, 293
86. M.J. Peel, M. Preuss, A. Steuwer, M. Turski, and P.J. Withers, *Proc. Sixth Int. Conf. on Trends in Welding Research,* April 2002 (Phoenix, AZ), ASM International, p 273–278
87. D. Lawrjaniec, A. Abisror, C. Decker, M. Kocak, and J. dos Santos, Numerical Simulation of Friction Stir Welding, *Mater. Sci. Forum,* Vol 426–432, 2003, p 2993– 2998
88. R.V. Preston, H.R. Shercliff, P.J. Withers, and S.D. Smith, Finite Element Modeling of Tungsten Inert Gas Welding of Aluminum Alloy 2024, *Sci. Technol. Weld. Join.,* Vol 8 (No. 1), 2003, p 10–18
89. R.V. Preston, H.R. Shercliff, P.J. Withers, and S.D. Smith, Microstructurally-Based Constitutive Modeling of Residual Stress Development in Welding of Aluminum Alloy 2024, *Acta Mater.,* Vol 52 (No. 17), 2004, p 4973–4983
90. A. Steuwer, M.J. Peel, and P.J. Withers, Dissimilar Friction Stir Welds in AA5083-AA6082, Part III: Residual Stresses, *Metall. Mater. Trans. A,* in press

## CHAPTER 11

# Robots and Machines for Friction Stir Welding/Processing

Christopher B. Smith, Friction Stir Link, Inc.

FRICTION STIR WELDING (FSW) and its variants of friction stir processing (FSP) and friction stir spot welding (FSSW) have numerous equipment solutions for production and development applications, like most manufacturing processes. There are three basic categories of production equipment solutions for most processes: manual, fixed automation, and robotic solutions. Production FSW solutions are similar, with the exception that a manual solution is generally not possible due to the high forces required for FSW and its variants.

Typically, the decision for the type of production solution is based on economic and technical factors. The economic factors include cost and productivity, for example, parts per unit of time the machine is capable of producing, while there are several technical factors for FSW that affect the choice for the production equipment solution. These technical factors for FSW include the force requirements, the stiffness requirements, the intelligence or sensing requirements, and the flexibility requirements. For FSW applications, the force, stiffness, intelligence, and flexibility requirements can be vastly different depending on the application. Thus, the equipment solution can vary depending on the specific application characteristics.

This chapter first reviews the various FSW application characteristics (material thickness, alloy, etc.) and how they affect each of these technical categories (force requirements, stiffness requirements, intelligence requirements, and flexibility requirements). These application characteristics ultimately dictate the equipment solution that is required for any one application. This chapter also reviews the basic equipment solutions and their relative ability with respect to these technical categories. Lastly, peripheral equipment is discussed that provides solutions for special applications.

## 11.1 Application Characteristics

Each FSW, FSP, and FSSW application has basic characteristics that affect the force requirements, stiffness requirements, intelligence requirements, and flexibility requirements for the application. These characteristics are application dependent and dictate the type of machine solution that should be employed.

**Part Geometry.** There are several geometrical characteristics of the part/application that affect the force, stiffness, intelligence, and flexibility requirements of the production machine. These include the following.

*Part thickness* most significantly affects the force and stiffness requirements of the machine:

- As thickness increases, force requirements increase.
- For thin material (<1 mm, or 0.04 in.), stiffness requirements can increase due to increased sensitivity of the FSW process.
- For thin material (<1 mm), intelligence or sensing requirements can increase to overcome increased sensitivity of the FSW process.

*Weld Path.* The weld path mainly affects the flexibility required from the machine. This flexibility is related to the number of axes the machine must possess:

- One-dimensional (1-D) paths typically require the least flexibility (fewest axes of motion), although a 1-D path can still require a five-axis machine. One such case is welding of tailor-welded blanks, where dissimilar-thickness butt welds are required. A typical 1-D application requiring the fewest number of axes is the joining of long extrusions.
- Two-dimensional (2-D) paths require significantly more flexibility due to the need to maintain work and travel angles along the path, in most applications. This will typically require at least a five-axis machine, unless the FSW or FSP tool is held perpendicular to the path. A typical application with 2-D paths is an FSP application on a flat surface of a casting.
- Three-dimensional (3-D) paths require the most flexibility and always require the machine to have at least five axes of motion. Typical applications requiring 3-D paths are FSP of castings on a complex surface or FSW on complex surfaces.
- Circumferential paths (e.g., tank ends) require a moderate level of flexibility. A single-axis machine can be used for circumferential welding, with the aid of an external rotary positioner.
- Multiple welds are required in many applications at multiple orientations, which affect the flexibility requirements of the FSW machine. In this case, a machine with six axes is often the most suitable for economic and technical reasons, although machines with fewer axes can be used in special cases where external positioners could be used. If machines with fewer than six axes are employed, it often means that multiple setups are required. This significantly affects productivity in a negative manner with machines having less than six axes.

*Part Size/Weld Lengths.* Weld length basically affects the required working envelope of the machine. For example, welding of long extrusions requires a long machine, whereas small welds on small parts only require a small machine.

*Lack of Access to Both Sides.* For applications where there is no access to the back of the part, a self-reacting tool (bobbin tool) can be used. The self-reacting tool is special peripheral equipment and is described in detail in section 11.5, "Special Peripheral Equipment." The use of the self-reacting tool can significantly increase the stiffness and intelligence requirements of the application.

**Joint Type.** There are several basic joint configurations where FSW can be applied. These different joint configurations can affect the requirements of the FSW machine. The following lists the joint configurations and their effects on the machine requirements.

*Full-Penetration Butt Weld.* The full-penetration butt weld requires the highest relative level of force.

*Partial-Penetration Butt Weld.* The partial penetration butt weld requires less force than a full-penetration butt weld in the same thickness. However, the intelligence or sensing requirements may be increased, due to increased sensitivity of the process. That is, the range of force over which quality welds can be produced may be smaller than for a full-penetration weld.

*Lap-Penetration Weld.* The lap-penetration weld typically requires less force than a butt weld. Additionally, the lap-penetration weld is insensitive to the location of the FSW tool with respect to the joint line. This decreases intelligence and stiffness requirements.

*Dissimilar-Thickness Butt Weld.* The dissimilar-thickness butt weld places the most constraints on the machine. Major constraints include:

- Because the FSW tool must be tilted backward (travel angle) and sideways (work angle) in a dissimilar-thickness butt weld application, the flexibility requirements of the machine are greatly increased. Without a five-axis machine, welding of dissimilar-thickness butt welds is very difficult. Another solution is to employ complex fixturing that allows for tilting of the parts. This alternative is cumbersome and limits the ability to optimize the work and travel angles.
- As the thickness difference increases or the work angle increases, the process becomes more sensitive to off-seam conditions. This can place added stiffness or intelligence requirements (e.g., seam tracking) on the machine.
- As the thickness difference increases or the work angle increases, the process becomes more sensitive to flash generation. The flash

generation can be caused by an off-seam condition or a small difference in work or travel angle. Thus, increased thickness differences also require the machine to be more flexible and have better control over the work and travel angles.

*Lap Fillet Joint.* The lap fillet joint has similar requirements to the dissimilar-thickness butt weld, due to the need for both a work and travel angle.

*For FSSW and FSP,* there are no joint types. However, in relation to FSW, FSP has characteristics similar to a partial-penetration butt weld. For FSSW, the joint type can be equated to a lap-penetration joint. However, with the speeds at which friction stir spot welds must be made in most production applications, the force requirements for FSSW are significantly higher than for FSW in the same thickness and alloy combination.

**Material and Alloy.** The material and alloy can significantly affect the requirements of the FSW, FSP, or FSSW machine.

*Aluminum.* Friction stir welding, processing, or spot welding of aluminum alloys is the most common application of the FSW process. However, machine requirements vary significantly based on the alloy. The alloy affects the force requirements of the machine. For example, an FSW butt weld in 6 mm (0.24 in.) 1100 aluminum alloy can require 2.5 kN (0.28 tonf) or less welding force, whereas a butt weld in 6 mm 7*xxx* aluminum alloy can require five times or more force.

*Magnesium alloys* tend to require a little higher thrust force than an equivalent-thickness aluminum alloy.

*Copper alloys* require some additional thrust force and a moderate increase in torque. Bronze alloys tend to require similar force levels to 6*xxx*-series aluminum but require additional torque.

*Steel* requires the most significant level of force as well as very high level of machine stiffness, due to current FSW tool material technology. The current FSW tool materials are sensitive to vibration and runout and thus dictate the requirement for a very stiff machine.

*Other materials* are weldable, including lead, titanium, and so on. As a broad generalization, the force and stiffness requirements tend to correlate with the melting point and the extrudability of the material that is to be welded. However, specific alloys within a material type can also significantly affect force and stiffness requirements.

***Tool Design.*** The design of the FSW tool affects the technical requirements of the machine. On any one application, a variety of tool designs can be considered (see Chapter 2, "Friction Stir Tooling: Tool Materials and Design"). A list of tool features and how they affect the machine requirements follows.

*Shoulder.* As the tool shoulder increases, the required welding force and torque increase.

*Pin Diameter.* As the pin diameter increases, the required welding force and torque increase.

*Pin Length.* As the pin length increases, the required welding force and torque increase.

*Shoulder-to-Pin-Diameter Ratio.* As this ratio decreases, the process becomes more sensitive. That is, the range of welding force over which a quality weld is produced decreases. Thus, low shoulder-to-pin-diameter ratios require increased stiffness and intelligence requirements from the machine.

*Conical pins* decrease the welding force and torque. Additionally, the welding thrust force trace is more desirable during the plunge. That is, the thrust force tends to continually increase until the shoulder contacts the material. This can allow for improved error-proofing strategies or strategies where the traverse can be initiated based on the thrust profile during the plunge. To the contrary, a cylindrical pin will have a thrust force profile during the plunge where the peak force can occur prior to the contact of the shoulder.

*Threads on Pin.* Threads alone tend to require the highest level of force due to the pumping action that they create. Increasing pitch tends to increase the welding force but makes the process more robust. That is, the process is less sensitive or variable, so sensing and intelligence requirements are reduced with increased thread pitch.

*Flats on Pin.* The addition of flats on the pin tends to decrease the welding force and torque.

*Spirals on Pin.* Spirals tend to generate a higher level of thrust force.

*Shoulder Features and 0° Travel Angle.* Through the use of special features on the tool shoulder, it is possible to perform FSW on flat surfaces at a 0° travel angle. This has the benefit of decreasing flexibility requirements (number of machine axes). However, the process is more sensitive using a 0° travel angle. As a consequence, this increases stiffness and intelligence requirements of the machine.

**The welding parameters** of the process can also affect the technical requirements of the machine.

*Rotation Speed.* Increases in rotation speed generally decrease the required welding force. The reduction in welding force is not directly proportional to increases in rotation speed.

*Travel Speed.* Increases in travel speed increase the required welding force.

*Travel Angle.* Increases in travel angle increase the required welding force. Additionally, above and below a certain range of angles, the process becomes more sensitive to flash generation. For similar-thickness materials, optimal travel angles, in terms of process robustness, tend to be in the 1.5 to 3° range. Outside this range, the stiffness and intelligence requirements of the machine may need to be increased to develop or maintain a consistent process. Travel angles of 0° can be achieved, which allows for the minimum welding force, fastest travel speeds, and minimum flexibility requirements from the machine. However, the consequence is increased process sensitivity; that is, the range of forces over which acceptable weld quality is achieved is quite small. Forces that are too low generate surface or internal voids, and forces that are too high cause flash to be generated. Thus, the machine stiffness and intelligence requirements must be increased if 0° travel angles are to be used.

*Work Angle.* The work angle has little effect on the machine requirements, other than the fact that a nonzero work angle makes a five-or-more axis machine highly desirable.

*Plunge Rate.* The plunge rate affects the force during the plunging operation. In FSW, the plunge rate can be set such that it is not the controlling factor in the maximum force. However, with FSSW, the plunge rate directly affects the required welding force. Increases in plunge rate increase the required welding force.

**Control Types (Force or Position Control).** Force or position control strategies affect the machine intelligence requirements. Depending on the application, force or position control, or both, could be required. A position-controlled machine requires the least intelligence, and a machine with a combination of position and force control requires the most intelligence. Each of the solutions has merit in different applications.

*Position control* is a viable control strategy, given certain application characteristics, such as the following applications:

- Where the FSW tool produces acceptable results over a wide range of forces
- Where the material thickness and position of the material will be very consistent from part to part
- Having partial-penetration butt welds, lap-penetration welding, or FSP applications. Many FSW tools can have a characteristic where, once the tool penetrates to a certain depth (e.g., shoulder below surface of material), it takes less and less force to plunge the tool. Thus, there is an unstable mode in the FSW process where the tool can potentially "dig" into the material, if operating in a force-controlled manner. In these cases, position or a combination of force and position control may be more desirable.

*Force control* is most desirable in the following conditions:

- Application requires full-penetration butt welds. This is helpful to guarantee sufficient penetration. Additionally, full-penetration butt weld applications tend to have a large range of acceptable forces.
- Application where material may vary in thickness, or position is liable to change from part to part

*A combination of force and position control* can be performed when force control is used as the master and the second as a servant. An application for force and position control may be FSP of a casting where the surface varies somewhat from part to part. The force control can be used to measure and react to surface variations, while the position is monitored and, if outside certain limits, the system can shut down or create warning messages.

**Other Application Characteristics.** Technical requirements of the machine are also affected by other characteristics of the production application.

*Quality requirements* include:

- *Weld strength:* In many applications, optimization of weld strength may not be required. This allows more freedom on weld parameters. As such, lower strengths may be acceptable. This, for example, can allow for higher rotation speeds, which lower force requirements.
- *Visual quality:* In some applications, the appearance of flash is not an issue. This may allow for faster travel speeds or overplunging

when using position control. Higher travel speeds require more force, but use of only position control reduces the intelligence requirements.

*Exit Hole.* There may be applications where the exit hole of the FSW is not acceptable. There are multiple solutions to this problem, each having different consequences on the technical requirements of the machine:

- *Run on/off tabs:* These tabs can be used to eliminate or avoid having the start and stop of the weld within the part of interest. Generally, these do not affect the technical requirements of the machine.
- *Placing the hole in a more desirable location:* It is often possible to place the exit hole in a nondetrimental location. However, this is often not along the joint line or the original processing path. Thus, this will require the machine to have additional axes or increased flexibility.
- *Plug:* It is possible to perform a postweld or post-FSP operation where the hole is plugged with a friction plug. This requires the machine to have additional capability and intelligence.
- *Retractable pin tool:* In cases where there is no other solution, a retractable pin tool can be considered. The retractable pin tool is described in section 11.5, "Secial Peripheral Equipment." This is a special peripheral solution that allows the pin of the tool to retract up into the shoulder over time. This can be used in circumferential welding or FSP applications. The use of this peripheral equipment increases the stiffness and intelligence requirements of the machine. To be effective, the retractable pin technology requires the retraction of the pin to occur in a prewelded area or in an area of parent material away from the joint line. This may require the FSW machine to have additional axes.

## 11.2 FSW and FSP Machines

There are several different categories of equipment solutions for FSW and FSP. Each of these categories of equipment has different technical capabilities in the areas of force capability, stiffness, intelligence, and flexibility. Equipment solutions for FSW and FSP are relatively similar, because both processes involve the plunge and traversing of an FSW tool through material.

There are three basic equipment solutions that can be considered: custom-built machines, robots, and modified machining centers. Each of these machines has different capabilities in the technical categories (force capability, stiffness, intelligence and sensing capability, and flexibility), as discussed previously.

**Custom-built machines** are available in many sizes and shapes and can have a very large range of technical capabilities, as discussed in the previous sections. As the name implies, they tend to be built exactly to the requirements of the application. They tend to have the highest force capability and highest stiffness but can have a large range in these categories. However, their intelligence is very application-specific, ranging from being very simple to very complex. Their flexibility also covers a great range, from single-axis to multiaxis machines. As a consequence, their cost also has a large range, from under $100,000 to multimillions of dollars.

As of publication, there are several suppliers of this type of equipment, including ESAB, AB (Sweden), General Tool (Cincinnati, OH), MTS (Minneapolis, MN), Novatech (Seattle, WA), TTI (Elkhart, IN), and Hitachi (Japan), among others.

A good example of a custom-built machine is one that is used for welding long extrusions to fabricate paneling. Figure 11.1 shows an ESAB, AB machine welding extrusions and a Friction Stir Link, Inc. machine used for marine paneling. Using these machines, multiple extrusions are welded together to create a panel. The extrusions are welded in a mode where one weld is made per setup. The process is as follows:

1. Load parts (unwelded extrusion and a partially welded panel, one or more previously welded extrusions)
2. Clamp
3. Weld (single long weld)
4. Retract machine (return to start position)
5. Unclamp
6. Shift welded panel or unload
7. Return to step 1

This type of machine is used because this application requires high travel speeds and high force capability for optimal productivity. These extrusion welding machines have high force capability, moderate intelligence (force control capability), and limited flexibility (single axis).

Another example of a machine in this category is an MTS five-axis gantry machine, shown in Fig. 11.2. This type of machine welds airplane fuselage sections at Eclipse Aviation. This particular machine is used for welding relatively thin material on complex surfaces. It has moderate force capability, high flexibility, high stiffness, and significant intelligence, all designed specifically for the application.

These types of machines tend to be quite large and the most costly, especially machines with multiaxis capability. From an economic perspective, they are more difficult to justify, especially if the application is only replacing an alternative process. These types of machines are more likely to be economically justifiable in applications where:

- The process is being combined with other items that help eliminate other operations. For example, for fabrication of large panels out of multiple long extrusions, it is often possible to integrate other functions into the extrusions (e.g., mounting surfaces), elimi-

(a)

(b)

**Fig. 11.1** Custom-built machines for welding long extrusions nate downstream operations (e.g., grinding), and reduce distortion.

- Machine stiffness is very important
- A high-value-added operation is being replaced, for example, riveting.
- There is a high scrap rate, or the cost of scrap is high.
- There is no alternative, that is, where the welding is an enabling technology, and no other joining process can compete with the weld strength provided by FSW.
- The material is very thick, where the application would normally require many welding passes.

An exception to the high cost of these machines is a small machine with a limited number of axes. There are some applications, especially on small parts having short linear welds, where this may be all that is necessary. In these cases, it is possible to develop a small custom machine that is less expensive than the

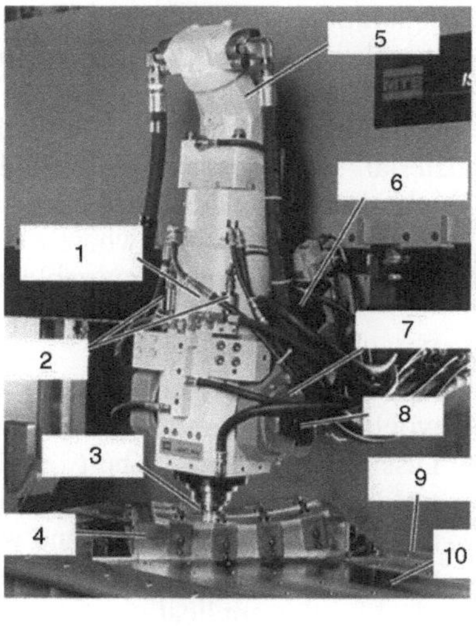


1. Forge and pin tool actuation system
2. Forge pressure transducers
3. Pin/shoulder tools
4. Curved anvil (not included)
5. Spindle drive
6. Pitch actuators (2)
7. Gimbal assembly
8. Pitch axis encoder
9. Specimen welding table
10. Wear p

**Fig. 11.2** MTS I-STIR 5 axes process development system (PDS) lower-cost solutions discussed in the following sections.

Custom-built machines are most suited to the following applications:

- Welding long parts
- Welding thick parts
- Applications where high stiffness is required, for example, welding very thin material, welding with a 0° travel angle, or welding with low shoulder-to-pin-diameter ratios
- Single- or multiaxis applications
- Applications where the alternative solution does not exist or is very expensive

**Robotic FSW or FSP Machines.** One alternative to using custom-built machines is robotic-based FSW systems. As with other manufacturing processes, the availability of robotic solutions has allowed for improved flexibility and significantly lower fabrication costs. Historically, many manufacturing processes have transitioned from the use of custom-built machines to robotic-based solutions. This transition has occurred at different times with other processes, due to the four technical requirements discussed in this chapter and the relative ability of robots in these categories. As discussed, FSW generally requires a high level of force, moderate-to-high levels of stiffness, and a significant amount of intelligence. With this in mind, robots are now available that have sufficient force, stiffness, and intelligence for some FSW and FSP applications.

Robots have two main advantages that allow them to eventually be a preferred solution for many applications. The first is cost, and the second is flexibility. Because they are produced in moderate production volumes, their cost is significantly less than a custom-built machine. Additionally, they typically have much improved flexibility. This flexibility allows for significant productivity improvements. As an example, consider a part with welds on multiple sides. A robotic solution can allow for welding on multiple sides of the part in a single setup. This reduces non-value-added materials handling applications and can yield 100% or more improvements in productivity. This, of course, reduces net welding cost.

Industrial robots, since their advent in the 1970s, have continually experienced improvements in force capability, intelligence, and stiffness. For this reason, robots have become the preferred solution for many production processes. Because of the different force, stiffness, and intelligence requirements of any process, this transition to robotics has occurred at various times for each process throughout the last 30 years. As an example, simple material handling applications were the first to be automated with robots, due to their low force, intelligence, and stiffness requirements. On the contrary, laser cutting has been one of the more recent applications for robots, because laser cutting requires moderate stiffness, high precision, and significant control technology.

Friction stir welding, due to its high force requirements and moderate stiffness and intelligence demands, has been impossible to perform with a robot until recent years. There are now robots available that can generate in excess of 4500 N (1000 lb) of thrust force, making them capable of performing FSW on material of thin-to-moderate thickness. Given that the force capability and stiffness of robots are improving by a factor of 2 every 5 years or so, it stands to reason that robots will eventually become the dominant machine solution for FSW, as robots have done with many other processes.

The robotic-based solutions are available in two basic categories: articulated arm robots and parallel-kinematic robots. Articulated arm robots are the most common and widely used. A typical articulated arm robot is shown in Fig. 11.3. These robots tend to have six axes and six degrees of freedom, with all motion axes being situated in a serial fashion. Compared to custom-built machines, these types of robots have relatively low stiffness but moderate force capability. Their intelligence and flexibility can be significantly better than custom machines. Furthermore, they are low in cost. Given their flexibility and low cost, they can be the lowest-cost solution by far but have a limited range of materials on which they can perform FSW or FSP. As a general rule, they are capable of welding up to 6 mm (1/4 in.) thick aluminum material. Their capability in higher-melting-point materials tends to be somewhat less.

Example applications where a robotic-based solution would be more favorable include:

- Relatively thin material
- Applications having multiple welds that would otherwise require multiple setups
- Dissimilar-thickness butt welds (tailor-welded blanks). Dissimilar-thickness welds require both a travel angle and work angle (a minimum of five axes of motion). Robots are ideal for this application.
- Applications where multiaxis capability is required

- Higher-volume applications where productivity is more important

The other basic robotic configuration is the parallel-kinematic robot. They differ from the articulated arm robots in that the axes of motion are in parallel instead of series. A photograph of one such parallel-kinematic robot is shown in Fig. 11.4. Their benefit is that they can generate more force and have significantly more stiffness than an articulated arm robot. However, their cost can be significantly higher, and their work envelope (flexibility) is significantly less. They are more suited to applications where the parts are relatively small, multiaxis capability is required, and the force requirements are a little higher than what the articulated arm robot is capable of generating.

Parallel-kinematic robots should be considered in similar applications to the articulated arm robots, with the following exceptions:

- The work envelope of the part is relatively small.
- The part can be welded near or close to the horizontal plane.
- The force or stiffness requirements are somewhat higher.

Robot-based solutions are available from Friction Stir Link (Waukesha, WI) and GKSS (Hamburg, Germany), although GKSS provides only prototyping and application development services.

**Modified Machining Centers.** Another alternative to the custom-built machine is modified machining centers. Friction stir welding and processing are similar in nature to machining at a high level. Thus, there are potential opportunities to modify existing equipment to perform FSW. There are several items that must be considered before deciding whether or not to modify an existing machining center:

- Friction stir welding and processing can require relatively more force than machining. The base equipment (ways, guides, rails, motors, spindles, etc.) must be investigated to determine the capability of the machine prior to any decision.
- Friction stir welding typically requires more intelligence than machining. For example, force control may be needed for FSW. This means that the base machine must possess a controller with an open architecture; if not,

**Fig. 11.3** Articulated arm robot

**Fig. 11.4** Parallel-kinematic arm robot

then a separate controller may be required. If a separate controller is required, then some communication with the base controller will likely be required. The communication capability of the base machine should be considered.

- In most applications, FSW or FSP requires at least a nonzero travel angle and perhaps a nonzero work angle. Unless the machining center has five-axis capability, this may pose a challenge. Mechanical fixed solutions can be implemented to apply a travel and/or work angle to overcome this limitation.
- Friction stir welding and processing produce heat that can transfer into spindles, which are not designed to handle high temperatures. Thermal management must be considered.

A modified machining center can also be used for FSW. Providers of modified machining center equipment include General Tool (Cincinnati, OH), among others. Modification of existing machining centers can be an economical means of implementing FSW or FSP, but the considerations list given previously must be investigated prior to implementation.

## 11.3 FSSW Equipment

Friction stir spot welding is a variant of FSW, where the traverse part of the FSW process is eliminated. This means that the equipment requires only two axes of motion (rotary and vertical). Like FSW, it requires significant force. However, one major benefit is that fixturing need not be as robust as with FSW. Friction stir spot welding is very similar to resistance spot welding (RSW) and riveting in that they are all "point" processes, require a significant level of thrust force, and have similar fixturing requirements. However, the intelligence and stiffness requirements of FSSW are increased because of more precise vertical position control requirements. Due to the similarities to RSW, the equipment solutions are quite similar.

For FSSW, the equipment solutions come in four basic categories: pedestal units, benchtop units, C-frame units, and a poke solution. A typical pedestal unit is pictured in Fig. 11.5. The pedestal unit is a self-contained stand-alone solution. An operator or robot can be used to manipulate the parts under the pedestal machine. These units are controlled with servo drives that communicate with an operator interface, robot, or programmable logic controller (PLC). The weld schedule is stored internally or in the external control equipment. In the manual mode, the system is activated via push button or operator interface. In the automatic or robotic mode, the system is activated via communication from a PLC or robot.

The second type of machine is a tabletop or benchtop system, as shown in Fig. 11.6. This is a smaller stand-alone system that will sit atop a stiff table. It has all of the other features and can be operated in the same manner as the pedestal-type unit.

The third type of FSSW machine is a C-frame unit. The purpose of the C-frame is to contain the welding forces internal to the unit. This means that the robot or operator does not have to generate any of the forces required for the process. Thus, smaller robots can be used for C-frame FSSW than for FSW. The robot arm only manipulates the C-frame unit through space to the part that is to be welded. Robots that are used for RSW can also be used for FSSW. Typical C-frame FSSW units are shown in Fig. 11.7.

To perform the process, the robot first places the C-frame against the backside of the part. The robot then activates the spot welder. The

Fig. 11.5 Pedestal-type friction stir spot welding unit. Courtesy of Friction Stir Link, Inc.

rotary axis rotates, and the vertical axis forces the FSSW tool down into the material, creating an FSSW. Then, the FSSW tool is retracted. After the tool is retracted, the robot moves the C-frame to the next weld position. Note that the C-frame can also be used in a manual mode, where the C-frame hangs from a counterbalance unit, and the operator manually moves the unit up to the part.

The FSSW process can also be used in the "poke" mode, where there is lack of access to the backside of the part. In this variant, a robot is typically used to poke the part. The robot forces the FSSW tool down into the part. This means that the robot must generate the force required for the FSSW process. Thus, robots capable of generating the high forces required for FSSW must be used. A robot that is used for FSW can be used for poke FSSW.

The FSSW equipment is supplied by several companies, including Friction Stir Link (Waukesha, WI) and Kawasaki (Japan).

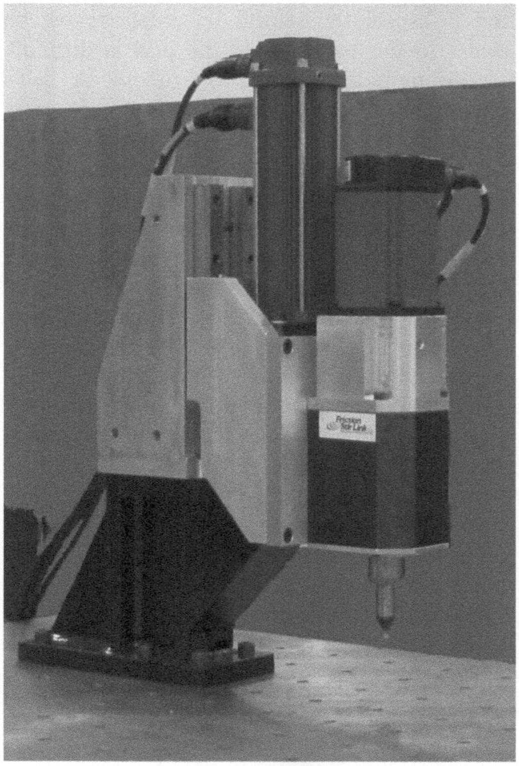

Fig. 11.6 Benchtop friction stir spot welding unit. Courtesy of Friction Stir Link, Inc.

Friction stir spot welding is most suited to applications where RSW or riveting is employed. This would include:

- Relatively thin material (<3 mm, or 1/8 in.)
- Where joint strength requirements are lower, as with other spot joining processes
- Where parts are contoured
- Where flanges or other local flat areas are available in locations of the spots
- Where access to the backside of the part is available (not an absolute requirement)

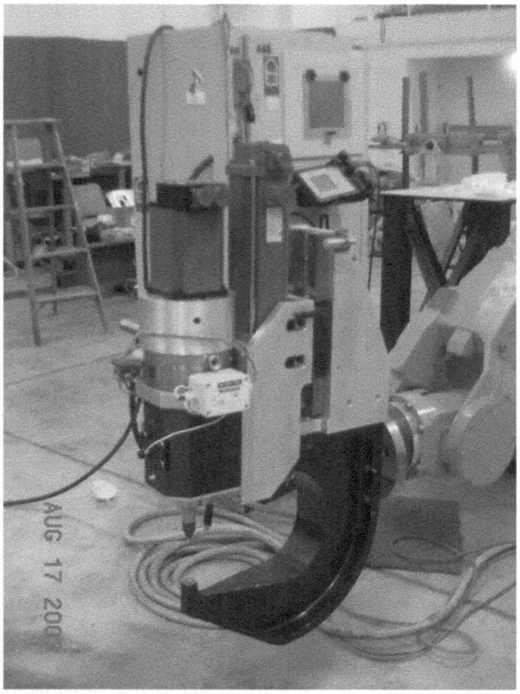


(a)

(b)

Fig. 11.7 (a) and (b) C-frame friction stir spot welding unit. Courtesy of Friction Stir Link, Inc.

There exists another variant of FSSW where the FSSW tool is traversed a short distance. This may be referred to as "stitch" FSSW. However, it eliminates the fixturing benefits that would normally be accompanied by FSSW, but if the weld is long enough, a stitch FSSW will have higher strength than an FSSW.

## 11.4 Fixturing

Proper fixturing is critical to the success of FSW and FSP. Because these processes have very high forces, the fixturing must be built to withstand the forces. Additionally, the part must be fully supported on the backside. Furthermore, the fixturing must be designed to prevent parts from moving relative to one another. These considerations are a function of the joint design.

With the butt weld configuration (similar thickness and dissimilar thickness), there are very high splitting forces in the joint. This means that not only must the fixture support the thrust force, but it must prevent the parts from spreading apart relative to one another. Because of the high forging forces, material near the joint line can lift, causing deformation on the backside if the fixturing does not hold the parts down sufficiently. This is more of a concern with thinner or softer material.

Lap welds are generally easier to fixture. If the weld is close to the edge of the lap, then the material can deform if fixturing does not prevent this. With lap welds, there are also lifting forces, especially at the start of the weld. This is caused by the material attempting to extrude between the faying surfaces of the joint. This condition must be prevented by the fixturing.

As noted, the part must be fully supported because of the high forces of FSW. This can be accomplished by having backing support behind the part. The backing can be a fixture itself, or a rib, or some other feature within the part itself. This means that special considerations must be given for open or hollow sections. Open or hollow sections can be welded if the weld is on a rib or other feature in the part, or if a mandrel is built to support the part.

## 11.5 Special Peripheral Equipment

There are special FSW tool solutions that can be implemented for certain applications. These include retractable pin tools and self-reacting tools. These are special FSW tools with actuation or other features that allow them to overcome some of the concerns of FSW.

The retractable pin tool is essentially an FSW tool where the pin can retract up into the shoulder. This can be used in applications where the exit hole of the FSW process is not acceptable. In most situations, the exit hole is not an issue, because it is no worse than a start or stop in other welding processes. If the exit hole is a potential issue, it can often be placed in an area where it is not an issue. The retractable pin requires a more complex spindle, with the ability to shift the FSW tool pin with respect to the shoulder. Thus, it adds cost and complexity to the application and requires a machine with more intelligence capability (additional control). The retractable pin can be considered for circumferential applications where the exit hole may not be acceptable. An example of a retractable pin tool is shown in Fig. 11.8.

The self-reacting tool is a dual-sided FSW tool that has two shoulders. One shoulder contacts the top surface, and another contacts the bottom surface. It is referred to as a self-reacting tool because the net thrust force on the machine is theoretically zero. It is self-reacting similar to a C-frame in the FSSW application. This solution is especially helpful in situations where access to the backside of the part is difficult (e.g., longitudinal welds on tubes). A self-reacting tool is shown in Fig. 11.9.

The self-reacting tool must be operated with the tool vertical to the material surface (zero travel angle). Therefore, the intelligence and stiffness requirements for the machine are higher when employing the self-reacting tool. Additionally, some self-reacting tools have individual force control capability and ability to move the shoulders with respect to one another, to overcome the effects of material thickness variation. This capability adds complexity and requires additional intelligence in the machine, although this allows for improved control. However, recent developments in tool design help mitigate some of these issues and allow the self-reacting tool to be less sensitive to variation. (Ref 1). Other factors may also need to be considered when using a self-reacting tool, including:

- Potential need for assembly and disassembly of the self-reacting tool at the start and end of the operation
- Potential need for a hole to be drilled into the part at the start of the weld

**Fig. 11.8** Retractable pin tool. Courtesy of NASA

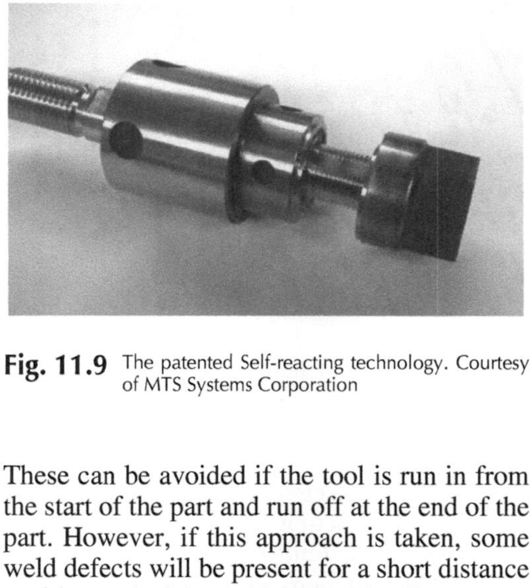

**Fig. 11.9** The patented Self-reacting technology. Courtesy of MTS Systems Corporation

These can be avoided if the tool is run in from the start of the part and run off at the end of the part. However, if this approach is taken, some weld defects will be present for a short distance at either end of the weld. This side effect can be overcome with the use of run-on and run-off tabs. Thus, one should consider run-on and run-off tabs when self-reacting tools are used, to avoid unnecessarily increasing the complexity of employing the self-reacting tool concept.

## 11.6 Control and Process Monitoring

As with all joining technologies, there are sensing, control, process monitoring, and error-proofing strategies that are specific to the technology. Because FSW and its variants tend to have fewer variables, these strategies are less expansive than other technologies.

**Sensing.** There are several variables that should be sensed on any FSW machine and some other variables that may be required to be sensed. The following are the variables that can be sensed.

*Thrust Force.* This is the force on the tool in the axial direction of the tool. In most applications (FSW and FSP), it should be a requirement to measure force. Forces that are too high or too low lead to undesirable welding results. It can be used for force-control strategies (recommended in many applications) or for error proofing. In FSSW, measuring thrust force is not required but can be used for error proofing.

*Traverse Force.* The traverse force is the force to push the FSW tool through the material in the direction of travel. It can be used for control (travel speed is varied, based on maintaining constant traverse force), or it can be used for error proofing (or monitoring). Measuring the traverse force is not required.

*Lateral Force.* This is the force perpendicular to the welding direction. It is typically only used for monitoring and is not necessarily required.

*Welding Torque.* This is primarily used for monitoring and can indicate such items as tool wear or tool failure. It can be sensed directly with sensors or through motor current. Direct measurement is typically more accurate.

*Rotation Speed.* Most often, FSW and its variants use motors with internal control through drives. The motor speed is almost always controlled and is often monitored via output from motor drives. Separate sensors are not required.

*FSW Tool Orientation.* With limited-axis machines, the angles (travel angle and work angle) can be controlled mechanically and do not need to be measured, except for simple error-proofing means. In a multiaxis machine, the internal software often controls and maintains these angles. These angles affect the FSW and FSP processes quite significantly. Strategies should be in place to ensure that the correct angles are used.

*Seam Position.* As with other joining technologies, tracking of the seam may be desirable. This can be accomplished with standard off-the-shelf technologies. It can be simpler to implement for FSW because of the more benign operating environment (no arc, dust, or spatter) found with FSW compared to other processes.

**Control Strategies.** There are two basic control strategies found with FSW and FSSW: force and position control. Force control can be very desirable in many applications. It is often the case that FSW has a larger operating range with respect to thrust force than vertical position. Thus, it can be a more robust control strategy. This is especially the case for butt welds. Additionally, force control tends to be preferred for less stiff machines (e.g., robots).

Position control can also be an effective strategy, especially for applications where thermal effects and geometries vary significantly over time or position. For example, a part that has welding in an area that transitions from a large thermal mass to a small thermal mass may require significant changes in actual welding force. The same can be true for processing in cases where the temperature of the part changes significantly over time.

**Process Monitoring/Error Proofing.** Several of the variables that can be sensed are also recommended for monitoring and/or error proofing.

*Thrust Force.* For FSW and FSP, it is highly recommended to implement at least a force-monitoring strategy. Force monitoring can have the following benefits:

- Detect changes in FSW tool condition (wrong tool, wear, etc.)
- Detect wrong parts
- Detect missing parts
- Detect weld-quality changes

*Traverse Force.* Monitoring of traverse force can be beneficial, but most of the same errors can be detected via thrust-force monitoring.

*Torque Monitoring.* This has similar benefits to thrust-force monitoring.

*Angle (Work and Travel Angles).* Strategies should be implemented to ensure that these angles are maintained or set up properly for each application. This can be performed with simple strategies, such as proximity sensing on mechanically adjusted machines.

*Rotation speed* should be controlled. Monitoring of this variable can be performed but will provide only confirmatory results.

Other error-proofing strategies are similar to other joining processes (e.g., part sensing). In some cases, these can be easier with FSW due to its relatively benign environment.

## REFERENCE

1. K. Colligan, Concurrent Technologies Corp., Tapered Friction Stir Welding Tool, U.S. Patent 6,669,075, Dec 30, 2003

## SELECTED REFERENCES

- B. Christner et al., Friction Stir Welding System Development for Thin-Gauge Aerospace Structures, *Proceedings of the Fourth International Symposium on Friction Stir Welding,* May 2003 (Park City, UT), TWI
- R.J. Ding, Retractable Pin-Tool Technology for Friction Stir Welding, *Proceedings of the Fifth International Trends in Welding Research Conference,* June 1998 (Pine Mountain, GA), ASM International
- R.J. Ding, Force Characterization on the Welding Pin of a Friction Stir Welding Retractable Pin-Tool Using Aluminum-Lithium 2195, *Proceedings of the Second International Symposium on Friction Stir Welding,* June 2000 (Gothenberg, Sweden, TWI
- R.J. Ding and P.A. Olgetz, Mechanical Property Analysis in the Retracted Pin-

Tool Region of Friction Stir Welded Aluminum Lithium Alloys, *Proceedings of the First International Symposium on Friction Stir Welding,* June 1999 (Thousand Oaks, CA), TWI

- G. Engelhard et al., Orbital Friction Stir Welding of Aluminum Pipes, *Proceedings of the Third International Symposium on Friction Stir Welding,* Sept 2001 (Kobe, Japan), TWI
- S. Hirano et al., Development of Three-Dimensional Type Friction Stir Welding Equipment, *Proceedings of the Third International Symposium on Friction Stir Welding,* Sept 2001 (Kobe, Japan), TWI
- C. Jones and G. Adams, Assembly of a Full-Scale External Tank Barrel Section Using Friction Stir Welding, *Proceedings of the First International Symposium on Friction Stir Welding,* June 1999 (Thousand Oaks, CA), TWI
- Z. Loftus, Friction Stir Welding Tooling Development for Application on the 2195 Al-Li-Cu Space Transportation System External Tank, *Proceedings of the Fifth International Trends in Welding Research Conference,* June 1998 (Pine Mountain, GA), ASM International
- Z. Loftus et al., Development and Implementation of a Load-Controlled Friction Stir Welder, *Proceedings of the First International Symposium on Friction Stir Welding,* June 1999 (Thousand Oaks, CA), TWI
- M.W. McLane et al., Free-Form Friction Stir Welding, *Proceedings of the Fourth International Symposium on Friction Stir Welding,* May 2003 (Park City, UT), TWI
- J.E. Mitchell et al., Force Sensing in Friction Stir Welding, *Proceedings of the Sixth International Trends in Welding Research Conference,* April 2002 (Pine Mountain, GA), ASM International
- R. Sakano et al., Development of Spot FSW Robot System for Automobile Body Members, *Proceedings of the Third International Symposium on Friction Stir Welding,* Sept 2001 (Kobe, Japan), TWI
- M. Skinner and R. Edwards, Improvement to the FSW Process Using the Self-Reacting Technology, *Mater. Sci. Forum,* Vol 426–432, 2003, p 2849–2854
- C.B. Smith, "Robotic Friction Stir Welding, Phase I: Initial Feasability Study," Report APPT-1493, Tower Automotive Internal Report, May 1997
- C.B. Smith, "Robotic Friction Stir Welding, Phase II: Robot Performance Comparison," Report APPT-1494, Tower Automotive Internal Report, May 1998
- C. Smith, Robotic Friction Stir Welding Using a Standard Industrial Robot, *Proceedings of the Second International Symposium on Friction Stir Welding,* June 2000 (Gothenberg, Sweden), TWI
- C. Smith, Robotic Friction Stir Welding: The State of the Art, *Proceedings of the Fourth International Symposium on Friction Stir Welding,* May 2003 (Park City, UT), TWI
- W.M. Thomas, Friction Stir Welding Developments, *Proceedings of the Sixth International Trends in Welding Research Conference,* April 2002 (Pine Mountain, GA), ASM International
- W.M. Thomas et al., Friction Stir Butt Welding, U.S. Patent 5,460,317
- J. Thompson, FSW for Cost Savings in Contract Manufacturing, *Proceedings of the Sectond International Symposium on Friction Stir Welding,* June 2000 (Gothenberg, Sweden), TWI
- A. Von Strombeck et al., Robotic Friction Stir Welding—Tool Technology and Applications, *Proceedings of the Second International Symposium on Friction Stir Welding,* June 2000 (Gothenberg, Sweden), TWI

# CHAPTER 12

# Friction Stir Spot Welding

Harsha Badarinarayan, Frank Hunt, and Kazutaka Okamoto
Hitachi America Ltd., R&D

THE USE OF ALUMINUM in the automotive industry is increasing. To date, aluminum has been used predominantly for closure panels such as hoods, decklids, and lift-gates to reduce weight and improve vehicle fuel economy. Current closure panel welding techniques include resistance spot welding (RSW), self-piercing rivets (SPR), and clinching. The disadvantages of these methods include weld electrode dressing, high energy consumption, and the use of consumables. In the case of RSW, higher electric power source and electrode dresser are required because of the physical properties of the aluminum alloy. The SPR also require rivets that add to the cost of assembly manufacturing via consumables.

The welding method used for aluminum sheet assembly is one of the key technology drivers to enhance weight reduction in the automotive industry, and hence, friction stir spot welding (FSSW) was evaluated as an alternative welding technique (Ref 1). Weight reduction is an important challenge in the automotive industry in order to improve fuel economy. Lightweight materials such as aluminum and magnesium, when properly designed, can be used to replace equivalent steel assemblies with approximately half the weight.

Over the past few years, there have been developments in the process of spot friction stir welding (FSW). Friction stir spot welding can be broadly classified into three main categories:

- Pure spot FSW
- Refill FSSW
- Swing FSSW

The pure spot FSW technique was invented by Mazda (Ref 2). In this case, a rotating tool is plunged into the workpiece, held for a certain duration of time, and then retracted, hence creating a spot FSW. This technology was first used in the Mazda RX-8 rear door panel spot welding in 2003. Mazda claimed to have reduced the energy consumption by 99% of that used by the conventional earlier process (Ref 3).

Conventional friction stir spot welding leaves behind a keyhole (exit hole) after the weld has been done. In order to avoid this, GKSS of Germany invented a process that would fill the keyhole (Ref 4). This method was called the refill FSSW process. The joined region consists of a spot of material that has been plasticized, displaced in a process similar to a back extrusion, and then replaced, forming a fully consolidated weld that is nominally flush with the original surface.

The third variation of FSSW, developed by Hitachi, is called swing FSSW. Unlike the conventional spot technique, where spot geometry is a perfect circle, swing FSSW produces a spot that is elliptical in shape (elongated spot) (Ref 5). Because the area of contact is larger for an elongated spot, the strength offered by swing FSSW may be higher.

Friction stir spot welding is still an evolving technology. There are various aspects of this technology that are still being worked on by researchers around the world—be it as simple as designing a jig/fixture for welding or a more challenging aspect of trying to use existing or emerging nondestructive testing techniques to evaluate the integrity of the weld. There has been a steady growth in the knowledge base for this technology, and as people continue to devote their research focus to FSSW, there will be more insight into this complex process, and many questions will be answered.

## 12.1 FSSW Methods

**Pure Spot FSW.** Sakano et al. (Ref 6) illustrated a newly developed spot FSW robot system for lap joints of aluminum plates. The system was comprised of a specially designed spot FSW gun and a multiarticulate robot. The gun had an FSW probe with rotational and axial movements individually executed by servomotors; therefore, the entire welding sequence was controlled by the central processing unit (CPU) of the robot system. Not only did the spot FSW lap joint have equal or superior mechanical properties to the conventional RSW, it also showed significantly lower energy consumption and maintenance cost in comparison with current RSW systems.

A schematic illustration of the spot FSW process is shown in Fig. 12.1. The process is applied to a lap joint consisting of upper and lower sheets. A rotating tool with a probe is plunged into the material from the top surface for a certain time to generate frictional heat. At the same time, a backing plate contacts the lower sheet from the bottom side to support the downward force. Heated and softened material adjacent to the tool causes a plastic flow. In addition, the tool shoulder gives a strong compressive force to the material. After the tool is drawn away from the material, a solid-phase bond is made between the upper and the lower sheets. Figure 12.2 shows the appearance and the cross-sectional configuration of a spot friction stir weld. The upper surface of the weld looks like a button with a hole, and the bottom surface is kept almost flat. In the cross section, there is a hole that is made by the probe and reaches into the lower sheet.

A spot FSW gun was designed and manufactured to make these welds. Figure 12.3 shows the appearance of the gun design. It has a C-shaped frame structure similar to conventional RSW guns and consists mainly of a tool rotation unit and an axial loading unit. An induction motor was used to rotate the tool, and the gun weighed approximately 80 kg (176 lb). The spot FSW gun was attached to a multiarticulate Kawasaki robot with six motion axes, as shown in Fig. 12.4. In this system, a CPU of the robot controller also controls the axial motion and rotation of the tool. The robot controller has a welding sequence program that executes the precise sequential change of the tool rotational speed during the weld.

Static strengths of spot FSW lap joints were examined to evaluate the joint properties. A 6000-series aluminum was used for welding. Lap-shear and cross-tension tests were performed. As a general observation, strength is higher at higher revolutions per minute and shorter weld time. Mechanical properties of these spot welds are discussed in detail later in this chapter. In another test, multiple spots were made on a large sheet of aluminum to demonstrate that the distortion seen in FSSW is much smaller than that seen in RSW. Figure 12.5 shows an example of one such sheet.

Mazda estimated that the investment of the spot FSW system was approximately 50% less than the equivalent RSW system, because several pieces of equipment, including a large electric power supply, a cooling unit, an electrode

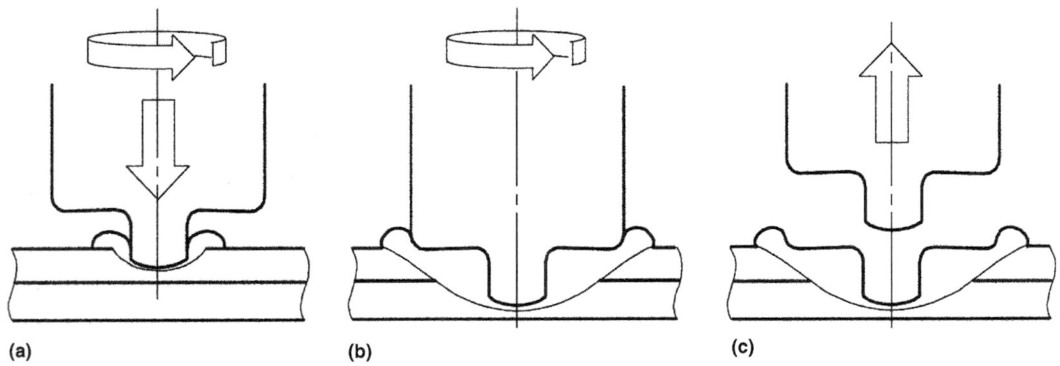


**Fig. 12.1** Spot friction stir welding illustration. (a) Plunging. (b) Bonding. (c) Drawing out. Source: Ref 6

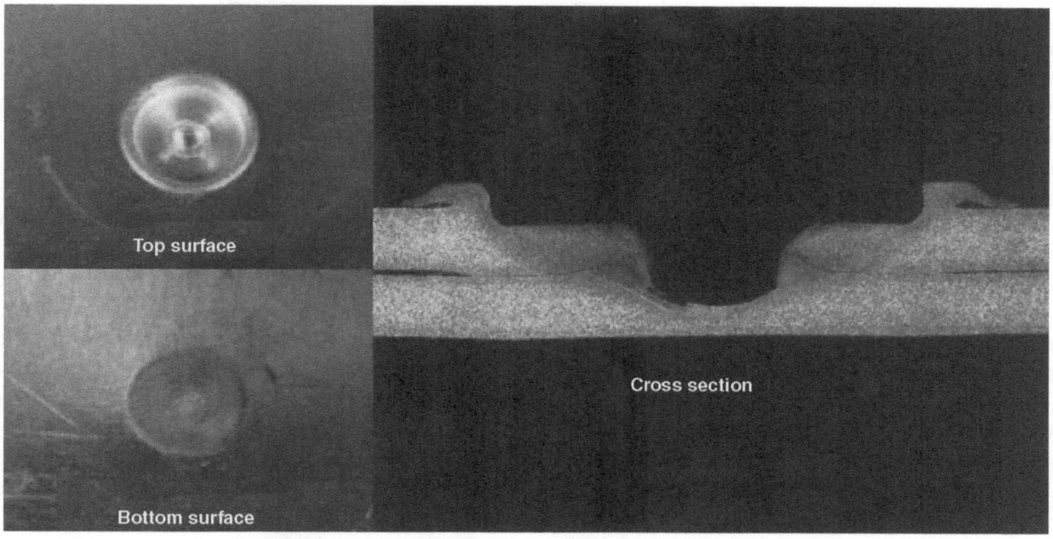


**Fig. 12.2** Spot friction stir welding appearance and cross section. Source: Ref 6

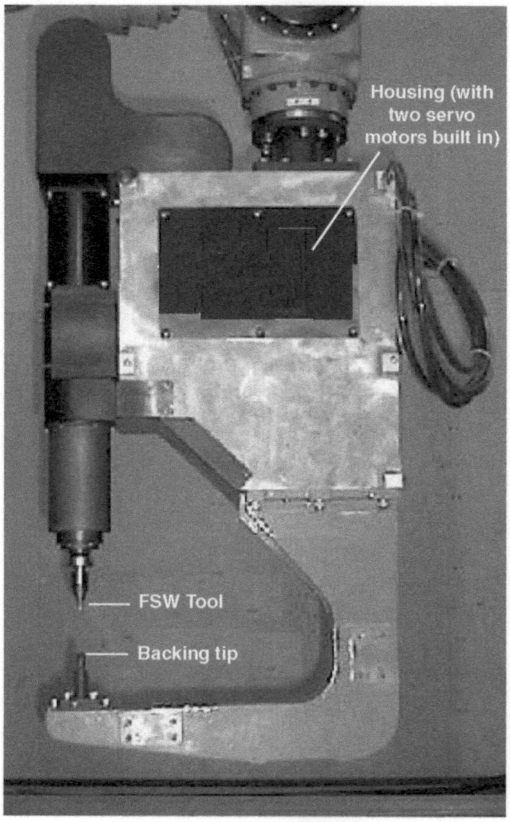


**Fig. 12.3** Spot friction stir welding (FSW) gun design. Source: Ref 6

dresser, and others, were not necessary. The cost per single spot estimation showed that the cost of the spot FSW system is 85% less than that of the RSW system. This drastic cost reduction was brought about by cutting the utility cost and the consumables. Based on cost evaluation analysis, the spot FSW system was considered to be a very viable welding process for the automotive industry.

**Refill FSSW.** The refill FSSW is a patented process of GKSS (Germany) that joins two or more sheets of material together in the lap configuration (Ref 2). The joined region consists of a spot of material that has been plasticized, displaced in a process similar to a back extrusion, and then replaced, forming a fully consolidated weld that is nominally flush with the original surface.

The refill FSSW process is performed using a three-piece tool system consisting of a clamp ring, outer shoulder, and inner pin (Fig.12.6) (Ref 7). Each of these three components is contained on a separate actuation system such that each can be moved in and out independently of the other. The pin and shoulder rotate at the same revolutions per minute in the same direction. The stationary clamping ring holds the workpiece in the proper position during processing. The inner pin and outer shoulder are rotated at a specified revolutions per minute and moved to the surface to fictionally preheat the workpiece. When the workpiece is sufficiently heated and begins to plasticize, the inner pin continues to plunge to the faying surface between the upper and lower sheets, while the outer shoulder retracts to form a reservoir to capture the displaced material.

A typical FSSW process sequence initiates with the clamping ring moving into position to hold the workpiece firmly in place (Fig. 12.7).

During the full retract phase, the inner pin is retracted, and the outer shoulder is extended to extrude the reservoir material back into the weld zone. Assuming no material loss, this process sequence leaves the hole completely refilled with minimal or no surface indentation. The clamping ring holds the upper and lower sheets firmly in contact during the process and prevents sheet lifting, separation, and expulsion and spitting of material. The microstructure of the weld region shows a dynamically recrystallized zone, thermomechanically affected zone (TMAZ), and a heat-affected zone. This particular type of refill FSSW is known as shoulder-first refill FSSW.

The aforementioned technique, although innovative, has some material sticking issues. The larger-diameter shoulder displaces a significant volume of material and requires the smaller-diameter pin to retract to a greater distance to maintain constant volume exchange. This large pin retraction distance draws the plasticized material into cooler regions of the shoulder, where it subsequently adheres to the inner walls. This causes the pin to periodically stick and become lodged within the shoulder between spot weld cycles. Hence, a modification to this was suggested in which the rotating pin and the shoulder are initially plunged in a fixed position relative to each other (Fig. 12.8).

During stage 1, the pin is extended past the shoulder to a distance that ensures a constant volume exchange between the material displaced by the pin and that accepted beneath the

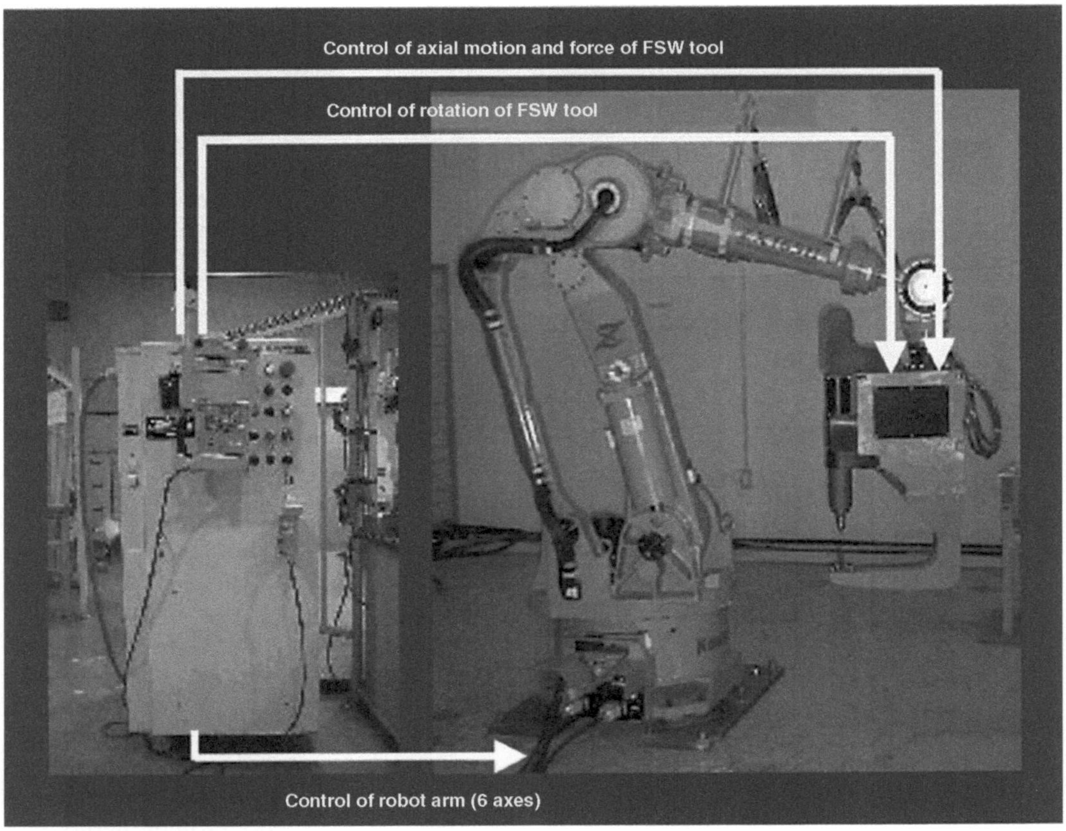


**Fig. 12.4** Spot friction stir welding (FSW) robot system. Source: Ref 6

shoulder during the stage 2 plunge to the desired depth. After penetration to the desired depth (stage 2) under constant plunge rate, the pin is retracted into the shoulder under position control (stage 3), while the shoulder is placed into forge control mode and extrudes the material back into the void left as the pin is retracted. At full retracted position, the shoulder and pin are nominally flush with the workpiece surface. During stage 4, the rotation speed is stopped, and a reforge cycle may be employed, where the pin and shoulder are commanded to a preset forge load to enhance consolidation of the materials within the stirred zones prior to removing the spot weld system from the workpiece.

The properties of the refill joints are discussed later in this chapter. The refill FSSW process has been shown to produce high joint strengths with minimal indentation and internal void formation.

**Swing FSSW.** Hitachi developed the technique of swing FSW (Ref 5). In the conventional spot FSW, the tool plunges into the workpiece, creates the weld, and retracts. However, in the technique of swing FSW, the tool, after plunging, traverses a short linear distance before retracting. The advantage of such a process is that the contact area is larger, which may result in higher strength.

Figure 12.9 shows the tool movement of various FSW processes and a corresponding top view of the welds. In the FSSW process, the rotating tool is plunged, momentarily held, and then extracted. In this process, the squeezed material is lumped around the shoulder indentation. Stitch FSW and swing FSW result in short-

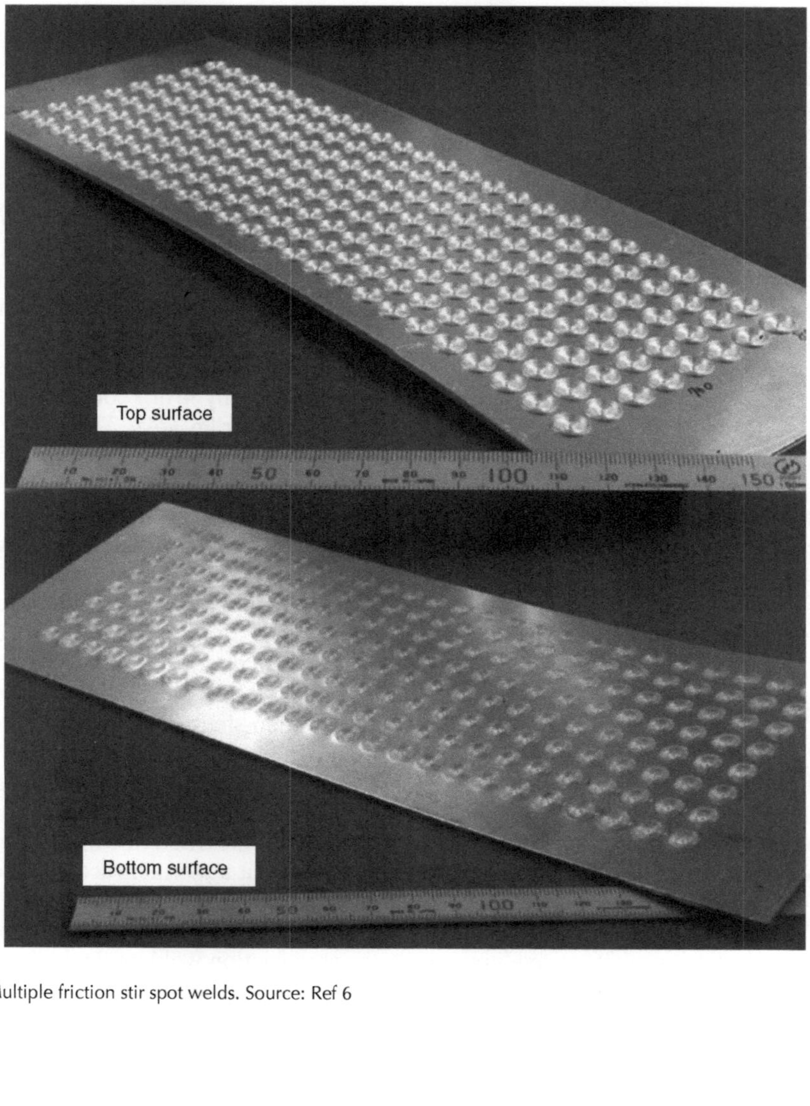


**Fig. 12.5** Multiple friction stir spot welds. Source: Ref 6

distance linear welds. In these processes, the tool is plunged and fed for a short distance, so that a small amount of burr is formed, such as in linear FSW. Furthermore, the joint area is greater than spot welding (FSSW), which may lead to higher joint strength.

In order to validate this technique, Hitachi developed a prototype C-frame gun. In terms of C-frame gun design, FSSW requires the simplest gun with spindle motor and tool plunge motor. However, an additional motor that drives the tool horizontally is necessary for a typical stitch FSW, which leads to complex and heavy C-frame gun design. In the case of swing FSW (an extension of stitch FSW), shown in Fig. 12.9(c), the tool is pivoted at one end, and the other end (which is in contact with the workpiece) is made to move in a swing motion with a very large radius and small angle, which practically results in a linear motion. This movement is controlled by a push/pull mechanism around a rotating axis on the C-frame head.

Swing-Stir, shown in Fig. 12.10, is a specially designed gun for FSSW and swing FSSW. The gun is made of an aluminum C-frame with the following features: anvil, spindle motor, tool plunge motor, workpiece clamping jig, and additional swing axis and drive mechanism to move the tool in arc (swing) motion. With this design, both welding speed and weld length are adjustable. This gun is designed for aluminum welding with up to 3 mm (0.12 in.) tool penetration depth. The spindle motor is 3.5 kW, and the unit weighs approximately 170 kg (375 lb). The C-frame gun is mounted onto a multiarticulate robot.

The properties of the swing FSW joints are discussed later in this chapter. The swing FSSW process, using the C-frame gun, was developed to improve the joint performance for spot FSW joints, which could then be applied for automotive closure panel applications.

## 12.2 Mechanical Properties and Microstructure of Friction Stir Spot Welds

Similar to other joining techniques, the quality of FSSW is measured by evaluating the mechanical properties of the joint. There are several mechanical tests that are conducted to study both the static as well as endurance strength of the joints. Some of these tests are widely used in the industry and more or less represent a test standard, while some of the tests are very specific to a particular industry. Some of the static strength tests employed are lap shear, coach peel, and cross tension, wherein the direction of application of load on the joint varies, consequently resulting in different stress concentration areas around the weld. The endurance (dynamic) tests are usually employed by the automotive and aerospace industry, where the final product is expected to undergo cyclic (or noncyclic) fluctuations in the applied load.

The microstructural observations provide valuable information regarding the metallurgy of the joint. Nugget/stir zone size, actual weld depth, and hook formation (thinning of top sheet) are some of the geometrical information that are possible to visualize through the cross-sectional images. Other properties that have been evaluated are grain size, hardness profile, and texture.

### *12.2.1 Pure Spot FSW Properties (AA6111-T4)*

**Static Strength Evaluation.** Lin et al. (Ref 8) investigated the microstructures and failure mechanisms of spot friction welds in aluminum 6111 lap-shear specimens. In this investigation, aluminum 6111-T4 sheets with a thickness of 0.9 mm (0.035 in.) were used. The lap-shear specimens were made by using two 25.4 by 101.6 mm (1 by 4 in.) coupons with a 25.4 by 25.4 mm overlap area. The welds were made by using a spot friction welding gun made by Kawasaki robot.

The lap-shear specimens were then tested by using an Instron Model 4502 testing machine at a monotonic displacement rate of 1.0 mm/min (0.04 in./min). The load and displacement were simultaneously recorded during the test. Tests were terminated when the two sheets of the specimen were separated. Figure 12.11(a) shows a lap-shear spot friction weld specimen. Figure 12.11(b) shows a close-up top view of the spot friction weld on the upper sheet. As shown in the top view, the top surface of the weld looks like a button with a central hole. The squeezed-out material is accumulated along the outer circumference of the shoulder indentation. Figure 12.11(c) shows a close-up back view of the spot friction weld on the lower sheet. In the back view, the contact mark due to the backing plate can be seen.

*Microstructure*. In order to understand the failure mechanisms of spot friction welds under lap-shear loading conditions, cross sections of spot friction welds before and after failure were obtained. Figure 12.12(a) shows the cross section of a spot friction weld before testing, and Fig. 12.12(b) shows close-up views of regions I, II, III, and IV, as marked in Fig. 12.12(a).

In Fig. 12.12(a), there is an indentation with a profile that reflects the shape of the probe pin and the flat tool shoulder. As shown in the figure, the bottom surface is kept almost flat, except near the central hole. Near the outer area of the central hole, there is a gray area that represents the stir zone, where the upper and lower sheets are bonded. Two notch tips can be seen near points "C" and "D." The notch tips extend into the weld and appear to be formed from the unwelded interfaces between the two sheets. Note that the weld joint has no defects in the stir zone, compared with the porosity reported in the aluminum resistance spot welds (Ref 9, 10). In Fig. 12.12(b), a close-up view of region I shows relatively coarse grains in the base metal. A close-up view of region II shows finer grains in the TMAZ. A close-up view of region III shows very fine equiaxed grains in the stir zone. The equiaxed grains in the stir zone are formed due to stir and recrystallization. The fundamentals of microstructural evolution are similar to linear

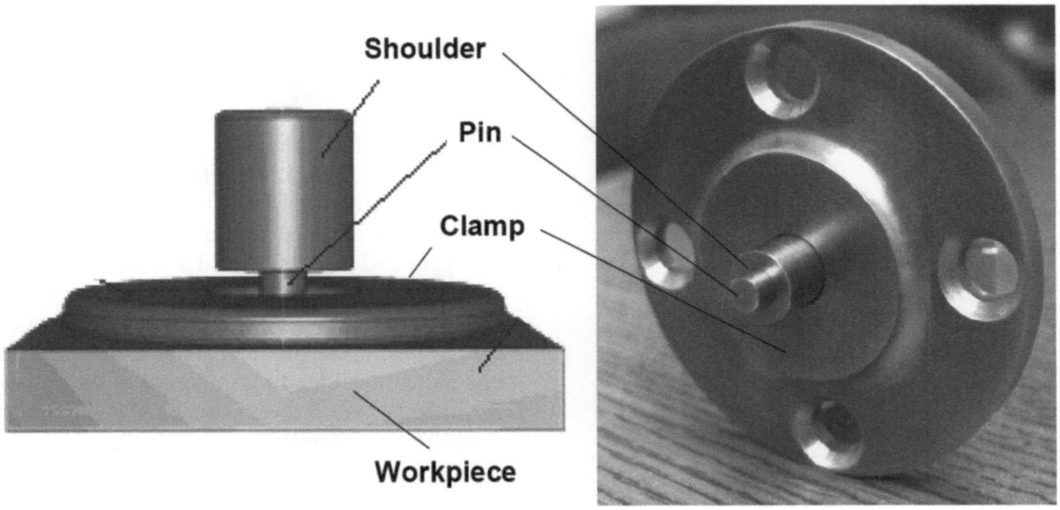


**Fig. 12.6** Refill friction stir spot welding tooling components. Source: Ref 7

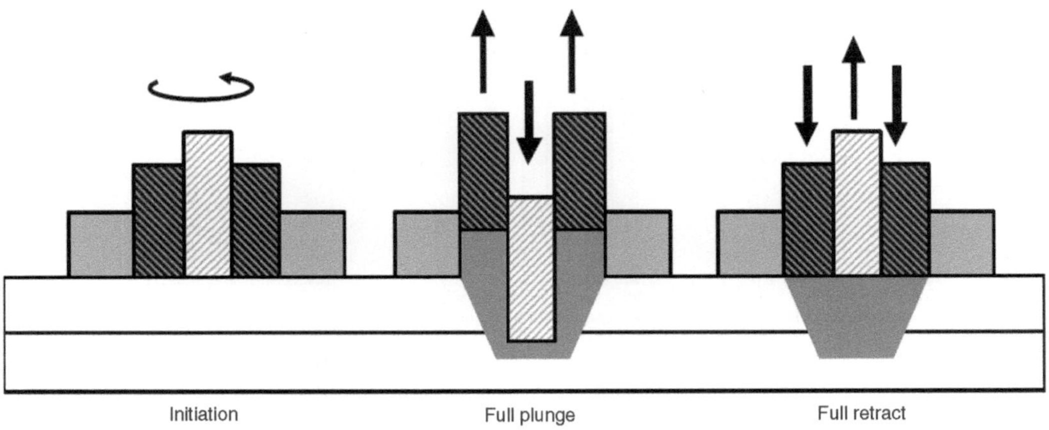


**Fig. 12.7** Refill friction stir spot welding process schematic. Source: Ref 7

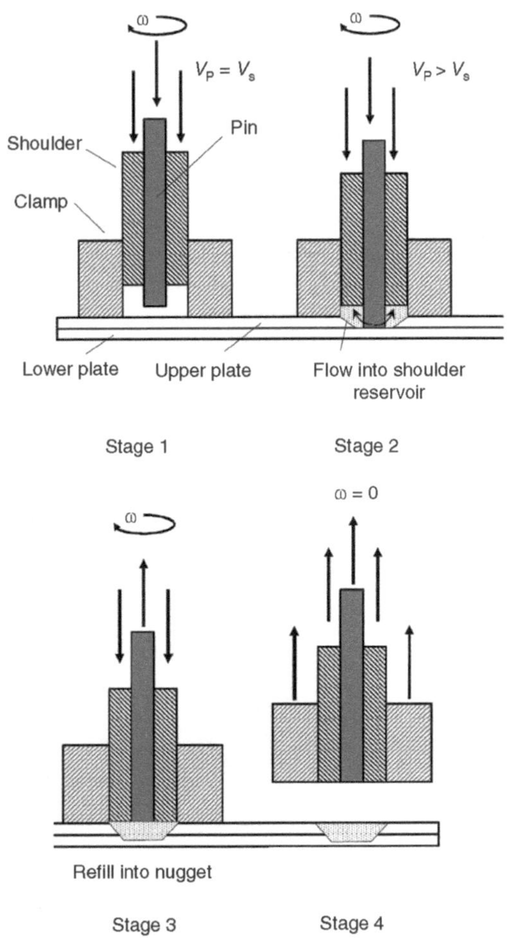


**Fig. 12.8** Schematic for the fixed-position refill friction stir spot welding process. Source: Ref 7

FSW and are covered in Chapter 4. As shown in Fig. 12.12(a), the interfaces horizontally pass through the TMAZ of the weld and rise up near the stir zone due to the deformation of the lower sheet from the indentation of the probe pin. In Fig. 12.12(b), a close-up view of region IV shows that the curved interface becomes vague and disappears close to the stir zone.

As the tool continues to rotate and plunge into the upper and lower sheets, the material under the tool shoulder near the probe pin is stirred. Outside the stir zone, the interfacial surface of the upper and lower sheets is distorted into a macroscopic curved interface, as shown in region IV in Fig. 12.12(b). The shoulder indentation squeezes out a portion of the upper sheet material, and consequently, the thickness of the upper sheet material decreases under the shoulder indentation. The reduction of thickness under the shoulder indentation results in a radial expansion of the upper sheet along the outer circumference of the shoulder indentation. However, due to the constraint of the neighboring material, the sheet is therefore bent along the outer circumference of the shoulder indentation. The bending of the sheet creates a gap between the upper and lower sheets. The bend is marked by "A" and "B," and the gap is marked by "C" and "D" in Fig. 12.12(a). The squeezed-out material from the shoulder indentation forms a ring along the outer circumference of the shoulder indentation on the top surface of the upper sheet. The squeezed-out material can be seen in Fig. 12.12(a).

*Failure Mode.* Figure 12.13 shows a failed lap-shear spot friction weld specimen and close-

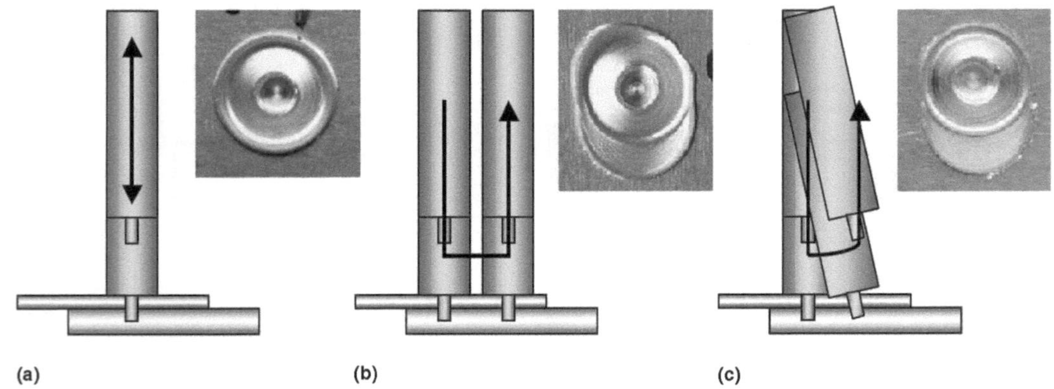


**Fig. 12.9** Tool movement and top view of variant of friction stir welding (FSW). (a) Friction stir spot welding. (b) Stitch FSW. (c) Swing FSW. Source: Ref 5

up views of the spot friction weld in the failed lap-shear specimen. The circumferential failure mode or the nugget pullout failure mode can be seen on the lower sheet of the failed specimen in Fig. 12.13(a). Figure 12.13(b) shows a top view of the failed spot friction weld. As shown in this figure, the hole diameter is much smaller than the indentation diameter or the tool shoulder diameter. Figure 12.13(c) shows a top view of a spot friction weld on the lower sheet of the failed specimen. As shown in Fig. 12.13(a) and (c), a small portion near the right side of the remaining weld nugget is removed, possibly due to tearing and rubbing of the upper sheet. The hole in the upper sheet, as shown in Fig. 12.13(b), is bent, distorted, and enlarged due to the tearing process. Therefore, the area of the hole, as shown in Fig. 12.13(b), is larger than the area of the remaining weld nugget, as shown in Fig. 12.13(c). The rough region surrounding the remaining weld nugget is possibly due to contact and rubbing from the upper sheet during the welding process.

The circumferential failure mode or the nugget pullout failure mode was observed. The experimental results suggest that under lap-shear loading conditions, the failure is initiated near the stir zone in the middle part of the nugget, and the failure propagates along the circumference of the nugget to final fracture. The initial shear failure emanated from the original curved notch tip. The failures of both spot friction welds were initiated and fractured through the upper sheet in the indentation zone near the weld nuggets.

**Effect of Paint Bake on a 6111-T4 Spot Weld.** Blundell et al. (Ref 11) studied the effects of paint bake cycles on the static perfor-

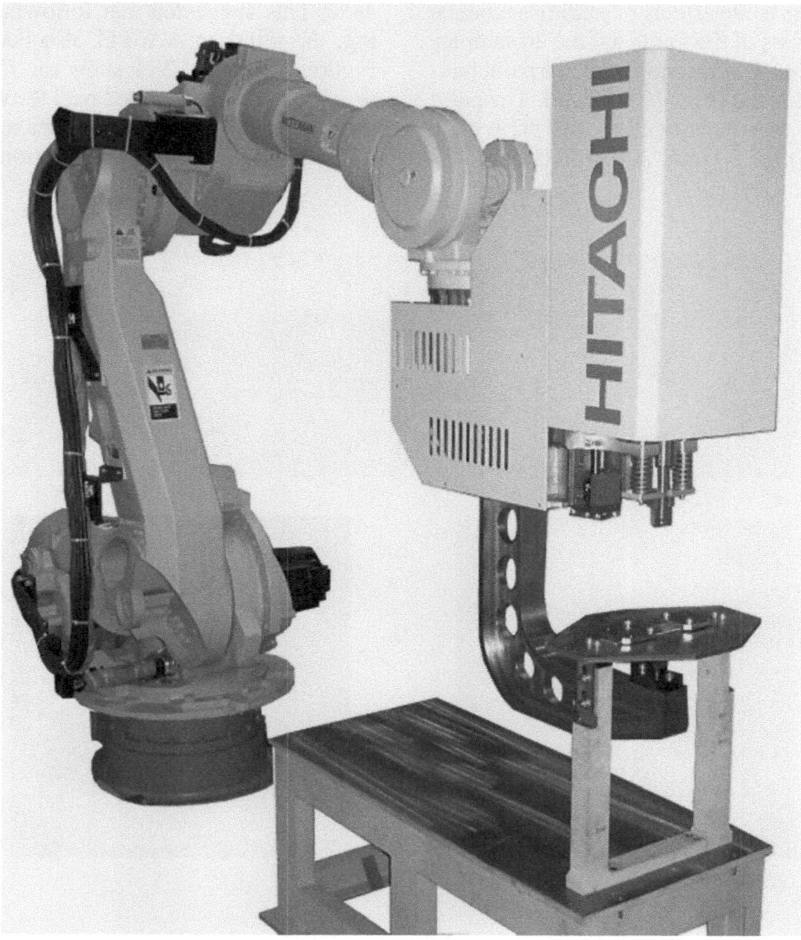


**Fig. 12.10** Swing, friction stir, welder Swing-Stir. Source: Ref 1

mance of AA6111-T4 FSSW. Welded samples were subjected to mechanical testing and exposed to a typical paint bake cycle of 180 °C (355 °F) for 30 min. The joint mechanical properties with and without the paint cycle were evaluated. The failure modes obtained from the testing were also examined. In a typical automotive production line, assembled components are subjected to heat treatment processes such as the paint bake cycle. This may significantly influence the physical properties of the base material and may, as a consequence, have a direct influence on the strength of a joint made within the material. Previously published research (Ref 12) has reported the effect of the paint baking cycle on SPR joints. The motivation of this study was to evaluate if such a phenomenon exists with FSSW.

The AA6111 sheet was received in the T4 condition. A total of 30 samples were welded, from which 20 were chosen randomly in order to avoid the possible effects of joining sequence on the properties of the joints. Of the 20 samples for each group, 10 were subjected to a paint baking cycle, while the other 10 were not. The paint baking cycle was performed at 180 °C with ± 10 °C (18 °F) for 30 min. A thermocouple was used to monitor the baking temperature. Approximately 48 h after manufacturing or the paint baking cycle, samples were tested under shear and peel conditions. At least five samples were tested at each condition in terms of sample group and baking condition.

Figure 12.14 shows the shear and peel test results. The mean maximum shear load was 3.2 kN (0.36 tonf) for the unbaked samples and 3.1 kN (0.35 tonf) for the baked samples. Following paint baking, a 3.1% reduction in shear strength was observed. In peel testing, mean maximum loads of 0.6 and 0.5 kN (0.067 and 0.056 tonf) were obtained for the unbaked and baked samples, respectively. This represented a 17% reduction, attributed to the paint bake cycle. The graph in Fig. 12.15 also shows a reduction in extension at maximum load following paint baking. In lap-shear tests, the extension reduced by approximately 38%. In peel tests, the extension reduced by approximately 44%. This suggested that following paint baking, the joints in AA6111 also became brittle. Figures 12.16 to 12.19 show the failure modes that occurred in shear and peel tests. Fracture of the nugget dominated the failure mechanism in the shear test, while the separation of coupons

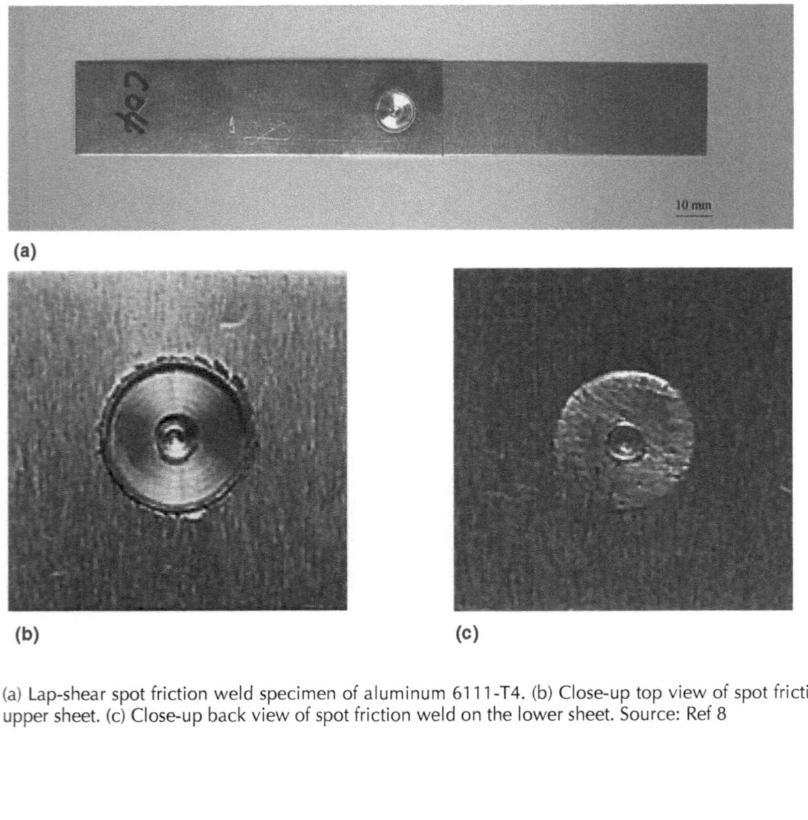


**Fig. 12.11** (a) Lap-shear spot friction weld specimen of aluminum 6111-T4. (b) Close-up top view of spot friction weld on the upper sheet. (c) Close-up back view of spot friction weld on the lower sheet. Source: Ref 8

from the nugget boundary was the only failure mode for peel test.

The conclusion based on the data obtained was that for FSSW of A6111-T4, there was no significant change in the static strength in both coach peel and lap shear of specimens that were as-welded and those subjected to a paint bake cycle.

**Fatigue Life.** Lin et al. (Ref 13) investigated fracture and fatigue mechanisms of spot friction welds in aluminum 6111-T4 lap-shear specimens. A concave tool was used to make the spot welds. Optical and scanning electron micrographs of spot friction welds before and after failure under quasi-static and cyclic loading conditions were examined. The failure mechanisms of spot friction welds under quasi-static, low-cycle, and high-cycle fatigue loading conditions were also investigated by Lin et al. Aluminum 6111-T4 sheets with a thickness of 0.94 mm (0.037 in.) were used. Lap-shear specimens were made by using two 25.4 by 101.6 mm sheets with a 25.4 by 25.4 mm overlap area.

Lap-shear specimens were first tested by using an Instron 4502 testing machine at a monotonic displacement rate of 1.0 mm/min. The tests were

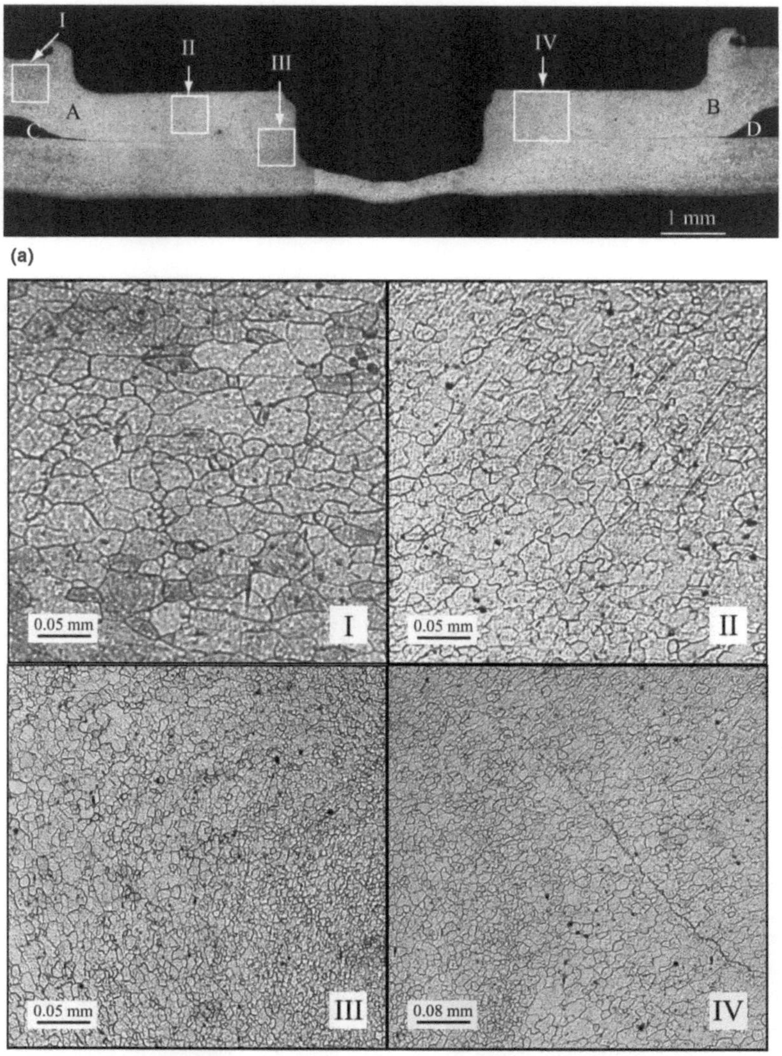


(a)

(b)

**Fig. 12.12** (a) Micrograph of the cross section of a spot friction weld. (b) Close-up views of regions I, II, III, and IV. Source: Ref 8 terminated when specimens separated. The loads and displacements were simultaneously recorded during the tests. The failure loads were then used as the reference loads to determine the loads applied in the fatigue tests. Lap-shear specimens were then tested by using an Instron servohydraulic fatigue testing machine with a load ratio $R$ of 0.2. A lap-shear specimen and the fixture are shown in Fig. 12.20. The test frequency was 10 Hz. The tests were terminated when specimens separated or nearly separated when the displacement of the two grips of specimens exceeded 5 mm (0.2 in.). Figure 12.21 shows the load range as a function of the life for spot friction welds made by the concave tool in lap-shear specimens under cyclic loading conditions.

*Fatigue Failure Mode.* The experiments conducted by Lin et al. were conducted under cyclic loads that resulted in the fatigue life of spot friction welds from $10^3$ to $10^5$. Based on the experimental observations, the failed spot friction welds with fatigue lives from $10^3$ to $10^4$ show one failure mechanism, while the failed spot friction welds with fatigue lives from $10^4$ to $10^5$ show another failure mechanism. They classified it as fracture mechanism under quasi-static loading conditions and fatigue mechanisms under loading conditions of low-cycle fatigue (lives of $10^3$ to $10^4$) and high-cycle fatigue (lives of $10^4$ to $10^5$).

Figure 12.22(a) shows a schematic plot of the cross-sectional symmetry of a lap-shear specimen made by the concave tool, with the sheet thickness $t$ under an applied load (shown as the bold arrows). Figure 12.22(b) shows a schematic plot of the cross section near the spot friction weld. In these figures, the shadow represents the stir zone, the dashed line represents the unwelded interfacial surface, and the thin solid line represents either the fracture surface or fatigue crack. Figure 12.22(c) shows a table that lists the failure mechanisms of the spot friction welds under quasi-static, low-cycle fatigue, and high-cycle fatigue loading conditions.

(a)

(b)

(c)

**Fig. 12.13** (a) Failed spot friction weld lap-shear specimen. (b) Top view of a spot friction weld on the upper sheet of the failed specimen. (c) Top view of a spot friction weld on the lower sheet of the failed specimen. Source: Ref 8

As shown in Fig. 12.22(b) and summarized in Fig. 12.22(c), under quasi-static loading conditions, a necking failure is initiated at location "A"; the failure then propagates along the nugget circumference, and finally, the upper sheet is torn off at location "B." Under low-cycle fatigue loading conditions, the experimental observations suggest that one fatigue crack (marked by "C") appears to emanate from the original crack tip, and then another fatigue crack (marked by "D") appears to emanate from the surface of the bend. The experimental observations suggest that the fatigue crack (marked by "C") appears to be the dominant crack that propagates through the sheet thickness. Without the support of the lower sheet near the stretching side of the nugget, the nugget is rotated clockwise, and the sheets near the nugget are therefore bent. Eventually, the stir zone is separated through the interfacial surface

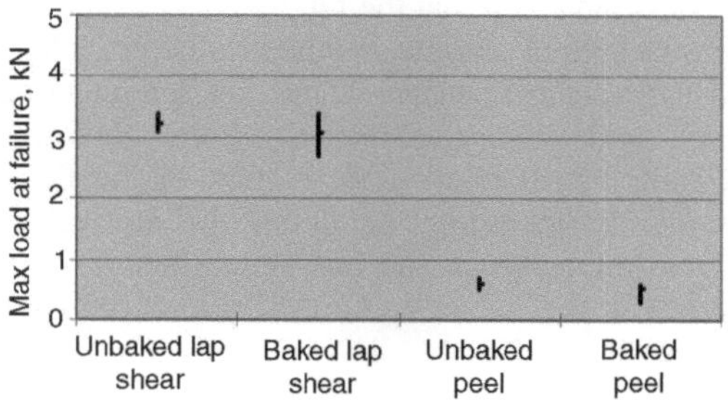


**Fig. 12.14** Lap-shear and t-peel results. Source: Ref 11

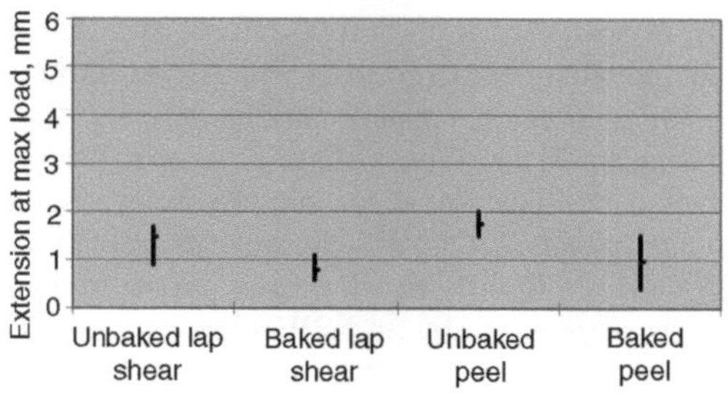


**Fig. 12.15** Extension at maximum load. Source: Ref 11

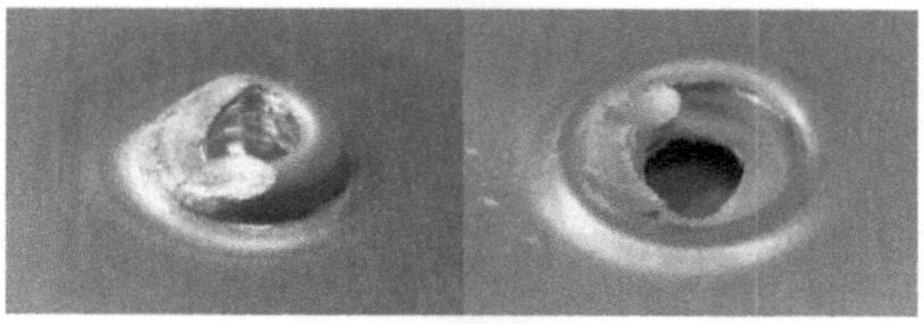

**Fig. 12.16** A6111 unbaked. Typical lap shear failure. Source: Ref 11

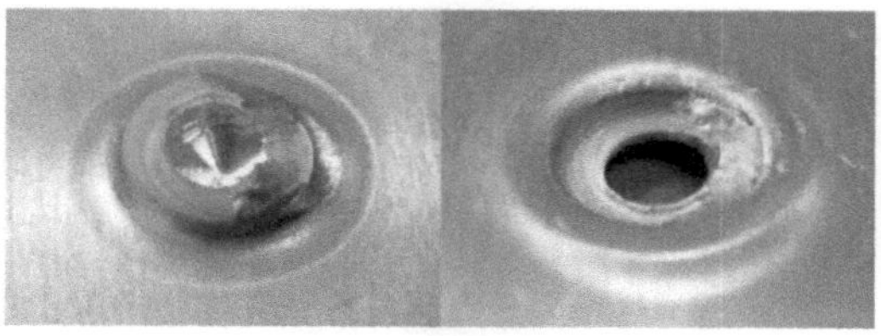

**Fig. 12.17** A6111 paint baked. Typical lap shear failure. Source: Ref 11

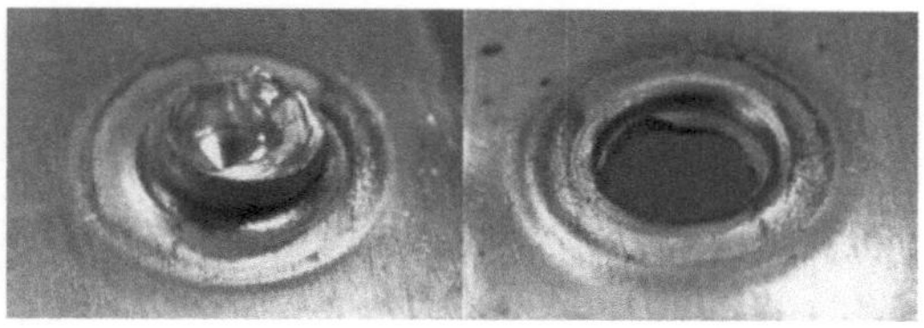

**Fig. 12.18** A6111 unbaked. Typical t-peel failure. Source: Ref 11

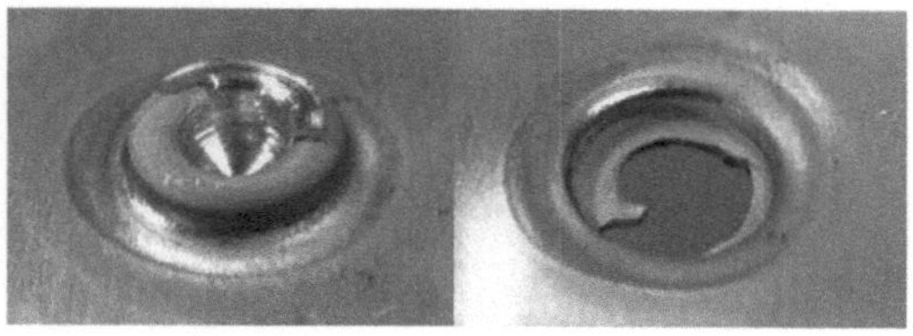

**Fig. 12.19** A6111 paint baked. Typical t-peel failure. Source: Ref 11

**Fig. 12.20** A lap-shear specimen and the fixture are mounted in an Instron fatigue testing machine. Source: Ref 13

(marked by "E"), and the upper sheet is torn off. Under high-cycle fatigue loading conditions, the experimental observations suggest that one fatigue crack (marked by "C") appears to emanate from the original crack tip, and another fatigue crack (marked by "D") appears to emanate from the surface of the bend. Both fatigue cracks propagate through the sheet thickness, then become transverse cracks growing toward the width direction of the specimens and finally cause the fracture of the specimen.

*Empirical Model for Fatigue Crack Growth.* Lin et al. (Ref 14, 15) also proposed a fatigue crack growth model based on the Paris law for crack propagation. Furthermore, the global and local stress-intensity factors for kinked cracks were adopted to predict the fatigue lives of the spot friction welds. The global stress-intensity factors and the local stress-intensity factors based on previous work (Ref 16, 17) were used to estimate the local stress-intensity factors for kinked cracks with experimentally determined kink angles. Their results indicated that the fatigue life predictions based on the Paris law and the local stress-intensity factors as functions of the kink length agree well with the experimental results obtained. Detailed mathematical derivation of the Paris equation (including obtaining the equivalent stress-intensity factor) has been illustrated in depth in the above-mentioned references (Ref 14, 15).

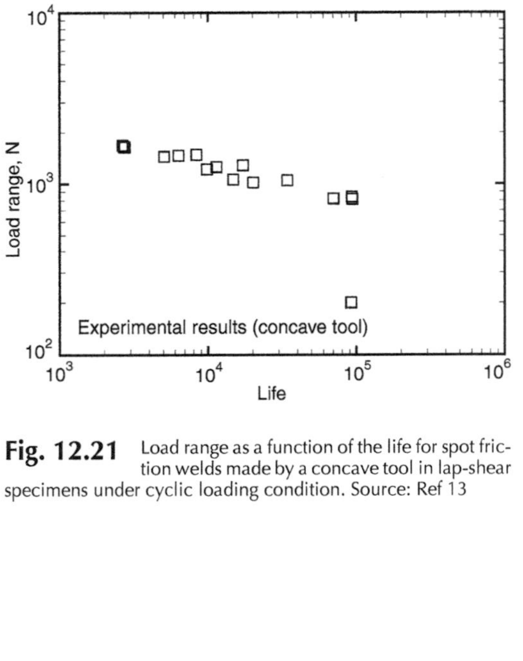


**Fig. 12.21** Load range as a function of the life for spot friction welds made by a concave tool in lap-shear specimens under cyclic loading condition. Source: Ref 13

### *12.2.2 Pure Spot FSW Properties (AA5754)*

**Static Strength Evaluation.** Arul et al. (Ref 18) investigated the microstructures and failure mechanisms of spot friction welds in aluminum 5754 lap-shear specimens. In this investigation, aluminum 5754 sheets with thickness of 1.0 mm were used. The lap-shear specimens are made by using two 25.4 by 101.6 mm coupons with a 25.4 by 25.4 mm overlap area. Spot friction welds were made by an FSW system manufactured by MTS Systems Corporation. The lap-shear specimens were tested to obtain the shear strength by using an Instron Model 4502 testing machine. The crosshead displacement was set at a rate of 10.0 mm/min (0.4 in./min). In this

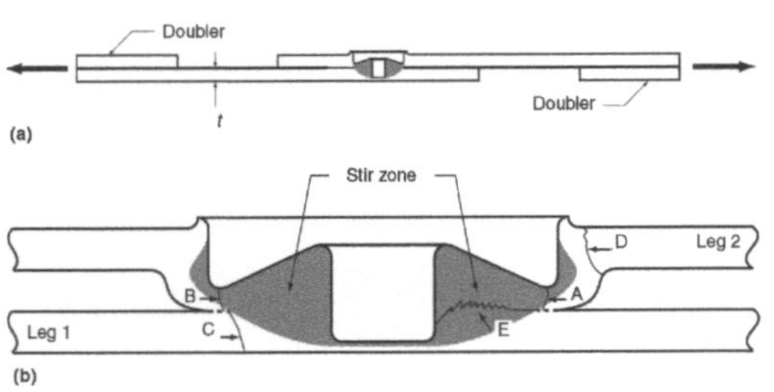


| Loading condition | Failure mechanism |
|---|---|
| Quasi-static | A → B |
| Low-cycle fatigue | C, D → E |
| High-cycle fatigue | C, D → transverse cracks |

**Fig. 12.22** (a) Schematic plot of the cross-sectional symmetry of a lap-shear specimen made by a concave tool, with a sheet thickness *t* under an applied force (shown as bold arrows). (b) Schematic plot of the cross section near the spot friction weld made by a concave tool. (c) Failure mechanisms of spot friction welds made by a concave tool under quasi-static, low-cycle fatigue, and high-cycle fatigue loading conditions. Source: Ref 13

investigation, a tool with a concave shoulder and a tool with a flat shoulder were used. In order to study the effect of the penetration depth, the specimens were made with two different depths of 1.85 and 1.95 mm (0.073 and 0.077 in.). With a tool having a concave shoulder, the maximum load increases by approximately 4.7% (3.06 versus 2.92 kN, or 0.34 versus 0.33 tonf) when the depth increases from 1.85 to 1.95 mm. However, with a flat shoulder tool, the maximum loads stay the same (2.88 kN, or 0.32 tonf) for the depths of 1.95 and 1.85 mm. For the depth of 1.85 mm, the maximum load for the concave tool is larger than that for the flat tool by 1% (2.92 versus 2.88 kN). For the depth of 1.95 mm, the maximum load for the concave tool is larger.

**Microstructure.** Figure 12.23(a) shows a micrograph of the cross section of a spot friction weld made by a tool having a concave shoulder. Near the center, the shape of the indentation matches the profile of the probe pin and the shoulder. With a concave shoulder, the shoulder squeezes a lot of material from the upper sheet metal to the location near the probe. The light-gray area around the pin and the shoulder represents the stir zone, and the slightly darker area surrounding the stir zone is the TMAZ. Two notch tips at the unwelded interface between the upper and lower sheets near the spot friction weld are denoted by "C" and "D".

A comparison of Fig. 12.23(a) and the results obtained by Lin et. al (Ref 8) for a flat-shoulder tool in aluminum 6111-T4, discussed earlier, shows that the stir zone for the concave tool (light-gray area around the probe and the shoulder) is much larger compared to that of the flat tool. Due to different flow patterns, the shapes of the interface between the upper and lower sheets under the shoulder indentation are quite different. The different flow patterns also result in different shapes of spot friction welds.

In Fig. 12.23(a), the boxed areas indicate where the grain structure samples are taken to show the details of the stir zone and TMAZ. A close-up view of the stir zone in Fig. 12.23(b) shows very fine equiaxed grains. This is due to stirring and recrystallization. A close-up view of the TMAZ in Fig. 12.23(c) shows fine grains. For comparison, a close-up view of the base metal in Fig. 12.23(d) shows coarse grains.

**Failure Mode.** Figure 12.24 shows a cross-sectional view and close-up views of a spot friction weld made by a concave tool with a depth of 1.95 mm in a partially failed lap-shear specimen (Ref 18). The two arrows in Fig. 12.24(a) schematically show the loading direction. In Fig. 12.24(a), near the upper right portion of the

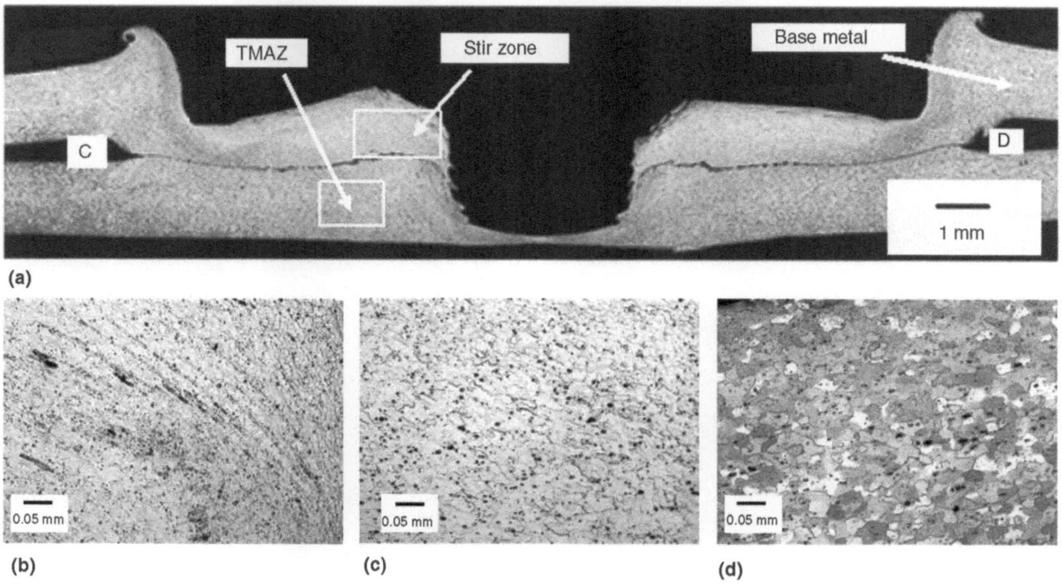


**Fig. 12.23** (a) Micrograph of the cross section of a spot friction weld made by a concave tool with a depth of 1.95 mm (0.077 in.). (b) Close-up view of the stir zone. (c) Close-up view of the thermomechanical affected zone (TMAZ). (d) Close-up view of the base metal. Source: Ref 18

spot friction weld, marked as leg 2, a necking and shearing failure appears at point "A." The necking and shearing failure mechanism is very similar to that of the failed resistance spot welds in lap-shear specimens (Ref 19). Note that the location of the necking and shearing failure is close to the outer circumference of the shoulder indentation near the crack tip. In Fig. 12.24(b), a close-up view of region I shows the necking failure. In Fig. 12.24(c), a magnified view of region II shows the microstructures near the crack tip. Note that the material in the lower portion of region II appears to be the base metal, and the material in the upper left portion of region II appears to be the TMAZ. The circumferential failure mode or the nugget pullout failure mode was observed. The experimental results suggest that under lap-shear loading conditions, the failure is initiated near the stir zone in the middle part of the nugget, and the failure propagates along the circumference of the nugget to final fracture. The initial shear failure was emanated from the original curved notch tip. The failures of both spot friction welds were initiated and fractured through the upper sheet in the indentation zone near the weld nuggets. The necking and shearing failure mechanism is the principal failure initiation mechanism, similar to the study for the spot friction welds in aluminum 6111-T4 sheets (Ref 20). The failure was initiated and fractured through the upper sheet under the shoulder indentation near the crack tip.

### 12.2.3 Pure Spot FSW Properties (AA5052)

Freeney et al. (Ref 21) evaluated the effect of process parameters on FSSW of AA5052 using a plunge-type FSW machine. Sheets with two different thicknesses were used. The dwell time and revolutions per minute were process variables. Lap-shear tests were performed in two-sheet and three-sheet configurations to determine the influence of processing parameters on the mechanical properties of lap-joint friction stir spot welds.

Due to the variation in material thickness being welded, two different conical pinned tools were used during this study. The first tool used for the single-overlap 1 mm sheet had a shoulder diameter of 12 mm (0.47 mm) and a 1.77 mm (0.070 in.) long conical pin, with a root diameter of 4.5 mm (0.18 in.) and tip diameter of 3 mm (0.12 in.). The second tool was used for both the single-overlap 1.6 mm (0.06 in.) and the double-overlap 1 mm coupon configurations. The tool had a conical pin and was machined from H13

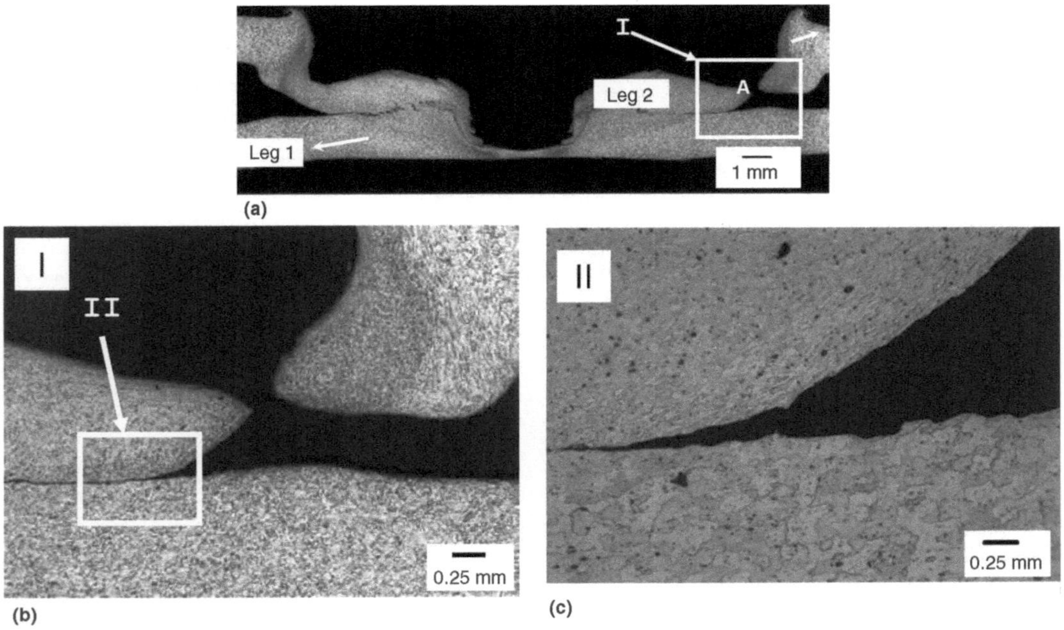


**Fig. 12.24** (a) Micrograph of the cross section of a spot friction weld made by a concave tool with a depth of 1.95 mm (0.077 in.) in a partially failed lap-shear specimen. (b) Close-up view of region I. (c) Close-up view of region II. Source: Ref 18

tool steel. The tool had a shoulder diameter of 12.5 mm (0.49 in.), a pin height of 2.65 (0.10 in.), a root diameter of 5 mm (0.20 in.), and tip diameter of 3.3 mm (0.13 in.). In the experiments, the plunge rate and dwell time were held constant at 2.5 mm/s (0.10 in./s) and 490 ms, respectively. When the minimum target depth was established, the plunge depth was increased in increments of 0.15 mm (0.006 in.), so that the shoulder penetrated slightly in the top sheet. Three different target depths were tested for each coupon arrangement.

Maximum load to failure was recorded by loading the welds in shear. Figure 12.25 shows load to failure for various tool rotation rates and plunge depths. Maximum weld strength was observed at lower tool rotation rates in all the welds made. Also, at higher tool rotation rates, varying plunge depths did not significantly influence the loads to failure. Further, it was observed that welds made on 1.6 mm thick sheets showed significantly lower weld strength than the 1 mm thick sheets. The thinner sheet showed better weld strength because of larger weld interface. In the thinner sheets, the material flow from the shoulder into the weld interface at lower tool rotation rates led to a larger weld zone. The frictional condition varies from sticking-dominated to slip-dominated with changing tool rotation rate (Ref 22). The sticking condition exhibited at lower tool rotation rates leads to higher material flow around the shoulder and hence to better interface strength in the spot welds.

### 12.2.4 Refill FSSW Properties

As discussed earlier, there are two types of refill methods: shoulder-first refill and fixed-position refill.

**Shoulder-First Refill.** Allen et al. (Ref 7) performed weld trials using the shoulder-first refill method in 2 mm (0.08 in.) thick upper and lower sheet 7075-T6 aluminum lap welds under forge control mode. A large effective shear area was formed and a high degree of refill achieved (Fig. 12.26).

As seen in Fig. 12.27, several effects of processing parameters were noted. The higher the forge load, the greater the expulsion of material between the shoulder and the clamping ring, resulting in loss of material and increased depth of indentation (lack of refill). The effective shear area at the faying surface was independent of forge load and a direct function of the shoul-

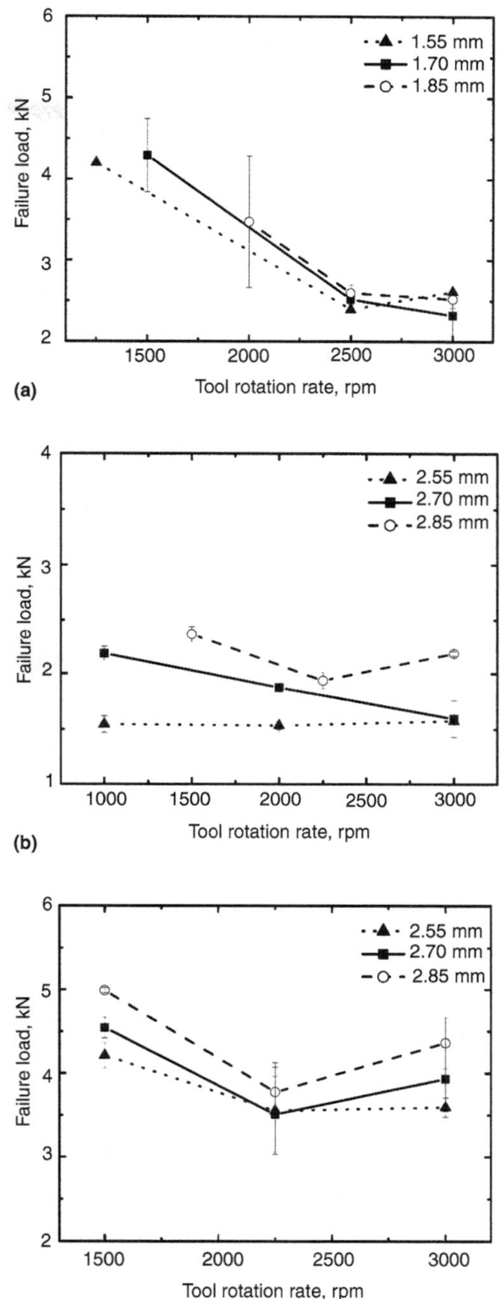


**Fig. 12.25** (a) Failure loads for friction stir spot welded 5052. (a) 1 mm (0.04 in.) single overlap. (b) 1 mm (0.04 in.) double overlap. (c) 1.6 mm (0.06 in.) single-overlap configuration. Mechanical properties are not significantly influenced by plunge depth with increasing tool rotation rate. The thicker sheet exhibits lower weld strength. Source: Ref 21

der diameter. Sheet lifting and separation was not observed, and material was sufficiently plasticized and reflowed easily.

The spot welds created with this shoulder-first method were characterized by high, unguided lap-shear strengths, with an average load-carrying capability per spot of over 8.9 kN (1.0 tonf). These specimens failed almost exclusively in the nugget pullout mode. The authors, however, point out that caution must be maintained while interpreting pullout geometry alone, because excessive indentation and lack of refill will result in tensile overload failure around the periphery of the spot rather than shear through the faying surface. Thus, nugget pullout and minimum pounds per spot should be used in evaluating joint quality.

**Fixed-Position Refill.** A process development matrix was obtained with a smooth cylindrical pin and shoulder profile using the fixed-position refill method (Fig. 12.28). The materials used for this study were 3.18 mm (0.13 in.) thick upper and lower sheet 2024-T3 aluminum lap joints. Characteristic measurements were made of surface indentation, effective shear area, void size, and lap-shear strength.

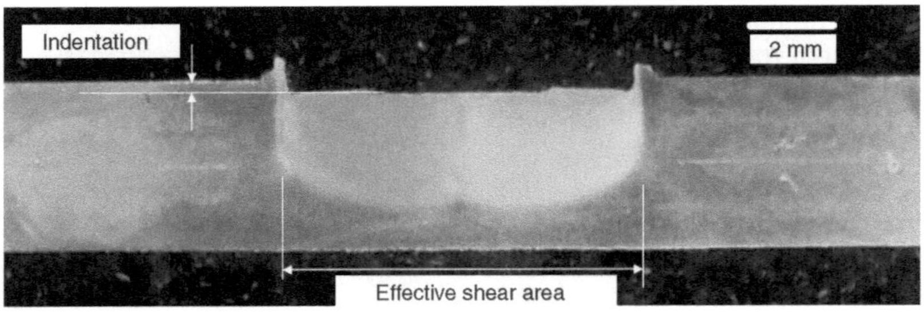


**Fig. 12.26** Typical structure of shoulder-first refill friction stir spot welding in 2 mm (0.08 in.) 7075-T6 lap joints. Source: Ref 7

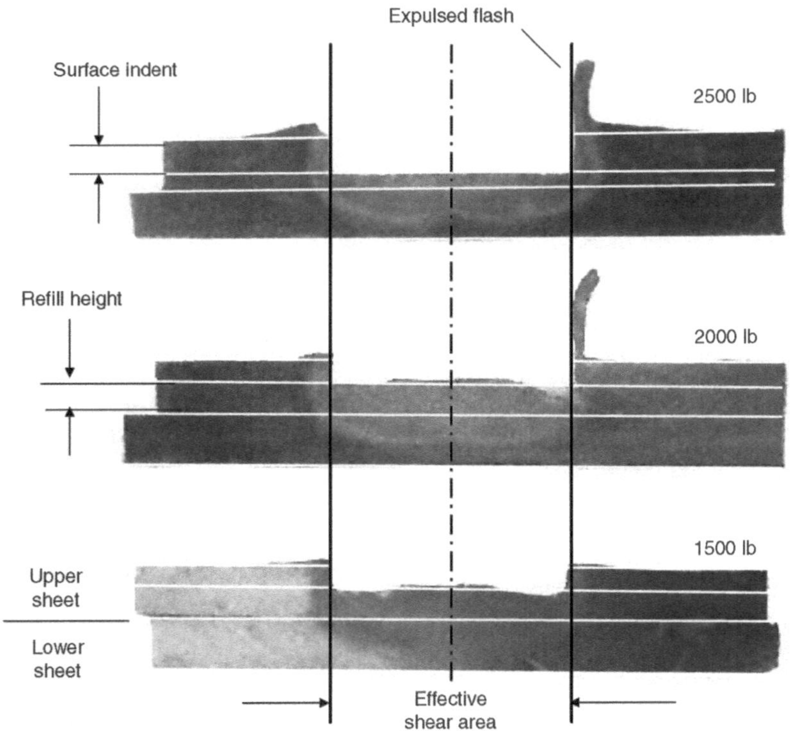


**Fig. 12.27** Effect of forge load on weld geometry for shoulder-first refill friction stir spot welding in 2 mm (0.08 in.) 7076-T6 lap welds. Source: Ref 7

Surface indentation arises from excessive flash being extruded through the clearance space between the shoulder and clamping ring.

The hotter parameters (higher revolutions per minute and higher shoulder forge loads) resulted in more material loss as flash and a greater surface indent. Internal void size increased with low revolutions per minute and low shoulder force levels (cold welds). The voids showed increased consolidation toward the upper-right quadrant of the matrix (hotter welds). Surface indentation showed a reverse trend, with the hotter welds showing the larger indentation. The potential to fill the void left by the retreating pin apparently increases with higher revolutions per minute and extrusion forces, because the material becomes easier to plasticize and extrude. The unguided lap-shear strength value of the fixed-position refill FSSW is shown at the bottom left side of each image on Fig. 12.28. This strength was maximum in the central combination of parameters, where a strength of 4.23 kN (0.48 tonf) per spot was seen. The lap-shear strengths were generally greater in the upper-right quadrant of the matrix, where the hotter welds and higher forge forces resulted in less internal void formation. There was, however, a drop in strength at the highest levels, where the surface indentation was greatest. This suggests that the strength of these joints involves a competing mechanism between the loss of effective shear area due to internal void formation and the reduction in tensile area around the spot periphery due to excessive indentation.

### 12.2.5 Swing FSSW Properties

**Static Strength.** Okamoto et al. (Ref 5) evaluated the mechanical properties of swing FSSW. In this study, the material welded was AA6022-T4 in lap configuration. This was chosen to mimic the automotive closure panel assembly process. Upper and lower sheet thicknesses were 0.8 and 1.5 mm (0.03 and 0.06 in.), respectively. Weld coupons of 150 mm (6 in.) in length and 40 mm (1.6 in.) in width were overlapped by 40 mm and lap welded for lap-shear specimens. For swing FSW, the effect of the welding length on lap-shear strength was studied. The tool rotating speed was 2500 rpm. The dwell time was 0.5 s. The swing length varied from 0 (pure spot) to 2.5 mm. The swing FSW direction was selected to be parallel to the lap-shear test direction of the coupons. The welding tool was made of tool steel, with a shoulder diameter of 8 mm (0.3 in.) and thread pin diam-

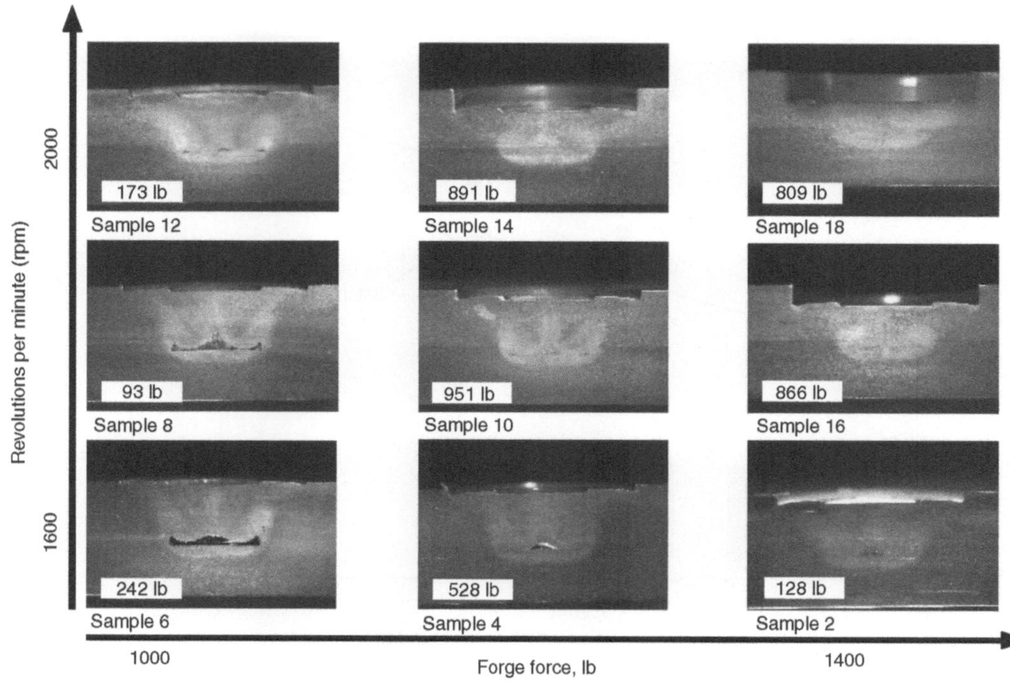


**Fig. 12.28** Effect process development matrix for the fixed-position refill friction stir spot welding method. Source: Ref 7

eter of 3 mm (0.12 in.). Figure 12.29 shows the effect of welding length on lap-shear strength for stitch FSW joints. A pure spot shows a static strength of 225 kgf. It was observed that the shear strength increases as a function of welding length.

Figure 12.30 shows the macrostructure of the cross section of FSSW and swing FSW joints. The weld area between upper and lower sheets is wider for welds with longer welding length. Figure 12.31 shows the failed FSSW specimen, swing FSW specimen, and the as-welded micrograph. In both specimens, failure occurred at the tool shoulder indentation. The shoulder indentation area has the lowest hardness and the minimum thickness in the upper sheet. The fracture seems to initiate at the hook horizontally and expand into the shoulder indentation.

**Fatigue Life.** Okamoto et al. (Ref 5) conducted preliminary work on fatigue strength for swing FSW. Figure 12.32 shows the shear fatigue strength of swing FSW. In the case of lap-shear tension, a swing FSW joint shows fairly high fatigue strength, especially at a lower cycle.

Figure 12.33 shows the failed specimens of static and fatigue lap-shear tests. In the case of static and fatigue test under higher applied load, the failure is pullout mode. Due to the large diameter of the hole size, swing FSW appears to have a higher static lap-shear and fatigue strength at low cycle. On the other hand, the crack initiated at the weld region and grew into the base metal in all the joints. This indicates that the high-cycle fatigue strength of the swing FSW is comparable to the other friction stir spot techniques. However, a detailed study is required to determine crack initiation in swing FSW.

## 12.3 Numerical Simulation of FSSW

Numerical simulation of FSSW has always been challenging, primarily because the weld sequence—comprised of the plunge, stir, and retract periods—is relatively short as compared to linear FSW. Modeling this dynamic phenomenon is a challenge for simulation engineers because of the numerous complexities involved in the process. Effective and reliable computational models of the FSW process would greatly enhance the study of material flow and microstructure evolution around a tool pin as well as temperature distribution along a weld line. Approaches for the computational modeling of the FSW process, however, are still under development, and a great deal of work is underway, particularly the application of explicit

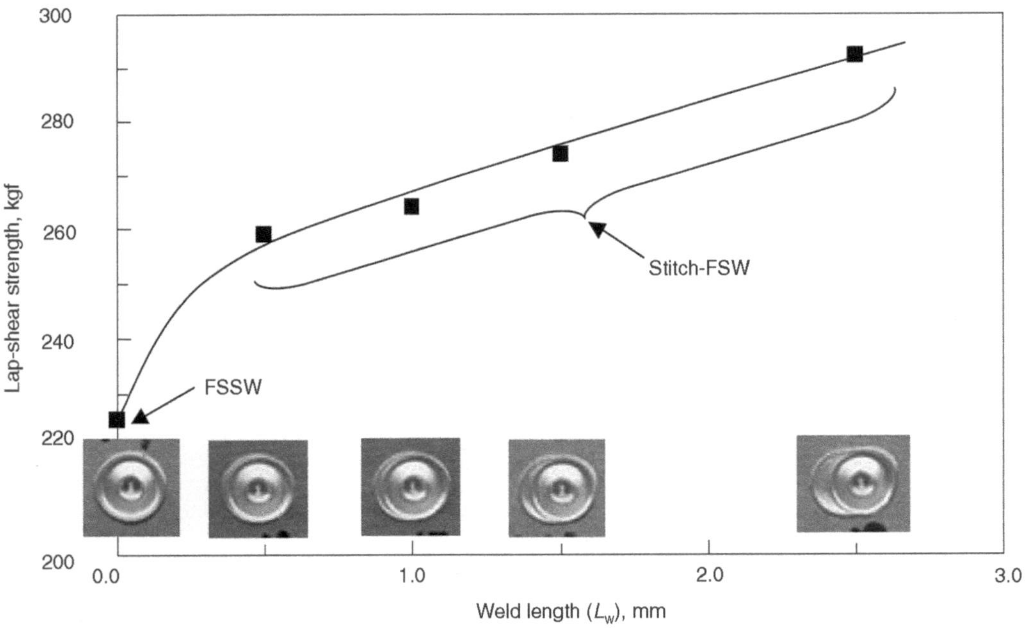


**Fig. 12.29** Effect of welding length on lap-shear strength of stitch friction stir welding (FSW) joints. FSSW, friction stir spot welding. Source: Ref 5

finite element codes for a verifiable simulation (Ref 23).

**Pure Spot FSW.** Awang et al. (Ref 23) presented some results on finite element modeling of FSSW using ABAQUS/Explicit (ABAQUS, Inc.) as a finite element solver. A three-dimensional (3-D) coupled thermal-stress model was used to calculate the thermomechanical response of the FSSW process. Adaptive meshing and advection schemes, which make it possible to maintain mesh quality under large deformations, were used to simulate the material flow and temperature distribution in the FSSW process.

The FSSW process simulation involved modeling the coupled thermoelastoplastic response of the tool-workpiece system, in which the constitutive model of the material and the nonlinear temperature-dependent transient heat-transfer response produce both plastic deformations and a temperature distribution as the material flows and stirs, forming the weld.

The finite element (FE) model of the FSSW process was done using ABAQUS/Explicit software. A 3-D dynamic fully coupled thermal-stress analysis was performed to obtain thermomechanical responses of the FSSW process. Two features in the FE package were deployed in order to obtain the results:

- The adaptive mesh scheme that automatically regenerates the mesh when the elements are severely distorted due to large deformation
- The mass scaling technique that modifies the densities of the materials in the model and improves the computational efficiency while retaining the accuracy of the results.

The FE analysis was conducted by prescribing displacement and angular velocity of the pin tool and by imposing appropriate boundary conditions. The rate of pin penetration was prescribed in two time steps, based on an actual experimental setup. In step 1, the pin was

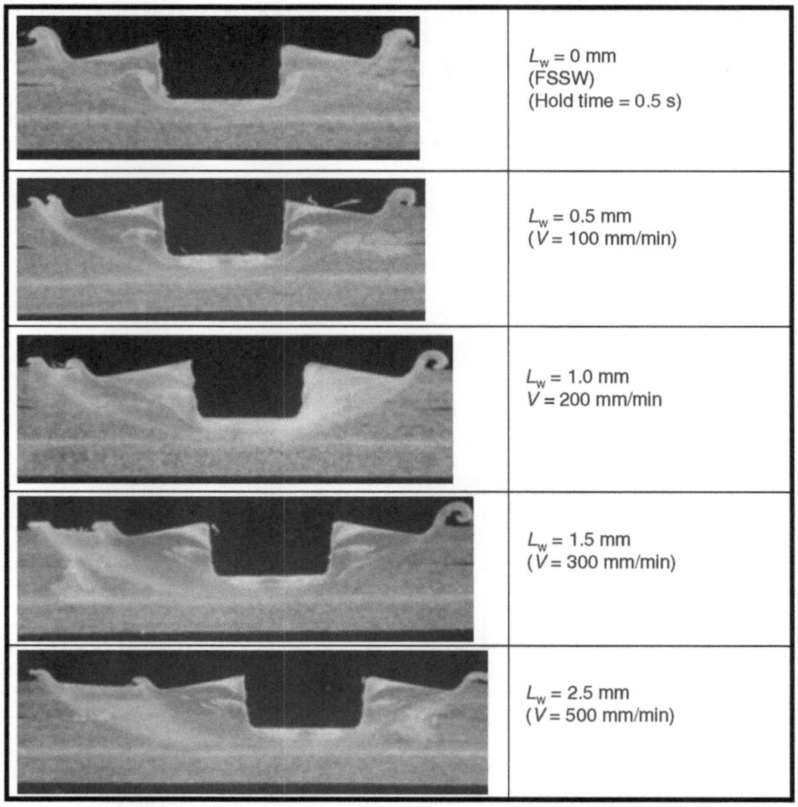


**Fig. 12.30** Micrographs of cross section of friction stir spot welding (FSSW) and stitch friction stir welding joints. $L_w$, weld length; $V$, velocity. Source: Ref 5

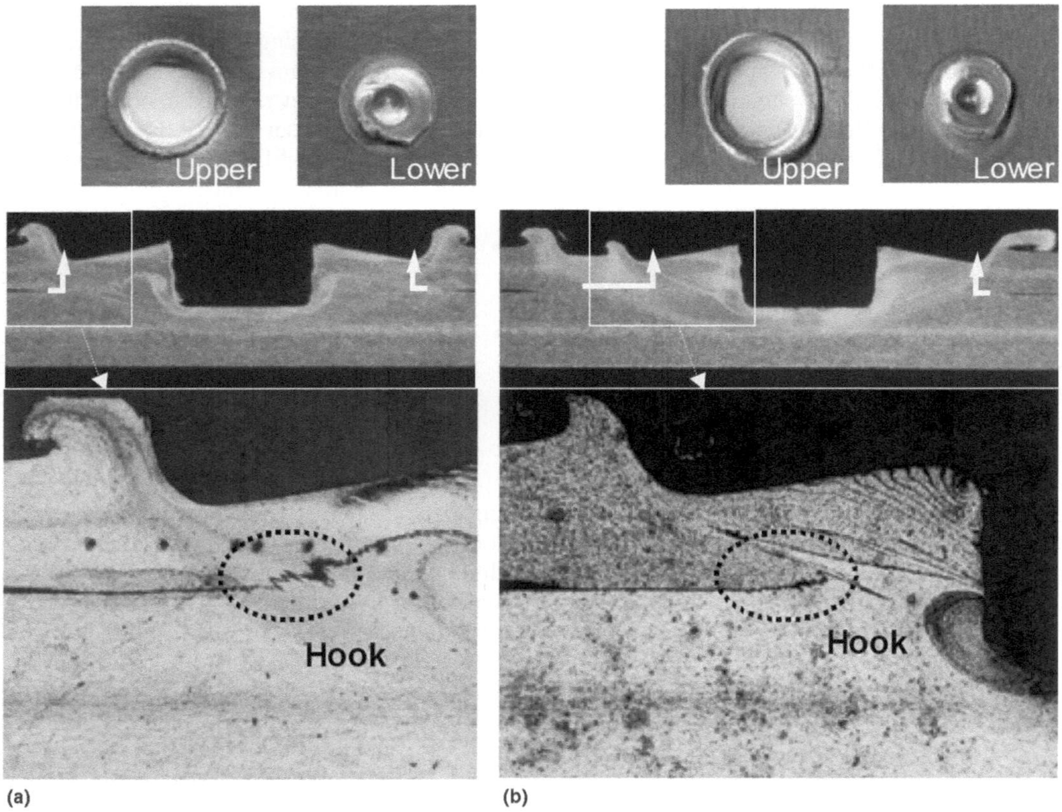


Fig. 12.31 Failed lap-shear specimens and as-welded micrograph of (a) friction stir spot welding and (b) stitch friction stir welding joints. Arrows show fracture path. Source: Ref 5

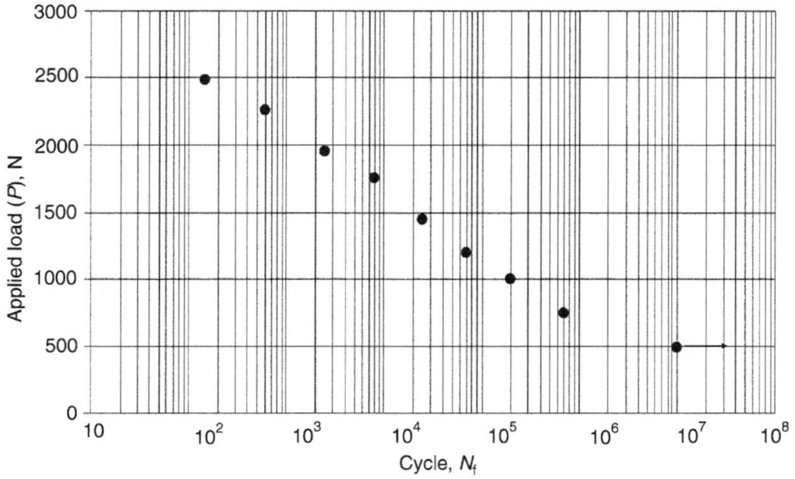


Fig. 12.32 Shear fatigue strength of swing friction stir welding. Source: Ref 5

plunged with a rate of 2.668 mm/s (0.105 in./s). In step 2, the plunge rate was set at 0.493 mm/s (0.019 in./s).

The workpieces were spot welded in lap-joint configuration, as shown in Fig. 12.34. The geometry of the workpieces had a dimension of 25 by 25 mm (1 by 1 in.), with a thickness of 1 mm (0.04 in.). They were meshed with eight-node trilinear displacement and temperature and reduced integration with hourglass control. A total of 80,000 elements and 102,010 nodes were generated in the model. The pin and the backing anvil were modeled as isothermal analytical rigid surfaces. This assumption would

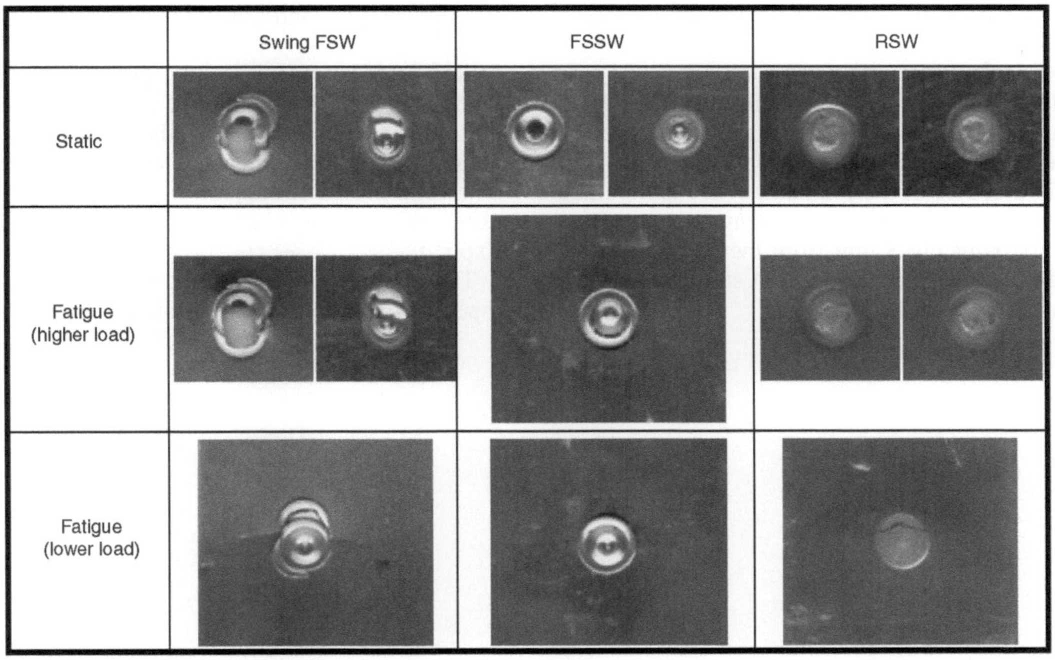


**Fig. 12.33** Failed specimens of static and fatigue lap-shear tests. FSW, friction stir welding; FSSW, friction stir spot welding; RSW, resistance spot welding. Source: Ref 5

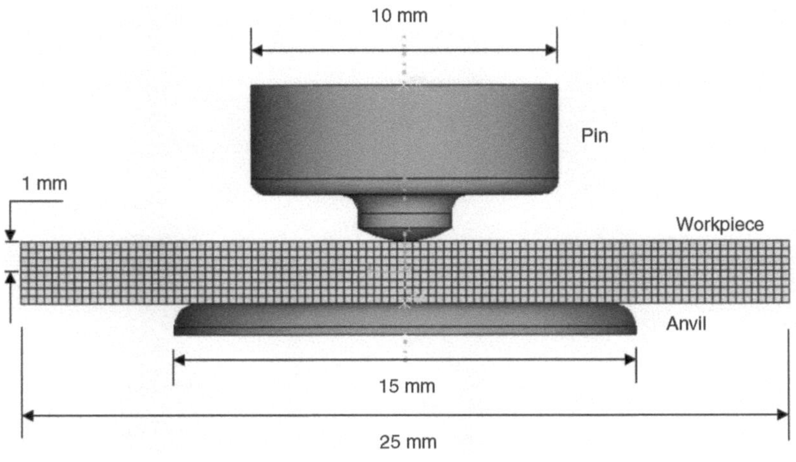


**Fig. 12.34** Mesh representation of two layers of workpiece with a pin and an anvil. Source: Ref 23

reduce the computational time, because the internal resistance of the rigid bodies to heat is negligible in comparison with the external resistance. Several assumptions were made and boundary conditions set accordingly (Ref 23).

Temperature distribution during the FSSW process is shown in Fig. 12.35. In this simulation, the maximum temperature after 1.235 s is 491.3 °C (916.3 °F). Prior experiments on linear FSW of aluminum alloys (Ref 24, 25) suggest that the actual temperatures in the stir region would be 80% of the melting temperature, which is 460 °C (860 °F) for aluminum 6061-T6. The result of maximum temperature is approximately 6.8% higher than the theoretical temperature due to the assumption of isothermal rigid bodies of the pin and anvil. Figure 12.36 shows that the maximum temperature occurs approximately 3 mm from the center point of the workpiece, after which it starts to decrease away from the center point.

The simulation results for the von Mises stress profile indicate that it is lowest in the nugget region and begins to increase and finally stabilize away from the center point of the workpiece (Ref 23). It is believed that further refinements to include the tool (pin) and the anvil as elements that absorb and release heat during the operation would enhance the accuracy of the model. This model, along with the adaptive remesh option, leads the way to simulate the complex and dynamic phenomenon of spot FSW.

**Refill FSSW.** Muci-Küchler et al. (Ref 26) reported results on a simplified isothermal 3-D finite element model (FEM) of the initial plunge phase of the FSSW process. The model, based on a solid mechanics approach, was developed using the commercial software ABAQUS/Explicit.

The reason to focus on the solid mechanics aspects first is that modeling the material as a

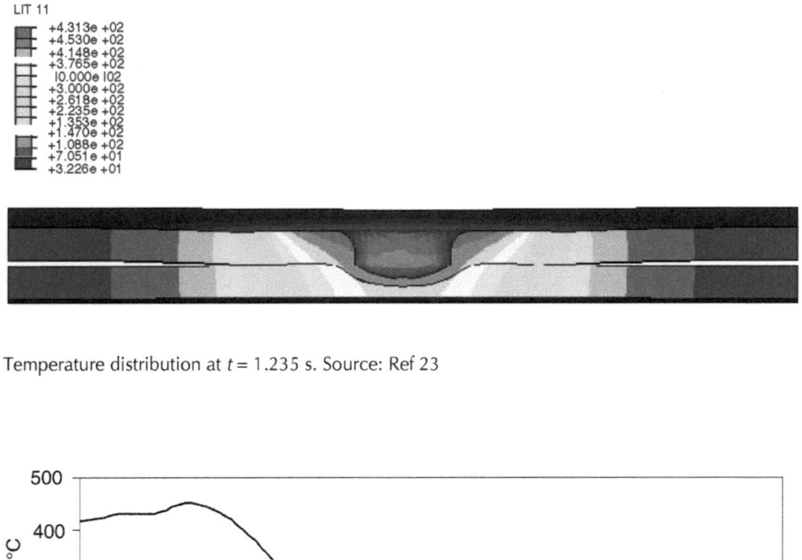


**Fig. 12.35** Temperature distribution at $t$ = 1.235 s. Source: Ref 23

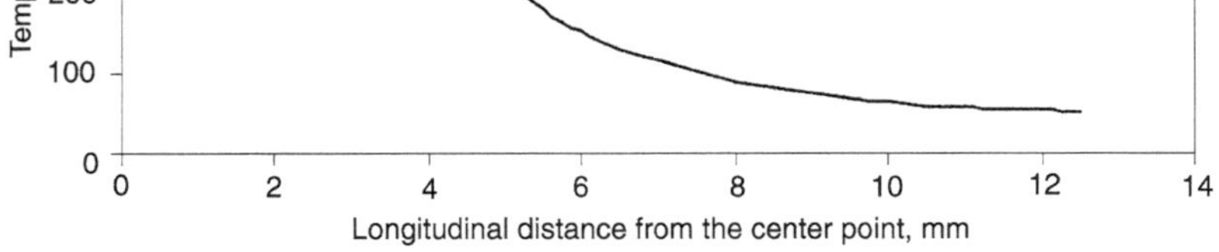


**Fig. 12.36** Graph of temperature versus radial distance from the center of the tool. Source: Ref 23

solid presents challenges when a Lagrangian or an arbitrary Lagrangian-Eulerian FEM formulation is used. The large deformations caused by the combined effect of the translation and rotation of the tool could lead to problems with the numerical method if the elements close to the tool become excessively distorted. Although using a code that models the plates employing an Eulerian approach could be a possible solution, the commercial FEM programs commonly used to solve solid mechanics problems do not offer that alternative (Ref 26). The mechanical behavior of the material of the plates is represented using an elastic/perfectly plastic constitutive relation in which the material properties correspond to the value of the temperature assigned to the plates.

In the simulations, a linear elastic/perfectly plastic constitutive relation was used for the material of the plate, and the effect of the strain rate on the mechanical properties was not taken into consideration. An adaptive meshing technique was also employed to reduce the distortion of the elements. Because the deformations of the pin, shoulder, and clamp are minimal compared to those of the plate, those components were considered as rigid, and the surface of each one was modeled using rigid shell elements. The general contact algorithm available in ABAQUS/Explicit was used to define the interaction between the components of the tool and the plate. The frictional contact has been modeled based on a modified Coulomb friction law. A maximum shear-stress value was defined that controls the stick/slip behavior of the material around the pin. For the boundary conditions, an independent reference node was defined for each component of the tool, and the boundary conditions corresponding to that component were applied to its reference node. The motion of the clamp was constrained in all directions, and it was in contact with the plate from the beginning of the simulation. The motion of the pin was constrained in all directions except the translation and rotation about the vertical axis. The bottom face of the plate was constrained in *Y*, the right and left faces in *X*, and the front and back faces in *Z*. For the simulations, the velocity control method was considered, and the pin was plunged with a constant velocity.

A square plate was considered for the simulation run. The plunge velocity of the pin was 25.4 mm/min (1.0 in./min), and its angular velocity was 800 rpm. The pin was 12.7 mm (0.5 in.) long and had a diameter of 4.75 mm (0.19 in.) with fillets; the shoulder was not included in the simulation. The material used for this model was aluminum 7075-T6, and information about its temperature-dependent material properties was taken from graphs provided in the MIL-HDBK-5H (Ref 27). Temperature-dependent data were extrapolated appropriately wherever required. The value assigned to the friction coefficient was 0.64, and the plunge depth was 0.3175 mm (0.0125 in.).

Figure 12.37 shows a minimum amount of flash generated during the plunge experiment. Figures 12.38 and 12.39 correspond to results from the numerical simulation for the same cross section as Fig. 12.37. Those figures also indicate minimal flash, which was in agreement with the experimental results. Figures 12.39 and 12.40 show the symmetric distribution of stresses obtained as the pin plunges through one-quarter of the top plate. The deformed geometry plot presented in Fig. 12.41 indicates the motion of the plate material obtained during the plunge of the pin. The arrow plot of the velocity vector shown in Fig. 12.42 provides a convenient way to visualize the material flow during the process. It also shows how the plate material tends to be stirred as the pin plunges. Based on the results of the simulation and the experiment, it can be inferred that, for the case under consideration, the rotation of the tool did not have a substantial effect on the material flow. The frictional force generated at the bottom of the pin is directly related to the plunge force. As the pin plunges into the material, the increase in the plunge force originates a corresponding increase in the frictional force. For small plunge forces, the material flow is very

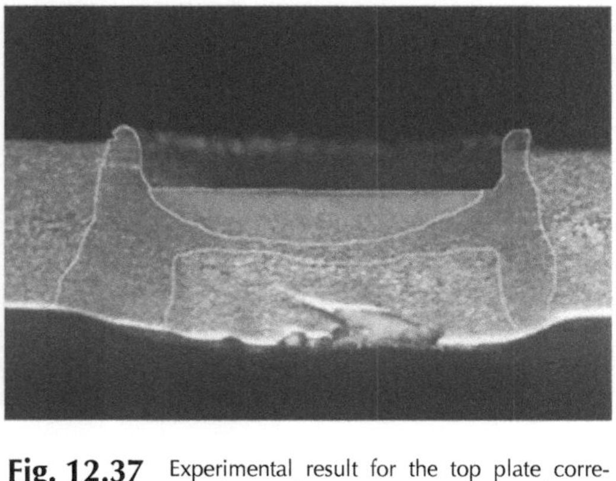

**Fig. 12.37** Experimental result for the top plate corresponding to the plunge test. Source: Ref 26

similar to one corresponding to a forging process. As can be seen in Fig. 12.43, the predicted values of the vertical plunge force as a function of plunge depth are close to the ones obtained from the experiment. The difference in the values could be attributed to the higher temperature at which the spot weld was done in the experiment. The average torque that was recorded during the experiment also compared well with the values obtained in the numerical simulation. The cyclic nature of the forces measured may suggest a stick/slip condition.

Subsequent simulations were run with varying process parameters, the results of which were in agreement with the experimental runs. Furthermore, Itapu et al. (Ref 28) reported a 3-D isothermal FEM of the plunge phase of a refill FSSW process using ABAQUS/Explicit.

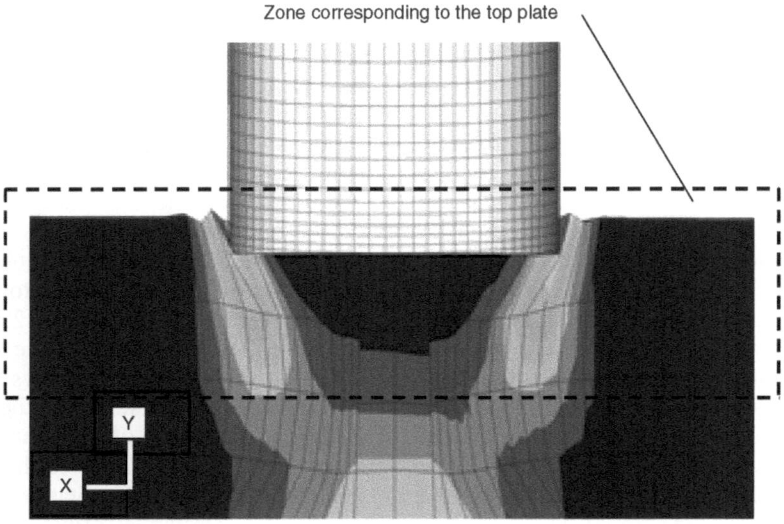


**Fig. 12.38** Equivalent plastic strains at 0.75 s. Source: Ref 26

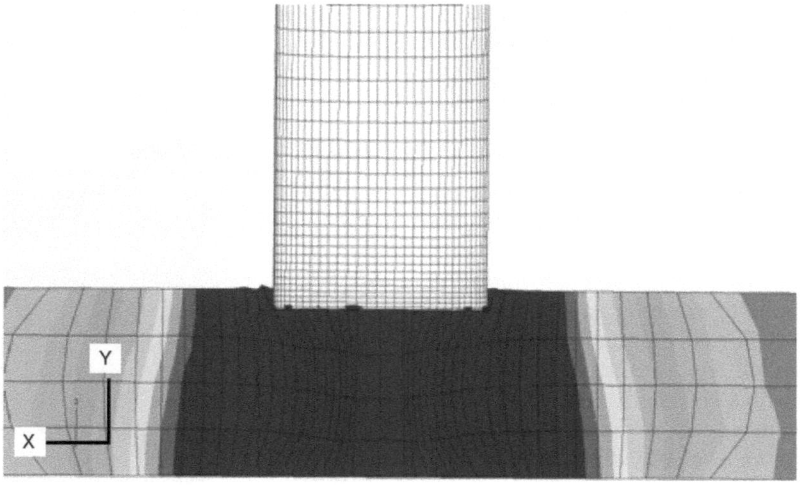


**Fig. 12.39** The von Mises stresses at 0.75 s. Source: Ref 26

Deformations, stresses, and strains induced in the plates being spot welded were computed. Virtual tracers were also incorporated in the simulation in an attempt to visualize the material flow near the tool. The authors reported a good correlation between the experimental and simulation results obtained.

## 12.4 Advancements in FSSW

**FSSW in Advanced High-Strength Steel.** The conventional electric RSW process can be problematic for many new high-performance lightweight structural materials such as aluminum alloys and advanced high-strength steels (AHSS) (Ref 29, 30). The great emphasis on safety and vehicle weight reduction to improve fuel efficiency has been driving the increased use of AHSS in automobile body construction. The biggest technology barrier inhibiting the use of RSW for AHSS is the profound weld property degradation (Ref 29, 31–33). Due to the extremely high cooling rate in RSW, the weld nugget region of AHSS would develop highly brittle microstructures and is prone to solidification-related weld cracks/defects. However, past work on linear FSW has shown that steels are much more difficult to friction stir weld than aluminum alloys (Ref 34). The technical difficulties arise from the very fundamental aspect of the FSW process: compared to aluminum alloys, FSW of AHSS must operate at much higher temperatures and requires much

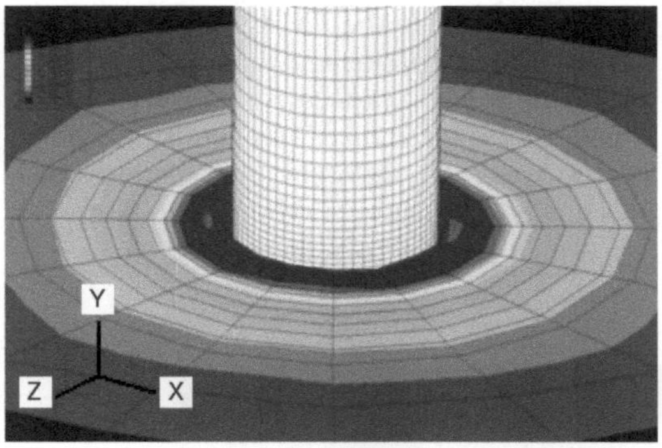


Fig. 12.40 The von Mises stress distribution at 0.75 s. Source: Ref 26

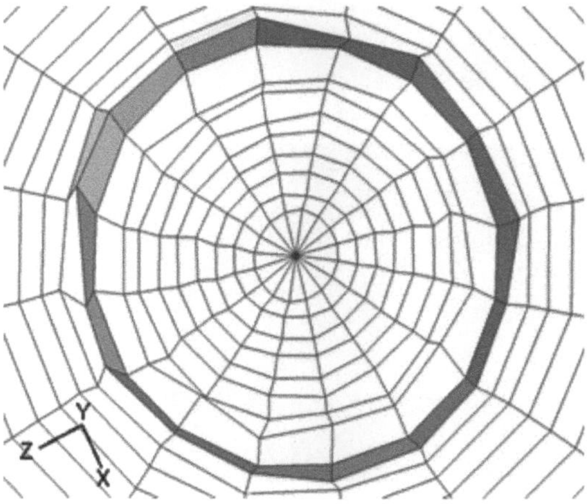


Fig. 12.41 Deformed geometry plot indicating the flow of material at 0.75 s. Source: Ref 26

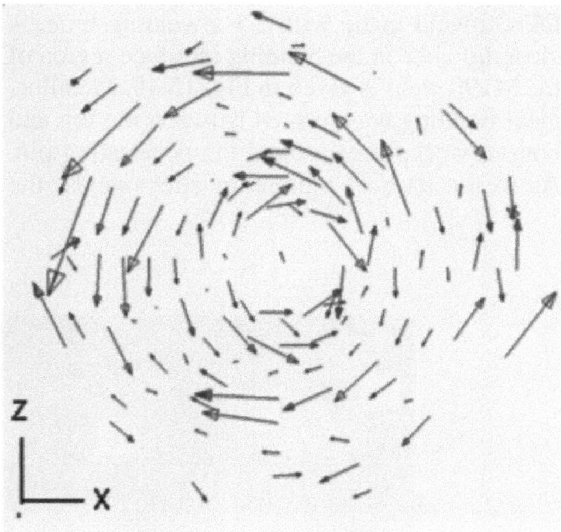


Fig. 12.42 Arrow plot of the nodal velocities on the plate surface at 0.75 s. Source: Ref 26

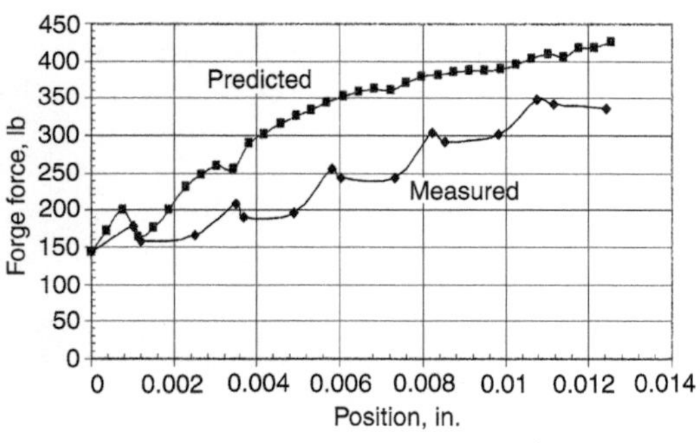


Fig. 12.43 Comparison of experimental and predicted plunge forces. Source: Ref 26

higher mechanical loading for plunging and stirring. These technical difficulties are also expected for FSSW.

Feng et al. (Ref 35), conducted a preliminary study to investigate the feasibility of FSSW of AHSS sheet metal. The objective was to weld 600 MPa (87 ksi) dual-phase steel and 1310 MPa (190 ksi) martensitic steel. A single tool, made of polycrystalline cubic boron nitride, survived over 100 welding trials without noticeable degradation and wear. The tool had a tapered pin, 2.0 mm (0.08 in.) long, and the shoulder was 10 mm (0.4 in.) in diameter. Solid-state metallurgical bonding was produced with welding time in the range of 2 to 3 s. Tensile-shear and cross-tension mechanical testing was performed for selected welding conditions to evaluate the mechanical strength of the joints produced.

Figure 12.44 shows the overall cross-sectional views of both the M190 weld and DP600 weld made with 2.1 s welding time. A close-up view in the bonding interface region of the M190 weld is given in Fig. 12.45. Metallurgical bonding was formed between the top and bottom workpieces around the penetrating pin. As in the case of aluminum alloy welds, the material from the bottom piece was pushed up by the plunging action of the rotating pin, causing the workpiece interface to bend upward and form a hook. The solid-state phase transformations that occur in carbon steels during cooling make it difficult to directly observe details of the stirring/mixing of the material between the two sheets. The width of the bonding ligament, a critical factor determining the strength of the weld, was relatively small in this study.

The martensitic M190 weld shows considerable softening outside the stir zone. The minimum hardness, approximately 200 HV, was located approximately 5 mm away from the weld center, corresponding to the shoulder radius of the tool. However, the hardness in the stir zone was fully recovered back to the 430 HV base-metal level. The minimum hardness location was located quite far away from the bonding region at the interface. The softened region was outside the TMAZ, where substantial plastic deformation and material flow occur during the welding process.

Due to the differences in chemistry, DP600 steel showed very different microhardness profiles under the same welding condition. The

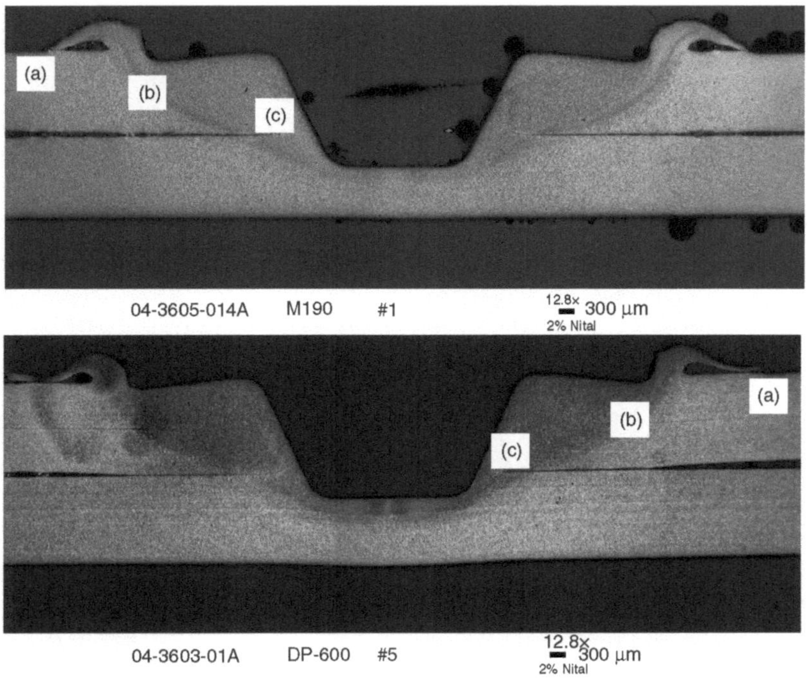


**Fig. 12.44** Cross section of friction stir spot weld. Top: M190; bottom: DP600. Welding time: 2.1 s. Source: Ref 35

softening was relatively insignificant compared to the base-metal microhardness level; the softening was mostly outside the shoulder diameter. On the other hand, the stir zone appeared to be hardened. The maximum hardness was approximately 250 HV, compared to the base-metal average of 210 HV. This variation of the microhardness can be related to the microstructural changes in the different regions of the weld. The resulting microstructure in the bonded region also suggests that the material flow and bonding takes place when the material is fully austenitized. Such information would be important for the future process and tool material development for FSSW of AHSS.

The lap-tensile test for the two materials showed similar value for shear strength. The shear strength increased with increase in weld time. It was also pointed out that the welding process conditions produced relatively small bonding ligament widths, thereby limiting the tensile strength levels of the joint. It is expected that substantial improvement in joint strength can be achieved if the bonding ligament width can be increased through further process development and modifications to the tool geometry.

**FSSW of Aluminum-Magnesium Alloys.** Fusion welding of magnesium-base alloys is complicated due to problems such as hydrogen porosity formation and solidification cracking in weld deposits and liquation cracking in heat-affected zone regions (Ref 36–39). Hence, Su et al. (Ref 40) evaluated FSSW as a joining technique to weld aluminum 5754 and AM60 sheet. The objective in this particular study involved determining the factors that determine joint mechanical properties.

**Fig. 12.45** Magnified section view of the bonding interface region. M190 steel. Welding time: 2.1 s. Source: Ref 35

The 1.5 mm (0.06 in.) thick sheets of aluminum 5754 and thixomolded AM60 base materials were used during this investigation. The tool was heat treated to a hardness of 46 to 48 HRC and coated with TiAlN to minimize wear during FSSW trials. Mathematical equations were used to calculate the energy that was produced when the pin was forced into the workpiece and the energy that was produced due to tool rotation. Tool revolutions per minute, plunge depth, and plunge speed were the process parameters that were varied. Joint mechanical properties were evaluated by measuring the peak fracture load during overlap-shear testing at a loading rate of 1 mm/min.

In aluminum 5754 spot welds, the stir zone had a fine equiaxed structure having a grain size <10 μm, while the TMAZ had a microstructure comprising a mixture of deformed and partially recrystallized grains. In AM60 base material, the stir zone comprised fine-grained (<10 μm) $\alpha$ phase, while the TMAZ contained elongated primary $\alpha$ particles and partially recrystallized $\alpha$ grains. The mode of specimen failure changed when welding parameters varied and the FSSW joints produced contained discontinuities, which affected test specimen fracture during overlap-shear testing. In addition, the tool shoulder produced increased thinning of the upper sheet when the penetration depth was increased during spot welding. This may have facilitated failure in some spot-welded joints.

Unbonded regions are formed when the oxidized surfaces of the two sheets contact each other but are not metallurgically bonded (Fig. 12.46a). Figure 12.46(b) shows the microstructural features observed at the extension of an unbonded region in a spot weld in aluminum 5754 base material. The $Al_2O_3$ oxide films originally present on the surfaces of the contacting aluminum alloy sheets are disrupted, producing a microstructure comprising unbonded regions, $Al_2O_3$ particles, and areas showing evidence of metallurgical bonding.

The influence of unbonded regions at the edges of completed welds on sample failure during overlap-shear testing is illustrated in Fig. 12.47. Failure initiated from unbonded regions located on either side of the spot weld, and as the fracture propagated, the transition from

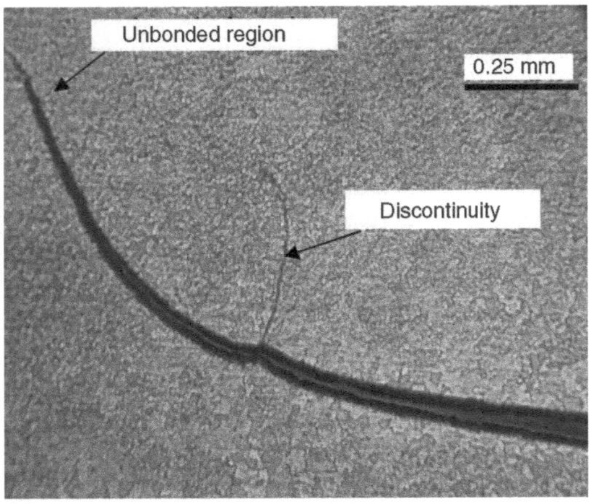


(a)

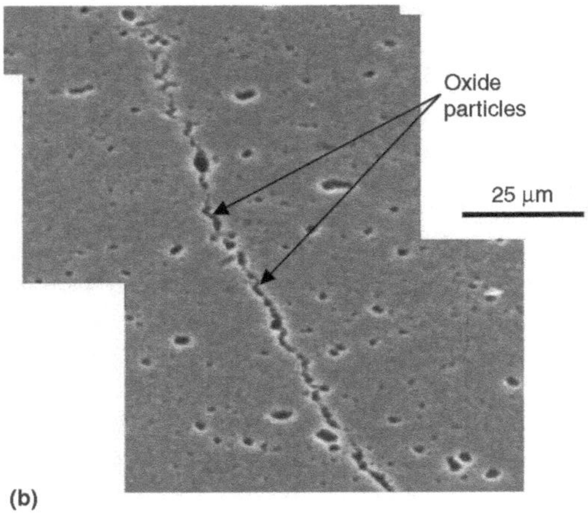


(b)

**Fig. 12.46** (a) An unbonded region and a discontinuity on periphery of unbonded region in friction stir spot welded aluminum 5754 base material. (b) Oxide particles located at the extension of the unbonded region formed in an aluminum 5754 spot weld. Source: Ref 40

shear to peel-mode loading promoted crack propagation in the upper and lower sheets. Failure in this particular test sample may have also been facilitated due to thinning of the upper sheet by the tool shoulder.

Higher fracture loads were produced when the energy input increased for spot welds in both AM60 and aluminum 5754 base materials. The energy input during FSSW also influenced the mode of fracture during mechanical testing. It has already been shown that the eutectic temperature is attained when AZ91D base material is FSSWed, and therefore, the rotating tool will be coated with an adhering eutectic that leads to the formation of intermingled lamellae adjacent to the stir zone when the tool is then used during welding of aluminum 5754 material (Ref 41). The intermingled regions formed in aluminum 5754 were found to be comprised of aluminum 5754 and a $Mg/Mg_{17}(Al,Zn)_{12}$ eutectic lamellae. Magnesium contamination markedly decreased the energy input during aluminum 5754 spot welding and hence reduced the projected bonded area adjacent to the keyhole periphery.

From the study, it was found that the joint mechanical properties were determined by the energy input during welding and by the projected area of the bonded region immediately adjacent to the keyhole periphery. The fracture load during overlap shear testing of welds increased when the projected bonded area immediately adjacent to the keyhole periphery and the energy input during welding increased. Partial pullout failure involving crack propagation from unbonded regions located on either side of the welded joint occurred in welds produced using high energy inputs. Also, magnesium contamination of FSSW tools had a markedly detrimental effect on the mechanical

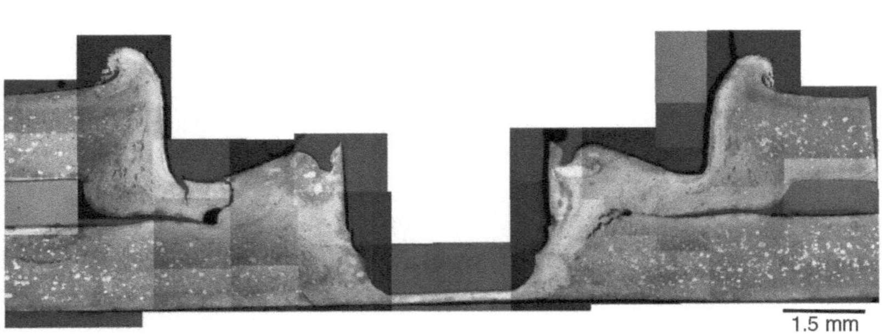


**Fig. 12.47** Partially failed overlap-shear specimen of AM60 base material, showing failure propagation into the upper and lower sheet materials. Source: Ref 40

properties of spot-welded aluminum 5754 base material.

**Energy Savings by Mazda.** Mazda Motor Corporation became the first auto manufacturer to apply FSW to the manufacture of aluminum body assemblies (Ref 3, 42). Mazda used FSW for the rear doors and hood of their RX-8 models. Figure 12.48 illustrates the welding setup for the rear door panel. An RX-8 rear door panel with friction stir spot welds is shown in Fig. 12.49.

Traditional resistance welding requires that a large current be instantaneously passed through the aluminum. This approach not only uses a large amount of electricity but also requires

**Fig. 12.48** Body panels welded together using friction stir spot welding. Source: Ref 3

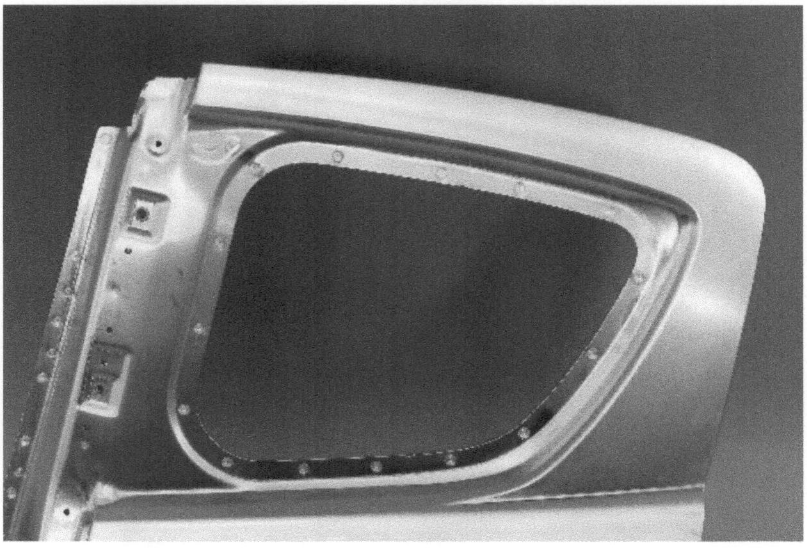

**Fig. 12.49** Close-up photo of a completed friction stir spot weld on an RX-8 aluminum rear door. Source: Ref 42

large, specialized equipment. Through the new spot joining method used by Mazda, they were able to overcome the disadvantages of RSW. Mazda reports that it has achieved a 40% reduction in equipment investment compared to that of resistance welding for aluminum (roughly the same level of investment is required for FSW of standard steel).

The only energy consumed using the friction welding technology is the electricity needed to rotate and apply force to the welding tool in order to create frictional heat. Because the process eliminates the need for the large current and coolant/compressed air required for conventional resistance welding, Mazda reported that the energy consumption was reduced by approximately 99% in the case of aluminum (and approximately 80% for steel) (Ref 3). This significantly reduces the impact on the environment while achieving the same or greater level of joint strength. Additionally, the welding method has simplified the overall joining system, because, unlike in resistance welding, a large current source and specialized joining equipment are not required. Figure 12.50 shows the cell layout for welding the Mazda RX-8 rear door panel.

**Energy Generation in FSSW.** Su et al. (Ref 43) investigated the energy generation and use during FSSW of aluminum 6061-T6 and AM50 sheet metals. With no dwell time, the rotating pin accounts for the majority of the energy generated when 6.3 mm (0.25 in.) thick aluminum 6061-T6 and AM50 sheet materials are spot welded. However, the contribution made by the tool shoulder increases significantly when a 4 s long dwell period is incorporated. The increased contribution made by the tool shoulder is due to the tool shoulder remaining in contact with stir zone material for a much longer period during the FSSW operation. Furthermore, only a small percentage of the total energy generated during tool rotation (approximately 4%) is required for stir zone formation during plunge testing of aluminum 6061-T6 and AM50 sheets. The remainder of the energy generated by tool rotation dissipates into the sheets being welded, the tool assembly, anvil support, clamp, and surrounding atmosphere. The presence of a thread on the rotating tool has negligible influence on the amount of energy generated during spot welding.

Three different tool designs were used in order to see the contribution of the tool geome-

**Fig. 12.50** A robot controls the friction spot welds in an aluminum door. Source: Ref 42

try on the heat input to the welds. The plunge depth was kept the same for all tool designs. Plunge speed and rotational speed were varied to see the effect of these parameters on the energy input. Earlier, Su et al. (Ref 44) used simple calorimetry to determine how much of the energy produced during tool rotation dissipated in the aluminum alloy sheets being spot welded. It was found that only 12.6% of the energy resulting from tool rotation dissipated into the aluminum alloy sheet material during FSSW of aluminum 6061-T6 sheet with a steel tool, clamp, and anvil support.

The stir zone dimensions of aluminum 6061-T6 and AM50 sheet materials are largely unaffected when the tool rotational speed increases from 1500 to 3000 rpm (using a plunge rate of 1 mm/s). The observation was similar when the rate of tool penetration increased from 1 to 10 mm/s (using a tool rotational speed of 3000 rpm). The tool shoulder accounts for approximately 30 and 34% of the energy generated during spot welding of 6.3 mm thick aluminum 6061-T6 and AM50 sheets without a dwell period (when the tool has a shoulder diameter of 10 mm, a pin diameter of 4 mm, and the rotational speed and plunge rate are 3000 rpm and 2.5 mm/s). In contrast, when a dwell time of 4 s is applied, the tool shoulder accounts for approximately 48 and 65% of the energy generated during spot welding of aluminum 6061-T6 and AM50 sheets. The increased contribution resulting from the tool shoulder is explained by the longer time of contact with stir zone material during the spot welding operation. The cross sections are shown in Fig. 12.51. It is evident that a longer dwell time results in a larger stir zone. Only a small percentage of the total energy generated during the FSSW operation is required for stir zone formation. The highest percentage utilization values during stir zone formation are approximately 4% during plunge testing of 6.3 mm thick aluminum 6061-T6 and AM50 sheets. The remainder of the energy resulting from tool rotation dissipates in the sheets being welded and in the tool assembly, anvil support, clamp, and surrounding atmosphere.

Gerlich et al. (Ref 45) examined the tool penetration phenomenon in detail. They concluded that this can be readily explained as a progression of wear events, from mild (delamination) wear through severe wear and finally to melt wear in material beneath the base of the rotating pin. Melt wear can occur under the rotating tool shoulder when there is sufficient penetration of the upper sheet produced during spot welding. Furthermore, during the experiments, the highest temperatures attained during FSSW of aluminum 6111 and AZ91 base materials were found to be close to the solidus temperatures of each base material.

**Design of Experiments on FSSW.** Hunt et al. (Ref 1) carried out a design of experiments on the effects of weld parameters on swing FSSW. Lap-shear specimens of automotive aluminum alloy A6022-T4 were welded using a C-frame welder. Welds were made with numerous weld parameters, such as tool pin length, tool rotating speed, tool plunging speed and depth, hold time, welding speed, and weld length. The results of this study showed that transverse welds have higher lap-shear strength than longitudinal welds, and lap-shear strength increases linearly with weld length. Increasing hold time also increased shear strength. The effects of pin length and revolutions per minute needed more investigation and would be studied further. The effects of eight factors on the shear

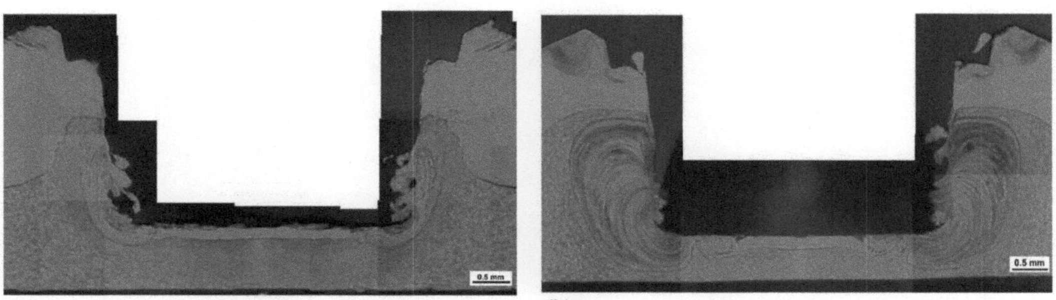


(a) (b)

**Fig. 12.51** Stir zone profile produced in aluminum 5754/aluminum 6111 spot welds with and without a dwell time. (a) With no dwell time applied. (b) With a dwell time of 2 s. Source: Ref 44 strength were studied using Taguchi methods with an L18 orthogonal array.

**Digital Shearography for Nugget Size Measurement.** Yang et al. (Ref 46) used an innovative optical technique of digital shearography to measure the nugget size. Digital shearography, a laser measuring technique based on digital data processing, phase-shifting techniques, and interferometry, has shown a great potential for nondestructive testing of spot welds. Digital shearography has a very high measuring sensitivity, and any anomaly in deformation of approximately 100 nm can be detected. This technique, however, measures relative deformation and not the absolute deformation, as holography does. Consequently, it is insensitive to rigid body movement and well suited for an on-line inspection.

## 12.5 FSSW Commercial Applications

**Friction Stir Spot Welding Aluminum Steel.** Mazda Motor Corporation says it has developed the world's first direct spot joining technology to weld aluminum and steel (Ref 47). Up until now, welding two different metals, such as aluminum and steel, has been a difficult task. However, by optimizing the rotating tool shape and joining characteristics, and by using galvanized steel on one side, joining aluminum and steel is possible. Figure 12.52 shows Mazda's aluminum-to-steel friction stir welded deck lid. Use of galvanized steel helps prevent the galvanic corrosion that would otherwise result from the contact of the two different types of metal. Mazda claims that this technology improves the potential of coupling aluminum parts to steel in vehicle bodies and helps lower the costs of production. The company adds that the technology contributed significantly to its vehicle weight-reduction efforts during the development of the new MX-5, where each gram of weight shed was counted.

**Innovative Use of Backing Plate.** Pan et al. (Ref 48) included an embodiment on the surface of the anvil to make a decorative imprint on the surface of the lower sheet during FSSW. This feature could be used as a design feature or an identification mark.

Friction stir spot welding uses a stationary anvil on the opposite side of the spinning tool. After the weld has been done, a flat dimple is left on the surface of the lower sheet. Pan et al. suggested that an embodiment could be

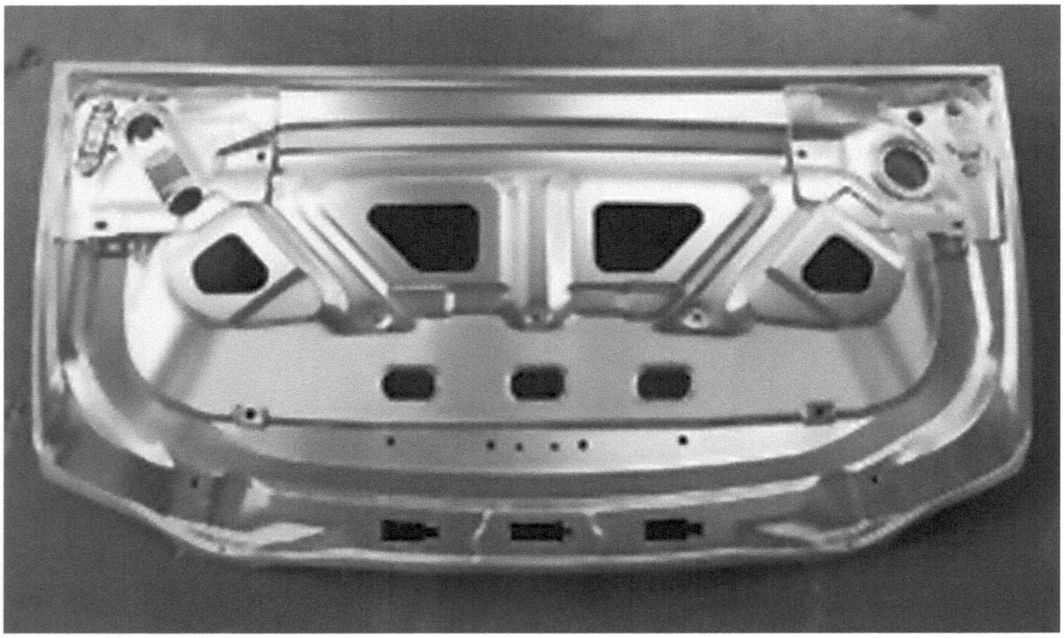

**Fig. 12.52** Mazda has used friction welding to join the aluminum deck lid to the steel bolt retainer on the new Mazda MX-5. Source: Ref 47

included on the surface of the anvil (Fig. 12.53) to make a decorative imprint or logo on the friction stir spot joint, as shown in Fig. 12.54. Decorative spot joints can be added as design features (as desired by the end user) or could be used for identification purposes (such as the imprint of vehicle identification numbers, on cars). Based on a previous study (Ref 49), it is claimed that higher joint strength may also be achieved with patterns on the anvil.

## 12.6 Conclusion and Future of FSSW

This chapter reviewed the current knowledge base and understanding in the developments of FSSW process, microstructure and properties, computer modeling, and its applications. With recent advancement in the research of this process, there has been steady progress made in capturing the physics of this complex phenomenon, both experimentally and numerically.

**Fig. 12.53** Example of decorative anvil for the spot friction welding process. Source: Ref 48

Unlike linear FSW, which is predominantly used for butt welding, most of the spot friction welding is done in the lap configuration. Spot FSW, a key contender to compete with existing spot welding techniques such as RSW, SPR, and TOX (Pressotechnik), has evolved strongly since the beginning of this decade when Mazda introduced it for the first time on a production line. The usual cycle time for a typical spot weld is on the order of a few seconds. It is during this short interval that the tool has to plunge into the workpiece, stir and metallurgically bond the material, and retract. So far, research has been done on optimizing key welding parameters such as tool rotation rate, plunge speed, target depth, and dwell time to better understand the influence of each parameter on the weld quality (which is typically mechanical strength of the joint).

Perhaps the critical parameter for spot welding is accurately controlling the plunge depth. There has been little study done on capturing the plunge phenomenon as the tool penetrates the workpiece. The plunge depth not only influences the appearance of the weld but strongly controls the joint strength and failure mode. Some preliminary work has been done in trying to capture the dynamics involved during the plunge period (Ref 50). This study looks at the formation of the nugget zone, first sheet thinning and hook formation for different plunge depths, eventually giving an insight as to how the weld zone is formed and grows.

Another critical issue relating to plunge depth control is the thermal expansion of the tool.

**Fig. 12.54** Spot friction welded sample with the normal pin hole on the top sheet but with a decorative imprint on the bottom surface of the lower sheet. Source: Ref 48

Starting from a tool at room temperature, spot welds subsequently cause the tool temperature to rise until it attains a steady-state value. During this transition, the tool "grows" due to thermal expansion. This, in turn, will cause the quality of welds that are produced in this transition period to vary drastically. An investigation into the effect of thermal expansion on spot welds has been carried out to determine the steady-state tool temperature and tool "growth" (Ref 51). A finite element analysis model was developed to numerically predict tool thermal expansion.

Having stated the aforementioned, FSSW has already found its place in commercial applications. Apart from Mazda, Toyota has implemented FSSW on the rear door hatch of its popular Prius hybrid vehicle. With several advantages of FSSW over conventional spot welding techniques, more original equipment manufacturers are now taking a serious look at implementing this technology in the production line. With fuel costs on the rise, the energy-saving potential of FSSW gives it a significant competitive edge over other welding techniques. Up until now, it was commonly believed that only lightweight materials (aluminum, magnesium) could be joined by using FSSW; however, that has been proved wrong with results already available for friction spot welding AHSS (Ref 35) and also joining aluminum steel, which, until now, was a technically challenging process. Nearly a decade and a half after it was invented, FSW is now finding its way as a potential joining technique in many applications, and it is believed that this will be one of the forefront joining techniques in the manufacturing sector in the years to come.

## REFERENCES


1. F. Hunt, H. Badarinarayan, and K. Okamoto, "Design of Experiments for Friction Stir Stitch Welding of Aluminum Alloy 6022-T4," Paper 2006-01-0970, 2006 SAE World Congress (Detroit, MI), Society of Automotive Engineers, 2006
2. T. Iwashita. Method and Apparatus for Joining, U.S. Patent 6,601, 751, 2003
3. Mazda the First to Use Friction Stir Welding for Aluminum Body Assembly, *Aluminum Now Online,* Vol 5 (No. 3), May–June 2003, The Aluminum Association, http://www.aluminum.org/ANTamplate.cfm?IssueDate=05/01/2003&Template=/ContentManagement/ContentDisplay.cfm&ContentID=7156 (accessed 8/17/06)
4. C. Shilling and J. dos Santos, Method and Device for Joining at Least Two Adjoining Work Pieces by Friction Welding, U.S. Patent Application 2002/0179682
5. K. Okamoto, F. Hunt, and S. Hirano, "Development of Friction Stir Welding Technique and Machine for Aluminum Sheet Metal Assembly," Paper 2005-01-1254, 2005 SAE World Congress (Detroit, MI), Society of Automotive Engineers, 2005
6. R. Sakano, K. Murakami, K. Yamashita, T. Hyoe, M. Fujimoto, M. Inuzuka, Y. Nagao, and H. Kashiki, Development of Spot FSW Robot System for Automobile Body Members, *Proceedings of the Third International Symposium of Friction Stir Welding* (Kobe, Japan), TWI, Sept 27–28, 2001
7. C.D. Allen, and J.A. Arbegast, "Evaluation of Friction Spot Welds in Aluminum Alloys," Paper 2005-01-1252, 2005 SAE World Congress (Detroit, MI), Society of Automotive Engineers, 2005
8. P.-C. Lin, S.-H. Lin, J. Pan, T. Pan, J.M. Nicholson, and M.A. Garman, "Microstructures and Failure Mechanisms of Spot Friction Welds in Lap-Shear Specimens of Aluminum 6111-T4 Sheets," SAE Technical Paper 2004-01-1330, Society of Automotive Engineers
9. P. Thornton, A. Krause, and R. Davies, Aluminum Spot Weld, *Weld. J.,* Vol 75, 1996, p 101s–108s
10. A. Gean, S.A. Westgate, J.C. Kucza, and J.C. Ehrstorm, Static and Fatigue Behavior of Spot-Welded 5182-0 Aluminum Alloy Sheet, *Weld. J.,* Vol 78, 1999, p 80s–86s
11. N. Blundell, L. Han, R. Hewitt, and K. Young, "The Influence of Paint Bake Cycles on the Mechanical Properties of Spot Friction Joined Aluminum Alloys," SAE Technical Paper 2006-01-0968, Society of Automotive Engineers
12. L. Han, Y.K. Chen, A. Chrysanthou, and J.M. O'Sullivan, The Influence of Paint-bake Cycle on the Mechanical Behavior of Self-Piercing Riveted Aluminum Alloy Joints, *Sheet Metal 2003: Proceedings of the International Conference,* 2003

13. P.-C. Lin, J. Pan, and T. Pan, "Fracture and Fatigue Mechanisms of Spot Friction Welds in Lap-Shear Specimens of Aluminum 6111-T4 Sheets," SAE Technical Paper 2005-01-1247, Society of Automotive Engineers
14. P.-C. Lin, J. Pan, and T. Pan, "Investigation of Fatigue Lives of Spot Friction Welds in Lap-Shear Specimens of Aluminum 6111-T4 Sheets Based on Fracture Mechanics," SAE Technical Paper 2005-01-1250, Society of Automotive Engineers
15. P.-C. Lin, J. Pan, and T. Pan, "Fatigue Failures of Spot Friction Welds in Aluminum 6111-T4 Sheets under Cyclic Loading Conditions," SAE Technical Paper 2006-01-1207, Society of Automotive Engineers
16. S. Zhang, Stress Intensities at Spot Welds, *Int. J. Frac.,* Vol 88, 1997, p 167–185
17. D.-A. Wang, P.-C. Lin, and J. Pan, Geometric Functions of Stress Intensity Factor Solutions for Spot Welds in Lap-Shear Specimens, *Int. J. Solids Struct.,* Vol 42 (No. 24–25), Dec 2005, p 6299–6318
18. S.G. Arul, T. Pan, P.-C. Lin, J. Pan, Z. Feng, and M.L. Santella, "Microstructures and Failure Mechanisms of Spot Friction Welds in Lap-Shear Specimens of Aluminum 5754 Sheets," SAE Technical Paper 2005-01-1256, Society of Automotive Engineers
19. S.-H. Lin, J. Pan, T. Tyan, and P. Prasad, A General Failure Criterion for Spot Welds under Combined Loading Conditions, *Int. J. Solids Struct.,* Vol 40, 2003, p 5539–5564
20. P.-C. Lin, S.-H. Lin, and J. Pan, "Modeling of Plastic Deformation and Failure near Spot Welds in Lap-Shear Specimens," SAE Technical Paper 2004-01-0817, Society of Automotive Engineers, 2004
21. T.A. Freeney, S. R. Sharma, and R.S. Mishra, "Effect of Welding Parameters on Properties of 5052 Al Friction Stir Spot Welds," SAE Technical Paper 2006-01-0969, Society of Automotive Engineers
22. R.S. Mishra and Z.Y. Ma, Friction Stir Welding and Processing, *Mater. Sci. Eng. R,* Vol 50, 2005, p 1
23. M. Awang, V.H. Mucino, Z. Feng, and S.A. David, "Thermo-Mechanical Modeling of Friction Stir Spot Welding (FSSW)," SAE Technical Paper 2006-01-1392, Society of Automotive Engineers, 2006
24. J.C. McClure, Z. Feng, W. Tang, J.E. Gould, L.E. Murr, and X. Guo, A Thermal Model for Friction Stir Welding, *Fifth International Conference on Trends in Welding Research,* ASM International, 1998, p 590–595
25. H. Schmidt, J. Hattel, and J. Wert, An Analytical Model for the Heat Generation in Friction Stir Welding, *Model. Simul. Mater. Sci. Eng.,* 2004, p 143–157
26. K.H. Muci-Küchler, S.S.T. Kakarla, W.J. Arbegast, and C.D. Allen, "Numerical Simulation of the Friction Stir Spot Welding Process," SAE Paper 2005-01-1260, Society of Automotive Engineers
27. *Metallic Materials and Elements for Aerospace Vehicle Structures,* MIL-HDBK-5H, *Military Handbook,* U.S. Department of Defense, Dec 1998
28. S.K. Itapu and K.H. Muci-Küchler, "Visualization of Material Flow in the Refill Friction Stir Spot Welding Process," SAE Paper 2006-01-1206, Society of Automotive Engineers
29. S.G. Shi and S.A. Westgate, "Resistance Spot Welding of High-Strength Steel Sheet," Corporate Research Program Report 767/2003, The Welding Institute, Cambridge, U.K., 2003
30. D.J. Spinella, J.R. Brockenbrough, and J.M. Fridy, Trends in Aluminum Resistance Spot Welding for the Automotive Industry, *Weld. J.,* Vol 84 (No. 1), 2005, p 34–40
31. S. Ferrasse, P. Verrier, and F. Meesemeacker, Resistance Spot Weldability of High-Strength Steels for Use in Car Industry, *Weld. World,* Vol 41 (No. 2), 1998, p 177–195
32. K. Yamazki, K. Sato, and Y. Tokunaga, Static and Fatigue Strength of Spot Welded Joint in Ultrahigh-Strength, Cold-Rolled Steel Sheets, *Weld. Int.,* Vol 14 (No. 7), 2000, p 533–541
33. W. Peterson, Dilution of Weld Metal to Eliminate Interfacial Fractures of Spot Welds in High and Ultra-High Strength Steels, *Proc. Int. Conf. Advances in Welding Technology* (Columbus, OH), EWI/AWS/NJC/SME/TWI, Sept 17–19, 1997, p 331–346

34. J.E. Gould, T.J. Lienert, and Z. Feng, "Recent Developments in Friction Stir Welding," SAE Technical Paper Series 981875, Society of Automotive Engineers, 1998
35. Z. Feng, M.L. Santella, S.A. David, R.J. Steel, S.M. Packer, M. Kuo, and R.S. Bhatnagar, "Friction Stir Spot Welding of Advanced High-Strength Steels—A Feasibility Study," SAE Technical Paper 2005-01-1248, Society of Automotive Engineers
36. K. Nakata, S. Inoki, Y. Nagano, T. Hashimoto, S. Johogan, and M. Ushio, Friction Stir Welding of AZ91 Thixomolded Sheet, *J. Jpn. Inst. Light Met.*, Vol 10 (No. 51), Oct 2001, p 528–533
37. M. Avedesian, *Magnesium and Magnesium Alloys,* ASM International, 1999, p 106–118
38. S. Lathabai, K.J. Barton, D. Harris, P.G. Lloyd, D.M. Viano, and A. McLean, Welding and Weldability of AZ31B by Gas Tungsten Arc and Laser Beam Welding Processes, *Magnesium Technology 2003,* H.I. Kaplan, Ed., TMS, 2003
39. A. Stern, A. Munitz, and G. Kohn, Application of Welding Technologies for Joining of Mg Alloys: Microstructure and Mechanical Properties, *Magnesium Technology 2003,* H.I. Kaplan, Ed., TMS, 2003
40. P. Su, A. Gerlich, and T.H. North, "Friction Stir Spot Welding of Aluminum and Magnesium Alloy Sheets," Paper 2005-01-1255, 2005 SAE World Congress (Detroit, MI), Society of Automotive Engineers, 2005
41. A. Gerlich, P. Su, and T.H. North, Friction Stir Spot Welding of Mg-Alloys for Automotive Applications, *Magnesium Technology 2005,* N.R. Neelameggham, H.I. Kaplan, and B.R. Powell, Ed., TMS, 2005
42. R. Hancock, Friction Welding of Aluminum Cuts Energy Costs by 99%, *Weld J.,* Vol 83 (No. 2) Feb 2004, p 40
43. P. Su, A. Gerlich, T.H. North, and G.J. Bendzsak, "Energy Generation and Stir Zone Dimensions in Friction Stir Spot Welds," SAE Technical Paper 2006-01-0971, Society of Automotive Engineers
44. P. Su, A. Gerlich, T.H. North, and G.J. Bendzsak, Energy Utilization and Generation during Friction Stir Spot Welding, *Sci. Technol. Weld. Joining,* Vol 11 (No. 2), March 2006, p 163–169
45. A. Gerlich, P. Su, and T.H. North, Tool Penetration during Friction Stir Spot Welding of Al and Mg Alloys, *J. Mater. Sci.,* Vol 40, 2005, p 6473–6481
46. L. Yang, P.R. Samala, S. Liu, K.W. Long, and Y. Lee, Measurement of Nugget Size of Spot Weld by Digital Shearography, *Optical Diagnostics, Proc. of SPIE,* Vol 5880, 2005, p 50–57
47. A Friction Heat First for Mazda, Friction Stir Welding Joins Aluminum and Steel on MX-5, *Aluminum Now Online,* Vol 7 (No. 6), Nov–Dec 2005, The Aluminum Association, http://www.aluminum.org/ANTemplate.cfm?IssueDate=11/01/2005&Template=/ContentManagement/ContentDisplay.cfm&ContentID=9294 (accessed 8/17/06)
48. T. Pan, A. Joaquin, D.E. Wilkosz, L. Reatherford, J.M. Nicholson, Z. Feng, and M.L. Santella, Spot Friction Welding for Sheet Aluminum Joining, *Proc. Fifth Int. Symposium on Friction Stir Welding* (Metz, France), TWI, 2004
49. O.O. Popoola, J.B. Skogsmo, V.L. Reatherford, and D.E. Wilkosz, Ultrasonic Welding Apparatus, U.S. Patent 6523732B1, Feb 25, 2003
50. H. Badarinarayan, F. Hunt, and K. Okamoto, "Evaluation of Plunge Motion Study during Friction Stir Spot Welding," to be presented at TMS Annual Meeting, 2007
51. H. Badarinarayan, F. Hunt, and K. Okamoto, "Tool Thermal Expansion during Friction Stir Spot Welding," to be presented at SAE Congress, 2007

# CHAPTER 13

# Application of Friction Stir Welding and Related Technologies

William J. Arbegast
NSF Center for Friction Stir Processing (CFSP) & Advanced Materials Processing and Joining Center (AMP), South Dakota School of Mines and Technology

FRICTION STIR WELDING (FSW) is an innovative solid-state welding process invented in 1991 by The Welding Institute (TWI) (Ref 1). Friction stir welding can arguably be said to represent one of the most significant developments in joining technology over the last half-century (Ref 2). The initial development by TWI and its industrial partners under various group-sponsored projects focused on single-pass, complete joint penetration of arc-weldable and unweldable aluminum alloys up to 25 mm (1 in.) thick.

By 1995, FSW had matured to a point where it could be transitioned and implemented in the U.S. aerospace aeronautics, marine, ground transportation, and automotive markets. The many advantages of FSW compared to conventional arc welding have repeatedly been demonstrated with both improved joint properties and performance. Often, production costs are significantly reduced. Other times, FSW enables new product forms to be produced or skilled labor to be freed to perform other tasks. Research and development efforts over the last decade have resulted in improvements in FSW and the spin-off of a series of related technologies.

In the 1920s and 1930s, arc welding replaced rivets as the joining method for pressure vessels. Weld usage expanded through the 1940s with application to buildings, structures, and ships. By 2006, arc welding had evolved into an international industry, complete with welder education and certification programs and governed by extensive specifications, design criteria, and standards. A 2002 survey by the American Welding Society (AWS) estimated that U.S. manufacturing industries spend over $34.4 billion annually on arc welding of metallic materials, with an anticipated growth rate averaging 5 to 15% per year (Ref 3). The construction, heavy manufacturing, and light manufacturing industries make up the majority, with $25 billion in annual expenditures. Industry-wide repair and maintenance of welded structures is estimated to cost $4.4 billion annually. In doing so, these industries are a major consumer of energy and a producer of airborne emissions and solid waste.

An excellent state-of-the-art review of FSW technology is provided by Mishra and Ma (Ref 4) and is described in the other chapters of this book. Conventional arc welding of metals creates a structural joint by local melting and subsequent solidification. This normally requires the use of expensive consumables, shielding gas, and filler metal. The melting of materials is energy-intensive, and solidifying metals are often subject to cracking, porosity, and contamination. Undesirable metallurgical changes can occur in the cast nugget due to alloying with filler metals, segregation, and thermal exposure in the heat-affected zones (HAZs). These may result in degraded joint strengths, extensive and costly weld repairs, and unanticipated in-service structural failures. Solid-state (nonmelting) joining avoids these undesirable characteristics of arc welding.

## 13.1 Implementation Incentives

Friction stir welding is one such nonmelting joining technology that has produced structural joints superior to conventional arc welds in aluminum, steel, nickel, copper, magnesium, and titanium alloys. Friction stir welding produces higher strength, increased fatigue life, lower distortion, less residual stress, less sensitivity to corrosion, and essentially defect-free joints compared to arc welding. Because melting is not involved, shielding gases are not generally used, although argon gas may be used during the FSW of the higher-temperature alloys, mainly to protect ceramic and refractory pin tools from oxidation. Other expensive consumables and filler metals are not required. Simple argon environmental chambers and trailing shields are used during the FSW of titanium alloys to minimize interstitial pickup and contamination. Additional performance benefits are described in other chapters of this book.

The FSW researchers and producers (Ref 5) estimate that if 10% of the U.S. joining market can be replaced by FSW, then $1.28 \times 10^{13}$ Btu/year energy savings and 500 million lb/year greenhouse gas emission reductions can be realized. Hazardous fume emissions during the FSW of high-temperature and chromium-containing alloys are eliminated. Rockwell Scientific (Ref 6) reports emission levels of chromium, copper, manganese and $Cr^{6+}$ (<0.03, <0.03, <0.02, and <0.01 mg/mm$^3$, respectively) during FSW of ferrous alloys to be considerably lower than those measured during gas tungsten arc welding (0.25, 0.11, 1.88, and 0.02 mg/mm$^3$, respectively). The simplified processing, higher structural strength, increased reliability, and reduced emissions of FSW are estimated to create an annual economic benefit to U.S. industry of over $4.9 billion/year.

ESAB Welding, Inc. has compared the production costs of aluminum joints made using FSW and gas metal arc welding (GMAW). This study (Ref 7) identifies high-production applications with long, straight runs or applications using nonweldable aluminum alloys (2*xxx* and 7*xxx*) as ideally suited to FSW. The faster speed of FSW, lower distortion, elimination of consumables, and elimination of solidification-related defect repairs are cited as factors that reduce the production costs from $2.11/ft for GMAW to $1.27/ft for FSW. Other factors contributing to these lower production costs are reduced preweld preparation time and minimal postweld finishing and grinding. Simplified personnel training and reduced personnel safety considerations also reduce total production costs. The study does, however, indicate that the cost of FSW equipment and fixturing is approximately twice that of GMAW systems and can be a barrier to extensive FSW implementation.

## 13.2 Barriers to Implementation

The aeronautic and aerospace industries represent less than 1% ($300 million) of the total U.S. annual welding expenditures, because mechanical fastening is the joining method of choice. However, the bulk of FSW development dollars has been spent by these sectors. As a result, the broader automotive, marine, heavy manufacturing, light manufacturing, and construction markets for FSW implementation have been neglected.

As of January 2005, the FSW licenses granted by TWI (Fig. 13.1) were almost equally split between North America (36), Europe (37), and Asia (41), with no reported licensees in South America. Overseas, 68% of the licensees are industrial. In North America, only 36% of the licensees are industrial, with the remaining 64% being held by government laboratories, equipment manufacturers, and academic and research institutes (Ref 8). This suggests that industrial implementation of the FSW process in the United States is lagging behind the overseas industries. Several overriding issues have been identified as barriers to more extensive FSW implementation in U.S. markets:

- Lack of industry standards and specifications
- Lack of accepted design guidelines and design allowables
- Lack of an informed workforce
- High cost of capital equipment

### *13.2.1 Industry Standards and Specifications*

To address the lack of industry standards and specifications, in 1998, the AWS D17 Subcommittee began development of a specification for FSW. Other AWS and ISO committees have also begun preparation of industry standards and specifications (Table 13.1). These specifications, when released, will provide postweld acceptance criteria for both continuous friction stir welds and friction stir spot welds, design requirements, and equipment, operator, and procedure qualification and certification requirements. In the meantime, most FSW users have developed internal specifications for application to their products. In 2002, AJT, Inc. secured American Bureau of Shipping approval to use FSW in marine applications.

**Pin Tools, Process Parameters, and Essential Variables.** Standardization efforts should address pin-tool designs, process parameters, and essential variables. A wide variety of pin tools are currently being used, depending on the nature of the parts being joined. The three basic pin-tool categories are fixed, retractable, and adjustable self-reacting (Fig. 13.2). Within each category, there is considerable diversity in the pin-to-shoulder diameter ratios, thread pitch, pin frustum shape, and pin tip and shoulder feature designs. Pin-tool design affects the process forces, processing speeds, metal flow paths, and resultant joint quality and performance.

Fixed-pin-tool configurations are generally used for both straight and complex curvature joints, where sufficient access to the backside of the joint is available to assemble tooling to resist the downward process forces. The retractable-pin tool provides for processing of varying-thickness materials but again requires a backside anvil to react the process forces. In those instances where backside access is limited, the self-reacting-pin tool provides a solution where the material to be joined is pinched between the upper- and lower-pin-tool shoulders.

As an example of how pin-tool selection affects the FSW process, one study by Toskey et al. (Ref 10) investigated the effects of the pin-tool design on the fixturing and tooling requirements, process forces, essential variables, and joint quality during the fabrication of square box beam extruded aluminum "C" sections. In this study, two fixed-pin-tools and one self-reacting-pin-tool configurations were investigated (Fig. 13.3). The fixed-pin-tool configurations included a standard threaded (28 UNJF) cylindrical pin tool with a concave shoulder and a 3 to 1 shoulder-to-pin diameter ratio. The tapered-pin tool incorporates a 10° taper in the

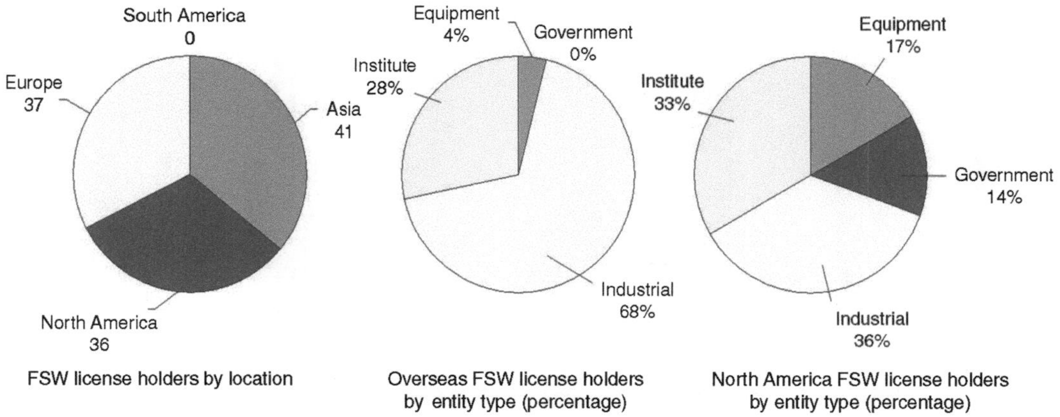


**Fig. 13.1** Demographics of friction stir welding (FSW) licensees as of January 2005

**Table 13.1 Current industry standards and specifications in development, as of 2006**

| Organization | Specification |
|---|---|
| AWS D8 Committee on Automotive Welding | AWS D8.17, "Specification for Automotive Weld Quality—Friction Stir Welding" |
| AWS D17 Committee on Welding in the Aircraft and Aerospace Industries | AWS D17.3, "Specification for Friction Stir Welding for Aerospace Applications" |
| AWS C6 Committee on Friction Welding | AWS C6.2, "Recommended Practices for Friction Welding" |
| International Institute of Welding, Subcommission III-B, Resistance and Solid-State Welding and Allied Joining Processes | ISO 25239, "Friction Stir Welding of Aluminum—General Requirements" |

Source: Ref 9

28 UNJF threaded section. These fixed pins are tilted 3.0° into the direction of welding.

The adjustable self-reacting-pin tool has a protruding double-half scroll on the top and bottom shoulders. The central pin section is threaded, with three flats ground 120° apart. The scroll direction feeds the material inward under a clockwise direction of rotation. The material being joined is pinched between the upper and lower shoulders with a zero tilt angle to form the required extrusion die cavity and constrain the various material flow paths (Ref 11).

The fixturing and tooling requirements differ for both the fixed-pin and adjustable self-reacting (ASR) pin tools (Fig. 13.4). The fixed-pin configurations require an internal mandrel to react the downward ($z$-force) processing loads through the hollow box beam structure into the support table below. This internal mandrel is not required for the ASR-pin tool. Additional fixturing prevents separation of the channels as the fixed-pin tool is plunged into and traverses along the joint. A starting hole is drilled into the start region of the joint, the ASR is inserted, and the bottom shoulder is installed. As an alternative, the ASR-pin tool can be slowly run-on into the joint from the end of the tube. Both of these approaches require fixturing to prevent sliding of the assembly in the direction of welding ($x$-force).

Process development trials with the cylindrical and tapered fixed pins and the ASR pin on

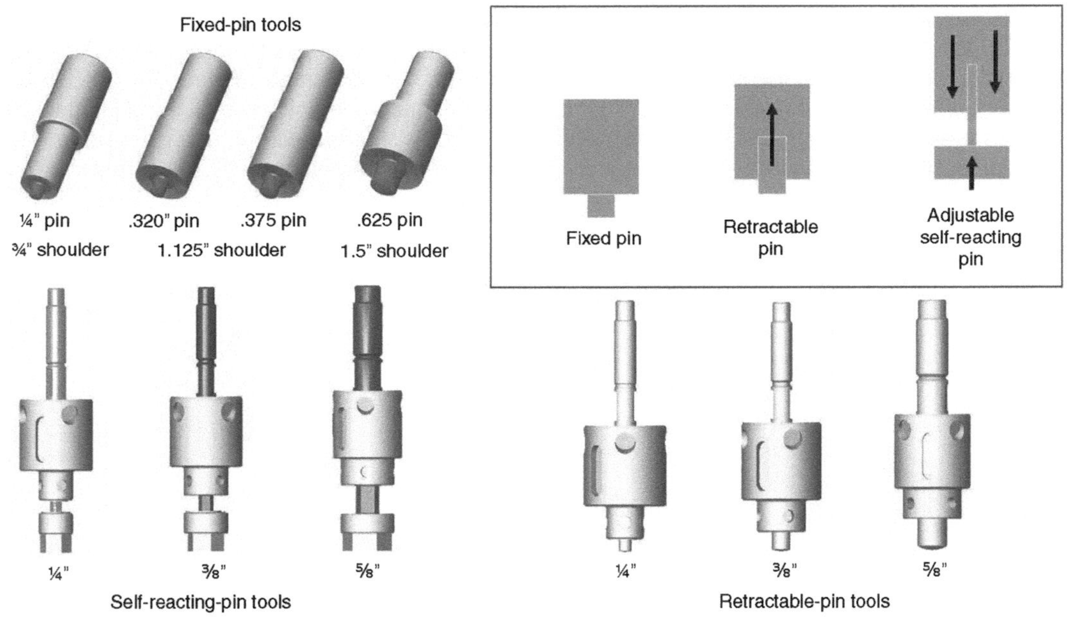


**Fig. 13.2** Schematics of fixed-, retractable-, and adjustable self-reacting-pin-tool configurations typical of current production applications

**Fig. 13.3** (a) Cylindrical fixed-pin tool, (b) adjustable self-reacting-pin tool, and (c) tapered fixed-pin tool used to join to the aluminum box beam sections

the 11 mm (0.43 in.) thick 5083-H111 box beam sections show the effects on essential variables, forces, and torques (Fig. 13.5). Three essential (controllable) variables are identified: rotation speed, travel speed, and forge or pinching ($z$) force for a selected pin tool, alloy, and joint type (Ref 12). Note that forge force applies when operating under load control, and this is substituted with shoulder plunge depth when position control methods are used (Ref 13). System responses include sliding ($x$) and separation ($y$) forces and spindle torque.

The ASR-pin tool optimal rotation speed was slower than the fixed-pin tools due to the added heat contribution of both the upper and lower shoulders. Excessively high rotation speeds result in extensive softening of the material outside the pin-tool footprint, loss of extrusion die cavity, and improper flow pattern formation. The cylindrical fixed-pin tool required higher rotation speeds than the ASR-pin tool to compensate for the added thermal loss due to the presence of the internal mandrel. Forward travel speeds for the standard fixed-pin and the ASR-pin tool were a factor of 10 slower than the tapered fixed-pin tool. The limiting material parameter for maximum forward travel speed is the flow stress at temperature and maximum allowable extrusion strain rate. These are governed by the die cavity formation rate and volume of material being swept around the pin with each cycle through each of the processing zones necessary to completely fill this cavity. The processing forces and torques are higher for the fixed-pin tools due to the lower processing temperatures. The very high traversing ($x$) and lateral ($y$) forces for the tapered fixed-pin tool result from the higher forging ($z$) forces necessary to maintain extrusion die cavity integrity at these high forward travel speeds.

Comparison of the resultant FSW nugget shape for the ASR-pin tool and the two fixed-

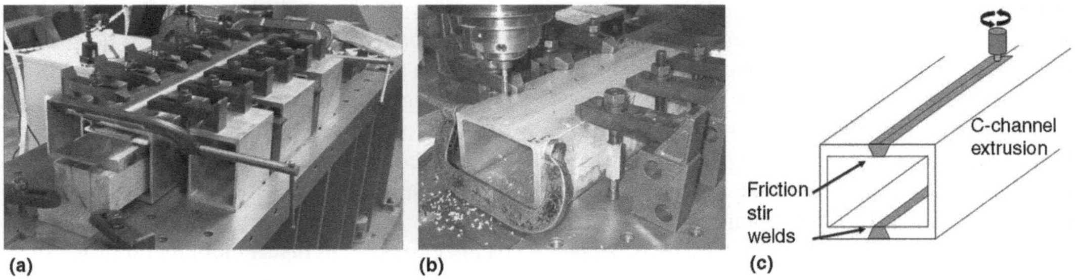


Fig. 13.4 Fixturing with internal backside anvil support for (a) fixed pin and (b) adjustable self-reacting pin without internal supports. (c) Two "C" channels of 11 mm (0.43 in.) 5083-H111 extrusions are joined to form a box beam section.

| | MTS System Corp. ASR pin tool | AMP ASR pin tool | Standard fixed pin | Fast tapered fixed pin |
|---|---|---|---|---|
| Thickness, mm | 6 | 11 | 11 | 11 |
| Rotation speed, rpm | 250 | 250 | 300 | 450 |
| Travel speed, (in./min) | 4 | 4 | 2 | 13.5 |
| Heel plunge, in. | 0.010 | 0.008 | 0.008 | 0.008 |
| Attack angle, degree | ... | ... | 3 | 3 |
| Tool $x$-force, lb | 750 | 210 | 630 | 3000 |
| Tool $y$-force, lb | 800 | 500 | 490 | 900 |
| Tool $z$-force, lb | 0 | 0 | 6400 | 15,000 |
| Spindle torque, in.-lb | 1000 | 500 | 760 | 1100 |
| Pinching force, lb | 2000 | 1450 | ... | ... |

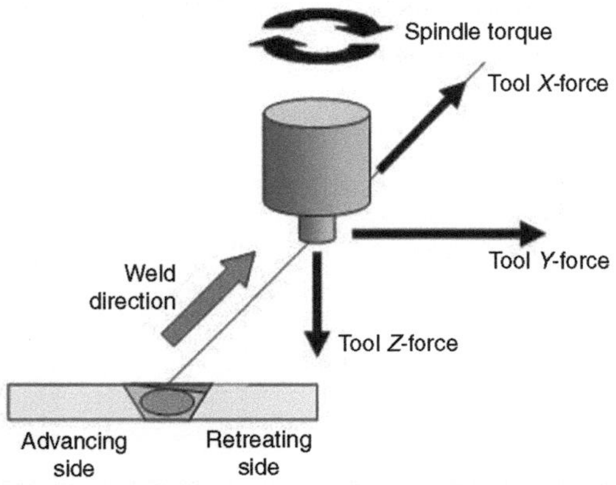


Fig. 13.5 Process forces and torques developed during friction stir welding of 5083-Hill extrusions "C" channels with fixed and adjustable self-reacting-(ASR) pin tools. AMP, Advanced Materials Processing and Joining Center

pin tools can also explain the differences observed in process forces (Fig. 13.6). For the ASR-pin tool, the larger pin diameter and higher and more uniform through-thickness heat input result in a larger nugget width with multiple horizontal flow-zone formation (lobes) with little vertical flow. This results in lower traversing forces (*x*) and lower spindle torque. The cylindrical fixed-pin tool shows a single lobed nugget zone width with more vertical flow and a smaller width than seen in the ASR-pin tool. Heating in this case is primarily from the flowing region below the shoulders and the extrusion of material directly around the pin tool. The depth of penetration of the nugget is influenced by the chill effects of the backside anvil, resulting in a thermal gradient within the joint and higher processing forces and torques. The tapered fixed-pin tool shows the smallest nugget width and complete penetration to the backside. This indicated a more localized heat input and less effect of backside chilling. The higher processing forces and torque for this pin-tool configuration is related to the faster forward travel speeds and colder processing conditions.

The influence of fixturing and clamping for these box beam welds is shown in mechanical testing of joints produced with the cylindrical fixed-pin tool FSW. The transverse tensile specimens were 25 mm (1 in.) wide, with no reduced section. Ultimate strength of the FSW averaged 307 MPa (44.5 ksi), with a parent-metal ultimate strength of 326 MPa (47.2 ksi), showing joint efficiency of 94%. The FSW yield strength averaged 150 MPa (21.8 ksi) compared to parent-metal yield strength of 190 MPa (27.5 ksi). The FSW elongation in a 51 mm (2 in.) gage length averaged 22.5% compared to the base-metal elongation of 21.0%. All samples failed outside the weld nugget in the HAZ on the retreating side of the weld. Each of the standard fixed-pin FSW in the 11 mm 5083-H111 box beams was measured for peaking and mismatch prior to tensile testing. For these FSWs, the high processing forces resulted in varying degrees of mismatch and peaking due to inadequate part restraint. This reduces the apparent yield strength of the joint due to induced bending stresses (Fig. 13.7). Specifications and standards for FSW should include an acceptable degree of peaking and mismatch.

**Process Control Algorithms.** Friction stir welding has been described as a "controlled-path metalworking process" consisting of distinct metallurgical processing regions (preheating, initial deformation, extrusion, forging, and cooldown region) ahead of, adjacent to, and behind the pin tool (Ref 12). Specifications and standards are more easily realized when the process is considered in this light. The cyclical flow patterns of material around the pin tool are constrained within the die cavity by the pin-tool upper shoulder, lower anvil, and the sidewall material, where the state of stress and temperature is insufficient to cause metal flow. The typically threaded and rotating pin tool acts as the extrusion die, with the volume of material flowing through the extrusion zone per revolution a function of pin tool geometry, processing parameters, temperature, and material flow stress. A theoretically optimal set of processing parameters can be calculated that maintains mass balance to prevent insufficient metal flow (volumetric void formation) and excessive flow (expulsion, nugget collapse, and flash formation).

Five distinct metal flow zones (Figure 13.8) have been identified within the transverse section of the FSW nugget dynamically recrystallized zone. Zones I and II represent the advancing and retreating side extrusion zones,

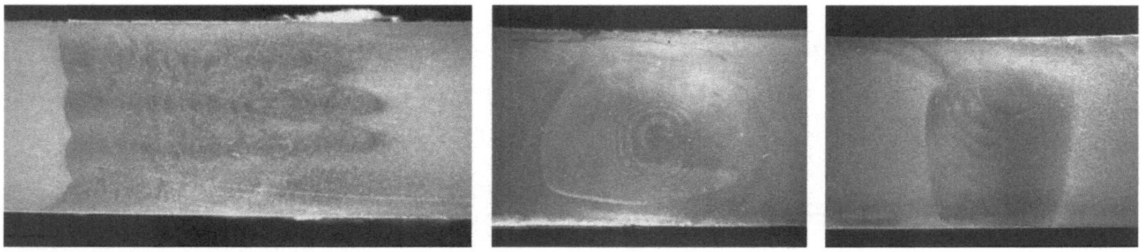

(a) (b) (c)

**Fig. 13.6** Metallurgical comparison of (a) adjustable self-reacting-pin-tool, (b) cylindrical fixed-pin tool, and (c) tapered fixed-pin tool FSW nugget formation in 11 mm (0.43 in.) 5083-H111 butt joints respectively, while zone III is the flow arm where material was dragged across the nugget top by the pin-tool shoulder. Zone IV is the swirl zone of material processing near and beneath the pin-tool tip. Zone I is filled in an interleaving pattern by material passing through the other zones. A zone V (recirculation zone) may form under very hot processing conditions, where the downward motion of material is greater than that which can be accommodated by the space behind the pin tool (excess flow), with the material changing direction and circulating back up toward the top surface, forcing increased deformation in the thermomechanically affected zone (TMAZ) located just outside the nugget. The presence of these distinct flow zones is readily apparent when aluminum FSW samples are subjected to high temperatures for short times and undergo time-incremented abnormal grain growth, which acts as in situ flow markers (Ref 14).

As with any metalworking process approximated by metal flow through converging channels, the flow rate and direction along slip lines are governed by the deformation zone geometry, hydrostatic stress state, and the local velocity vectors (Ref 15). Colegrove et al. (Ref 16) have shown the FSW process forces to be a function of the pin-tool geometry, stick-slip conditions, and the ratio of the tool area to swept

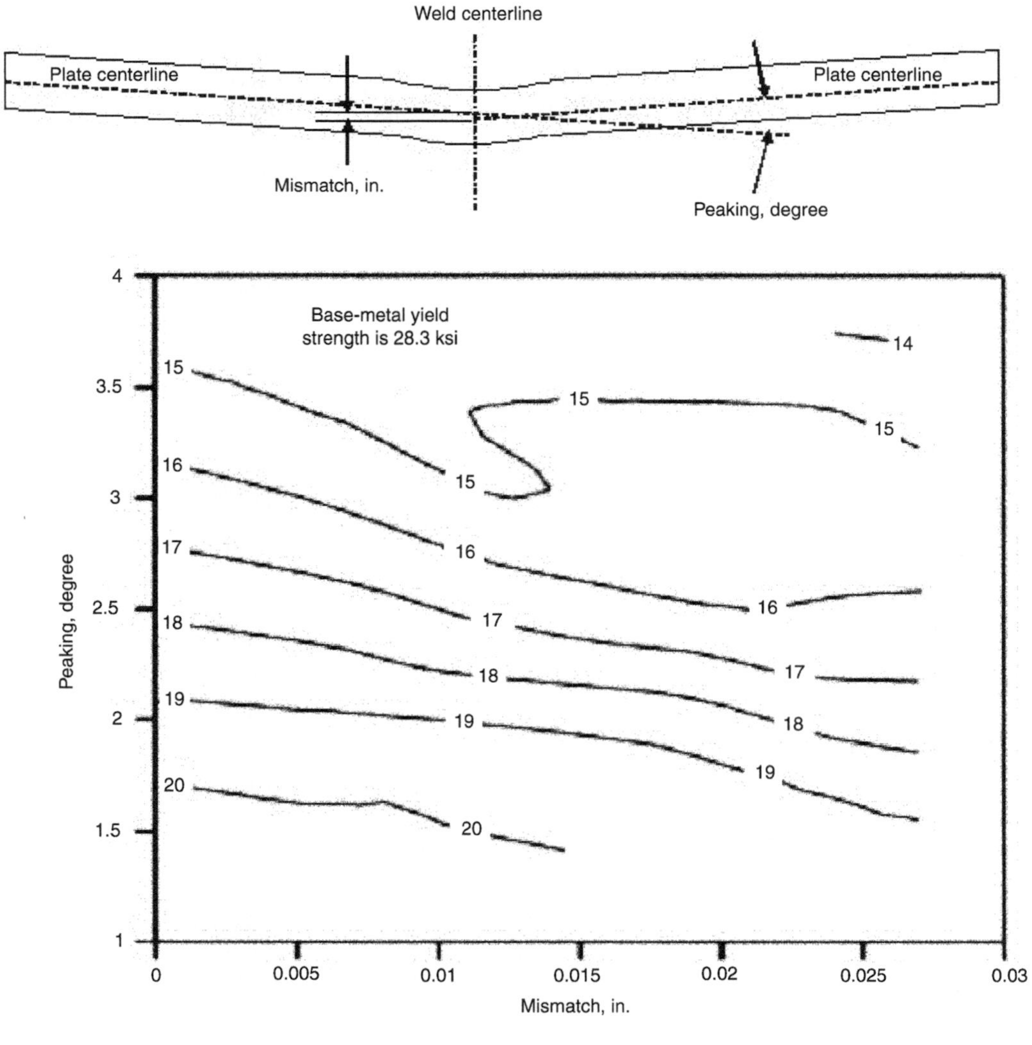


**Fig. 13.7** Effect of peaking and mismatch on the apparent yield strength of an 11 mm (0.43 in.) 5083-H111 friction stir weld

area. This author (Ref 11) has shown the FSW extrusion pressure ($P_e$) and time average strain rate ($\dot{e}$) to be a function of pin-tool geometry factors ($\alpha$ and $\beta$), temperature ($T$), material flow stress ($\sigma_1$), processing parameters (in./min and rpm), and the extrusion zone width ($W_r$), which includes the pin-tool swept area and that width of material outside the pin-tool projected area that also flows. Colligan (Ref 17) has shown that metal flow is highly cyclical and periodic in nature and results in the distinct metal flow patterns observed within each of the weld nugget flow zones.

The repeatable and cyclical nature of these flow patterns and their relationship to process forces provide an opportunity to develop intelligent path-planning algorithms, which include sensing and feedback/feed-forward control systems to monitor and control weld quality (Ref 18). As metal flows through each zone and converges again at the zone interfaces, perturbations in the metal flow patterns associated with defect formation are manifest in fluctuations in the magnitude and direction of the global processing forces ($x$-, $y$-, and $z$-axis) and torque and offer the opportunity to develop smart process control algorithms to monitor and control joint quality. The volumetric "wormhole" defect is the most common of the FSW defects. It is manifest by a lack of reonvergence of the materials flowing through each zone. When the die cavity geometry is constant (position control), the colder processing parameters, represented by slower rotation speeds and faster forward travel speeds (heat index = $\omega^2/V_f$), promote volumetric defect formation (Fig. 13.9). The magnitude of the forging force and pin-tool shoulder design provides boundary conditions and system constraints to ensure proper flow through each zone. Under force control, the size of these voids is directly related to the forge force (Fig. 13.10) and can be such that they extend completely to the surface (surface lack of fill), are embedded and continuous along the length of the FSW (wormhole), or embedded, discontinuous, and periodic along the length of the FSW (scalloping; lack of consolidation).

One FSW control algorithm approach monitors the periodic fluctuation in the global process forces and torques and adjusts the system parameters as necessary to maintain the proper temperature and metal flow to prevent volumetric defect formation. Several analytical methods to evaluate weld quality directly from process control variables and system torque and force responses have been investigated (Ref 12). In its simplest form, variations in process forces in frequency space are demonstrated to correlate well with volumetric defect formation, even down to the intermittent discontinuous

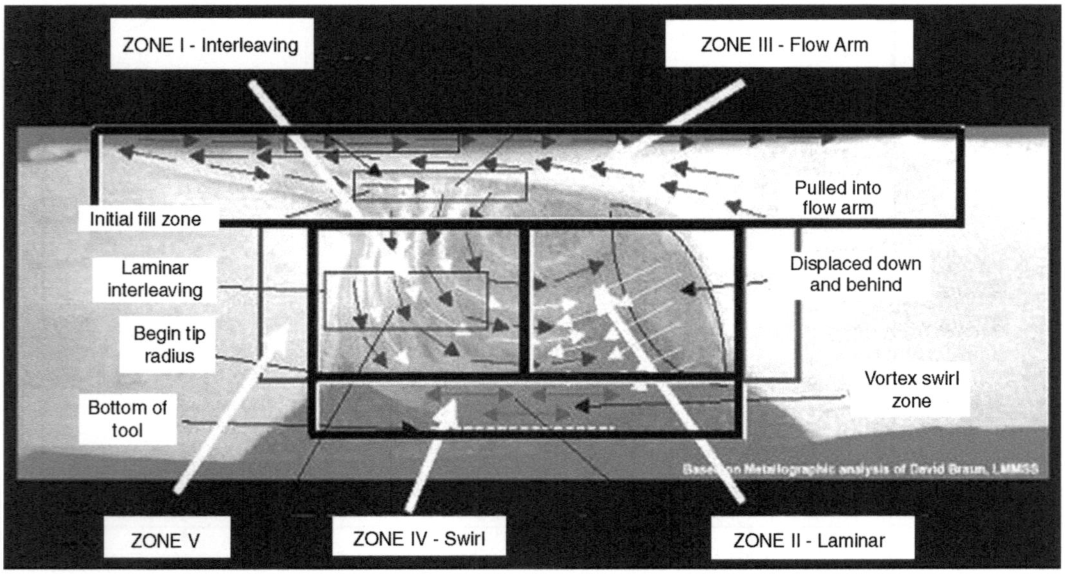


**Fig. 13.8** Metal flow zones developed during friction stir welding (transverse section view)

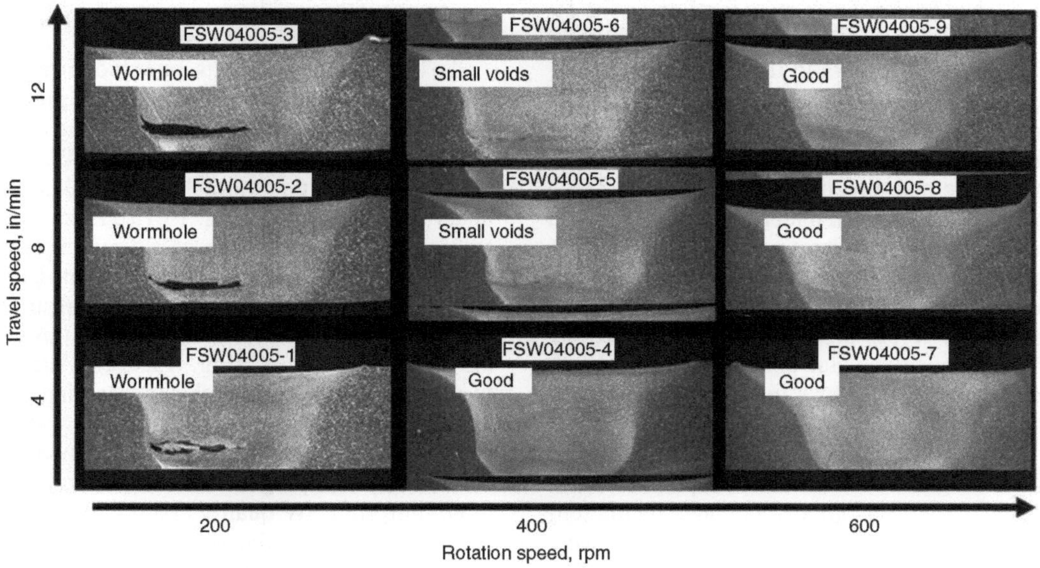


**Fig. 13.9** Colder processing parameters promote volumetric defect formation due to lower processing temperature and lower material flow stress

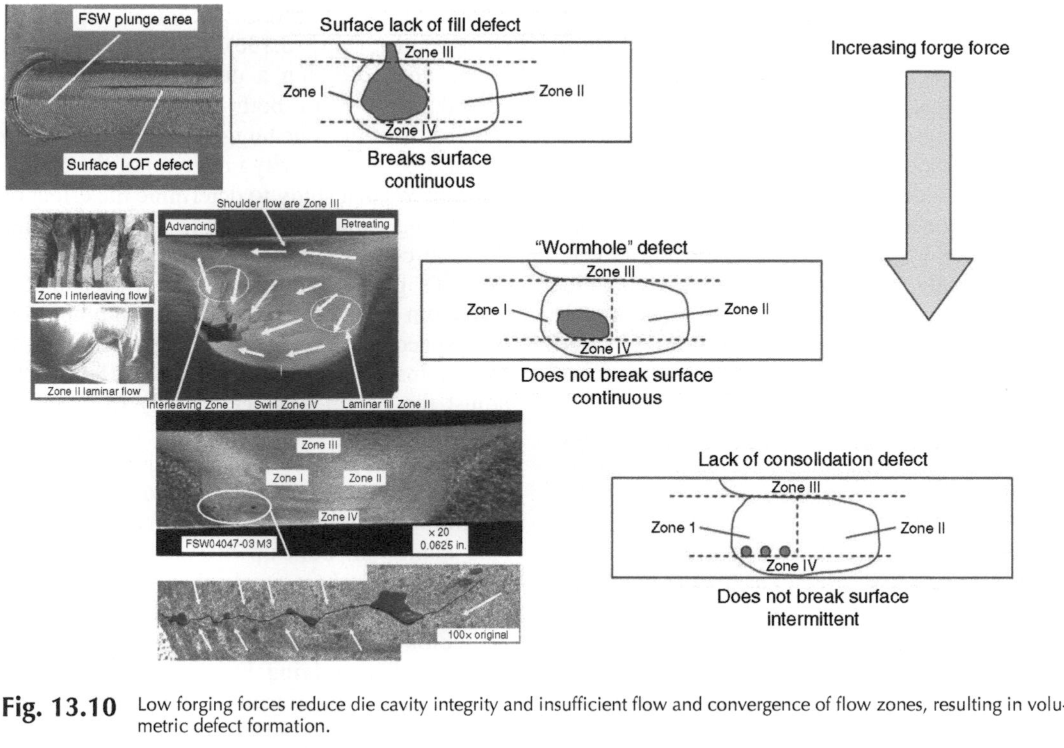


**Fig. 13.10** Low forging forces reduce die cavity integrity and insufficient flow and convergence of flow zones, resulting in volumetric defect formation.

microvoid formation in the zone I and zone IV convergence zone (Fig. 13.11). The increase in low-frequency events in the *y*-direction (transverse) represents the case where there is inadequate flow through zones II, III, and IV to completely fill the zone I region. The *y*-force magnitude and direction reflects the imbalance in the forces between materials flowing through the advancing and retreating sides of the joint. A larger negative *y*-force correlates with microvoid formation. Process control algorithms that maintain *y*-forces around a zero (balanced) or specified positive value correspond to void-free welds and provide for a real-time process control algorithm methodology.

**Intelligent FSW Path Planning.** Process control algorithms and FSW response to processing parameters are compounded by the sensitivity of the FSW process to support fixturing and tooling. The introduction of multiaxis FSW systems has enabled the welding and joining of more complex structures within three-dimensional (3-D) space (Ref 19). While these multiaxis systems provide for 3-D motion control under preprogrammed path plans, fixturing, support tooling, and clamping systems fabricated by the end user affect the resultant quality of the part being joined. Specifications and standards must address fixturing, clamping, support tooling, part geometry, machine control, and FSW process parameters to consistently produce high-quality joints (Fig. 13.12). Intelligent path-planning algorithms that integrate the virtual part geometry and weld path obtained from a computer-aided design/computer-aided manufacturing model into the motion-control systems of multiaxis FSW equipment are being developed. The algorithm logic includes automatic selection of approved process parameters and weld process schedules and employs process sensing and feedback and control systems to ensure weld quality (Ref 18).

For the various factors in FSW path planning (Fig. 13.12), tooling space is an important factor in the development of proper FSW practices for a particular application. While the welding parameter development and pin-tool designs are relatively straightforward to join most metals, the repeatability of the process is highly influenced by the heat transfer and restraint provided by the fixturing and tooling (Ref 20). In many cases, low-cost reconfigurable tooling has proven adequate to produce acceptable FSW. Variations in fit-up, restraint, and heat transfer along the length of the FSW can result in loss of processing forces, die cavity integrity, improper metal flow patterns, and potential defect formation.

To illustrate these effects, one study in 6061-T6 plate demonstrated the change in process forces due to clamping locations, welding direction, crossing over pre-existing FSW, and change in essential variables (rpm, in./min) under position control (Fig. 13.13). From these studies, it is seen that the process forces increase at the locations of discrete clamping, possibly due to increased die cavity sidewall restraint and increased resistance to metal flow through the processing zones (Fig. 13.13a). Alternatively, this may be due to colder processing temperatures and increased heat transfer at these locations. Subsequent studies have shown that continuous clamping methods can result in more uniform FSW quality along the length of the joint. Changing the welding direction into the retreating side increases the *y*-force. This may be due to the closure of the retreating side extrusion zone II. Changing direction into the advancing side of the weld results in a drop and change in sign (–) of the *y*-force (Fig. 13.13b). This may be due to the widening of the retreating side extrusion zone II.

Crossing of pre-existing FSW under position control also results in changes in process forces and compounds process control algorithm development (Fig. 13.13c). The drop in process loads when crossing a pre-existing FSW has contributions from both the softer dynamic recrystallization zone nugget and surface indentation (loss of die cavity integrity) of the underlying FSW. In a study to determine the effect of processing parameters (in./min and rpm) under position control on the FSW process forces in 7075-Tx plate (Fig. 13.13d), it is seen that a minimum in process forces occurs in both rotation speed and travel speed. Process control algorithms must consider these effects and adjust the parameters accordingly to maintain joint quality.

Equally important in ensuring joint quality is the path-plan sequencing. In a study to establish the fixturing and tooling requirements to fabricate aluminum built-up beams from standard extrusion and sheet stock materials, a loss of forging pressure and surface lack-of-fill defect formation was observed in those areas of overwelding of underlying FSW start-stop regions (Fig. 13.14 top). Starting (plunging) the end stiffener FSW lap welds near the exit keyhole of the previous underlying longitudinal butt joints resulted in a loss of forging pressure. Removal

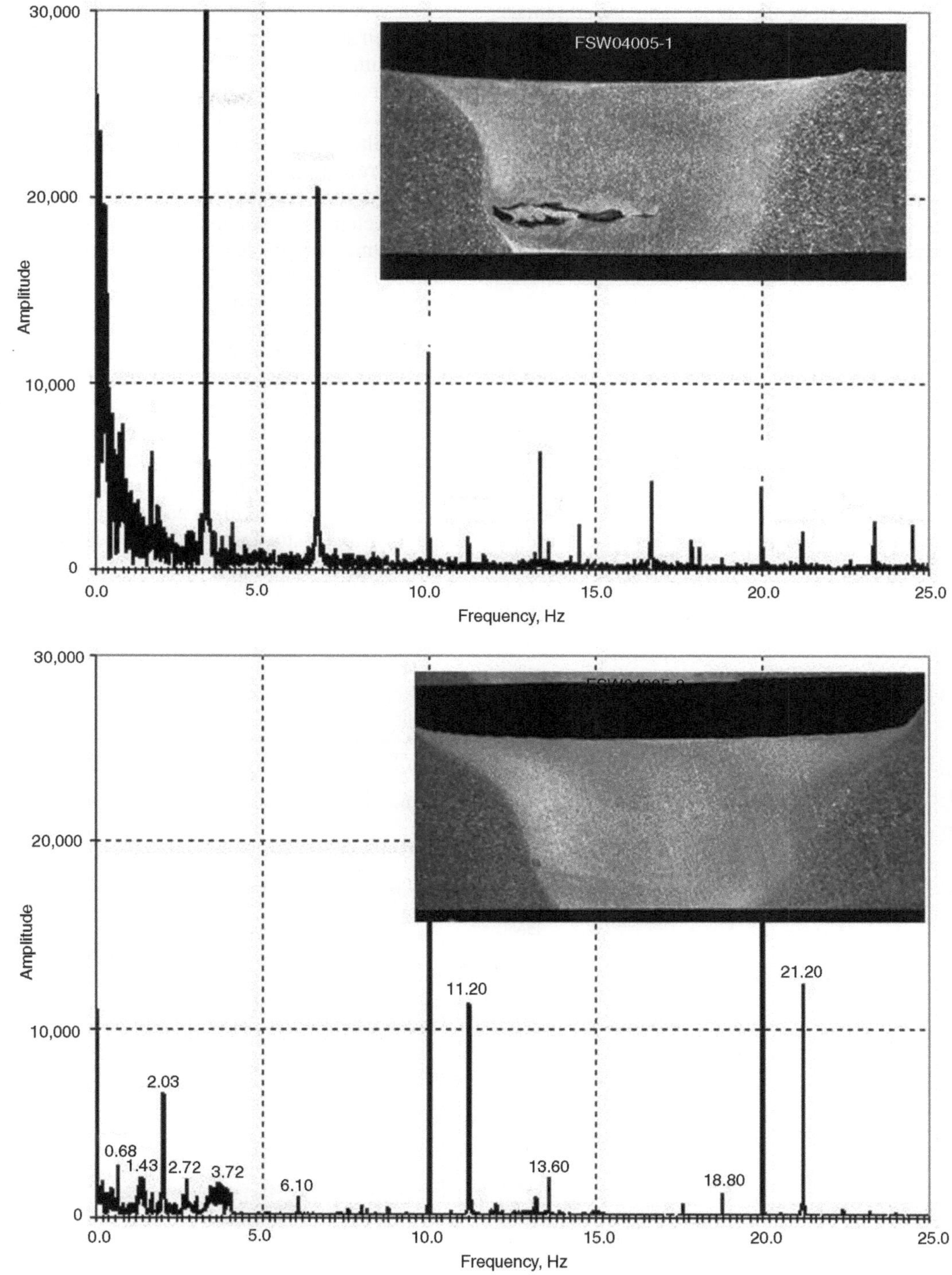


**Fig. 13.11** Fourier analysis of y-force fluctuations for 3.2 mm (0.13 in.) thick 2024-T3 sheet friction stir welded under position controls at (top) 200 rpm, 101 mm/min (4 in./min), and (bottom) 600 rpm, 202 mm/min (8 in./min). Note volumetric wormhole defect in top chart, showing large degree of low-frequency events.

of the flash from the underlying longitudinal FSW butt joints and stiffener FSW lap joints is essential to ensure part contact to each other and to the support tooling. This is also necessary to ensure consistent heat transfer.

Certification and qualification of the FSW process parameters and control algorithm should be done on production fixturing and tooling to ensure representative restraint and heat-transfer characteristics. Changes to production tooling should require requalification of the FSW process. Examples of the major elements of the fixturing and tooling that affect joint quality include the end, side, top (clamps and anvils), and antirotation restraints (Fig. 13.14 bottom).

**Joint Design Allowables and Service Life Assessments.** Friction stir welding has been demonstrated in a variety of joint designs (Fig. 13.15). The most commonly used joints are the full-penetration butt joints and the partial-penetration lap joints, followed by the edge joint, capture joint, and fillet joint, listed in order of ease of manufacture. Friction stir welding is not a "dropin" process, and existing riveted or welded structures should be redesigned to take full advantage of the process benefits and to accommodate the process limitations. New designs require innovative manufacturing approaches and special tooling to ensure that the FSW built-up assembly satisfies form, fit, and function requirements.

When a joint design, pin tool, and alloy have been selected, there are three essential variables (rpm, in./min, and forge force) that must be considered. During processing, sliding (*x*), separation (*y*), and forge (*z*) forces are introduced into the part by the rotating pin tool and flowing metal. A torque (M) is induced, which tends to rotate the part. Increasing the forward travel speed (in./min), rotational speed (rpm), and plunge depth generally increases the process forces (*x, y,* and *z*). Increasing the rotational speed and decreasing the travel speed increases the heat input to the weld. In addition to thermal expansion and distortion effects, the heating and plasticizing of the metal induces microstructural changes that govern the resultant mechanical properties.

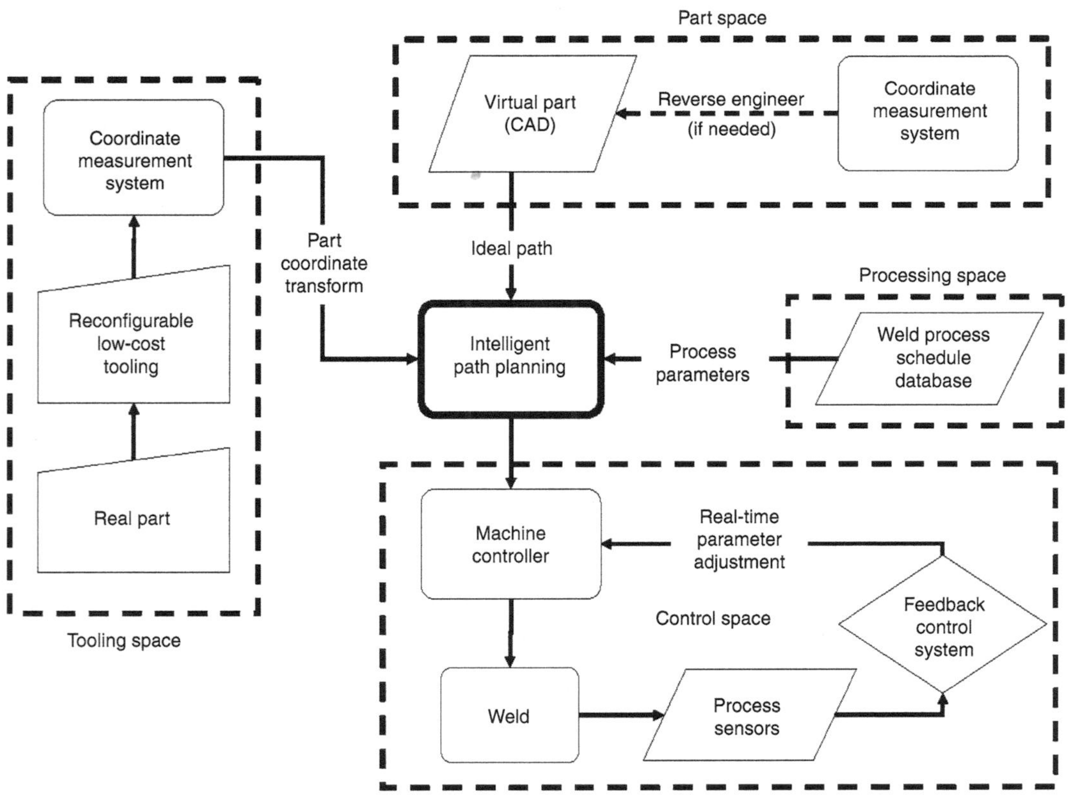


**Fig. 13.12** Factors and interactions for intelligent friction stir welding path planning

The strength of FSW butt and lap joints in aluminum alloys has been shown by many investigators to be a function of pin-tool design and processing parameters (Ref 20). The FSW butt joints typically exhibit 65 to 100% joint efficiencies when compared to the parent-metal ultimate strength (Table 13.2). For the heat treatable aluminum alloys, hotter welding parameters (high rpm, low in./min) generally result in lower joint static strengths (Fig. 13.16). At excessively cold processing parameters (low rpm, high in./min), the static strength is influenced and lowered by the formation of the characteristic "wormhole" or lack of consolidation defect (Ref 21). In thin-sheet 2*xxx*, 7*xxx*, and 5*xxx* partial-penetration lap joints, the pounds per inch of weld typically exceed those minima specified in the industry standards for resistance spot welds (Ref 22) and riveted structures (Ref 23). It is interesting to note, however, that this is not always true for lap joints in the 6*xxx* alloys (Fig. 13.17). For low flow stress materials such as 6061, under hotter processing parameters, the static lap shear strength is lowered by excessive formation of the characteristic zone V sheet-thinning defect (STD) on the advancing side of the joint, while at the colder processing parameters, the static strength is influenced by the formation of the characteristic cold lap defect (CLD) on the retreating side of the joint.

Based on static strength considerations, the use of FSW in either butt or lap joint configurations is a viable joining method and replacement for resistance spot welds and rivets in the design and development of built-up structures. It is recognized, however, that both the dynamic properties (fatigue and impact) and corrosion resistance of FSW joints compared to these conventional joining technologies must also be evaluated. Friction stir welding is readily adaptable to built-up design approaches (Fig. 13.18). In its simplest form, sheet and plate stock is welded to common extruded shapes using butt and lap joints. More complex designs using capture and fillet joints require machined details. To ensure low cost, simplified FSW joint types should be employed. Preference to butt and lap joints should be given, with other joint types used only in special situations.

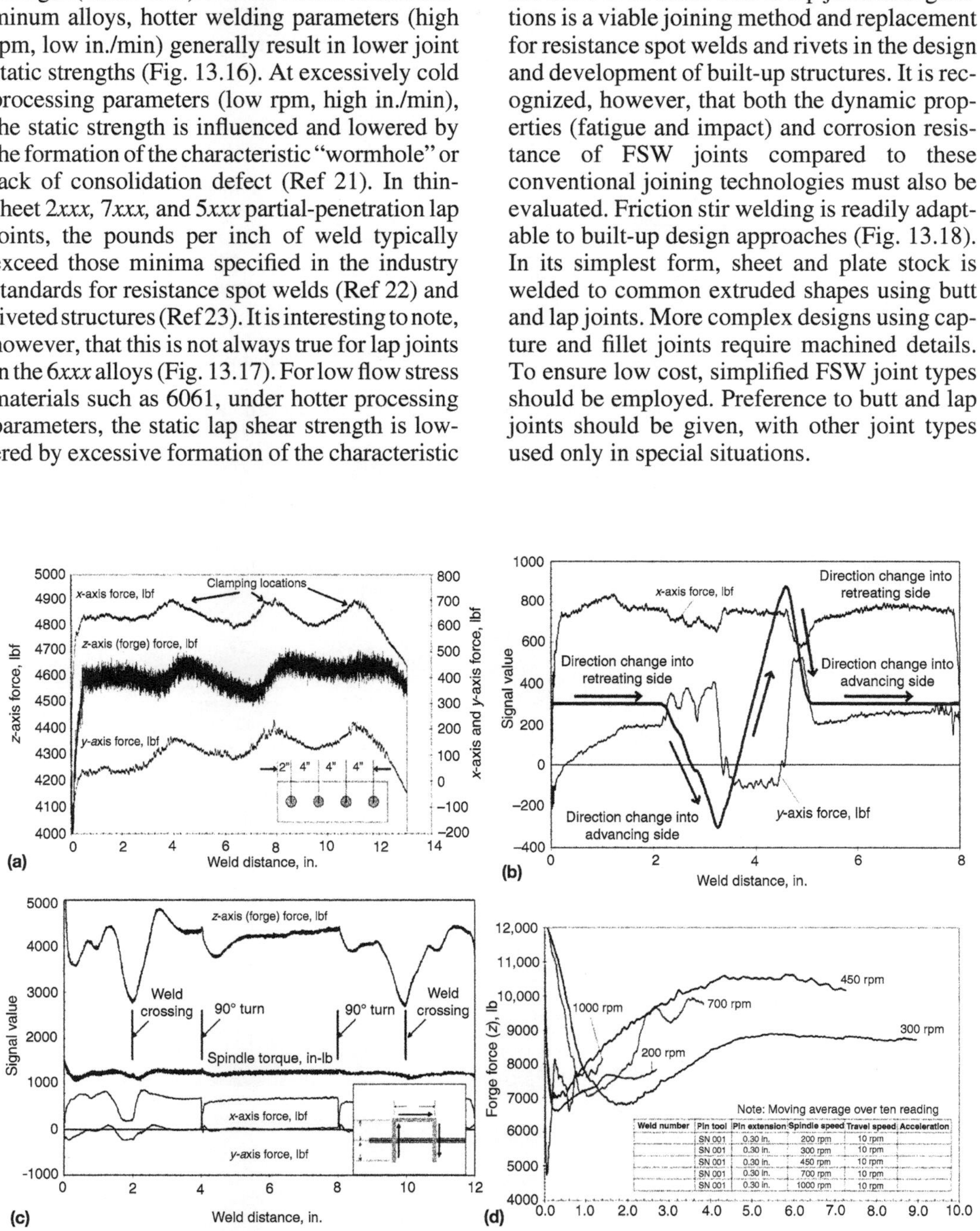


**Fig. 13.13** Effect of fixturing, tooling, and path planning on process forces. (a) Change in process forces due to clamping locations. (b) Change in process forces due to change in welding direction. (c) Crossing over pre-existing friction stir welding. (d) Effect of changing essential variables (rpm, in./min) under position control

Specifications and standards should specify the methodology to determine the static strength design allowables on welds made using production-like fixturing and tooling and the range of processing parameter adjustments allowed by process control algorithms. One such analysis, using the statistical lower-tolerance limit methods of MIL-HNBK-5 (Ref 23), shows a quadratic relationship between transverse as-welded strength and heat index to which the 99-95 and 90-95 probability and confidence factors can be applied to determine the effect of processing parameters on static strength allowables (Fig. 13.19).

The static strength (Fig. 13.20) and fatigue life (Fig. 13.21) of lap joints are influenced by the presence of the STDs and CLDs. The direction of welding is important to ensure that the

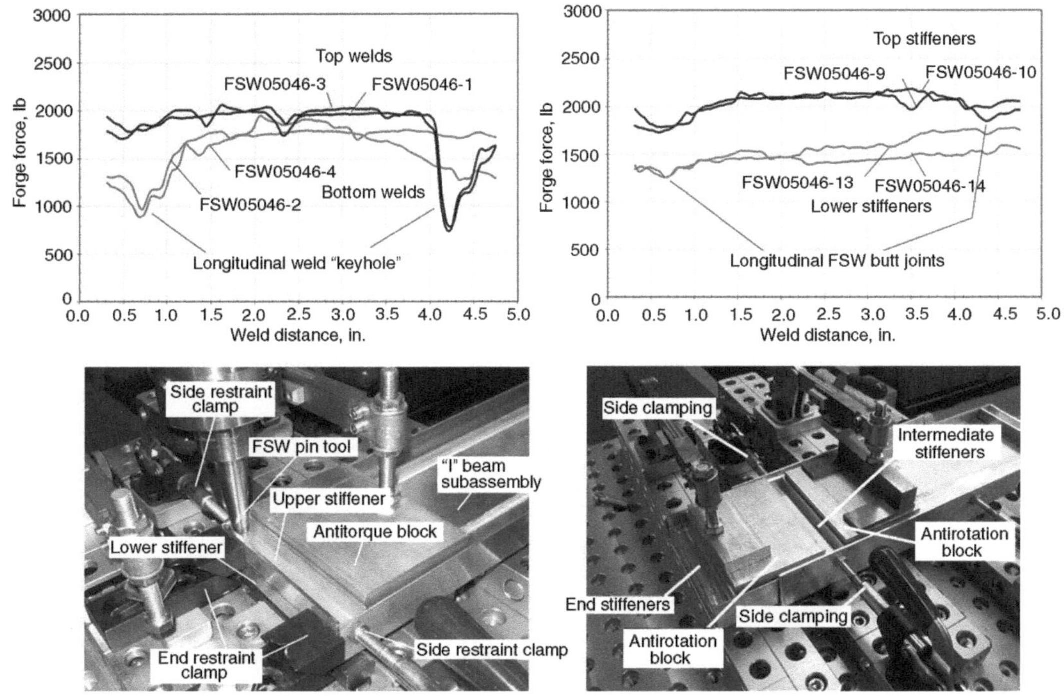


**Fig. 13.14** Major elements of fixturing and tooling to attached end (bottom left) and intermediate (bottom right) U-channel stiffeners to a built-up I-beam section. Effect on process forces with (top left) and without (top right) underlying friction stir welding start and stop features

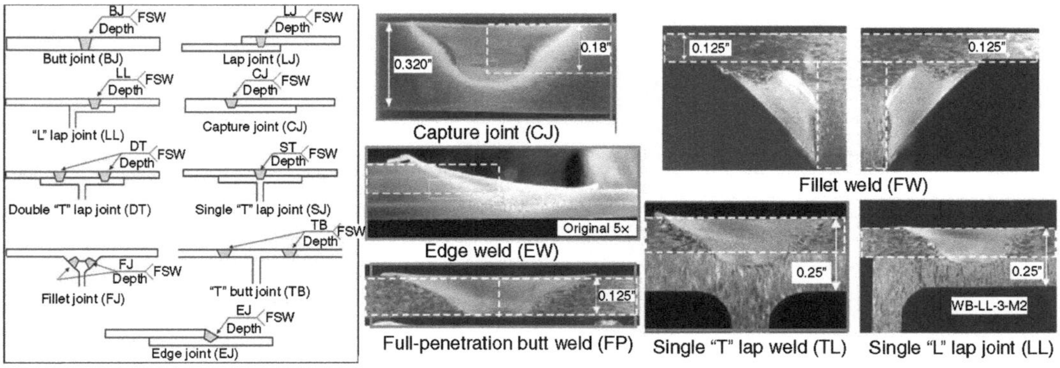


**Fig. 13.15** Typical friction stir weld joint designs

potential defect is placed on the nonload path side. If this is not possible, pin-tool selection and processing parameters must be optimized to minimize the STD, and the design allowables must be established assuming the presence of at least some degree of defect. It is noted here that the pin-tool design can greatly affect the STD and CLD formation, with those designs that promote more vertical flow increasing the tendency for STD formation, while those that promote horizontal flow increase the CLD formation. These two defects are competing. High-heat index welds show more STD and less CLD, while low-heat index welds show the opposite. Optimal allowables of STD and CLD must be determined for each application.

**Table 13.2 Typical aluminum alloy friction stir weld butt joint efficiencies (not for design purposes)**

| Alloy | Parent-metal UTS | | Friction stir weld UTS | | Joint efficiency, % |
|---|---|---|---|---|---|
| | MPa | ksi | MPa | ksi | |
| AFC458-T8 | 545 | 79.0 | 362 | 52.5 | 66 |
| 2014-T651 | 483 | 70.0 | 338 | 49.0 | 70 |
| 2024-T351 | 483 | 70.0 | 434 | 63.0 | 90 |
| 2219-T87 | 476 | 69.0 | 310 | 45.0 | 65 |
| 2195-T8 | 593 | 86.0 | 407 | 59.0 | 69 |
| 5083-O | 290 | 42.0 | 296 | 43.0 | 102 |
| 6061-T6 | 324 | 47.0 | 217 | 31.5 | 67 |
| 7050-T7451 | 545 | 79.0 | 441 | 64.0 | 81 |
| 7075-T7351 | 472 | 68.5 | 455 | 66.0 | 96 |

UTS, ultimate tensile strength

The static strength and fatigue life allowables for specific joint designs should be determined on test samples prepared on production-like tooling, using the range of approved processing parameters. One study of 2297-T87 FSW "T"-butt, single- and double-lap "T", and fillet joints (Fig. 13.22) shows that the direction of loading on these joints has a significant impact on fatigue life. Another interesting observation from these tests was that, depending on the joint configuration and specimen loading conditions, the fatigue life of the FSW may not be the limiting factor for fatigue, with specimen failure occurring in the parent metal away from the joint.

**Precleaning and Edge Preparation.** Friction stir welding is not as sensitive to preweld cleaning as are fusion welding methods. As such, simple abrasive cleaning of the mating surfaces is generally all that is required, followed by solvent wiping to remove debris. Care

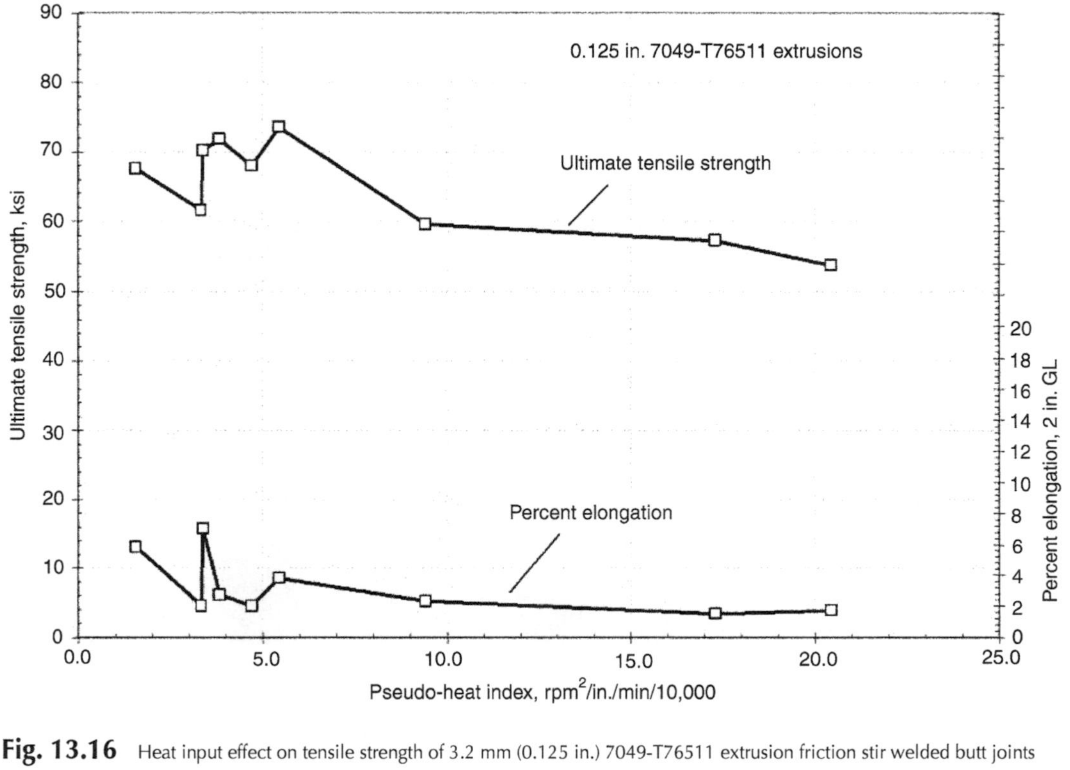


**Fig. 13.16** Heat input effect on tensile strength of 3.2 mm (0.125 in.) 7049-T76511 extrusion friction stir welded butt joints

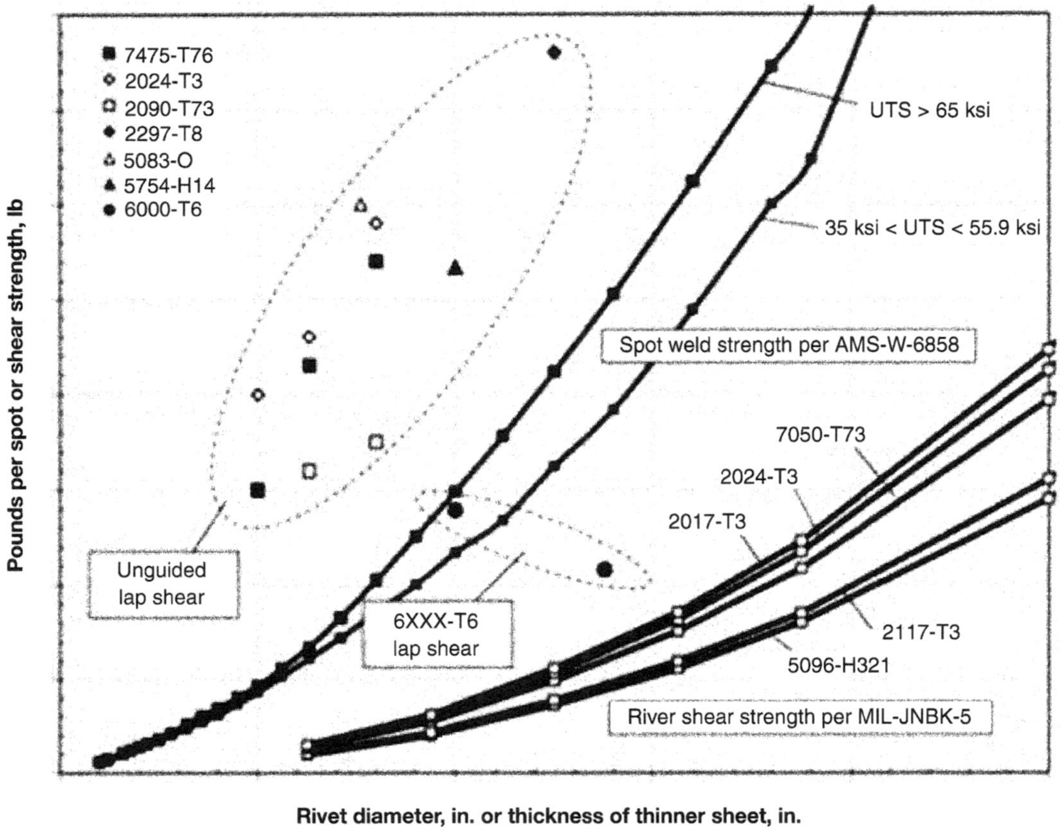


**Fig. 13.17** Unguided lap shear strength (pounds per inch of friction stir weld) for aluminum alloys compared to industry specification values for minimum spot weld strength and rivet shear strengths

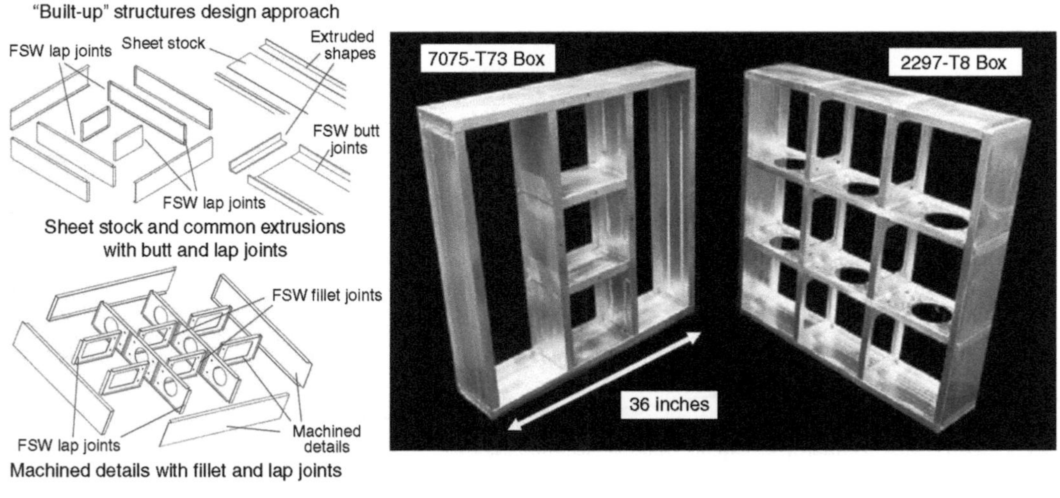


**Fig. 13.18** Built-up structure approach to fabricating detailed assemblies from simple sheet, extrusion, and machined details

must be taken to ensure that this abraded material is not entrapped within the joint line. Heavy oxide and debris entrapped within the FSW joint will become entrained within the periodic flow patterns and may prevent sound metallurgical bonding during reconvergence. Specifications and standards should identify the appropriate precleaning methods, and these should be used during process qualification trials. Acceptable edge-preparation methods include saw cutting, milling, and shearing. The tolerance for gap and fit-up are a function of pin-tool design, material thickness, and processing parameters and should be determined during process qualification trials.

The as-extruded edge of extruded shapes has been shown to produce acceptable FSW, provided the presence of abnormally large grains is not excessive and the pin-tool diameter completely consumes this region. For extrusions with excessively large grains on both the surface and at the edge (Fig. 13.23), low bend ductility in the FSW can result, with bend-test failures often occurring at a distance away from the joint line. During FSW, these abnormally large grains impede metal flow and resist recrystallization in the zone IV region beneath the pin tip and along the top of the joint in the zone III flow arm (Fig. 13.24).

This large grain size and lack of flow pattern formation beneath the pin tip contributes to the lack-of-penetration (LOP) defect formation. Very large amounts of these large grains that extend beyond the width of the FSW pin-tool-swept area may result in large grains remaining in the TMAZ, additional abnormal grain growth within the nugget DXZ, and loss of joint strength. Specifications and standards should address this issue and ensure that these large grains are removed from the ends of extrusions prior to FSW to ensure maximum joint strength and quality.

**Defect Formation and Acceptance Criteria.** The characteristic defects that typically occur in FSW are directly related to metal flow patterns and pin-tool geometric considerations (Fig. 13.25). Under hot processing conditions, an imbalance in the metal flow patterns may exist under which the nugget may collapse, caused by excessive material flow from the zone III flow arm filling the advancing-side zone I region. While no void is associated with this, it

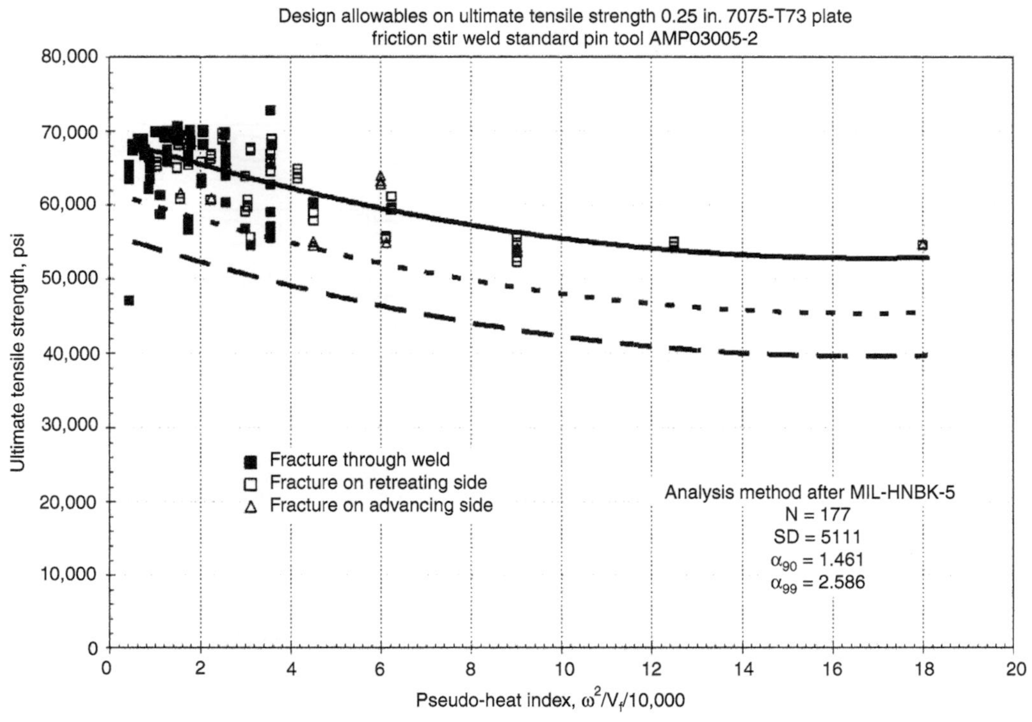


**Fig. 13.19** Statistically significant lower-tolerance limits for friction stir welding in 3.8 mm (0.15 in.) 7075-T73 plate

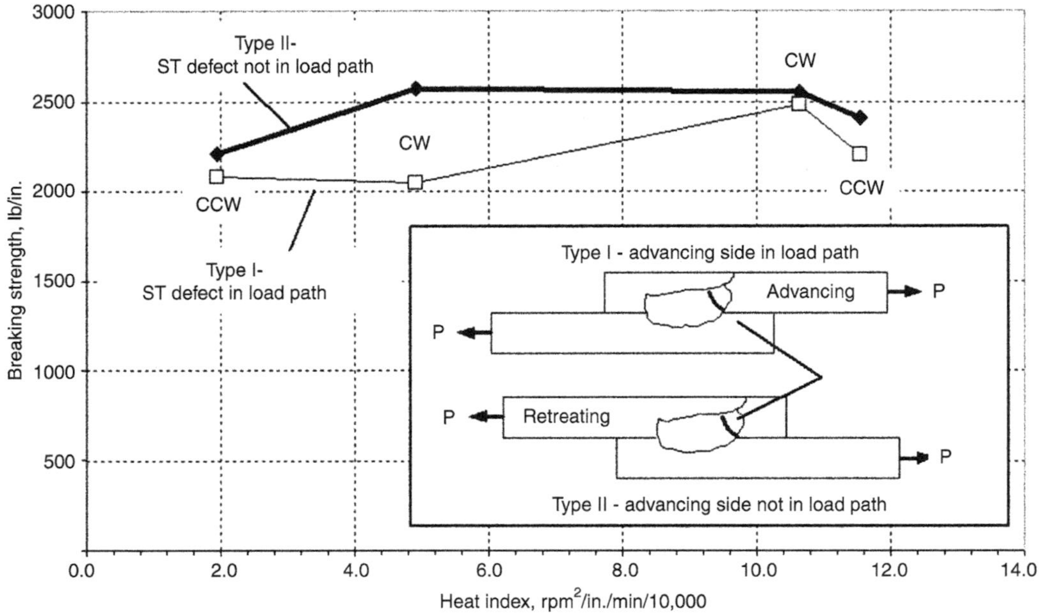


**Fig. 13.20** Effect of specimen orientation on breaking strength in 3.8 mm (0.150 in.) 7075-T73 sheet friction stir weld lap joints with sheet-thinning defects

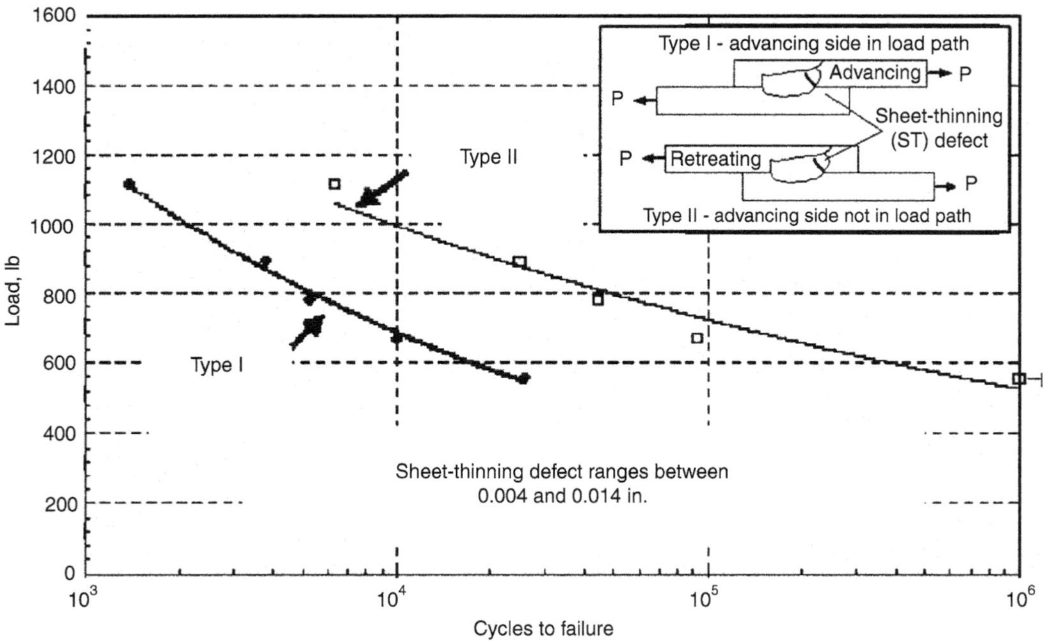


**Fig. 13.21** Effect of specimen orientation on fatigue life in 3.8 mm (0.150 in.) 7075-T73 sheet friction stir weld lap joints with sheet-thinning defects

does indicate overheating of the material and potential loss of parent-metal strength in the TMAZ or HAZ. The acceptability of this indication is determined by the required minimum joint strength.

Also associated with excessively hot processing parameters is the root-flow defect. This results from too much material flowing into the zone IV region and excessive penetration of the metal flow patterns to the backside beneath the pin-tool tip. This defect has been correlated with loss of joint strength, fatigue life, and bend ductility. Under extremely hot processing conditions, the top surface zone III flow arm materials may adhere to the pin-tool shoulder and result in severe galling and tearing of the metal. Also, under hot processing conditions, excessive zone III flow may result in expulsion of the

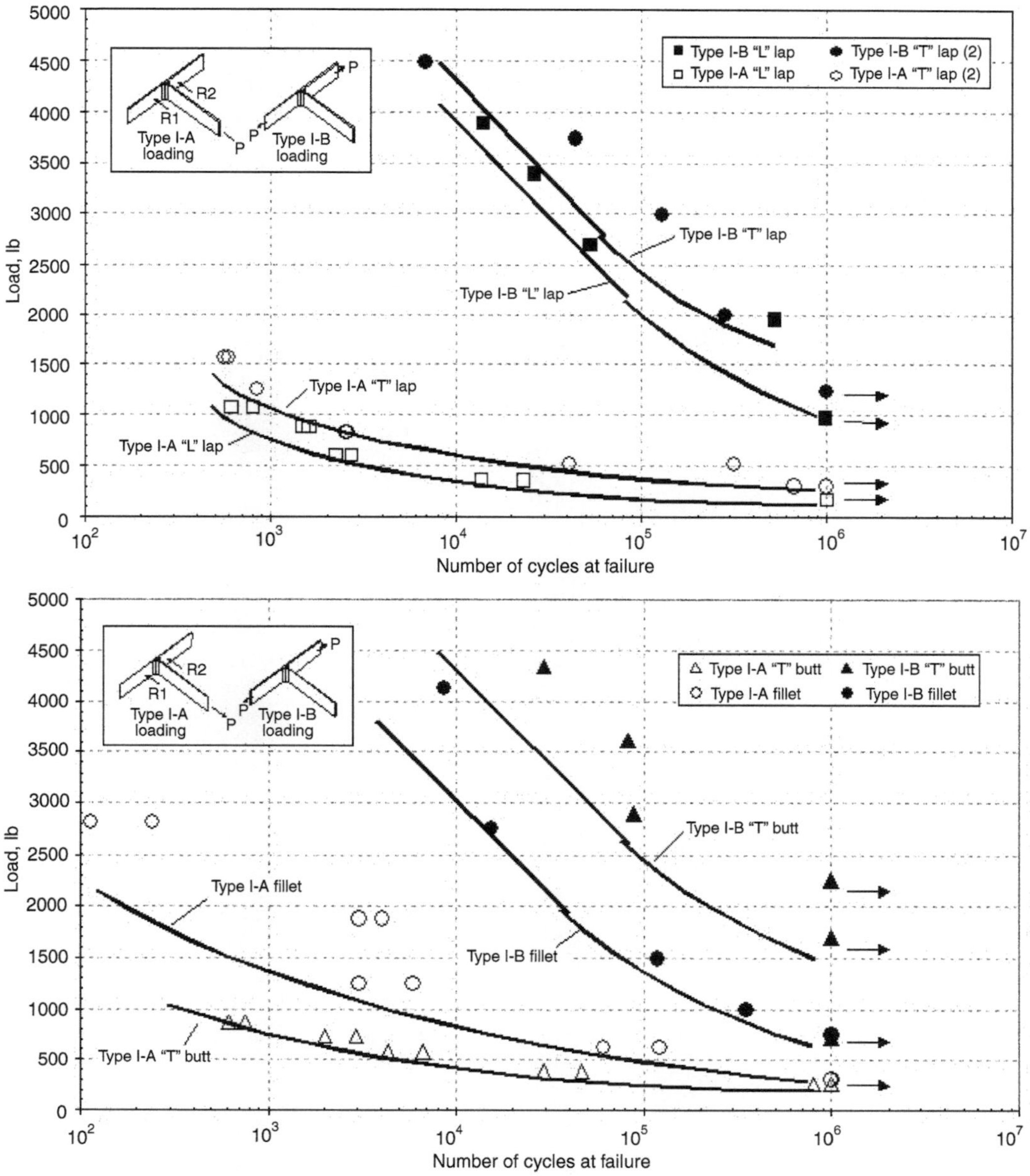


**Fig. 13.22** Effect of loading orientation on fatigue life of 2297-T87 "T"-lap, butt, and fillet joints

material from under the pin-tool shoulder and flash formation. Loss of material from the joint can affect the flow balance in the other zones and cause loss of forging pressure and volumetric void formation.

Under cold processing conditions, the volumetric wormhole defect may occur due to insufficient refill of the advancing-side zone I region. As described previously, the degree of this defect decreases as the forging force is increased. Under excessively low forging forces, the surface lack-of-fill defect may occur with penetration through the zone III flow arm to the top surface, and, under marginally low forging forces, the intermittent lack-of-consolidation (LOC) microvoids may occur at the zone I/zone IV interface. Each of these has an adverse effect on joint strength and fatigue life, depending on their size and degree. The presence of these defects is evident on the fracture surface of tensile specimens with the LOC defect, often manifesting in the form of scalloping fracture paths along the cyclical flow patterns.

In FSW butt joints, additional geometric defect indications are seen when the pin tool is too short or there is inadequate zone IV metal flow and recrystallization to completely consume the original faying surface on the backside of the joint. This is described as the LOP defect. Improper seam tracking results in the lack-of-fusion (LOF) defect, where the original faying surface remains off to one side of the full-penetration joint. A third geometric defect, known as excessive indentation, results from too high of plunge depth or forging force and severely reduces the section thickness in the joint region.

Two additional defects are seen in FSW of lap joints (Fig. 13.26). Under hot processing conditions, the zone V flow observed in the

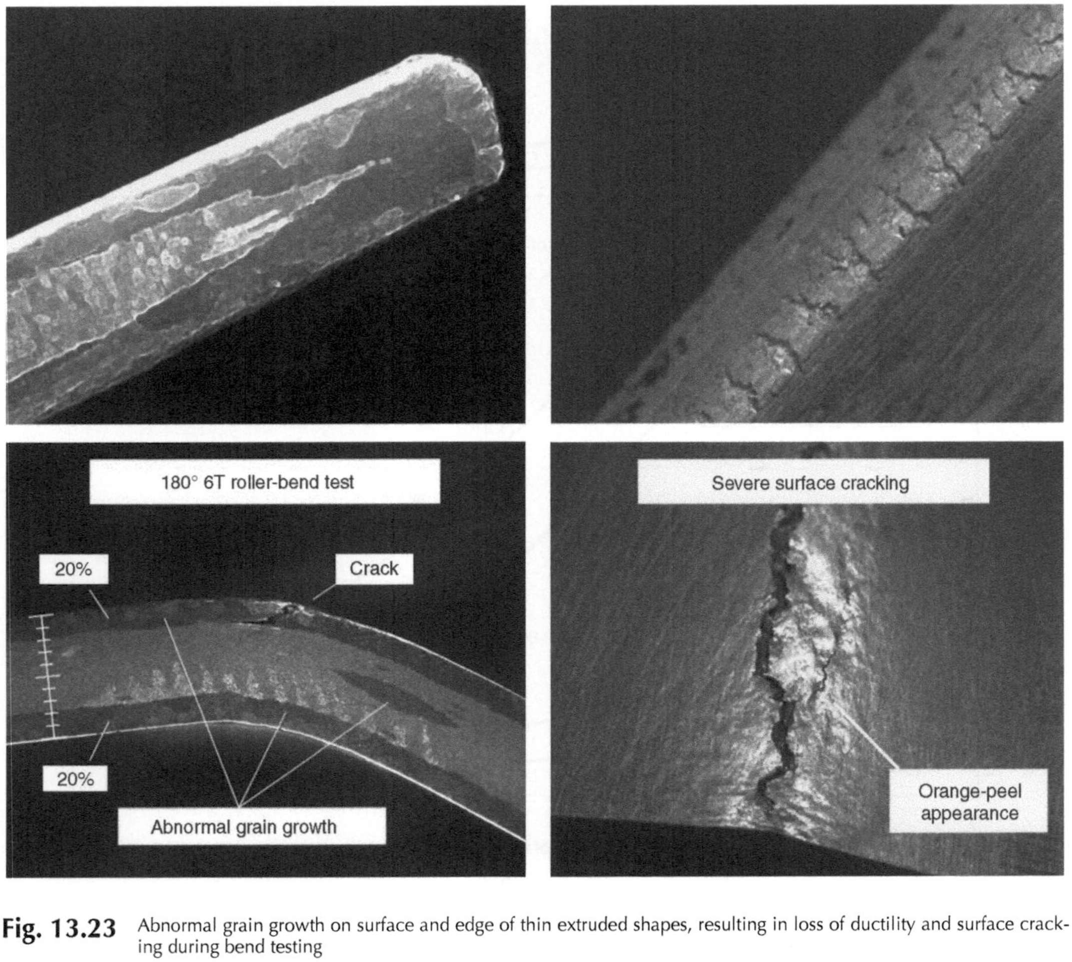


**Fig. 13.23** Abnormal grain growth on surface and edge of thin extruded shapes, resulting in loss of ductility and surface cracking during bend testing

TMAZ area can carry and uplift the horizontal faying surface of the joint. This is described as the sheet-thinning or hooking defect and has been correlated to reductions in static strength and fatigue life. Under cold processing conditions, this horizontal faying surface may be entrained in the zone II flow patterns and not completely reprocessed. The presence of this defect is described as the CLD and can affect the lap shear strength of lap joints.

Limits for each of these defects must be established and their acceptability determined based on testing for fitness for use. As with other joining processes, a certain degree of each is allowable, provided that strength and fatigue life are maintained for the intended criticality of the hardware. For example, the LOP defect still provides high static strengths but adversely affects bend ductility and fatigue life. Large volumetric defects reduce static strengths and fatigue life, but very small microvoids below a critical level do not. Excessive flash will reduce fatigue life, while excessive indentation will reduce static strength. Specifications and standards should establish guidelines and testing procedures to determine these limits.

### *13.2.2 Design Guidelines and Design Allowables*

The development of structural design guidelines and design allowables is being addressed through national FSW research programs and various successful industrial implementations. In 2004, the National Science Foundation Industry/University Cooperative Research Center for Friction Stir Processing (CFSP) was established to bring the South Dakota School of Mines and Technology (SDSMT), University of South Carolina (USC), Brigham Young Univer-

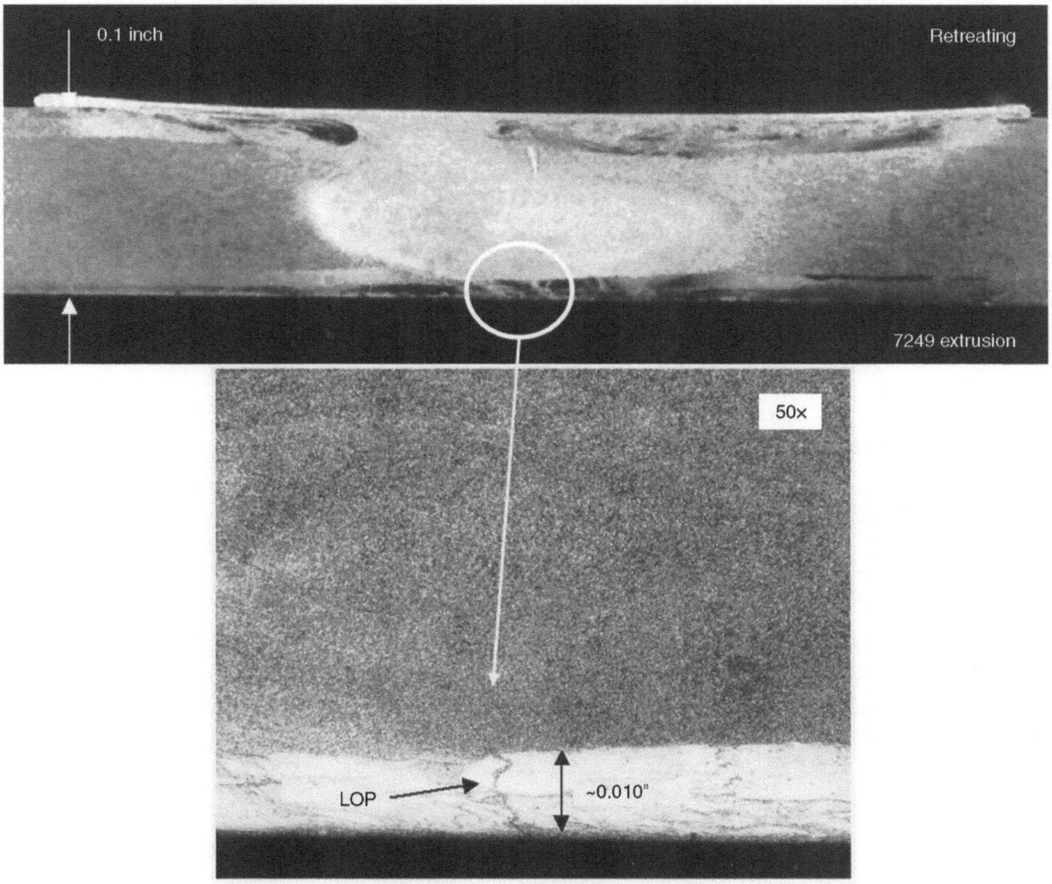


**Fig. 13.24** Unrecrystallized top surface (zone III) and bottom surface (zone IV) grains resulting from abnormally large grains on the as-extruded edge of aluminum extrusions. Note presence of lack-of-penetration (LOP) defect on bottom surface.

sity (BYU), the University of Missouri-Rolla (UMR) and Wichita State University (WSU) together in a collaborative FSW research program. The CFSP currently has 22 government laboratory and industrial sponsors. The center mission is to perform advanced and applied research, develop design guidelines and allowables, train scientists and engineers, and transfer the FSW technology into a broader base within the industrial sector. Current research programs at the CFSP include:

- Design allowables and analysis methodologies for FSW beam and skin-stiffened panel structures
- Intelligent FSW process control algorithms
- Thermal management of titanium and aluminum FSW for property control
- Microstructural modification of aluminum and magnesium castings
- FSW of high-strength low-alloy and 4340 steels
- FSW of austenitic steels and Inconel alloys
- Interactive database of FSW properties and processing parameters

The CFSP has also teamed with the Iowa State University Center for Nondestructive Evaluations to assess the effects of defects in aluminum alloy FSW. The probability of detection (POD) of various nondestructive evaluation (NDE) methods is being established for the volumetric and geometric characteristic discontinuities, and the relationship between flaw size and reduction in static strength and fatigue life is being determined. Statistical process control methods are being developed based on process force and torque responses in frequency space and are being compared to the POD of the NDE methods.

The Edison Welding Institute Navy Joining Center (NJC) has continued to develop and demonstrate FSW technologies in thick-section aluminum and titanium alloys for a variety of Department of Defense applications. One recent technology demonstration program at the NJC used a combination of FSW, GMAW, and hybrid laser welding to fabricate a large titanium structure from 12.7 mm (0.50 in.) thick Ti-6A1-4V plates (Fig. 13.27). In this assembly, the initial corner joints were friction stir welded from the outside of the structure to establish the basic shape, with the remaining structure assembled using GMAW and hybrid laser welding.

Under a recently completed Defense Advanced Research Projects Agency (DARPA) program, Rockwell Scientific and the Naval Sea

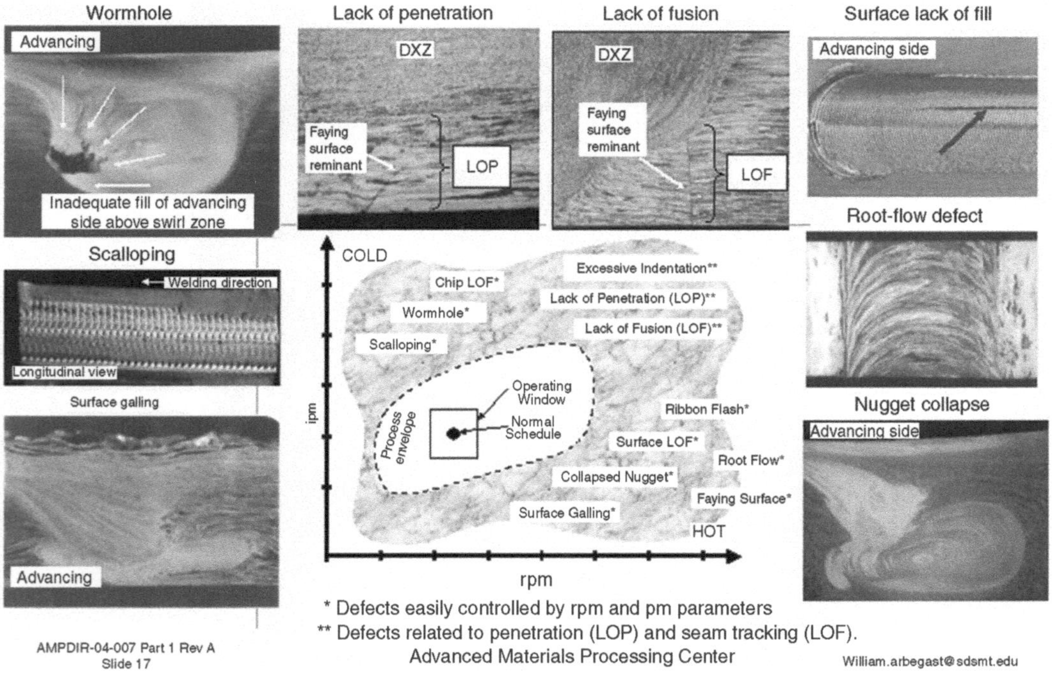


**Fig. 13.25** Characteristic defect types in friction stir welds

Systems Command (NAVSEA) Carderock Surface Warfare Center, in conjunction with 13 university and industrial partners, performed extensive development of friction stir processing on aluminum-, copper-, manganese-, and iron-base alloys. Within this program, MegaStir developed an advanced grade of polycrystalline cubic boron nitride (PCBN) capable of FSW of ferrous alloys up to 12.7 mm (0.500 in.) thick (Fig. 13.28). The fracture toughness of the PCBN is sufficiently high to allow features to be machined on the tool pin, thus accommodating material flow around the tool to fill the cavity in the tool wake.

Also, this same DARPA program demonstrated the ability to friction stir process large areas on the surface of complex-shaped propellers using large industrial robotic FSW systems provided by Friction Stir Link, Inc. (Fig. 13.29). Friction stir processing eliminates near-surface casting discontinuities, increases the yield strength (>2×), and increases fatigue life (>40%) compared to as-cast NiAl bronze. In addition, FSW equipment manufacturers (General Tool Corporation) are exploring alternatives to high-cost multifunctional FSW equipment by developing lower-cost, dedicated, single-purpose systems.

Concurrent Technologies Corporation (CTC), through the Navy ManTech National Metalworking Center (NMC), has advanced the development of FSW in thick-section 5083, 2195, and 2519 Al for ground and amphibious combat vehicles. Several large-scale prototypes have been completed. The work by CTC and NMC has provided a valuable transition of the technology from subscale laboratory work to full-scale prototype construction—the last major step before production implementation.

## 13.3 FSW Process Innovations

Innovations to the FSW process are ongoing. Since 1995, over 50 U.S. patents in FSW have been issued. Pin-tool designs have evolved from those originally developed by TWI to unique designs for thick-section, lap joint, high-

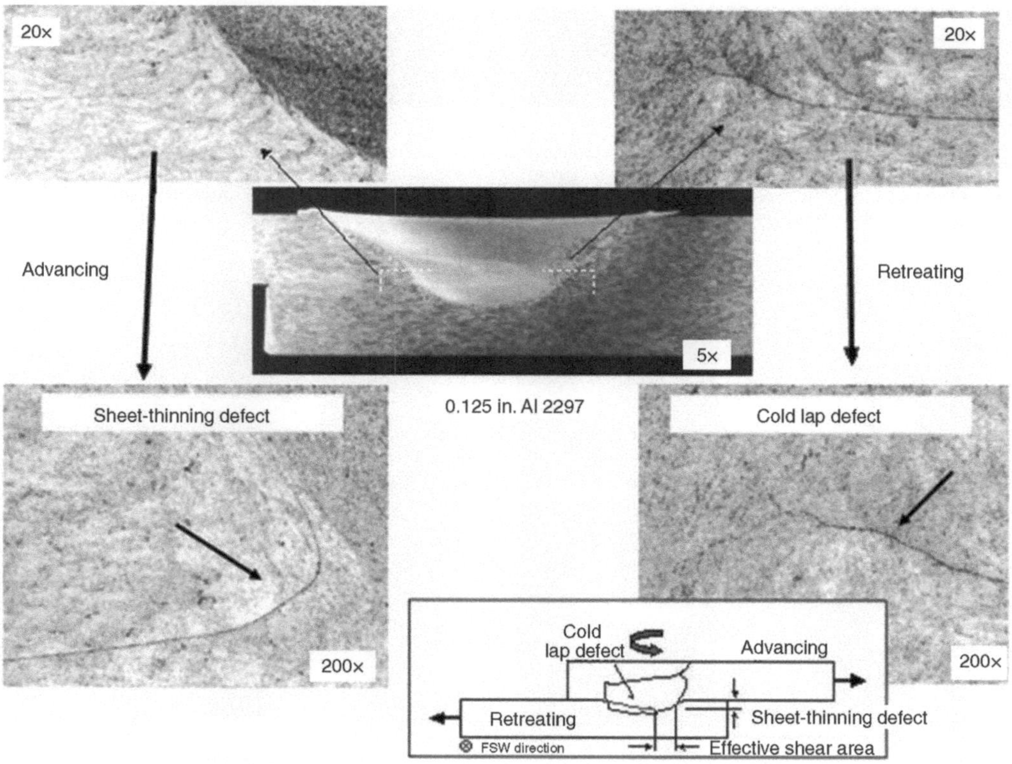


**Fig. 13.26** Characteristic defect types in friction stir lap weld joints

temperature, and fast travel speed joining. For example, in 2005, GKSS-GmbH reported that successful FSW at welding speeds in excess of 1980 cm/min (780 in./min) in thin-gage aluminum butt joints have been achieved. In 1999, the National Aeronautics and Space Administration (NASA) Marshall Space Flight Center (MSFC) and the Boeing Company developed the retractable pin tool (Ref 24) for the FSW of tapered-thickness joints. The MSFC is currently

**Fig. 13.27** The Edison Welding Institute used a combination of friction stir welding, gas metal arc welding, (GMAW), and hybrid laser welding to fabricate this demonstration article from thick-section titanium plates. Friction stir welding was used to join the 12 mm (0.50 in.) thick plates in a corner joint configuration (arrows) to establish the basic shape of the article, and GMAW and hybrid laser welding were used to complete the assembly. Courtesy of Edison Welding Institute

**Fig. 13.28** Photos of 6 mm (0.25 in.) tapered with flats (bottom left), 6 mm (0.25 in.) stepped-spiral (top left), and 12 mm (0.500 in.) stepped-spiral high-temperature polycrystalline cubic boron nitride friction stir weld pin tools. Courtesy of Mega Stir, Inc.

investigating the use of very high rotation speed (>50,000 rpm) FSW, thermal stir welding, and the integration of ultrasonic energy during FSW to enable portable hand-held devices. Other researchers are also evaluating modifications to the FSW process. The University of Missouri-Columbia is evaluating electrically enhanced FSW, where additional heat is applied by resistance heating through the pin tool. The University of Wisconsin is developing laser-assisted FSW of aluminum lap joints, where a laser is trained ahead of the pin tool to preheat the material.

Under a collaborative research program between the Army Research Laboratory and the SDSMT Advanced Materials Processing (AMP) and Joining Center, complex-curvature FSW, friction stir spot welding, dissimilar-alloy FSW, low-cost fixturing and tooling, and thick-plate titanium and aluminum FSW are being developed. Prototypes of advanced fuselage structures, helicopter beams, and naval gun turret weather shields have been built. The AMP Center is also developing induction preheated FSW using an Ameritherm 20 kW remote heat station to preheat thick-plate aluminum, steel, cast iron, and titanium alloys to increase travel speeds, reduce process forces, and reduce pin-tool wear (Fig. 13.30).

In 2001, the MTS Systems Corporation patented the self-reacting pin-tool technology (Ref 25). This innovation allows the FSW of tapered joints and eliminates the need for back-side anvil support to react the process loads. Lockheed Martin Space Systems and the University of New Orleans National Center for Advanced Manufacturing have demonstrated this self-reacting pin tool on the 8 m (27 ft) diameter domes of the Space Shuttle external tank. In this application, multiple gore sections of 8 mm (0.320 in.) thick 2195 Al-Li were joined along a simple curvature path to create the full-scale dome assembly.

## 13.4 FSW Industrial Implementations

The technology readiness level (TRL) for the FSW of aluminum alloys is high, with successful industrial implementation and space flight qualification by Boeing on the 2014 Al propellant tanks of the Delta II and Delta IV space launch

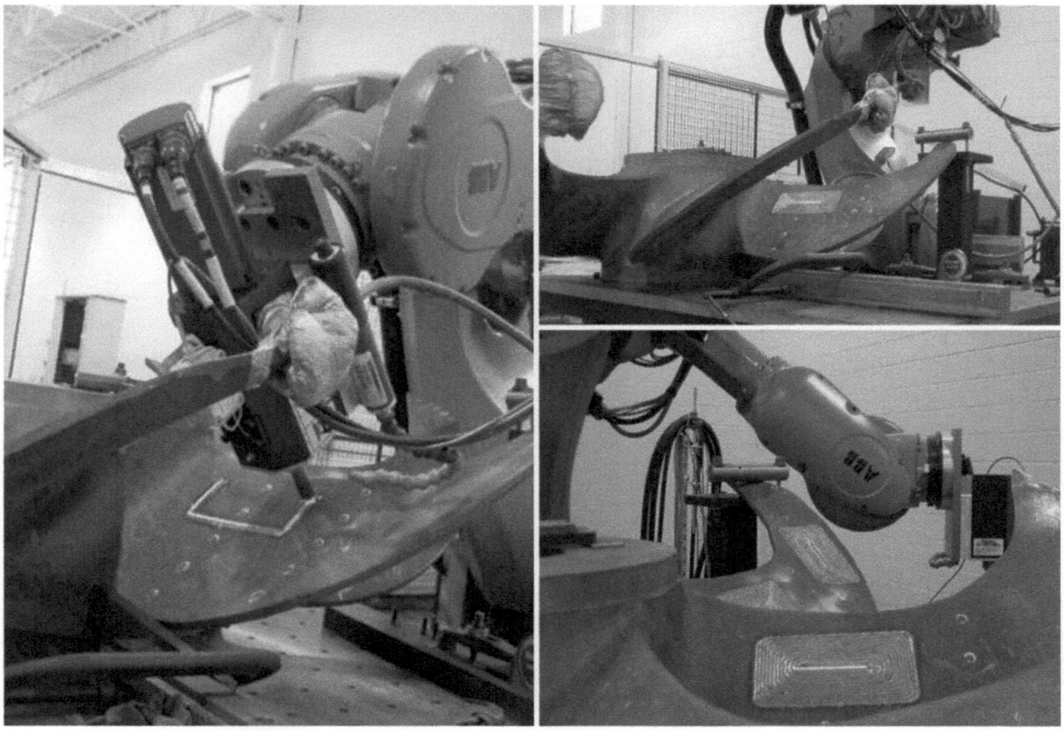

**Fig. 13.29** Friction Stir Link, Inc. robotic friction stir weld system processing large areas of NiAl bronze propellers to remove near-surface casting defects. Courtesy of Rockwell Scientific

vehicles. Lockheed Martin and NASA MSFC have developed and implemented FSW on the longitudinal welds of the 2195 Al-Li liquid hydrogen and liquid oxygen barrel segments of the external tank for the Space Shuttle (Fig. 13.31). Lockheed Martin Missiles and Fire Control and the SDSMT have developed square box beams for the High-Mobility Artillery Rocket System that are fabricated from thick-wall "C"-section extrusions joined by FSW to replace the current hollow, square tube extrusions. Airbus has announced the use of FSW in selected locations on the Airbus A350 and two new versions of the A340 (A340-500, A340-600).

In 2000, the Air Force Metals Affordability Initiative brought together a consortium of industry and university partners to develop FSW for a variety of Department of Defense applications (Fig. 13.32). Under task 1, joining of traditional aluminum assemblies, Lockheed Martin completed a development program that replaced the riveted aluminum floor structure of the C-130J air transport with an FSW floor structure. Under task 2, joining of complex aluminum assemblies, Boeing developed an FSW cargo slipper pallet and implemented an FSW cargo ramp toe nail on the C-17 transport. The toe nail is the only known friction stir welded part flying on a military aircraft. Under task 3, hard metals joining development, the Edison Welding Institute and General Electric developed high-temperature pin tools for the FSW of

**Fig. 13.30** MTS Systems Corporation ISTIR 10 friction stir weld system (left) with the Ameritherm 20 kW remote heat station and induction preheating coil (right). Courtesy of South Dakota School of Mines

**Fig. 13.31** Friction stir weld process development tool at the Marshall Space Flight Center (MSFC) shown with an 8.2 m (27 ft) diameter barrel segment of the 2195 Al-Li Space Shuttle external tank LH2 tank (left). Full-scale LH2 tank (right) at the National Aeronautics and Space Administration (NASA) Michoud Assembly Facility in New Orleans. Courtesy of NASA MSFC

steel, titanium, and inconel alloys for aircraft engine applications.

Eclipse Aviation is in final Federal Aviation Administration (FAA) certification for the Eclipse 500 business-class jet. First customer deliveries are scheduled for 2006. The FSW lap joints are used as a rivet-replacement technology to join the longitudinal and circumferential internal stiffeners to the aft fuselage section and to attach doublers at window and door cutout locations (Fig. 13.33). The use of FSW eliminates the need for thousands of rivets and results

**Fig. 13.32** Aircraft hardware items fabricated using friction stir welding under the Air Force Metals Affordability Initiative Program. C-17 cargo ramp (top left) and slipper pallet (bottom left). C-130 cargo floor (right)

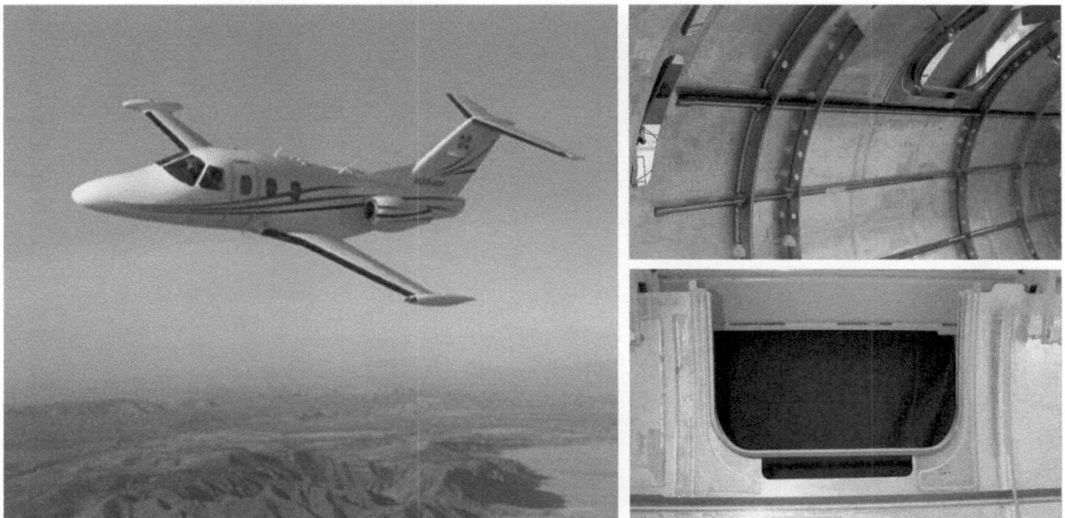

**Fig. 13.33** The Eclipse 500 business-class jet is currently in final Federal Aviation Administration certification trials (left). The internal longitudinal and circumferential aluminum stiffeners (top right) and window and door doublers (bottom right) are attached to the aluminum fuselage section with friction stir welded lap joints. Courtesy of Eclipse Aviation

in better quality and stronger and lighter joints at reduced assembly costs. MTS Systems Corporation designed and fabricated the custom-FSW equipment and production tooling for Eclipse Aviation. This equipment permits welding complex curvatures over many sections of the fuselage, cabin, and wing structures at travel speeds in excess of 51 cm/min (20 in./min) (Fig. 13.34). Because the process is faster than more conventional mechanical joining processes, the production cycle time is significantly reduced.

Over the last three years, the Ford Motor Company has produced several thousand Ford GT automobiles with an FSW central tunnel assembly (Fig. 13.35). This tunnel houses and isolates the fuel tank from the interior compartment and contributes to the space frame rigidity. The top aluminum stamping is joined to two hollow aluminum extrusions along the length of the tunnel, using a linear FSW lap joint. The use of FSW results in improved dimensional accuracy and a 30% increase in strength over similar GMAW welded assemblies.

The TRL for FSW of ferrous, stainless steel, nickel, copper, and titanium alloys is also high, with a variety of full-scale demonstration programs completed. MegaStir, Inc. has developed an improved grade of the PCBN high-temperature pin tool (HTPT) that has shown an acceptable service life for welding steels, nickel, and copper alloys. In 2004, MegaStir, Inc. completed a prototype oil field pipeline FSW demonstration program that successfully joined 30 cm diameter by 6 mm wall thickness (12 in. diameter by 0.25 in. wall thickness) X-65 steel pipe segments using an automated internal mandrel and external FSW tooling system (Fig. 13.36).

Chemical compatibility issues arise when welding titanium alloys with the PCBN pin

**Fig. 13.34** The friction stir welding (FSW) equipment used to attach the stiffeners and doublers to the Eclipse 500 fuselage sections was designed and fabricated by MTS Systems Corporation. It is capable of welding a variety of component geometries through the use of interchangeable holding fixtures located beneath the multiaxis FSW head and movable gantry frame. Courtesy of Eclipse Aviation

tools. The University of South Carolina has shown the suitability of tungsten-rhenium HTPT for most titanium alloys. However, issues with pin-tool wear and excessive metal adhesion still arise when welding Ti-6Al-4V. This is possibly due to reactions between the rhenium in the pin tool and the vanadium alloying elements in the titanium. Other refractory HTPT materials, such as tungsten-iridium, are under development at the Oak Ridge National Laboratory.

In 2005, Lockheed Martin performed FSW on 5 mm (0.20 in.) thick Ti-6Al-4V sheets using dispersion-strengthened tungsten HTPT that alleviated the sticking problem and allowed for many meters of welding (Fig. 13.37). They report that the joint efficiency ranged from 98 to 100% of base-metal strength at testing temperatures ranging from –195 to +260 °C (–320 to +500 °F). Titanium FSW produced at the CFSP using custom-designed environmental chambers and an argon atmosphere (Fig. 13.38) showed no evidence of surface discoloration or interstitial (oxygen, nitrogen, and hydrogen) contamination.

## 13.5 Friction Stir Spot Welding

If FSW is considered as a controlled-path extrusion rather than a welding process, several spin-off technologies can be realized. Friction stir spot welding (FSSW) has been in development over the last five years and has seen industrial implementation as a rivet-replacement technology. Currently, two variations to FSSW are being used. The plunge friction spot welding (PFSW) method was patented by Mazda in 2003 (Ref 26), and the refill friction spot welding (RFSW) method was patented by GKSS-GmbH in 2002 (Ref 27).

In the Mazda PFSW process, a rotating fixed-pin tool, similar to that used in linear FSW, is plunged and retracted through the upper and lower sheets of the lap joint to locally plasticize the metal and stir the sheets together. Even

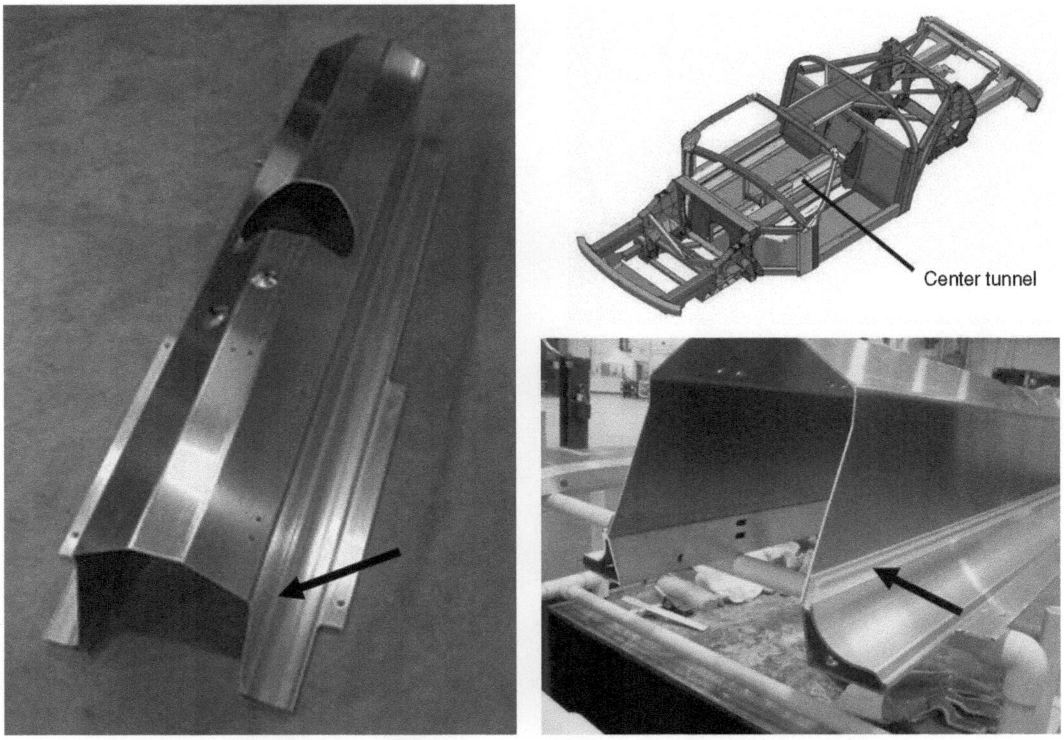


**Fig. 13.35** The central tunnel assembly of the Ford GT is a friction stir welded assembly made from aluminum stampings and extrusions. Courtesy of Ford Motor Company

though this approach leaves a pull-out hole in the center of the spot, the strength and fatigue life is sufficient to allow application at reduced production costs on the Mazda RX-8 aluminum rear door structure (Fig. 13.39). Since 2003, Mazda has produced more than 100,000 vehicles with this PFSW rear door structure. These PFSW doors provide structural stability against side-impact and impart five-star rollover protection.

The GKSS RFSW is being developed at the SDSMT AMP Center under license to RIFTEC-GmbH. This process uses a rotating pin tool with a separate pin and shoulder actuation system that allows the plasticized material initially displaced by the pin to be captured under the shoulder during the first half of the cycle and subsequently reinjected into the joint during the second half of the cycle. This completely refills the joint flush to the surface (Fig. 13.40). In addition to development as a rivet-replacement technology for aerospace structures, RFSW is also being developed as a tacking method to hold and restrain parts during overwelding by linear FSW.

## 13.6 Friction Stir Joining

Friction stir joining (FSJ) of thermoplastic materials uses the controlled-path extrusion characteristics of the process to join 6.3 mm (1/4 in.) thick sheets of polypropylene (PP), polycarbonate (PC), and high-density polyethylene (HDPE) materials (Fig. 13.41). Recent work at Brigham Young University has shown joint efficiencies for these materials ranging from 83% for PC to 95% for HDPE and 98% for PP. These joint efficiencies compare favorably with other polymer joining methods such as ultrasonic, solvent resistance, hot plate, and adhesive bonding. Current work at the SDSMT AMP Center in collaboration with the Air Force Research Laboratory-Kirtland is investigating the use of FSJ to join

**Fig. 13.36** Prototype pipe welding system showing external friction stir welded head and internal mandrel (inset). Courtesy of MegaStir, Inc.

fiber-, particulate-, and nanoparticle-reinforced thermoplastic materials.

## 13.7 Friction Stir Processing

Friction stir processing (FSP) uses the controlled-path metalworking characteristics of the process to perform metallurgical processing and microstructural modification of local areas on the surface of a part. In 1997, FSP was used by Lockheed Martin to perform microstructural modification of the cast structure of 2195 Al-Li variable polarity plasma arc (VPPA) welds to remove porosity and hot-short cracks. This also improved room-temperature and cryogenic strength, fatigue life, and reduced the sensitivity to intersection weld cracking by crossing VPPA welds (Ref 28).

In 1998, the Department of Energy's Pacific Northwest National Laboratory (PNNL) began investigating the processing of SiC powders into the surfaces of 6061 Al to increase wear resistance. Initial studies showed that both SiC and $Al_2O_3$ could be emplaced into the surface of bulk materials to create near-surface-graded metal-matrix composite structures. The University of Missouri-Rolla (UMR) has shown that a uniform SiC particle distribution can be achieved with appropriate tool designs and techniques, leading to significant increases in surface hardness.

In 2004, a PNNL/SDSMT AMP Center collaborative research program investigated increasing the wear resistance of heavy vehicle brake rotors by processing $TiB_2$ particles into the surface of class 40 gray cast iron. This resulted in a fourfold increase in the dry abrasive wear resistance when tested in accordance with ASTM G 65 (Fig. 13.42). The PNNL and Tribomaterials, LLC have performed subscale brake rotor/pad wear tests on FSP/$TiB_2$ cast iron rotors. These subscale brake tests have shown that FSP/$TiB_2$-processed brake rotors have

Fig. 13.37 Joining of long lengths of contamination-free Ti-6Al-4V is possible with out-of-chamber friction stir welding, using shrouds and trailing shoe shielding gas systems. Courtesy of Lockheed Martin Space Systems

improved friction characteristics and wear resistance over baseline heavy vehicle brake friction pairs.

## 13.8 Friction Stir Reaction Processing

Friction stir reaction processing (FSRP) was also investigated under the PNNL/SDSMT FSP/$TiB_2$ program. The FSRP uses the high temperatures and strain rates seen during processing to induce thermodynamically favorable in situ chemical reactions on the surface to a depth defined by the pin-tool geometry and metal flow patterns. This provides an opportunity for innovative processing methods to create new alloys on surfaces of materials and locally impart a variety of chemical, magnetic, strength, stiffness, and corrosion properties.

Studies performed at the University of Missouri-Rolla in conjunction with Rockwell Scientific have shown FSP to produce a fine-grain-size material and create low-temperature, high-strain-rate superplasticity in aluminum and titanium alloys. The PNNL is currently investigating the application of this FSP-induced superplasticity in the fabrication of large, integrally stiffened structures.

## 13.9 Summary

Friction stir welding has matured a great deal since its introduction into the U.S. market in 1995. The TRL for aluminum alloys is high, with several industrial implementations. While development efforts and property characterizations have shown that FSW can be used in fer-

**Fig. 13.38** Environmental chambers are used to provide an argon atmosphere and to minimize interstitial contamination in titanium friction stir welding. Courtesy of South Dakota School of Mines

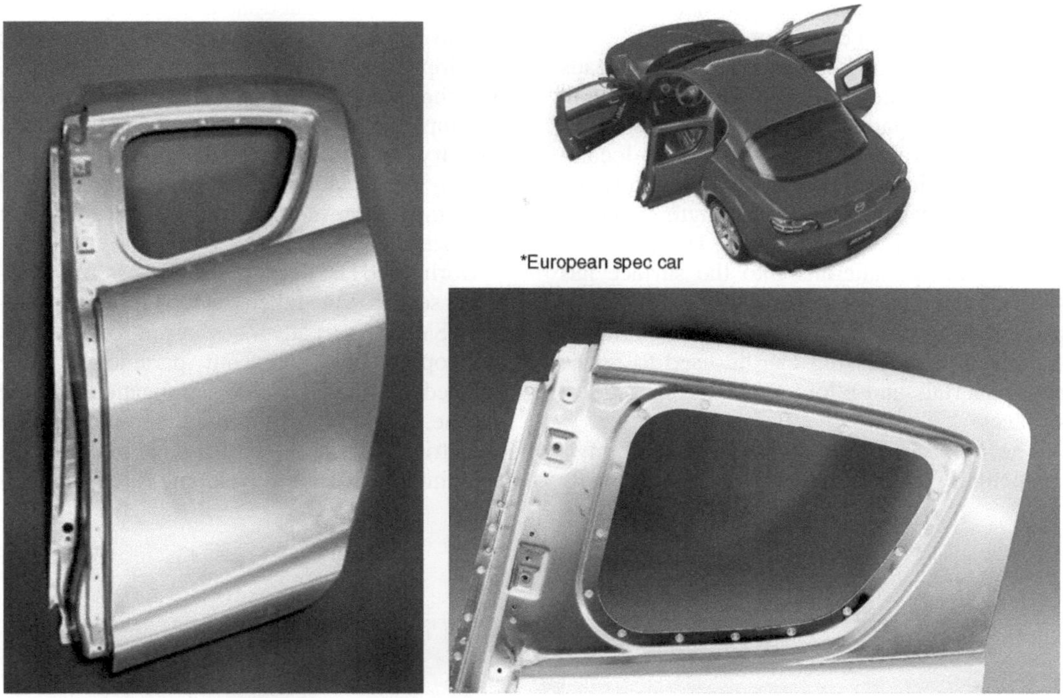


**Fig. 13.39** Use of the plunge friction spot welding method on the Mazda RX-8 rear door structure provides for structural stability against side impact and five-star rollover protection at reduced production costs. Courtesy of Ford Motor Company

**Fig. 13.40** Refill friction spot welding (RFSW) using MTS Systems Corporation ISTIR 10 system and custom-designed head adapter (left). The RFSW lap shear coupons (bottom right) and metallurgical cross section of RFSW showing complete joint penetration in 2 mm (0.080 in.) thick 7075-T73 Al (top right). Courtesy of South Dakota School of Mines

rous, stainless, nickel, copper, and titanium alloys, an industrial champion is needed.

The metalworking nature of the process leads to the plunge and refill FSSW methods, with properties comparable to riveted and resistance spot-welded joints. The use of FSP to locally modify the microstructure of arc welds and castings has shown to increase strength, improve fatigue life, and remove defects. Using FSP to stir particulate materials into the surface has shown increased wear resistance and creates particulate-reinforced surface layers. Friction stir reaction processing can be used to create new materials and alloy combinations on part surfaces.

The higher-strength, nonmelting, and environmentally friendly nature of the FSW process has shown cost reductions in a variety of applications and has enabled new product forms to be developed. Only a small percentage of the U.S. welding and joining market has been targeted for implementation. A variety of government, industry, and university collaborations are underway to accelerate the development and implementation of FSW and FSSW into these markets.

During the last decade, the defense and aerospace sectors have taken the lead in implementing FSW. Recent advances in pin-tool designs and optimized processing parameters have enabled FSW and FSSW applications in the marine, ground transportation, and automotive industries. Further innovations in low-cost equipment and the development of industry

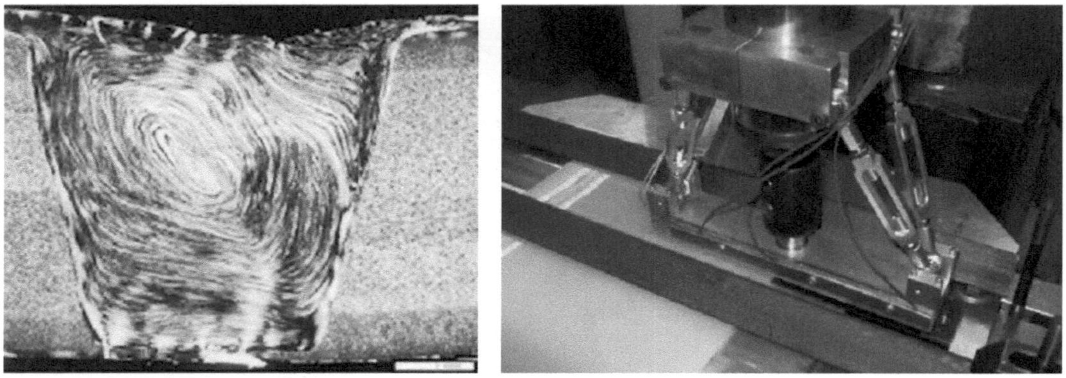

**Fig. 13.41** Cross section of polypropylene friction stir joining (FSJ) from Brigham Young University studies showing typical periodic flow patterns (left). Custom-designed thermoplastic FSJ system at South Dakota School of Mines (right)

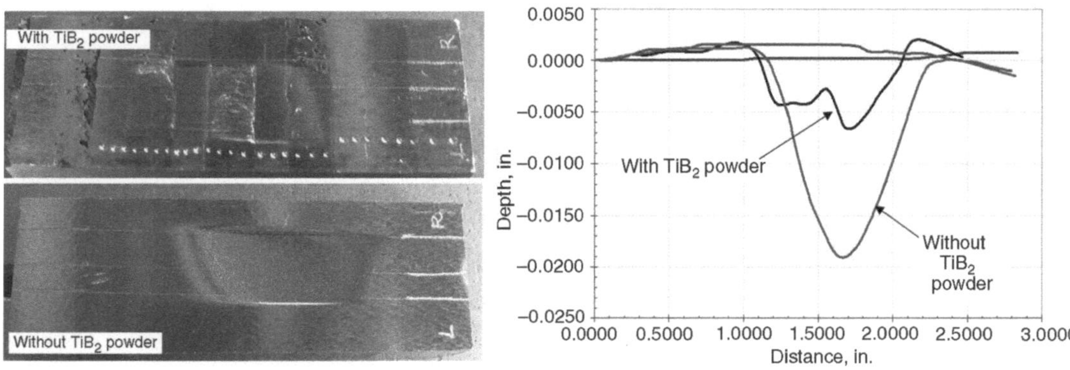


**Fig. 13.42** Grade 40 gray cast iron ASTM G 65 wear test results. Friction stir processing $TiB_2$ particles into the surface resulted in a fourfold increase in ASTM G 65 dry abrasive wear resistance over that seen in samples without $TiB_2$ particles. Courtesy of South Dakota School of Mines.

standards, design guidelines, and a trained workforce will enable the introduction of FSW and FSSW into the broader light manufacturing, heavy manufacturing, and construction industries during the next decade.


## ACKNOWLEDGEMENTS

Portions of this chapter were taken directly from W.J. Arbegast, "*Friction Stir Welding—After a Decade of Development,*" in the *Welding Journal,* March 2006, and from other writings of the author. Contributions to the *Welding Journal* article were received from Gilbert Sylva and Mike Skinner (MTS Systems Corporation), Glenn Grant (PNNL), Brent Christner (Eclipse Aviation), Doug Waldron (AJT, Inc.), Jeff Ding (NASA MSFC), Tim Trapp (EWI), Tracy Nelson (Brigham Young University), Carl Sorensen (Brigham Young University), Tony Reynolds (University of South Carolina), Zach Loftus (Lockheed Martin), Murray Mahoney (Rockwell Scientific), John Baumann (Boeing), Raj Talwar (Boeing), Dana Medlin (SDSMT), Anil Patnaik (SDSMT), Casey Allen (SDSMT), Rajiv Mishra (University of Missouri-Rolla), Chuck Anderson (ATI, Inc.), John Hinrichs (Friction Stir Link, Inc.), Kevin Colligan (CTC Corporation), Scott Packer (MegaStir), and Tsung-Yu Pan (Ford Motor Company).


## REFERENCES


1. W.M. Thomas et al., Friction Stir Butt Welding, U.S. Patent 5,460,317, Oct 24, 1995
2. W.J. Arbegast, Friction Stir Welding—After a Decade of Development, *Weld. J.,* March 2006, p 28
3. "Welding-Related Expenditures, Investments, and Productivity Measurement in U.S. Manufacturing, Construction, and Mining Industries," Internal Report, American Welding Society, Edison Welding Institute, and Office of Naval Research, May 2002
4. R.S. Mishra and Z.Y. Ma, Friction Stir Welding and Processing, *Mater. Sci. Eng.,* Vol 50, 2005, p 1–78
5. D. Waldren, Advanced Joining Technologies, unpublished data
6. M. Mahoney, Rockwell Scientific, private communication
7. J. Defalco, Friction Stir Welding vs. Fusion Welding, *Weld. J.,* March 2006, p 42
8. I. Smith, The Welding Institute (TWI), private communication
9. Friction Stir Welding Is a Hot Topic Worldwide, *Weld. J.,* March 2006, p 79
10. A Toskey, W.J. Arbegast, C.D. Allen, and A. Patnaik, Fabrication of Aluminum Box Beams Using Self-Reacting and Standard Fixed Pin Friction Stir Welding, *Friction Stir Welding and Processing III,* K.V. Jata et al., Ed., TMS (The Minerals, Metals and Materials Society), 2005
11. W.J. Arbegast, Modeling Friction Stir Joining as a Metalworking Process, *Hot Deformation of Aluminum Alloys III,* Z. Jin, Ed., TMS (The Minerals, Metals, and Materials Society), 2003
12. W.J. Arbegast, Using Process Forces as a Statistical Process Control Tool for Friction Stir Welds, *Friction Stir Welding and Processing III,* K.V. Jata, et al., Ed., TMS (The Minerals, Metals and Materials Society), 2005
13. Z.S. Loftus, W.J. Arbegast, and P.J. Hartley, Friction Stir Weld Tooling Development for Application on the 2195 Al-Li-Cu Space Transportation System External Tank, *Proceedings of the Fifth International Conference on Trends in Welding Research,* June 1–5, 1998 (Pine Mountain, GA), ASM International, p 580
14. W.J. Arbegast, "Using Grain Growth as an in situ flow marker during friction stir welding," presented at the Spring TMS (The Minerals, Metals, and Materials Society) Annual Meeting (San Diego, CA), 2003
15. W.A. Backofen, *Deformation Processing,* Addison-Wesley Publishing, 1972, p 88–115
16. P.A. Colegrove, H.R. Shercliff, and P.L. Threadgill, Modeling and Development of the Trivex Friction Stir Welding Tool, *Proceedings of the Fourth International Conference on Friction Stir Welding,* May 14–16, 2003 (Park City, UT), The Welding Institute (TWI)
17. K. Colligan, Material Flow Behavior during Friction Stir Welding of Aluminum, *Weld. J.,* July 1999, p 229
18. W.J. Arbegast and C.D. Allen, Friction Stir Welding of Complex Curvature Parts Using Rapid Configurable Tooling, *Proceedings of the Fifth International Con-*

*ference on Friction Stir Welding,* Sept 18–20, 2004 (Metz, France), The Welding Institute (TWI)

19. W.J. Arbegast and M. Skinner, "Multi-Axis Friction Stir Welding and Intelligent Laser Processing at the Advanced Materials Processing Center," presented at 13th Annual Advanced Aerospace Materials and Processes Conference and Symposium, June 9–12, 2002 (Orlando, FL), ASM International
20. W.J. Arbegast and A.K. Patnaik, Process Parameter Development and Fixturing Issues for Friction Stir Welding of Aluminum Beam Assemblies, *Proceedings of the 2005 SAE AeroTech Conference,* Oct 3–6, 2005 (Dallas, TX)
21. Z. Li and W.J. Arbegast, "Process Development of Friction Stir Lap Joints in AA7075 and AA2297 Alloys," presented at the TMS 2001 Annual Spring Meeting, Feb 11–15, 2001 (New Orleans, LA)
22. "Welding: Resistance, Spot and Seam," SAE-AMS-W-6858, April 2000
23. "Metallic Materials and Elements for Aerospace Vehicle Structures," MIL-HNBK-5H, Dec 1998
24. J. Ding and P. Oelgoetz, Auto-Adjustable Pin Tool for Friction Stir Welding, U.S. Patent 5,893,507, April 13, 1999
25. C.L. Campbell, M.S. Fullen, and M.J. Skinner, Welding Head, U.S. Patent 6,199,745, March 13, 2001
26. T. Iwashita et al., Method and Apparatus for Joining, U.S. Patent 6,601,751 B2, Aug 5, 2003
27. C. Schilling and J. dos Santos, Method and Device for Joining at Least Two Adjoining Work Pieces by Friction Welding, U.S. Patent Application 2002/0179 682
28. W.J. Arbegast and P.J. Hartley, Method of Using Friction Stir Welding to Repair Weld Defects and to Help Avoid Weld Defects in Intersecting Welds, U.S. Patent 6,230,957, May 15, 2001

# CHAPTER 14

# Friction Stir Processing

Rajiv S. Mishra, Center for Friction Stir Processing
University of Missouri-Rolla
Murray W. Mahoney, Rockwell Scientific Company

FRICTION STIR PROCESSING (FSP) is an adaptation of friction stir welding, and the following unique features of friction stir welding can be used to develop new processes based on the concept of friction stirring:

- Low amount of heat generated
- Extensive plastic flow of material
- Very fine grain size in the stirred region
- Healing of flaws and casting porosity
- Random misorientation of grain boundaries in the stirred region
- Mechanical mixing of the surface and subsurface layers

Therefore, the friction stir process can be used as a generic process to modify the microstructure and change the composition, at selective locations. At this time, FSP is the only solid-state processing technique that has these unique capabilities. In this chapter, examples of various FSP aspects are briefly presented to give readers an overview of the emerging trends. Many aspects of FSP are still in their infancy, and the examples covered are merely illustrative of the potential. Figure 14.1 shows a list of attributes and links to the FSP processes that build from those attributes.

## 14.1 Superplasticity

Superplastic forming is used to produce complex-shaped components and unitized structures. Superplasticity has emerged as an attractive, commercial, cost-effective, near-net shape forming process. Superplastic deformation and forming of materials have come of age in two decades of intensive research and technological development. Based on its potential impact in the manufacturing sector, a number of recent national reports have identified superplastic forming as a critical research area (Ref 1–4). At present, superplastic forming is used for a number of applications. In fact, it is an enabling technology for unitized structures (Ref 5). For example, the F-15E has implemented a superplastically formed and diffusion-bonded Ti-6Al-4V airframe structure as a replacement for built-up assemblies used in earlier models. This is a part of the U.S. Air Force Research Laboratory directed effort of the Metals Affordability Initiative Consortium to reduce the cost of metallic components in aircraft by 50% while accelerating implementation time (Ref 5). This initiative has resulted in a dramatic part-count reduction and demonstrated the successful use of unitized construction in service (elimination of 726 part details and 10,000 fewer fasteners).

In spite of such remarkable success stories of superplastic forming of aerospace components, the widespread use of superplastic forming in large-volume sectors, such as automotive, has been hampered by two factors: slow forming rates and the high cost of the starting material with a fine-grain superplastic microstructure. However, in the last ten years, high-strain-rate superplasticity has been developed. The emerging understanding continues to establish a relationship between grain size and superplastic strain rate and superplastic temperatures. For example, mechanically alloyed aluminum alloys (typical grain size 0.5 μm) exhibit superplasticity at strain rates >1 $s^{-1}$, comparable to conventional hot forming rates, whereas, for comparison, the usual (typical grain size 15 μm) superplastic strain rates are $10^{-4}$ to $10^{-3}$ $s^{-1}$ (Ref 6). The relationship between grain size and optimal strain rate of aluminum alloys is shown in Fig. 14.2(a) (data taken from Ref 6). Also, the superplastic temperature can change with grain size. Figure 14.2(b) shows the variation of superplastic temperature with grain size (data taken from Ref 7 and 8). Therefore, it is clear that by manipulating the grain size, it is possible to increase the superplastic strain rate and decrease the superplastic temperature. Both aspects have attractive technological significance.

**Microstructural Features for Enhanced Superplasticity.** Superplasticity is the ability of a metallic material to exhibit >200% uniform tensile elongation. The most important microstructural features that govern the overall superplastic behavior are:

- Fine grain size (<15 μm)
- Equiaxed grain shape
- Presence of very fine second-phase particles to inhibit grain growth
- Large fraction of high-angle grain boundaries

These requirements emanate from the mechanistic origin of superplasticity, that is, grain-boundary sliding. The high-temperature deformation based on grain-boundary sliding can be represented by a generic constitutive relation (Ref 9):

$$\dot{\varepsilon} = \frac{ADGb}{kT}\left(\frac{\sigma}{G}\right)^n\left(\frac{b}{d}\right)^p \quad \text{(Eq 14.1)}$$

where $\dot{\varepsilon}$ is strain rate, $G$ is shear modulus, $b$ is the Burgers vector, $\sigma$ is applied stress, $d$ is grain diameter (size), $D$ is appropriate diffusivity, $n$ is the stress exponent, $p$ is the inverse grain-size exponent, and $A$ is a microstructural- and mechanism-dependent dimensionless constant. Often, in superplasticity literature, the strain-rate sensitivity exponent ($m = \partial \log\sigma / \partial \log\dot{\varepsilon}$) is used instead of the stress exponent ($n$) shown in Eq 14.1. However, $m$ is just the reciprocal of $n$. Higher $m$ values mean a greater resistance to external neck formation and hence increased ductility. Generally, an $m$ value of ~0.5 and a $p$ value of 2 to 3 imply deformation by grain-boundary sliding.

Figure 14.3(a) shows a macrograph of 2024 Al with a friction stir processed nugget. The extent of grain refinement within the nugget is very apparent from Fig. 14.3(b) and (c). Friction stir processing has been generally found to be a very effective grain-refinement process. Table 14.1 gives a few examples of grain sizes obtained during friction stir welding and processing in several commercial aluminum alloys. A noteworthy feature is the 1 to 15 μm grain size range readily obtained by FSP. In addition,

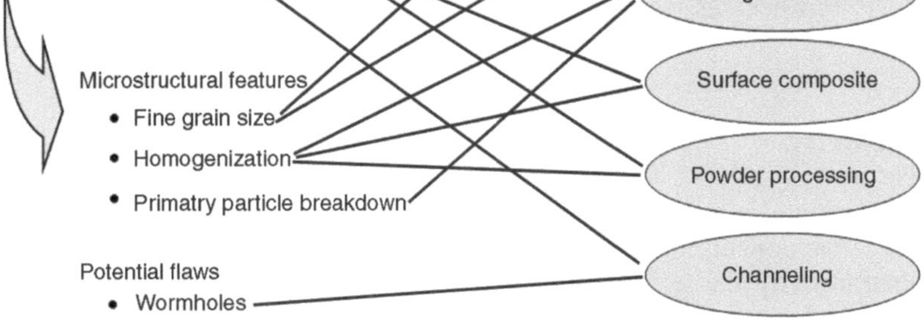


**Fig. 14.1** Schematic that lists attributes of friction stir processing and links to the friction stir processes

Fig. 14.3(d) shows the distribution of grain-boundary misorientation angles for a 7075 Al alloy. The friction-stirred material has a large fraction of high-angle grain boundaries (>0.95 in this example) after one pass. In comparison, conventional thermomechanical processing, involving rolling, produces a large fraction of low-angle grain boundaries. Equal-channel angular extrusion, an emerging severe plastic deformation technique, requires many passes to obtain a grain-boundary distribution similar to FSP. This illustrates the efficiency of FSP to convert the microstructure resulting from large processing strains and extensive mechanical mixing during FSP.

**Superplastic Behavior and Constitutive Relationships for FSP-Enhanced Superplasticity.** A number of studies have shown superplastic behavior after FSP (Ref 10, 24, 29–32, 34–55). Mishra et al. (Ref 34, 56) were first to report the possibility of using friction stir as a microstructural modification tool for enhanced superplasticity. There are three important aspects of superplasticity that are applicable to scientific and technological interests:

- *Flow stress:* Scientifically, the magnitude of flow stress provides insight to the difficulty of the deformation process, whereas technologically, it is important to keep the flow stress below 10 MPa (1.5 ksi) for gas forming.
- *Strain rate:* Scientifically, the strain rate is an indication of the flow kinetics of the deformation mechanism, whereas technologically, it has important implications for overall forming time. A strain rate of $10^{-2}$ $s^{-1}$ has been defined somewhat subjectively as the transition to high-strain-rate superplasticity (HSRS). The implication of HSRS is that components can be formed in minutes rather than the hours required at conventional forming rates.
- *Temperature:* Scientifically, the temperature for the onset of superplasticity (>200% uniform tensile ductility) is an indication of the efficiency of grain-boundary sliding-related

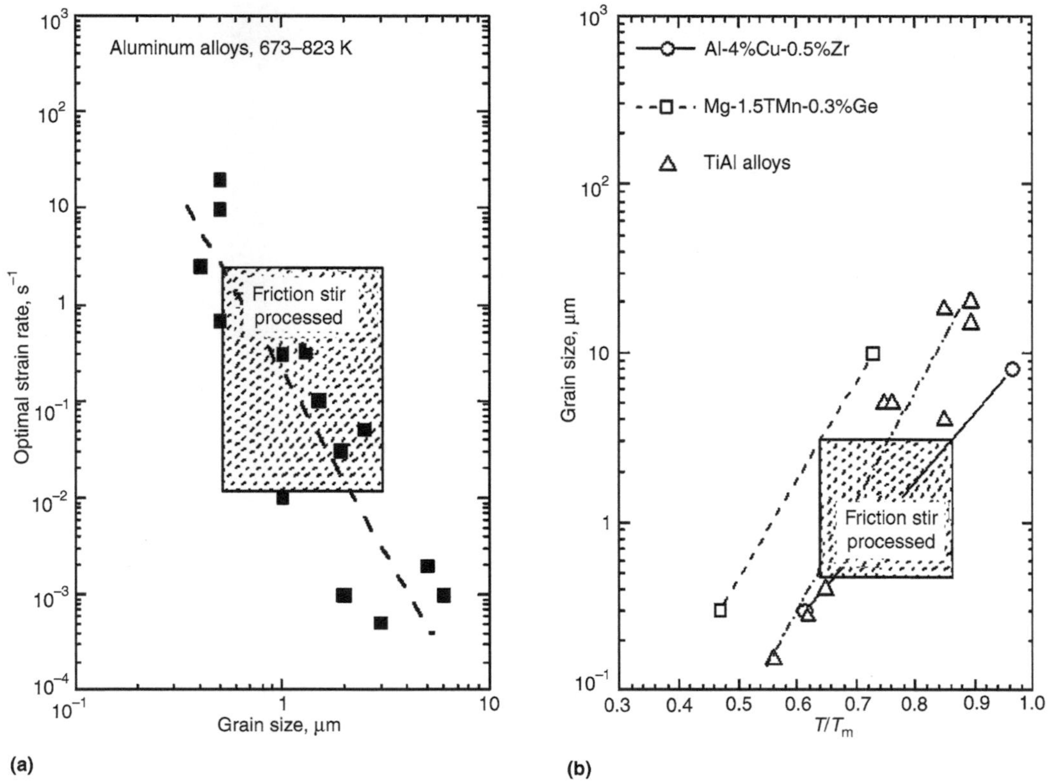


**Fig. 14.2** The influence of grain size on (a) optimal strain rate of aluminum alloys (Source: Ref 6) and (b) the superplastic temperature (Source: Ref 7, 8). Note the range of grain sizes obtainable by friction stir processing and the corresponding superplastic strain rate and temperature that are possible.

processes, whereas technologically, lower temperatures are preferable for multiple reasons, including energy efficiency of superplastic forming.

Figure 14.4 shows results on 7075 Al in this overall context. The results indicate that FSP lowers the flow stresses, increases the strain rates for superplasticity, and lowers the temperature range. Table 14.2 summarizes superplastic ductility in a number of alloys.

The stress-strain rate behavior is shown in Fig. 14.5(a) for three commercial aluminum alloys following FSP. As shown, the stress exponent is close to 2 and establishes grain-boundary sliding as the dominant deformation

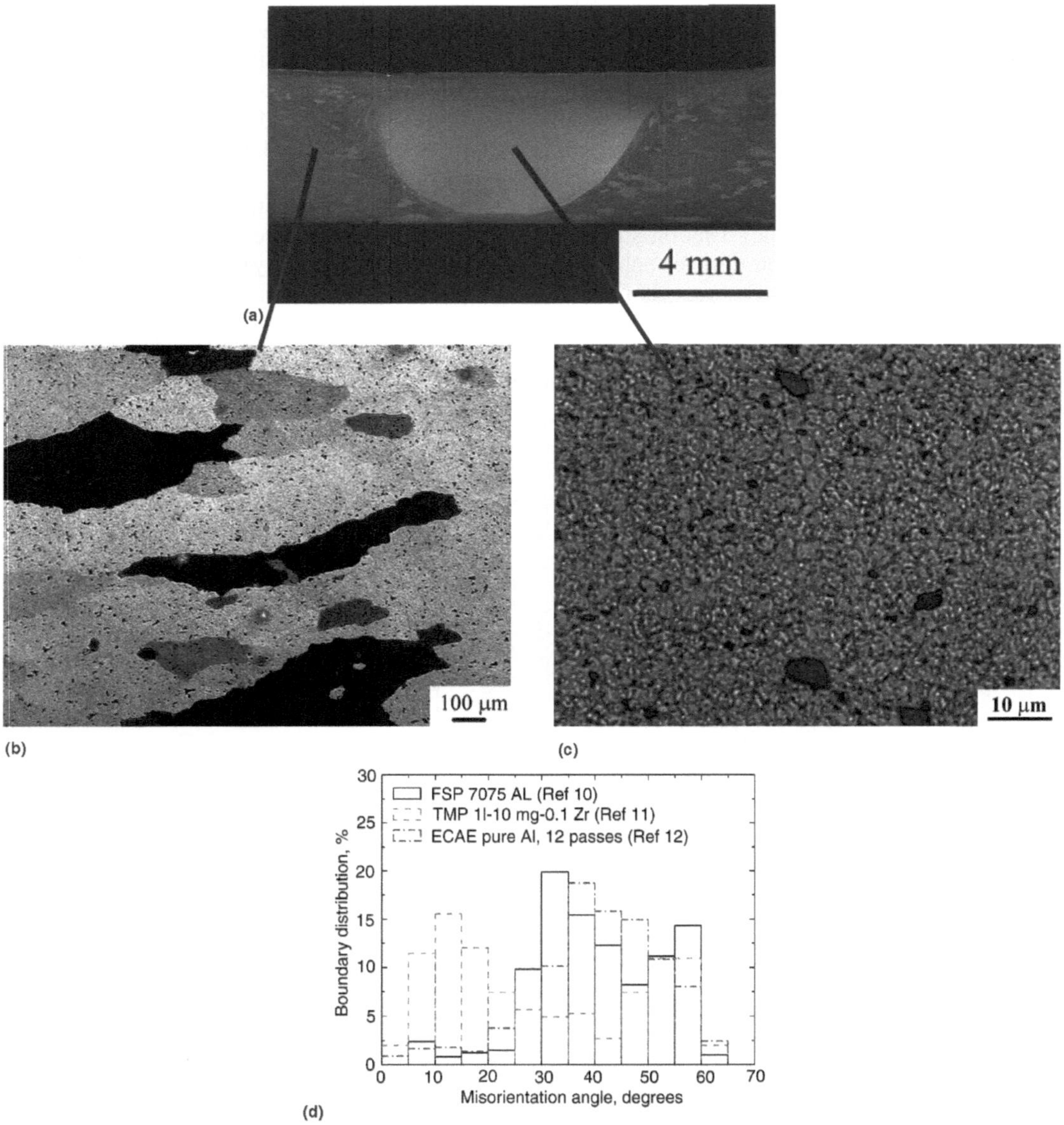


**Fig. 14.3** (a) Macrograph showing a friction stir processed (FSPed) nugget in a 2024 Al alloy. (b) and (c) Comparison of as-rolled and as-FSPed microstructure, showing grain refinement during FSP. (d) Comparison of grain-boundary misorientation distribution in FSP 7075 Al alloy with a distribution obtained by conventional thermomechanical processing (TMP) and equal-channel angular extrusion (ECAE)

process (Ref 57). Figure 14.5(b) shows a plot of grain size and temperature-compensated strain rate versus modulus-compensated flow stress for the FSP alloys. The constitutive relationship for superplasticity in aluminum alloys is given by Mishra et al. (Ref 6) as:

$$\dot{\varepsilon} = \frac{40D_0Gb}{kT}\left(\frac{\sigma}{E}\right)^2\left(\frac{b}{d}\right)^2\exp\left(\frac{-84000}{RT}\right) \quad \text{(Eq 14.2)}$$

The dimensionless kinetic constant for aluminum alloys is 40. The value observed for FSP 5083 Al is 279 (Ref 47) and that for 7075 Al is 750 (Ref 40). The implication is that the kinetics in FSP alloys are much faster than conventional aluminum alloys and by more than an order of magnitude in FSP 7075 Al. This raises the issue of the nature of grain boundaries after FSP. As highlighted earlier, FSP leads to a very high fraction of high-angle grain boundaries. The microstructure evolves through dynamic recrystallization during the friction stir process (see Chapter 4 of this book and a review in Ref 13 for more details). The current form of the constitutive relationship for superplasticity accounts for the grain size but not for the nature of grain boundaries. Although it is possible to comment that the nature of grain boundaries influences the kinetics of grain-boundary sliding, so far, a quantitative relationship has not been established.

**Friction Stir Processing as a Technology Enabler for New Concepts.** Apart from the opportunity for achieving high-strain-rate superplasticity in commercial alloys, FSP offers several new opportunities as a technology enabler (Ref 44, 56). Some of these possibilities are briefly described as follows (Ref 44):

- *Selective superplastic forming:* In many components, only selected regions require superplastic deformation. The concept of such a superplastically formed component is shown in Fig. 14.6. In essence, only the region undergoing superplastic deformation needs the fine-grained microstructure. However, conventional processing cannot be used to produce microstructural refinement on a selective basis. The FSP provides such an opportunity. Using FSP, a selected portion of the sheet can be processed for superplastic behavior. The difference in microstructure would result in selective deformation of the grain-refined region (Fig. 14.6). Recently, Wang et al. (Ref 58) have performed a finite element simulation of selective superplastic forming. Figures 14.6(c) and (d) show the results of the finite element simulation for sheet with two different grain sizes. The FSP

**Table 14.1 Summary of grain size in the friction stir nugget or processed zone in aluminum alloys**

| | Plate thickness | | | Traverse speed | | | |
|---|---|---|---|---|---|---|---|
| Material | mm | in. | Rotation rate, rpm | mm/min | in./min | Grain size, μm | Reference |
| 7075Al-T6 | 6.35 | 0.25 | ... | 127 | 5 | 2–4 | 14 |
| 6061Al-T6 | 6.3 | ¼ | 300–1000 | 90–150 | 3.5–6 | 10 | 15 |
| Al-Li-Cu | 7.6 | 0.3 | ... | ... | 9 | 16 | |
| 7075Al-T651 | 6.35 | 0.25 | 350, 400 | 102, 152 | 4, 6 | 3.8, 7.5 | 17 |
| 6063Al-T4, T5 | 4.0 | 0.16 | 360 | 800–2450 | 31–96 | 5.9–17.8 | 18 |
| 6013Al-T4, T6 | 4.0 | 0.16 | 1400 | 400–450 | 16–18 | 10–15 | 19 |
| 1100Al | 6.0 | 0.2 | 400 | 60 | 2.4 | 4 | 20 |
| 5054Al | 6.0 | 0.2 | ... | ... | ... | 6 | 21 |
| 1080Al-O | 4.0 | 0.16 | ... | ... | ... | 20 | 22 |
| 5083Al-O | 6.0 | 0.2 | ... | ... | ... | 4 | 22 |
| 2017Al-T6 | 3 | 0.12 | 1250 | 60 | 2.4 | 9–10 | 23 |
| 2095Al | 1.6 | 0.06 | 1000 | 126–252 | 5–10 | 1.6 | 24 |
| Al-Cu-Mg-Ag-T6 | 4.0 | 0.16 | 850 | 75 | 3 | 5 | 25 |
| 2024Al-T351 | 6.0 | 0.2 | ... | 80 | 3.15 | 2–3 | 26 |
| 7010Al-T7651 | 6.35 | 0.25 | 180, 450 | 95 | 3.7 | 1.7, 6 | 27 |
| 7050Al-T651 | 6.35 | 0.25 | 350 | 15 | 0.6 | 1–4 | 28 |
| Al-4Mg-1Zr | 10 | 0.4 | 350 | 102 | 4 | 1.5 | 29 |
| 2024Al | 6.35 | 0.25 | 200–300 | 25.4 | 1 | 2.0–3.9 | 30 |
| 7475Al | 6.35 | 0.25 | ... | ... | ... | 2.2 | 31 |
| 5083Al | 6.35 | 0.25 | 400 | 25.4 | 1 | 6.0 | 32 |
| 2519Al-T87 | 25.4 | 1.0 | 275 | 102 | 4 | 2–12 | 33 |

Source: Adapted from Ref 13

region with finer grain size undergoes superplastic deformation. This provides a versatile method to produce gas-formed components with an intricate design.

- *Superplastic forming of thick sheets:* This is described in the next section.
- *One-step processing for superplasticity from cast sheet or hot-pressed powder metallurgy sheet:* A conventionally cast microstructure can be converted to a superplastic microstructure in many steps. The present process of microstructural refinement can be used directly on cast sheets. This leads to very economical manufacturing. Ma et al. (Ref 46) have demonstrated superplasticity in A356 cast alloy. They were able to obtain a maximum superplastic elongation of 650% at 530 °C (985 °F) for an initial strain rate of 1 × $10^{-3}$ $s^{-1}$. Charit and Mishra (Ref 55) observed exceptional superplastic properties in an as-cast Al-8.9Zn-2.6Mg-0.09Sc (wt%) alloy. The FSP with a smaller pin tool led to ultrafine grains (0.68 μm grain size) from the as-cast state. The ultrafine-grained alloy exhibited superplasticity at relatively low temperatures and higher strain rates. An optimal ductility of 1165% at a strain rate of 3 × $10^{-2}$ $s^{-1}$ and 310 °C (590 °F) was obtained. Enhanced

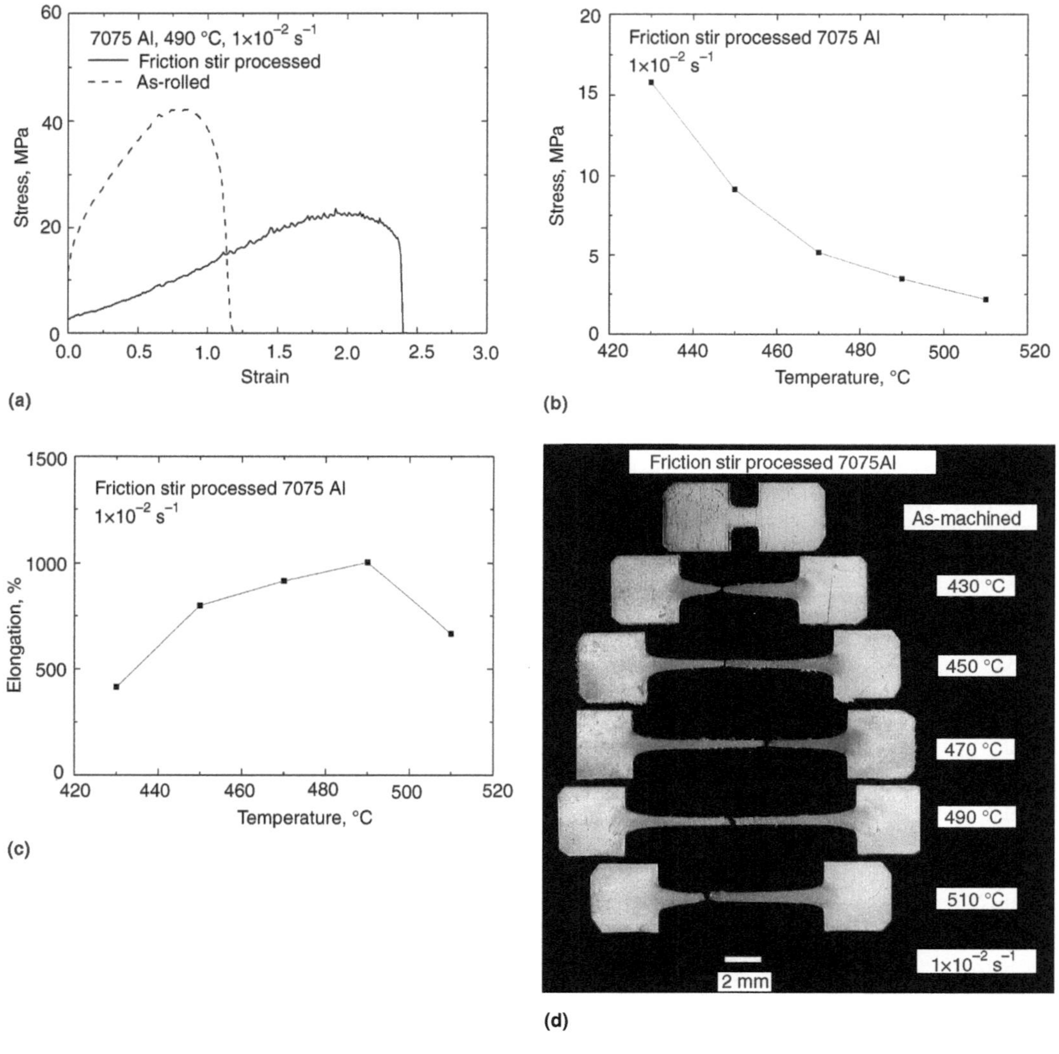


**Fig. 14.4** (a) Comparison of flow curves in as-rolled and as-friction stir processed conditions. (b) Variation of flow stress with temperature at a strain rate of $10^{-2}$ $s^{-1}$ and strain of 0.1. (c) Observation of exceptional ductility over a wide range of temperatures. (d) Photographs of deformed specimens show high uniform elongation, characteristic of superplastic flow.

superplasticity was also achieved at a temperature as low as 220 °C (430 °F). A similar approach can be taken for direct chill cast or continuous cast sheet, thereby eliminating several steps. Similarly, powder metallurgy processed aluminum alloys require extensive thermomechanical processing to break down the prior-particle boundaries that contain an alumina film. The friction stir process results in a very uniform microstructure from the hot-pressed sheet. For example, FSP of a nanophase Al-Ti-Cu alloy results in a remarkable combination of high strength and ductility (Ref 59). Again, the economical benefits of eliminating several steps are likely to be substantial. This approach will involve cast or powder metallurgy sheet + FSP + superplastic forming to produce high-strength, low-cost, unitized structures.

These concepts can be applied to many metallic materials and metal-matrix composites, but they have maximum impact on aluminum and magnesium alloys and components.

**Superplasticity in Very Thick FSP 7*xxx* Aluminum Alloys.** In conventional thermomechanical processing involving rolling of sheets, the sheet thickness reduces with every pass. To provide sufficient total strain for grain refinement, a number of passes are required, resulting in thin sheets (<3 mm, or 0.12 in.). For example, Grimes and Butler (Ref 60) have mentioned that a high-quality final product is currently available only as sheets having thickness less than 3 mm. On the other hand, when using FSP, the sheet thickness does not change. High-strain-rate superplasticity has been demonstrated in very thick-section (5 mm, or 0.2 in., thick) 7050-T7651 following FSP (Ref 36, 39, 42). High-strain-rate and thick-section superplasticity are two material properties never before demonstrated on a practical scale and are made possible only by FSP. For example, Mahoney et al. (Ref 39) demonstrated high uniform elongation (>500%) at strain rates >1 × $10^{-3}$ $s^{-1}$ at temperatures <460 °C (860 °F). These properties are possible because FSP produces a relatively small, uniform, and thermally stable grain size through the sheet thickness. This offers the potential to form complex-shaped parts at a higher strain rate and in section thickness never before possible. Figure 14.7 illustrates super-

**Table 14.2 Summary of superplastic elongation observed in a number of aluminum alloys**

| Alloy | Conditions | Elongation, % | Reference |
|---|---|---|---|
| 7075 Al | 3 × $10^{-3}$ $s^{-1}$, 480 °C (895 °F) | ~1450 | 40 |
| 2024 Al | 1 × $10^{-2}$ $s^{-1}$, 430 °C (805 °F) | 525 | 30 |
| 5083 Al | 3 × $10^{-3}$ $s^{-1}$, 530 °C (985 °F) | 590 | 47 |
| Al-4Mg-1Zr | 1 × $10^{-1}$ $s^{-1}$, 525 °C (980 °F) | 1280 | 29 |
| Al-Zn-Mg-Sc | 3 × $10^{-2}$ $s^{-1}$, 510 °C (950 °F) | ~1800 | 55 |

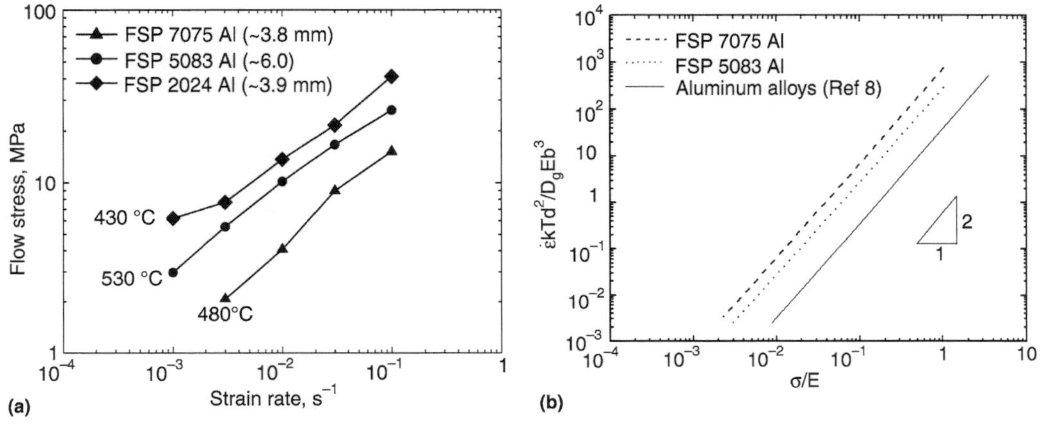


**Fig. 14.5** (a) Variation of flow stress with strain rate for three friction stir processed (FSPed) aluminum alloys. (b) Constitutive equations for FSPed 7075 (Ref 40) and 5083 (Ref 47) Al alloys compared to the equation for aluminum alloys proposed by Ref 8

plastic tensile elongation in 5 mm thick FSP 7050 Al, showing the initial tensile geometry (Fig. 14.7a), limited elongation and severe necking without FSP (Fig. 14.7b), and superplastic tensile elongation of ~800% at 460 °C following FSP (Fig. 14.7c). The thickness limit for superplasticity has not been established. In unpublished research, Mahoney et al. extended the material thickness to 12 mm (½ in.) and illustrated uniform elongations up to 500% (Fig. 14.8) (Ref 61). If the FSP tool and system are capable of greater depths and force, a uniform fine-grain microstructure should be possible to significantly greater depths, and a superplastic response can be anticipated for even thicker friction stir processed material.

The practical implications for enhanced superplasticity are illustrated in Fig. 14.9. Figure 14.9 illustrates results from gas pressure forming tests for different test conditions (Ref 42). Using a conventional superplastic 7475 Al alloy, that is, the sheet was processed to a fine grain size (~15 μm grain size) via special thermomechanical processing, Fig. 14.9(a) illustrates the ability to completely form the cone at 150 psi (1000 kPa) in 95 min. For the same test conditions but using FSP 7475 Al (~3 to 4 μm grain size), the time to completely form the cone was reduced to 18 min (Fig. 14.9b). Conversely, Fig. 14.9(c) illustrates the ability to reduce the internal gas pressure to 100 psi (690 kPa) from 150 psi (1000 kPa) and still reduce the time for forming to 49 min from 95 min. This demonstrates the high strain-rate enhancement associated with the very fine grain size created by FSP.

In a second example of a superplastic benefit attributed to FSP, Fig. 14.10 illustrates the ability to superplastically form a thick-section structure (5 mm, or 0.2 in.). Without FSP, the structure could not be fabricated, that is, the edges and corners could not being fully formed using a conventional superplastic 7475 Al alloy. Figure 14.10(a) shows the 5 mm thick 7475 Al sheet being locally friction stir processed to enhance strain just in the areas where the maximum strain is needed. It is not necessary to FSP the entire sheet. Figure 14.10(b) shows the part following superplastic forming. In this example, only one edge was friction stir processed to illustrate the difference in superplasticity following FSP. Figure 14.10(c) shows the inability to fully form the corners, whereas Fig. 14.10(d) illustrates complete forming in the corners where the sheet was friction stir processed. These results highlight an

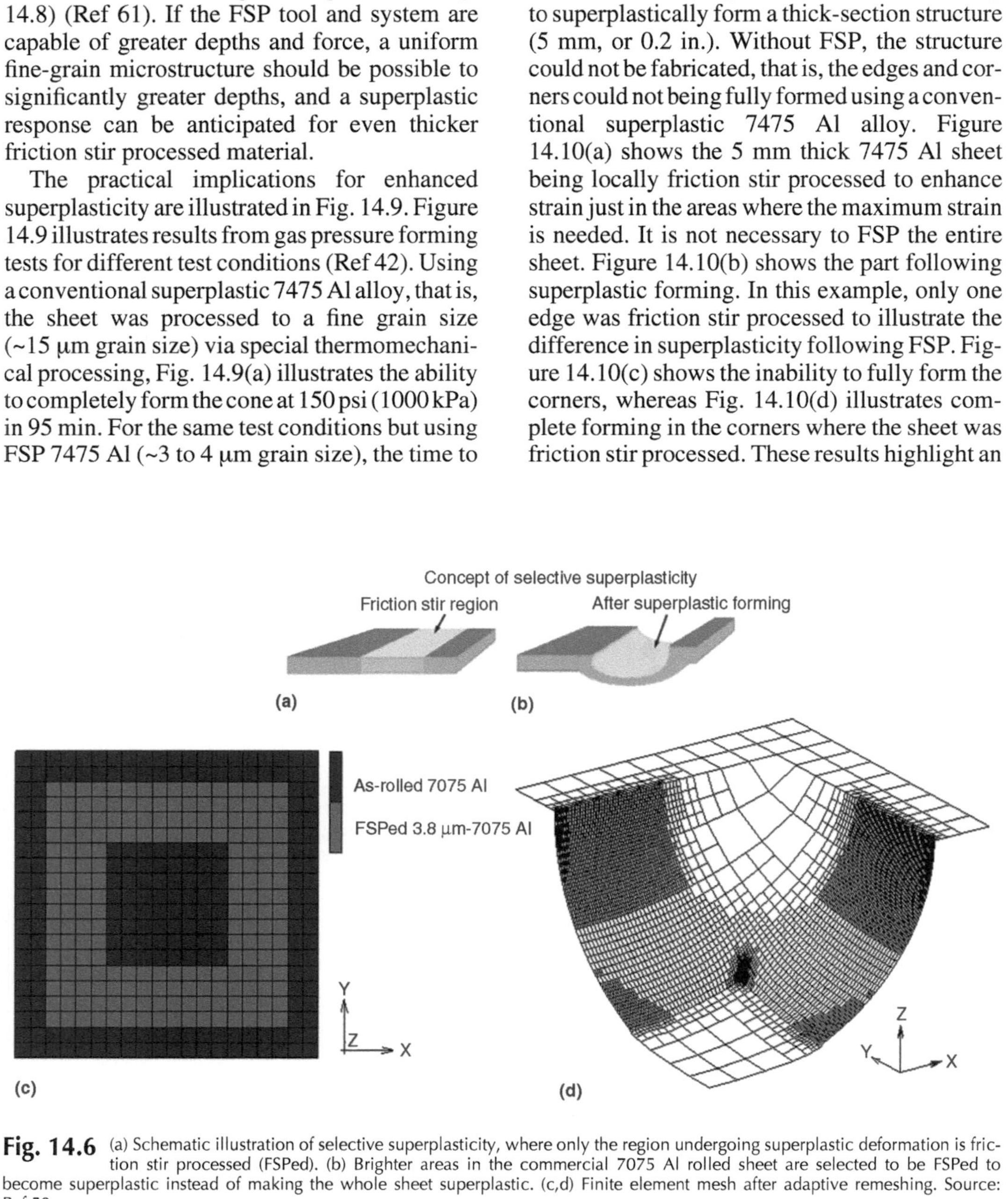


**Fig. 14.6** (a) Schematic illustration of selective superplasticity, where only the region undergoing superplastic deformation is friction stir processed (FSPed). (b) Brighter areas in the commercial 7075 Al rolled sheet are selected to be FSPed to become superplastic instead of making the whole sheet superplastic. (c,d) Finite element mesh after adaptive remeshing. Source: Ref 58.

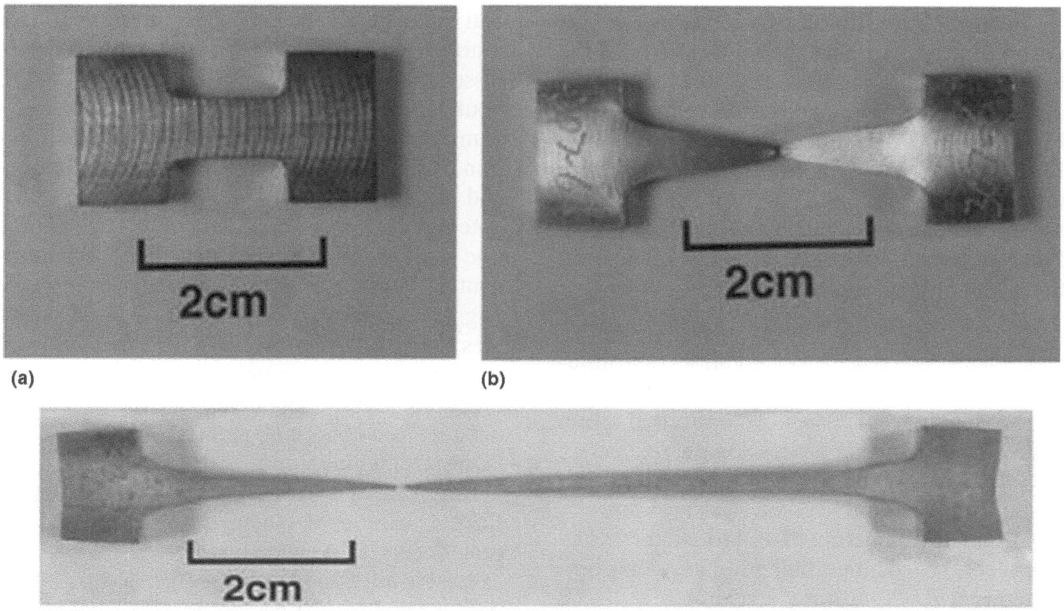


(a) (b) (c)

**Fig. 14.7** Superplastic tensile elongation. (a) 5 mm (0.2 in.) thick tensile sample. (b) Limited tensile elongation and severe necking without friction stir processing (FSP). (c) 800% superplastic elongation in 5 mm thick FSP 7075 Al

example where FSP combined with superplastic forming (SPF) results in the fabrication of a monolithic structure that could not be fabricated by any other means.

**Superplastic Forming of Multisheet Structures.** Multisheet structures are commercially fabricated by combining diffusion bonding and SPF of titanium alloys. The key issue that helps titanium alloys and hinders aluminum alloys is diffusion bonding. Because of the surface oxide layer, diffusion bonding of aluminum alloys is difficult. This has limited the development of aluminum alloy multisheet structures. The work of Grant et al. (Ref 62) has demonstrated the feasibility of making multisheet structures by combining friction stir welding (FSW) and friction stir spot welding (FSSW) with SPF. Figure 14.11 shows an example of a three-sheet structure created by FSW through two and three sheets. Fusion welding of aluminum alloys leads to a complete loss of superplasticity in the welded region, because of microstructural changes. As noted earlier, the FSW microstructure consists of fine grains, and superplastic properties are not degraded. A new opportunity involves microstructural tailoring by controlled heat input during FSW. The grain size can be varied by changing the thermal input. By controlling the microstructure, one can make the superplastic flow stress of the FSW region lower, higher, or equal to the parent sheet. Grant et al. (Ref 62) have also used FSSW to create different types of multisheet structures. Their work is opening up new possibilities of sandwich structures using aluminum alloy sheets.

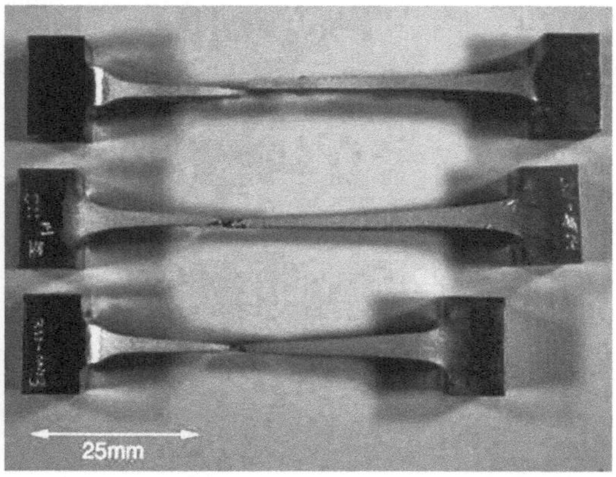


**Fig. 14.8** Superplastic strain in 12 mm (0.5 in.) thick friction stir processed 7475 Al at $2 \times 10^{-4}$ $s^{-1}$. (top) 460 °C (860 °F), 670% strain. (middle) 440 °C (825 °F), 630% strain. (bottom) 420 °C (790 °F), 470% strain

**Thermal Stability of the FSP Alloys at Elevated Temperatures.** Thermal stability of the FSP alloys has been investigated using elevated-temperature annealing experiments. Figure 14.12 shows optical macrographs of an FSP 7075 Al alloy (friction stir processed with different combinations of tool rpm and traverse speed) annealed at 490 °C (915 °F) for 1 h. Some important observations can be made from these sets of macrographs:

- Traverse speed and tool rotation rate influence abnormal grain growth.
- The location of onset of abnormal grain growth depends on processing parameters.
- Grain growth direction follows the nugget ring configuration in some cases.
- At some combination of tool rotation rate and traverse speed, abnormal grain growth is limited to a very thin surface layer.

Microstructural stability can be a critical issue for superplasticity in some FSP aluminum alloys, and it will define the upper limit for the SPF temperature range. As noted by Mishra and coworkers (Ref 30, 63), FSP alloys drastically lose their ductility over 450 °C (840 °F), accompanied by excessive growth of grains. A fine-grain microstructure has an intrinsic instability at elevated temperatures due to high grain-boundary driving forces. To resist grain growth, fine second-phase dispersions are often desirable. However, as Humphreys and Bate (Ref 64) have pointed out, abnormal grain growth can occur and destabilize the microstructure, even in the presence of pinning particles. Several factors contribute to the onset of abnormal grain growth: reduction of pinning forces due to dissolution of particles, anisotropy in grain-boundary energy and mobility, and thermodynamic driving forces from grain size distribution. It has been noted that anisotropy in energy and mobility of grain boundaries may not be a potential cause for abnormal grain growth in the fine-grained nugget because of the predominantly high-angle grain boundaries in the nugget region in an FSW 7075 Al. Two strategies can be adopted for SPF of FSW/FSP alloys: use processing parameters that eliminate abnormal grain growth, and determine the onset temperature for abnormal grain growth and work below that temperature.

**Cavitation during Superplastic Deformation.** During superplastic deformation, cavitation occurs in a wide variety of alloys, and extensive attention has been given to this aspect because of its influence on post-SPF properties (Ref 65, 66). It has been demonstrated that the post-SPF mechanical properties of the materials are significantly reduced when the cavity volume fraction exceeds approximately 1% (Ref 67). Ma and Mishra (Ref 41) have established the extent of cavitation during superplastic deformation of FSP 7075 Al alloy. Figure 14.13(a) shows the variation of cavity volume fraction with strain. The results for an FSP alloy are compared with a conventionally thermomechanical processed alloy. It is quite apparent that the FSP alloy shows lower cavitation at equivalent strain. Further, the grain size influences the onset and growth of cavities. Finer grain size shows lower cavitation at equivalent deformation strain. Figure 14.13(b) shows the

**Fig. 14.9** Gas pressure cone tests for different conditions including (a) conventional superplastic 7475 Al alloy, 150 psi (1000 kPa), 95 min, (b) friction stir processed (FSP) 7475 Al, 150 psi, 18 min, and (c) FSP 7475 Al, 100 psi (690 kPa), 49 min

critical strain for cavity nucleation in fine-grained FSP 7075 Al alloy. The values of critical strain are higher than ~1.0 in the high-strain-rate range. The technological implication of this is quite significant. It suggests that the FSP 7075 Al alloy can be formed to a total deformation of greater than 150% without any cavitation.

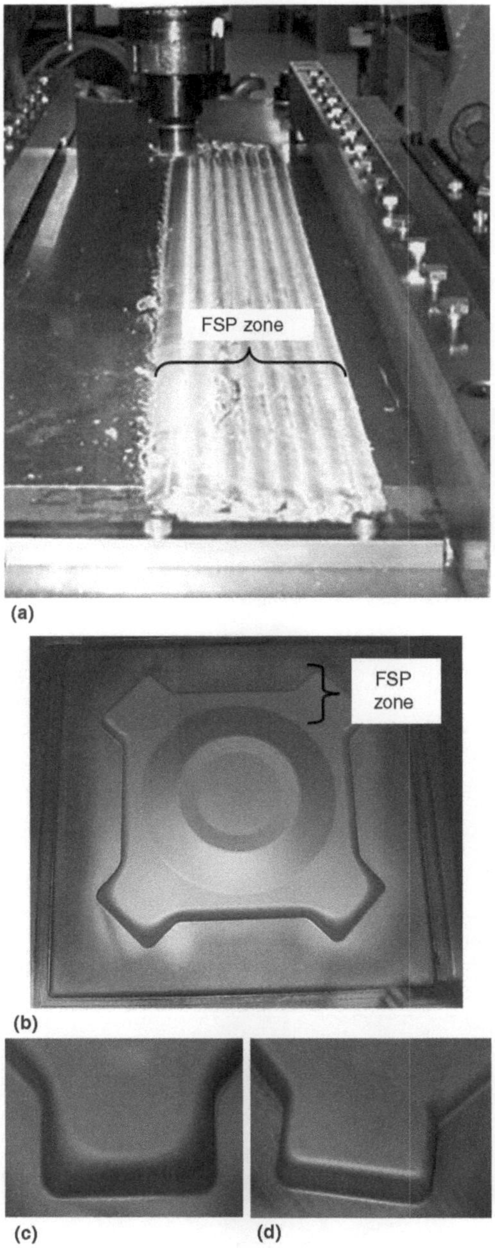


**Fig. 14.10** (a) Local friction stir processing (FSP) to enhance superplasticity. (b) Part after super plastic forming, where two corners were friction stir processed and two were conventional superplastic material. (c) Incomplete forming without FSP. (d) Complete forming following FSP

## 14.2 Enhanced Room-Temperature Formability via FSP

Thick-plate aluminum structures made using conventional fusion welding techniques result in built-up welded aluminum structures, such as plates, welded together to make enclosures. This inefficient, costly fabrication approach produces inferior properties, inasmuch as fusion welding creates a cast microstructure, high residual stresses, distortion, weld defects, and extensive precipitate overaging in the heat-affected zone of the precipitation-hardenable alloys. The ability to build monolithic or near-monolithic structures with improved properties and design flexibility is generally not possible with thick aluminum alloy plate. This is because forming or bending of thick conventional material is restricted to low angles, and even when forming or bending is possible, modestly elevated temperatures are required. Further, to final-machine a monolithic structure from a starting block may require first preforming or bending the block into a shape from which the final monolithic structure can be machined; that is, the available material size may be insufficient to fabricate the structure without first creating a shaped preform.

In addition to high-temperature formability, FSP can also be used to significantly enhance the room-temperature formability of aluminum alloys. The FSP locally anneals and creates a fully recrystallized fine-grain microstructure at selected areas within a thick aluminum plate, thus producing a selected region with low flow stress and enhanced formability. If the high-ductility surface is the tensile surface in bending, then bend limits can be extended, sometimes significantly. Superplastic forming requires high temperature and FSP through the entire sheet or plate thickness, that is, not just a shallow alteration of the surface microstructure. However, by using FSP as a surface modification procedure, room-temperature formability can be created in thick aluminum plate (Ref 69).

At times, it may be necessary to friction stir process a large surface area by rastering. Rastering refers to a pattern whereby the tool traverses a selected area on the surface and to a prescribed depth, wherein the microstructure is

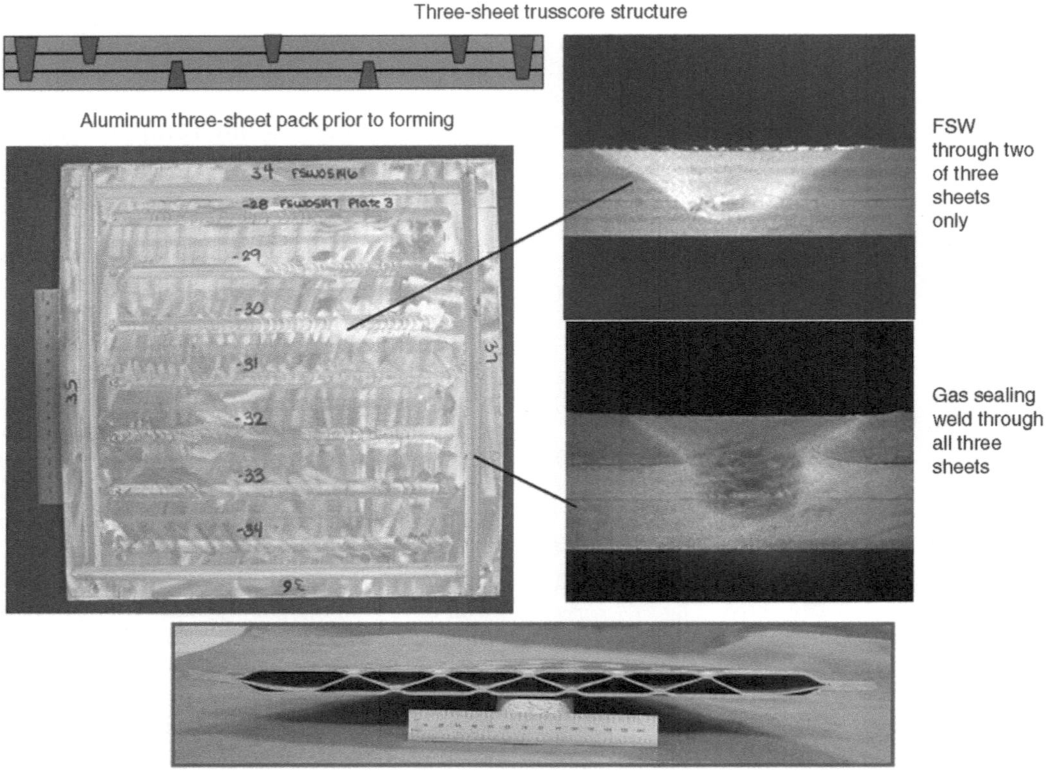


**Fig. 14.11** Example of multisheet structure created by a combination of friction stir welding (FSW) and superplastic forming. Courtesy of Glenn Grant

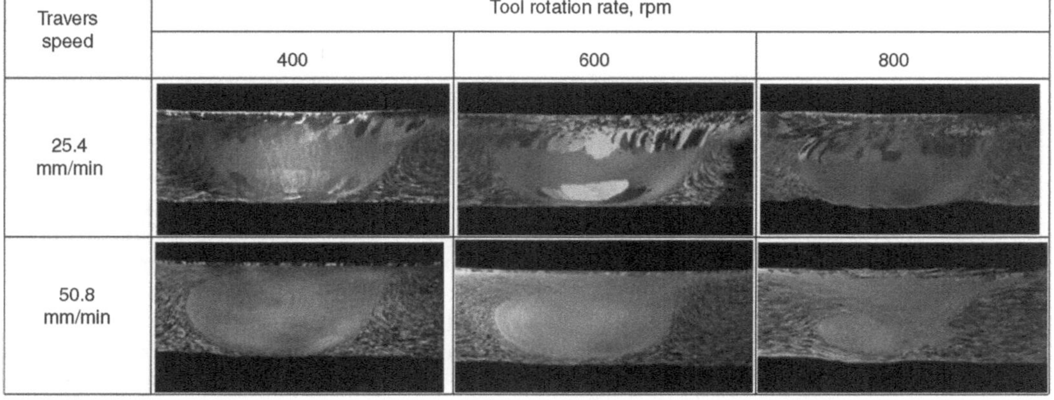


**Fig. 14.12** Collection of optical macrographs of friction stir processed (FSP) 7075 Al alloys (processed with different combinations of FSP parameters) heat treated at 490 °C (915 °F) for 1 h

modified by FSP. This is accomplished by traversing the FSP tool linearly forward and back or in some circular pattern until the surface area that subsequently experiences high tensile stresses during bending has been processed to create a fine-grained, fully recrystallized, annealed microstructure. Typically, to raster a large area, the FSP tool is moved a half-pin diameter to the advancing side of the previous pass to assure complete coverage. Because the advancing-side microstructure is most abrupt in friction stir processed aluminum alloys, moving the tool to the advancing-side direction creates a more homogeneous microstructure. Illustrations of different raster approaches are presented subsequently.

At the time of this writing, FSP-enhanced room-temperature formability is in its infancy, and most results are essentially qualitative. However, the early results illustrate an extremely promising new tool to create extended formability and thus are presented qualitatively to illustrate what can be accomplished using FSP as a surface-engineering approach. The FSP parameters have not been optimized to maximize either process efficiency or formability. For example, optimal penetration depth of the FSP tool for maximum travel speed (minimum cost) and subsequent maximum formability has not been established. Clearly, penetration of the tool beyond the neutral axis (approximately half the thickness) is not necessary. Further, it can be assumed that the requisite FSP penetration depth increases with an increased degree of bending, whereby the bend limit is restricted by the local ductility. A quantitative evaluation of this relationship has not been experimentally established. However, this penetration depth/bend radius relationship is important, especially for cost considerations, when large areas are rastered. Deeper penetration necessitates a slower travel speed. Further, the deeper the penetration, the greater the heat input. Higher heat input influences (reduces) other mechanical properties within the bulk of the structure. A computational model of this relationship has been attempted, with reasonable success (Ref 70).

Although results are limited, investigators have evaluated different aluminum alloys and different material thickness for room-temperature formability following FSP, including:

- 25 mm (1 in.) thick 2519-T87 (6.0Cu-0.3Mn-0.1Zr-Al) (Ref 71)
- 50 mm (2 in.) thick 7050-T7451 (2.3Cu-6.2Zn-0.12Si-2.3Mg-Al) (Ref 72)
- 150 mm (6 in.) thick 6061-T6 (0.6Si-0.7Fe-0.25Cu-0.15Mn-1.0Mg-0.2Cr-0.25Zn-Al) (Ref 72)

The 2519 Al alloy is an impact-resistant aluminum alloy, potentially for armor applications; the 7050 Al alloy is a high-strength nonweldable Al-Cu-Zn alloy of particular interest for aircraft applications; and the moderate-strength, weldable 6061 Al alloy is a versatile, commonly used aluminum alloy. Room-temperature formability results following FSP are presented as follows for these three alloys.

**FSP of 25 mm (1 in.) Thick 2519-T87 Aluminum.** Mahoney et al. (Ref 72) evaluated the ability to bend 25 mm thick 2519-T87 Al plate at room temperature following FSP. The tool

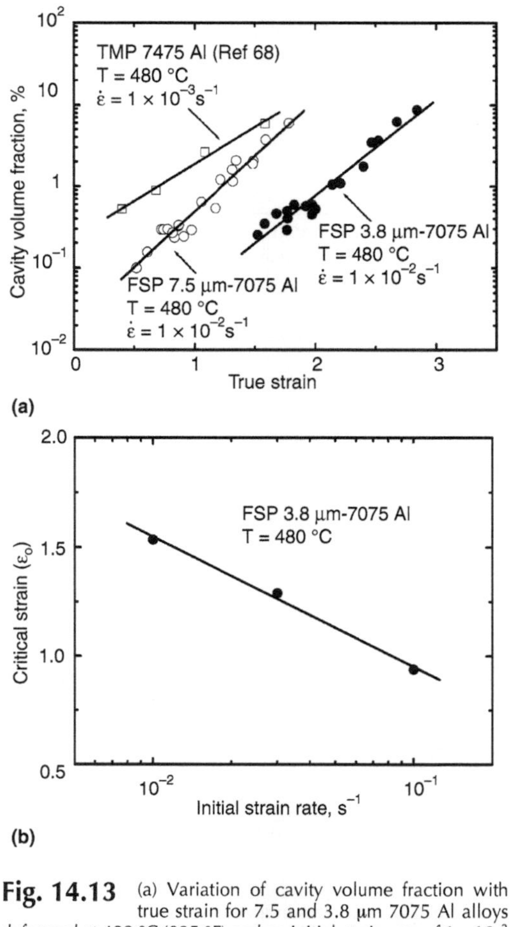


**Fig. 14.13** (a) Variation of cavity volume fraction with true strain for 7.5 and 3.8 μm 7075 Al alloys deformed at 480 °C (895 °F) and an initial strain rate of $1 \times 10^{-2}$ $s^{-1}$. (b) Variation of critical strain, $\varepsilon_0$, with initial strain rate for 3.8 μm 7075 Al alloy deformed at 480 °C. Source: Ref 41

penetration depth was 6.3 mm (¼ in.) using a standard threaded cylindrical pin. Process parameters of tool travel speed and rotation rate are dependent on tool design and can be considerably different, especially if a scroll shoulder tool with a different tool design is selected. Thus, these parameters are not reported. In addition, as discussed throughout the many chapters in this book, both FSP and FSW create inhomogeneous microstructures, for example, the advancing and retreating sides experience different strains, strain rates, and temperatures. This is especially true of the transition microstructures between the nugget and the thermomechanically affected zone on the two sides of the nugget. Thus, the FSP direction in relation to the bending direction can be important. Mahoney et al. showed that if the direction of tool travel during FSP is parallel to the eventual bend axis, the tensile surface experiences inhomogeneous flow, creating a ripple pattern and premature failure likely associated with strain localization. This is a clear example of the inhomogeneous nature of FSP. To achieve maximum strain, the FSP direction should be perpendicular to the bend axis.

Figure 14.14 illustrates a transverse cross section of the 25 mm thick 2519 Al plate following FSP to a depth of 6.3 mm. Again, this depth was chosen arbitrarily, and the same results may have been attained with less tool penetration. The FSP zone is essentially annealed, and below the FSP zone, there will be a heat-affected zone (HAZ) with reduced mechanical properties. The sample shown in Fig. 14.14 has been bent 85° (limit of the die) at room temperature without any indication of impending failure. Figure 14.15 illustrates flow properties of the parent metal and FSP metal, illustrating the significant reduction in flow stress and moderately enhanced ductility achieved in 2519 Al following FSP.

Hardness results as a function of depth below the surface show an extensive HAZ following FSP (Fig. 14.16). No attempt was made to limit the depth of the HAZ, such as increasing the FSP travel speed or by tool design. To regain full strength, the structure can be solution treated and reaged. To regain pre-FSP properties in the bulk of the structure without additional heat treatment, the surface layer can be machined to final shape, thus removing the annealed layer. Figure 14.17 illustrates this post-FSP machining approach for a structure where the final bend radius was moderate and where a sharp corner was required on the exterior. These results illustrate the ability to bend

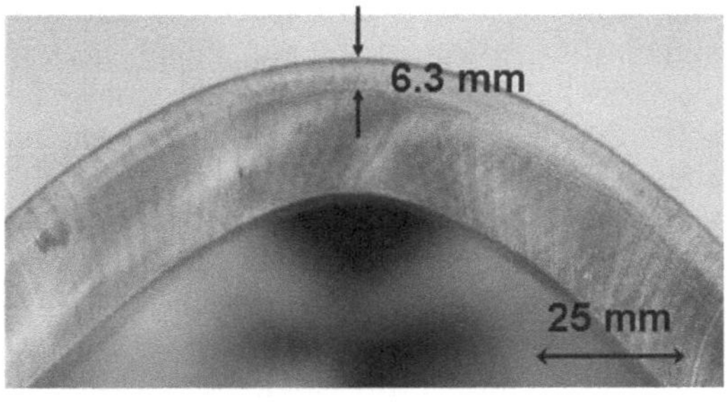


**Fig. 14.14** Illustration of the friction stir processing depth (6.3 mm, or 0.25 in.) and the ability to bend 2519-T87 Al ~85 ° at room temperature

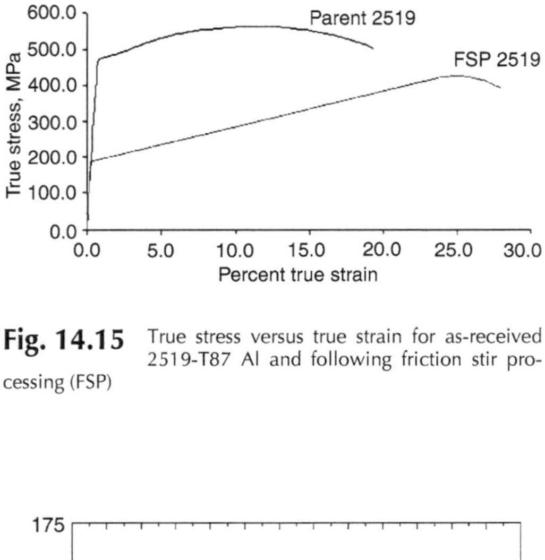


**Fig. 14.15** True stress versus true strain for as-received 2519-T87 Al and following friction stir processing (FSP)

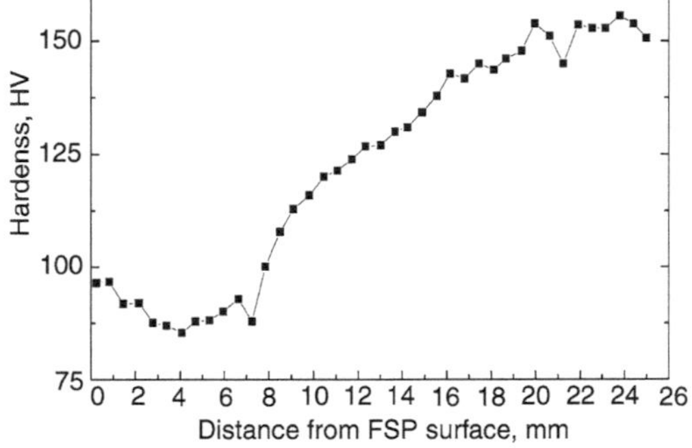


**Fig. 14.16** Hardness in friction stir processed (FSP) 2519 as a function of depth below the surface. Note the relatively deep heat-affected zone (15 to 18 mm, or 0.6 to 0.7 in.).

2519 Al to severe angles at room temperature following FSP.

**FSP of 50 mm (2 in.) Thick 7050-T7451 Al.** Friction stir processing was performed on 50 mm thick 7050-T7451 Al where the plate was friction stir processed to a depth of 6 mm (Ref 72). In this study, different FSP raster approaches were investigated, including linear and spiral-out raster patterns. For each raster pattern, FSP parameters included 350 rpm at 127 mm/min (5 in./min) with a tool translation of 3.3 mm (0.13 in.) per pass; that is, the tool was moved 3.3 mm toward unprocessed material after completion of a pass. For the linear raster pattern, this procedure produced an inhomogeneous processed zone, whereby alternating regions of advancing and retreating side zones are created. Conversely, for the spiral-out pattern, both the tool rotation and spiral were counterclockwise, resulting in a continuous movement of the tool to the advancing side of the previous pass. In each case, the raster was continuous; that is, the travel speed was maintained as the FSP tool reversed direction.

Another important aspect of FSP-assisted thick-section bending is the influence of the heat generated during FSP on subsequent parent-metal properties. With large aluminum blocks and the relatively low travel speed coupled with the conservative overlap per pass, considerable heat buildup occurs. The hardness profile for the processed 7050-T7451 plate is shown in Fig. 14.18 as a function of depth below the surface, and the microstructure is shown in Fig. 14.19, where the transition between the fine-grain processed material and the base material occurs at a depth of approximately 6 mm (0.2 in.). The hardness shows a small HAZ, with hardness equivalent to parent-metal hardness at a depth of ~10 mm (0.4 in.). Hardness results show a through-thickness gradient on the Rockwell hardness B scale of 63, 74, and 82 in the FSP zone, HAZ, and parent metal, respectively, for FSP 7050-T7451.

Similarly, a gradient in yield and tensile strength exists, with strength increasing through the plate thickness. Figure 14.20 shows tensile properties as a function of distance from the friction stir processed surface. To illustrate mechanical properties through the thickness, a series of tensile specimens were cut from different depths of the friction stir processed plate. Twelve tensile specimens, each one approximately 4 mm (0.16 in.) thick, were machined from the plate through its thickness. The first four layers of material on the processed side have lower yield and tensile strengths than layers through the remainder of the plate. Further, extended ductility is illustrated for the sample machined from all-FSP material (layer 1). These results correspond well with the hardness curve in Fig. 14.18. While some natural aging has occurred, for the FSP conditions used herein, there is considerable loss in strength in the FSP zone, that is, for the top 6 mm. Following FSP, natural aging will continue for years, with strength continuing to increase (Ref 73). Between the 6 and 10 mm depth, the slope changes considerably, and only

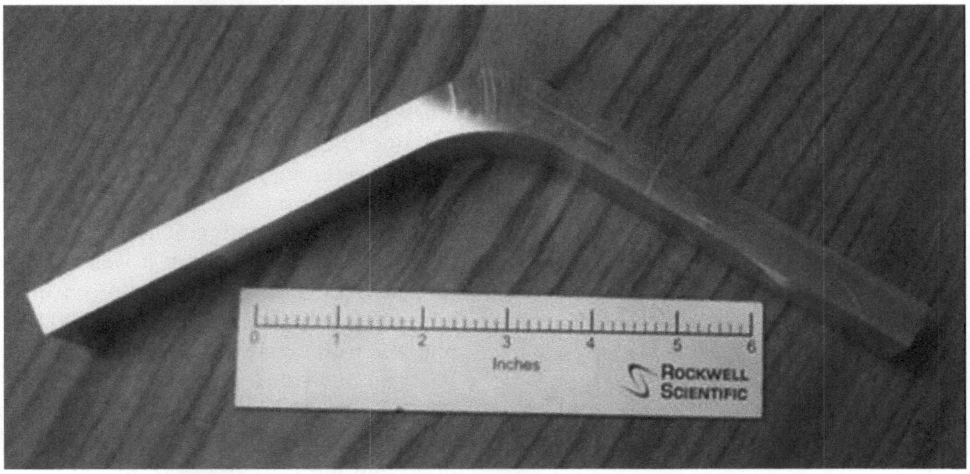


**Fig. 14.17** Final structure following room-temperature bending and machining to final thickness following friction stir processing

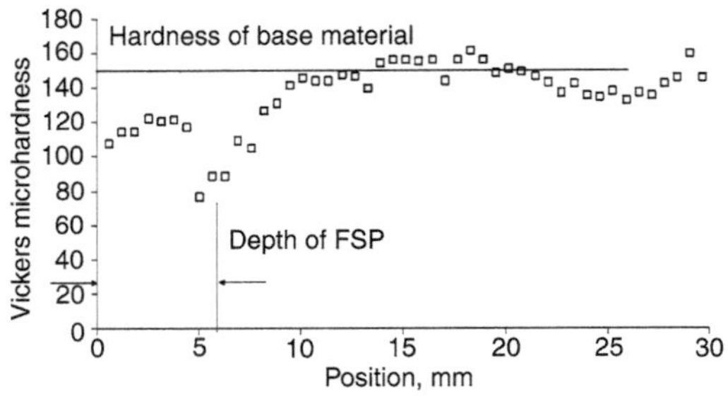


**Fig. 14.18** Hardness in friction stir processed (FSP) 6061 Al as a function of distance below the surface

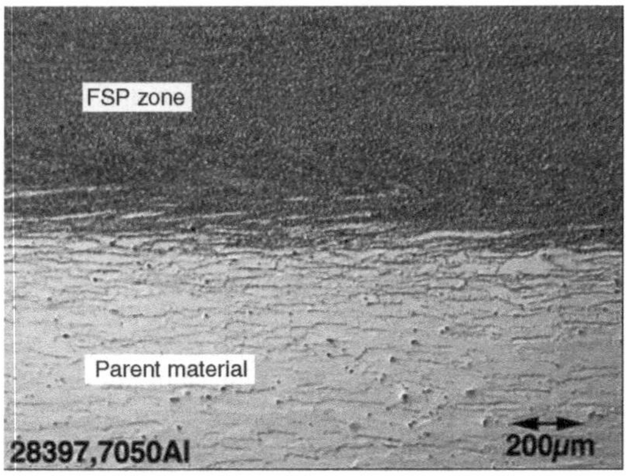


**Fig. 14.19** Micrograph of friction stir processed (FSP) 7050-T7451. The transition from the fine-grain microstructure produced by FSP into the unstirred zone is clearly evident.

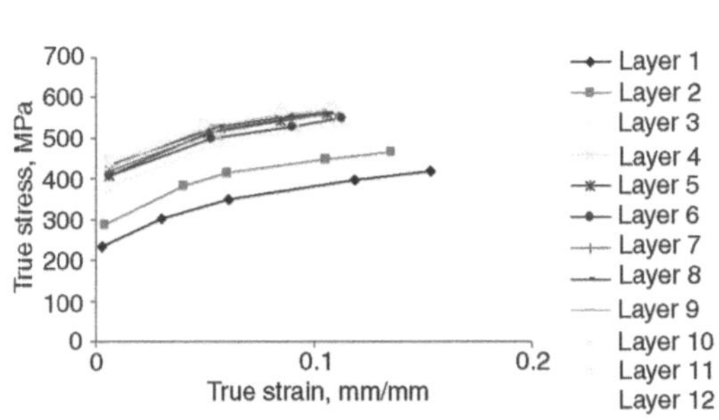


**Fig. 14.20** True stress-true strain tensile curves for layers taken through the thickness of a 50 mm (2 in.) thick 7075 friction stir processed plate. Layer 1 consists entirely of friction stir processed material, while layer 12 is on the opposite side of the plate. The area under the curve for layer 1 is 58 MPa (8.4 ksi), and the areas under the other curves decrease more or less uniformly to approximately 54 MPa (7.8 ksi) in layer 12.

slight strength reductions are measured for the next 20 mm (0.8 in.) of depth, with the loss decreasing with increasing depth. This region of lower rate of strength loss is presumably due to overaging. In this work, no attempt was made to minimize the heat input and thus the effects of overaging. However, approaches that can be easily introduced to reduce total heat input include increased tool travel speed, intermittent FSP with cooling to room temperature between passes, and continuous in situ active cooling. Alternatively, if the outer surface layer can be removed to fabricate the final structure, these results show that near-parent-metal strength can be retained in the bulk of the structure.

A longitudinal cross section of the FSP zone in the 7050 Al illustrates the uniform depth of the friction stir processed zone (Fig. 14.21). At low magnification, the FSP zone microstructure appears homogeneous. However, the processed zone contains the usual inhomogeneous microstructure typical of FSP; note the upward or vertical transition flow between FSP passes (Fig. 14.21). Initial bending trials demonstrated the ability to bend 50 mm thick FSP 7050-T7451 Al plate 14.5° without failure when the bend axis was perpendicular to the FSP direction. This is not the bend limit, and likely this plate could have been bent more, but this was sufficient for an initial trial. Another goal was to demonstrate the ability to bend the 50 mm thick 7050 Al into a compound curvature. However, using the same plate, planes of weakness were identified when the plate was rotated 90° and bending was applied parallel to the FSP direction. For example, Fig. 14.22(a) illustrates multiple cracks propagating in the plate in the FSP direction after only an 8° bend. Cracks bifurcated as they approached the FSP zone and did not penetrate into the HAZ (Fig. 14.22b).

Figure 14.22(b) shows a macrograph of a crack perpendicular to the tensile surface progressing in the direction of tool travel. This crack penetrates through the FSP zone, but when the parent metal is reached, the crack turns parallel to the surface. Metallography was used to determine the crack path, but it was not possible to determine if the crack followed either an advancing-side region or a retreating-side region. To eliminate this unidirectional aspect of FSP and to create a more homogeneous microstructure, a spiral raster pattern was applied to a similar plate. Figure 14.23 illustrates a 16° bend in 50 mm thick 7050-T7451 Al following FSP with a spiral-out raster. Even without surface machining, there was no cracking. The bend limit for this spiral-out FSP procedure was not explored. This ability to create a compound curvature can be useful for producing preshaped blanks to subsequently machine a monolithic structure when the required size of material cannot be attained by other cost-effective means. For example, Fig. 14.24 illustrates machining of a monolithic frame from a thick plate curved to first accommodate the final shape of the structure.

**FSP of Thick 6061-T6 Aluminum.** Bending trials were performed by Mahoney et al. using 25 mm thick 6061-T6 Al to illustrate the ability to bend under plane-strain conditions to very high bend angles, and secondly, to bend very thick plate (150 mm) (Ref 72). A plate of 25 mm thick 6061-T6 Al was friction stir processed to a depth of 6 mm. Prior to bending, the top layer of material was milled to provide a smooth surface, reducing the stir zone depth to ~3 mm.

Bending experiments were performed on as-received plate and on friction stir processed plates. For the 25 mm thick 6061-T6 plate, approximately 230 mm (9 in.) wide, the as-received plate failed at a bend angle of approximately 25°, while the friction stir processed plate failed at approximately an 80° bend angle (Fig. 14.25). Deformation on the plate surface was very close to plane strain, with 1% minor strain or less in the center of the plates on the crown. The increased ductility in the processed plate is due primarily to a decrease in hardness in the outer layer of material, resulting from the heat of processing. This is seen in the microhardness plot shown in Fig. 14.26. Hardness test results through the thickness in the 25 mm thick FSP 6061-T6 Al illustrate a 30 to 40% hardness decrease in the FSP zone, with a gradual increase in hardness until parent-metal hardness is reached near 12 mm penetration. No attempt was made during FSP to minimize the heat input.

Tension tests comparing as-received 6061-T6 Al and FSP 6061 Al illustrated the significant extended ductility and considerable reduction in flow stress following FSP (Fig. 14.27). To create specimens entirely of processed material, specimens 4 mm (0.16 in.) thick were

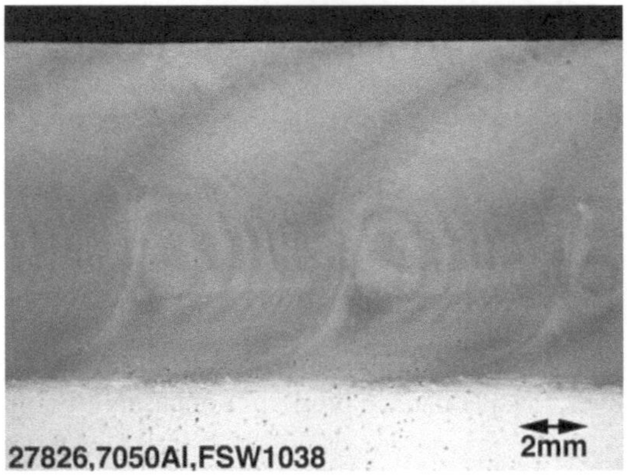


**Fig. 14.21** Longitudinal cross section illustrating depth and deformation pattern in the friction stir processed zone

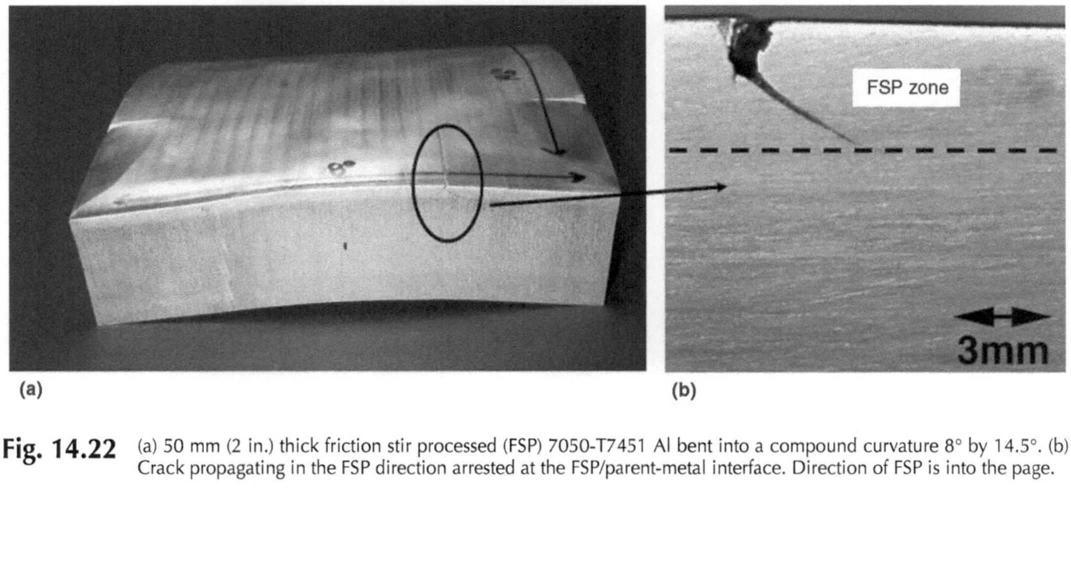


(a) (b)

**Fig. 14.22** (a) 50 mm (2 in.) thick friction stir processed (FSP) 7050-T7451 Al bent into a compound curvature 8° by 14.5°. (b) Crack propagating in the FSP direction arrested at the FSP/parent-metal interface. Direction of FSP is into the page.

machined from a 25 mm (1 in.) thick plate that had been processed to a depth of 5 mm (0.2 in.).

The thickness limits for room-temperature bending, enhanced via FSP, are not known. However, in a dramatic illustration of extreme-thickness bending, Mahoney et al. demonstrated the ability to bend 150 mm thick 6061-T6 Al plate following FSP (Ref 72). For the 150 mm thick 6061 Al, tool rotation rate was 500 rpm, travel speed was 50 mm/min, tool penetration depth was 25 mm, and the raster pattern followed a spiral-out path. Figure 14.28 illustrates FSP of this thick plate to a depth of 25 mm. This very thick plate was friction stir processed using a circular path, with the advancing side on the interior. Following FSP, ~6 mm was machined from the FSP surface to eliminate flash and other surface discontinuities associated with FSP. Thus, prior to bending, the depth of processed material was ~19 mm (0.75 in.). Again, the FSP penetration depth was not optimized, and it is likely that a more shallow depth could also provide enhanced room-temperature bending. Unprocessed material was bent to approximately 8°, at which time the tensile surface demonstrated an "egg-crate" appearance with small microcracks. In comparison, the FSP 6061 Al was bent to 30° (limit of the bending die) and still maintained a smooth surface, with no evidence of surface or subsurface cracking (Fig. 14.29).

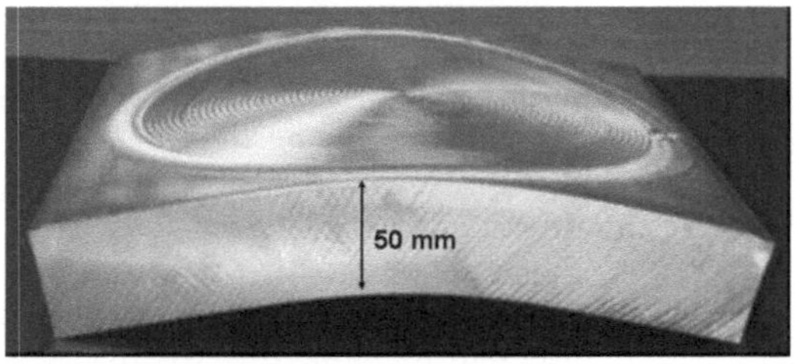


**Fig. 14.23** Spiral raster pattern in 50 mm (2 in.) thick friction stir processed 7050-T7451 Al bent 16° at room temperature

## 14.3 Casting Modification

Cast components are widely used because they provide a cost-effective manufacturing path for complex shapes and unitized substructures. The Al-7wt%Si-Mg alloys with magnesium contents in the range of 0.25 to 0.65 wt% (A356 and A357 alloys) are popular in the aerospace and automobile industries, because they offer a combination of high achievable strength (Ref 74–76) with good casting characteristics (Ref 77). However, the mechanical properties of cast alloys, in

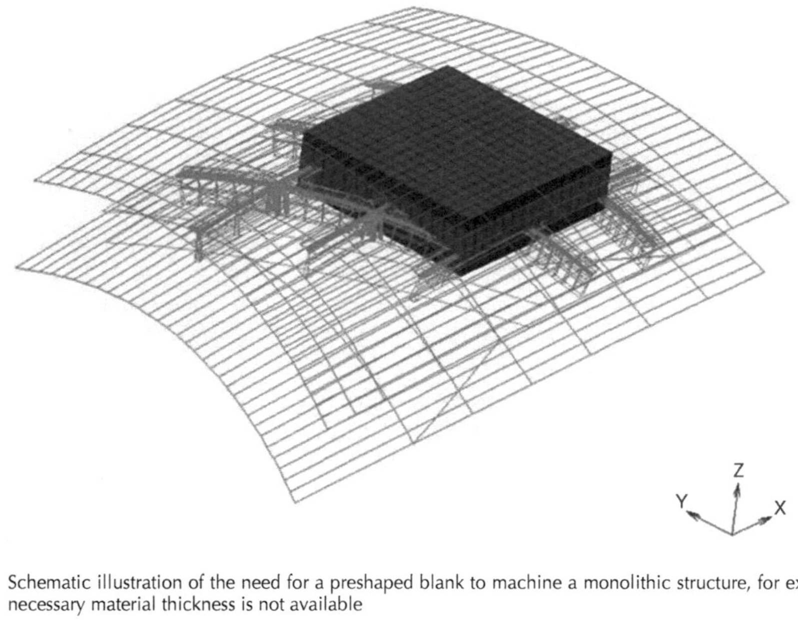


**Fig. 14.24** Schematic illustration of the need for a preshaped blank to machine a monolithic structure, for example, when the necessary material thickness is not available

particular, toughness and fatigue resistance, are limited by three drawbacks, that is, porosity, coarse acicular silicon particles, and coarse primary aluminum dendrites (Ref 78–81). Various modifications and heat treatment techniques have been developed to refine the microstructure of cast aluminum-silicon alloys. The conventional modification and heat treatment techniques pursued earlier cannot effectively eliminate porosity and redistribute the primary and constituent particles uniformly into the matrix. Therefore, a more effective modification technique is highly desirable for microstructural modification of cast components to enhance mechanical properties, in particular, ductility and fatigue. Very recently, a number of studies have reported the effectiveness of FSP for modification of cast microstructures (Ref 82–98). Some of the microstructural results and resultant mechanical behavior are reviewed subsequently, with A356 Al alloy as an illustrative example. Chapter 8 on copper alloys also highlights the microstructural changes in a cast NiAl bronze (Ref 82–84, 88, 89, 91, 93, 96), which are not included in this section. The basic influence of FSP on elimination of porosity and refinement of microstructure should be applicable for most metals and alloys.

**Fig. 14.25** Plane-strain bending in 50 mm (2 in.) thick 6061-T6 Al. (a) Parent metal bent to 27°, with cracks initiating on the tensile surface. (b) Friction stir processed 6061-T6 Al bent to 85° without cracking. Circle grid analysis of the surface strains showed that the negative minor strain at the crown was less than 1%.

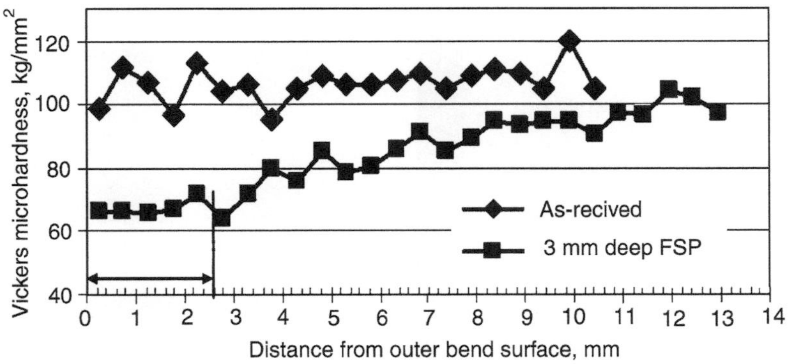


**Fig. 14.26** Microhardness through the thickness in 25 mm (1.0 in.) thick 6061-T6 Al following friction stir processing (FSP). The hardness in the 3 mm (0.12 in.) deep processed zone is uniform but then gradually increases through the thickness of the plate until the base-metal hardness is reached.

**Microstructural Evolution in A356 Al Alloy.** The effect of FSP parameters (tool rotation rate and traverse speed) on the microstructural evolution was examined. As shown in Fig. 14.30, lower tool rotation rates of 300 to 500 rpm resulted in generating FSP nuggets with a macroscopically visible banded structure (Ref 97). While a high density of fine silicon particles was uniformly distributed in most of the nugget zone, the banded zone was characterized by a low density of coarse particles (Fig. 14.31). The FSP at lower tool rotation rates did not result in a complete dispersion of silicon particles throughout the whole nugget zone. By comparison, at a higher tool rotation rate, a uniform microstructure with fine silicon particles was created.

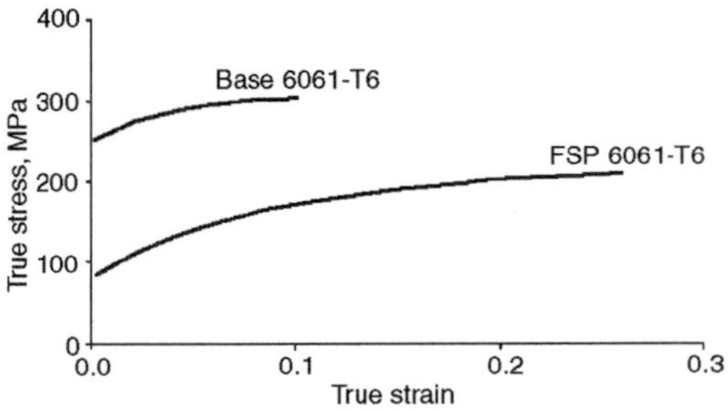


**Fig. 14.27** True stress-true strain tensile curves for base 6061-T6 and friction stir processed (FSP) 6061-T6 (5 mm, or 0.2 in., depth of processing). The area under the base-material curve is 29 MPa (4.2 ksi), while the area under the FSP-material curve is 45 MPa (6.5 ksi).

**Fig. 14.28** Friction stir processing of 150 mm (6 in.) thick 6061-T6 Al using a spiral-out raster path with the advancing side on the interior.

Figure 14.32 shows optical micrographs of as-cast A356 (12.5 mm, or 0.5 in., plate) and FSP A356 processed at 300 rpm for 0.85 mm/s (0.03 in./s) (Ref 98). Typical needle-shaped silicon particles were distributed within the as-cast A356 microstructure (Fig. 14.32a). The FSP resulted in the breakup of the needle-shaped silicon particles (Fig. 14.32b). Both particle size and aspect ratio are summarized in Table 14.3 for as-cast A356 and FSP A356 as a function of process parameters. Clearly, both particle size and aspect ratio were significantly reduced after FSP. Silicon particles in FSP samples processed at a lower tool rotation rate of 300 rpm exhibited a smaller size than at a higher rotation rate of 700 rpm for both the standard threaded pin and a trifluted pin.

Multiple FSP was reported on as-cast A356 plate using a triflute pin at a tool rotation rate of 700 rpm and a traverse speed of 3.4 mm/s (0.13 in./s) (Ref 95). Figure 14.33 shows the five-pass FSP samples with 50% overlapping. Although the cross section of the FSP sample shows flow lines between various FSP passes, optical microscopic examinations indicated that overlapping passes did not significantly influence the size and

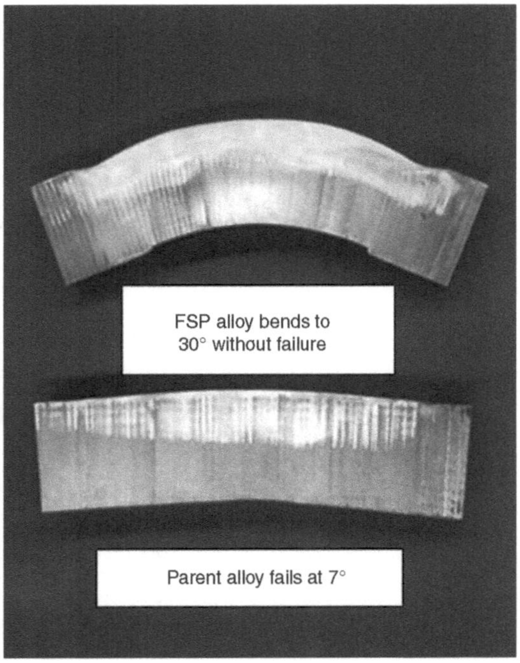


**Fig. 14.29** Friction stir processed (FSP) 150 mm (6 in.) thick 6061-T6 Al bent to 30° without cracking, compared to parent metal reaching a bend limit of 7°

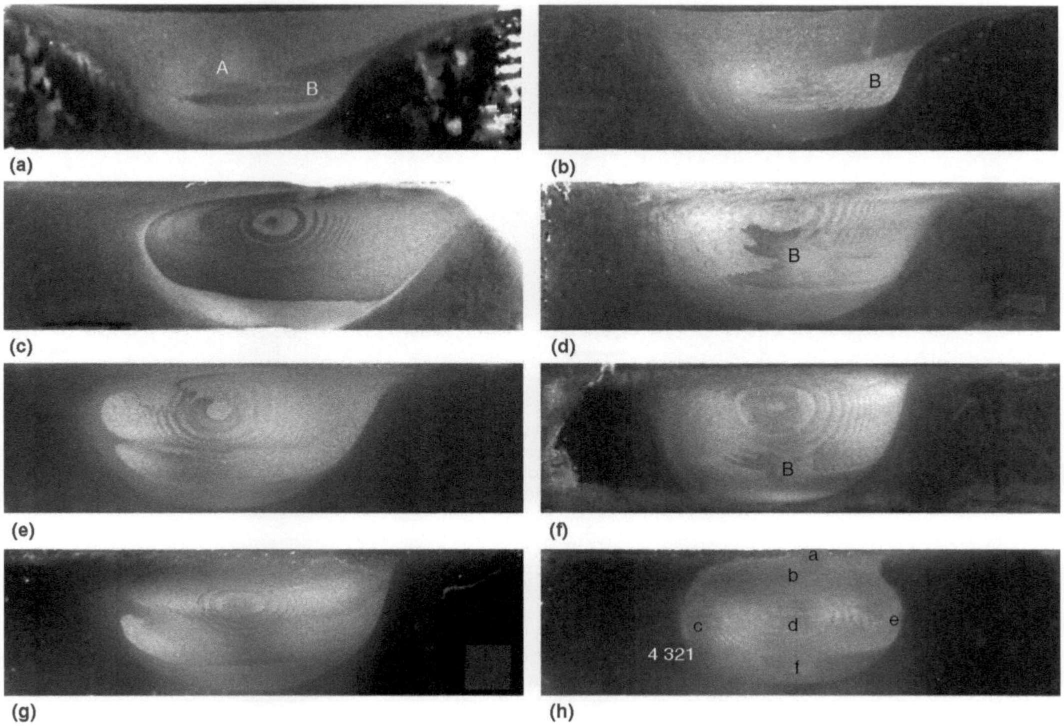


**Fig. 14.30** Macrographs showing stirred zone in friction stir processed A356 using processing parameter combinations of (a) 300 rpm, 0.85 mm/s (0.03 in./s), (b) 300 rpm, 1.7 mm/s (0.07 in./s), (c) 500 rpm, 0.85 mm/s, (d) 500 rpm, 1.7 mm/s, (e) 700 rpm, 1.7 mm/s, (f) 700 rpm, 3.4 mm/s (0.13 in./s), (g) 900 rpm, 1.7 mm/s, and (h) 900 rpm, 3.4 mm/s (samples were lightly etched). Source: Ref 97

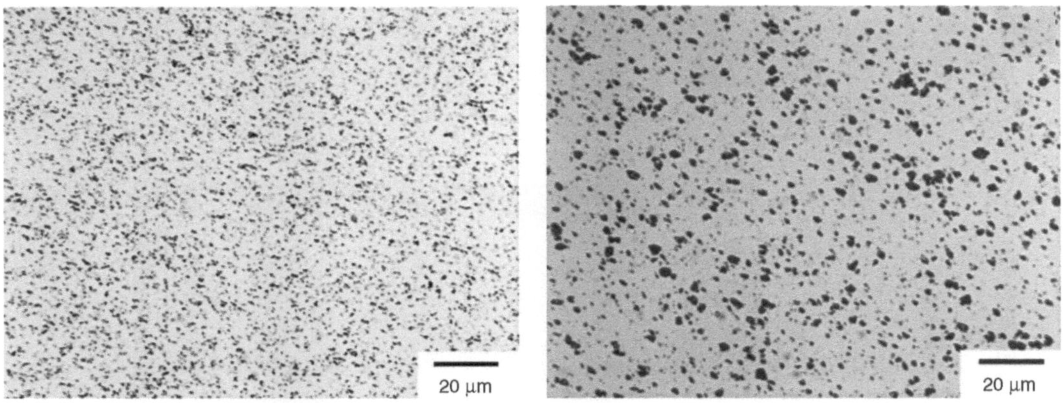


**Fig. 14.31** Optical micrographs showing (a) fine silicon particles in nugget center (region A in Fig. 14.30a) and (b) coarse silicon particles in banded structure (region B in Fig. 14.30a) of friction stir processed A356 sample (processing parameter: 300 rpm, 0.85 mm/s, or 0.03 in./s; sample was polished). Source: Ref 97

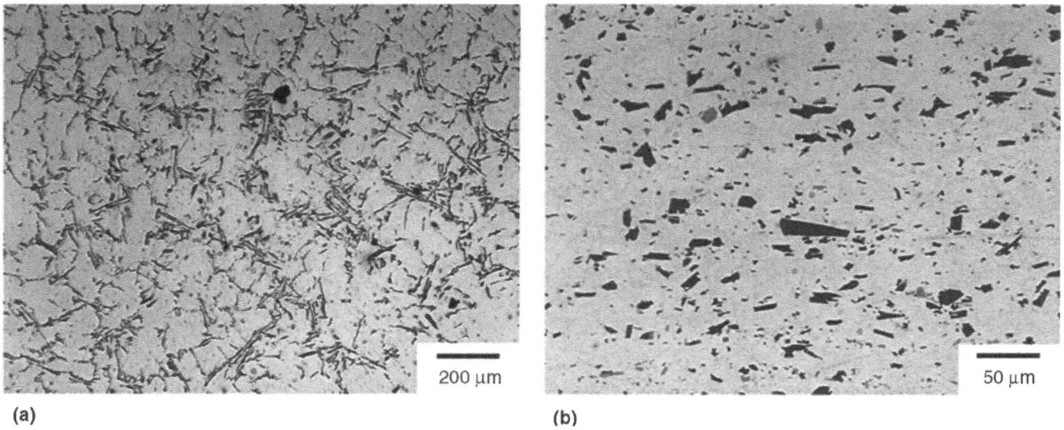


**Fig. 14.32** Optical micrographs showing the microstructure of (a) as-cast A356 12.5 mm (0.5 in.) thick plate, and (b) friction stir processed A356 at 300 rpm for 0.85 mm/s (0.03 in./s). Source: Ref 98

**Table 14.3 Size and aspect ratio of silicon particles in as-cast and friction stir processed (FSP) A356**

| Material | Particle size, $\mu m^2$ | Aspect ratio |
|---|---|---|
| As-cast | 7.28 ± 5.47 | 5.92 ± 4.34(a) |
| FSP—300 rpm/0.85 mm/s (standard tool) | 2.84 ± 2.37 | 2.41 ± 1.33 |
| FSP—700 rpm/3.4 mm/s (standard tool) | 2.62 ± 2.31 | 1.93 ± 0.86 |
| FSP—900 rpm/3.4 mm/s (standard tool) | 2.55 ± 2.21 | 2.00 ± 1.01 |
| FSP—1100 rpm/3.4 mm/s (standard tool) | 2.51 ± 2.00 | 2.04 ± 0.91 |
| FSP—300 rpm/0.85 mm/s (trifluted pin) | 2.70 ± 2.26 | 2.30 ± 1.15 |
| FSP—700 rpm/3.4 mm/s (trifluted pin) | 2.48 ± 2.02 | 1.94 ± 0.88 |
| FSP—900 rpm/3.4 mm/s (trifluted pin)-one pass | 2.50 ± 2.04 | 1.99 ± 0.94 |
| FSP—900 rpm/3.4 mm/s (trifluted pin)-two passes | 2.43 ± 2.02 | 1.86 ± 0.78 |
| FSP—1100 rpm/3.4 mm/s (trifluted pin) | 2.44 ± 2.00 | 1.86 ± 0.81 |
| FSP—300 rpm/0.85 mm/s (cone-shaped pin) | 2.90 ± 2.46 | 2.50 ± 1.35 |
| FSP—700 rpm/3.4 mm/s (cone-shaped pin) | 2.86 ± 2.32 | 2.09 ± 0.90 |

(a) The average aspect ratio in the as-cast condition is much higher than the computer software-generated number because of an artifact in the image processing. Source: Ref 85

**Fig. 14.33** Macrograph of cross section of five-pass friction stir processed A356 sample (triflute pin, 700 rpm for 3.4 mm/s, or 0.13 in./s). Source: Ref 95

distribution of silicon particles. It appeared that the size and distribution of silicon particles were uniform throughout the whole processed zone. Furthermore, porosity was eliminated within the whole processed zone. Ma et al. (Ref 95) also reported that the size and aspect ratio of silicon particles in various regions are quite similar, indicating that the overlapping FSP did not result in further breakup of silicon particles. Furthermore, the size and aspect ratio of silicon particles in various regions for the five-pass FSP sample were in good agreement with those achieved for the single-pass FSP sample.

Sharma and Mishra (Ref 99) have examined the effect of parameters on nugget shape and area of FSP in A356. Figure 14.34(a) shows a plot of nugget size as a function of pseudo-heat index. The nugget size varies with processing parameters. The nature of these curves indicates that during FSP the friction condition between the shoulder and the workpiece varies as a function of processing parameters. It is believed that the friction between the tool and workpiece changes from sticking friction to sliding friction with increasing heat input. This also has an influence on loads during processing. Initial results suggest that in the presence of sliding friction, process loads are lower when compared to the sticking friction condition, as can be seen in Fig. 14.34(b).

**Influence of FSP on Mechanical Properties.** Mechanical properties of FSP A356 samples have been reported by Ma et al. (Ref 85, 86) and Santella et al. (Ref 92, 94). To investigate

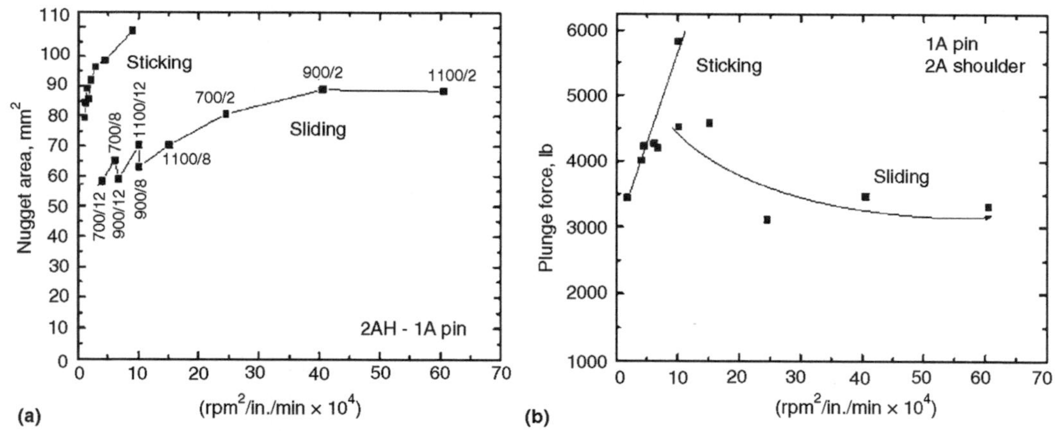


**Fig. 14.34** (a) Plot of nugget area as a function of pseudo-heat index. Notice the transition in the slopes of the two curves, suggesting transition from a sticking friction condition to a sliding friction condition. (b) Plot of plunge force as a function of pseudo-heat index. Notice reduction in plunge load as the processing parameters go from sticking to sliding friction. Source: Ref 99

**Table 14.4 Tensile properties of friction stir processed (FSP) A356 (12.7 mm, or 0.5 in., cast plate) at room temperature ($\dot{\varepsilon} = 10^{-3}$ s$^{-1}$)**

| | As-received or as-FSP | | | Aging (155 °C, or 310 °F, for 4 h) | | |
|---|---|---|---|---|---|---|
| Materials | UTS(a), MPa | YS(b), MPa | Elongation, % | UTS(a), MPa | YS(b), MPa | Elongation, % |
| As-cast | 169 ± 10 | 132 ± 5 | 3 ± 1 | 153 ± 7 | 138 ± 6 | 2 ± 1 |
| FSP—300 rpm/0.85 mm/s (standard pin) | 205 ± 6 | 134 ± 5 | 31 ± 2 | 206 ± 6 | 137 ± 9 | 29 ± 2 |
| FSP—700 rpm/3.4 mm/s (standard pin) | 242 ± 6 | 149 ± 10 | 31 ± 1 | 247 ± 7 | 169 ± 10 | 28 ± 2 |
| FSP—900 rpm/3.4 mm/s (standard pin) | 266 ± 4 | 171 ± 6 | 32 ± 1 | 288 ± 5 | 228 ± 9 | 25 ± 2 |
| FSP—1100 rpm/3.4 mm/s (standard pin) | 242 ± 3 | 157 ± 3 | 33 ± 1 | 265 ± 2 | 205 ± 8 | 23 ± 5 |
| FSP—300 rpm/0.85 mm/s (3A pin) | 202 ± 5 | 137 ± 4 | 30 ± 1 | 212 ± 5 | 153 ± 20 | 26 ± 3 |
| FSP—700 rpm/3.4 mm/s (3A pin) | 251 ± 5 | 171 ± 14 | 31 ± 1 | 281 ± 5 | 209 ± 3 | 26 ± 2 |
| FSP—300 rpm/0.85 mm/s (4A pin) | 178 ± 2 | 124 ± 5 | 31 ± 4 | 175 ± 2 | 119 ± 6 | 32 ± 1 |
| FSP—700 rpm/3.4 mm/s (4A pin) | 256 ± 5 | 169 ± 3 | 28 ± 2 | 264 ± 4 | 203 ± 10 | 21 ± 1 |

(a) UTS, ultimate tensile strength. (b) YS, yield strength. Source: Ref 85 the effect of heat treatment on tensile properties, FSP samples were subjected to post-FSP natural aging (room temperature for one month), post-FSP artificial aging (room temperature for one month + 155 °C, or 310 °F, age for 4 h), and a standard T6 (540 °C, or 1000 °F, solution treatment for 4 h, room-temperature water quench, and 155 °C age for 4 h) (Ref 85). Table 14.4 summarizes the effect of FSP parameters and heat treatment on the tensile properties of FSP A356. For the post-FSP natural-aged condition, in general, with increasing tool rotation rate as well as traverse speed, the strength of the FSP A356 increases and ductility decreases. Maximum strength is observed for this sample processed at 700 rpm and 3.4 mm/s. Post-FSP artificial aging also tends to increase the yield strength of FSP samples and decrease the ductility. The T6 heat treatment significantly increases the strength of FSP samples. Again, the sample processed at 700 rpm and 3.4 mm/s exhibits the optimal strength for the T6 condition. Compared to an as-received T6 A356 casting, FSP A356 samples exhibit significant increases in tensile strength while retaining the same ductility. Table 14.4 shows that higher tool rotation rates produce better tensile properties. At a lower tool rotation rate of 300 rpm, the tool geometry did not affect the tensile properties of FSP A356. However, at a higher tool rotation rate of 700 rpm, the triflute pin produces a higher strength than the standard pin, but ductility was not influenced. The aging treatment resulted in an increase in the strength of the FSP A356, in particular, yield strength, and a decrease in ductility. Figure 14.35 shows the appearance of failed tensile specimens. For as-cast A356, no necking occurred. The fracture propagates along the needle-shaped silicon particle/matrix interfaces. For the FSP specimen, obvious necking can be observed, indicating good plasticity.

Tensile properties of multiple-pass FSP A356 using minitensile specimens were established. The tensile properties in various microstructural regions are summarized in Fig. 14.36. For the as-FSP condition, the following observations can be made. First, the strength and ductility of the transitional zones, locations where the microstructure indicates overlapped passes, are slightly lower than those of center locations in the remnant nugget. Second, the strength of both the nugget and transitional zones decreases with increasing distance from the fifth-pass (last pass in this case) processed zone. Third, both strength and ductility of the fifth-pass FSP nugget zone are similar to those achieved in a single-pass FSP sample. These results suggest that in the as-processed condition, additional thermal cycles associated with subsequent FSP lowers the strength by 5 to 10%. For the T6-treatment condition, both strength and ductility are scattered within a band for the various microstructural zones, and no systematic variation is observed. However, the five-pass FSP sample, in various microstructural regions, exhibits increases in both strength and ductility, which compares favorably with results achieved in a single-pass FSP sample. This indicates that multiple-pass FSP with 50% overlapping is a feasible route to

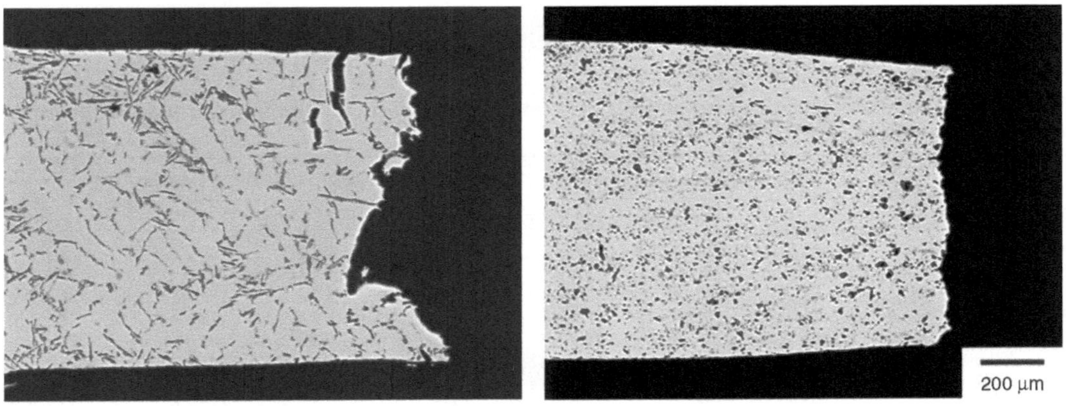


(a) (b)

**Fig. 14.35** Appearance of failed specimens. (a) As-cast A356 (12.5 mm, or 0.5 in., cast plate). (b) Friction stir processed A356 (triflute pin, 700 rpm for 3.4 mm/s, or 0.13 in./s) perform microstructural modification to cover larger regions of aluminum castings.

The tensile properties in the HAZ are equivalent to or lower than those of the as-received parent material, and the tensile and yield strengths decrease with increasing distance from the FSP zone boundaries (Fig. 14.37). In the HAZ, as expected, FSP did not break up the coarse silicon particles and aluminum dendrites, and conversely led to a coarsening of precipitates. In the HAZ, FSP did not result in an improvement in mechanical properties but actually resulted in a decrease. Control of microstructure in the HAZ during FSP will be critical for achieving mechanical properties equivalent to or better than the starting material.

Figure 14.38 illustrates fatigue results for A356 plates before and after FSP (Ref 87, 99). For processed plates, the samples were machined completely from the stir zone. The arrows in Fig. 14.38 indicate specimens that did not fail. As shown, the fatigue strength threshold stress increased by >80% after FSP. This fatigue strength improvement is attributed to both a reduction in silicon particle size and reduced porosity volume fraction. The fatigue life, $N_f$, has been related to the positive component of cyclic stress, $\sigma^*$, and the pore size, $a_0$, by:

$$\sigma^* = C(a_0 N_f)^{-1/m} \quad \text{(Eq 14.3)}$$

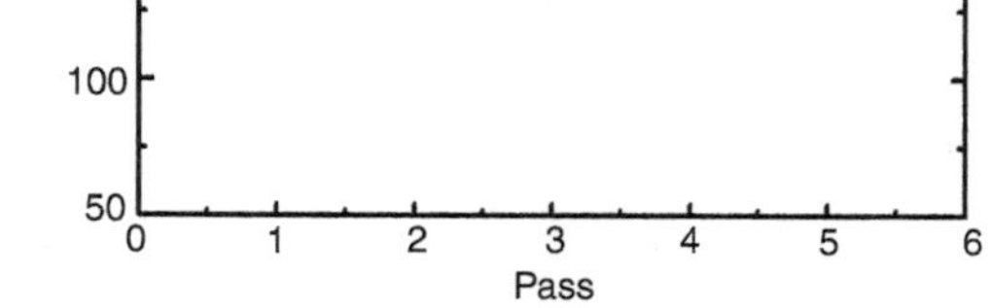


**Fig. 14.36** Effects of multipass friction stir processing (FSP) welds on mechanical properties. No significant change in percent elongation was observed. Note that the oval regions indicate properties in the transition region of the multipass FSP sample. UTS, ultimate tensile strength; YS, yield strength

where $m$ is the Paris exponent for fatigue crack growth, and $C$ is a constant that depends on the Paris pre-exponential constant and on the pore shape and position. From the previous analysis, it was concluded that fatigue life is influenced more by the size of the largest pore rather than porosity volume fraction or mean pore size. In addition to porosity, fracture characteristics of Al-Si-Mg castings are influenced by size, orientation and local distribution of silicon particles, as well as by the silicon-matrix interface strength. As stated by Lee et al. (Ref 100), fatigue failure in A356 occurs in four stages, including crack initiation at silicon or second-phase particles, crack growth, crack propagation across the aluminum-silicon matrix via linkage of microcracks generated as a result of decohesion and/or particle cracking, and high rate of crack growth, eventually leading to fracture of the aluminum matrix. Larger silicon particles present in the as-cast material accelerate crack nucleation due to stress-concentration effects. Murakami and Endo (Ref 101) have proposed the following equation for the fatigue limit in metals with three-dimensional defects:

$$\sigma_w = \frac{1.43(HV + 120)}{(\sqrt{A})^{1/6}} \quad \text{(Eq 14.4)}$$

where $\sigma_W$ is the fatigue limit (MPa), $A$ is the area obtained by projecting a defect or a crack onto the plate perpendicular to the maximum tensile stress ($mm^2$), and HV is Vickers hardness (kgf $mm^{-2}$) between 70 to 720 HV. Based on this equation, a 30% reduction in particle size alone would contribute to a 25% improvement in the fatigue limit. The FSP significantly refines the microstructure, leading to a homogeneous distribution of smaller silicon particles with smaller aspect ratios when compared to the as-cast microstructure. This refined microstructure also leads to increased plastic deformation in the aluminum matrix during cyclic crack tip propagation, resulting in a concurrent increase in crack energy dissipation and a consequent increase in crack growth resistance. Plastic deformation during fatigue leads to crack nucleation, either by separation of the silicon-aluminum interface, or by particle cracking, or both.

Crack growth studies were conducted using compact tension specimens machined from cast A356 and compared with friction stir processed regions (Ref 99). Figure 14.39 shows a comparison between the crack growth rates (*da/dN*) of different samples. To achieve similar crack

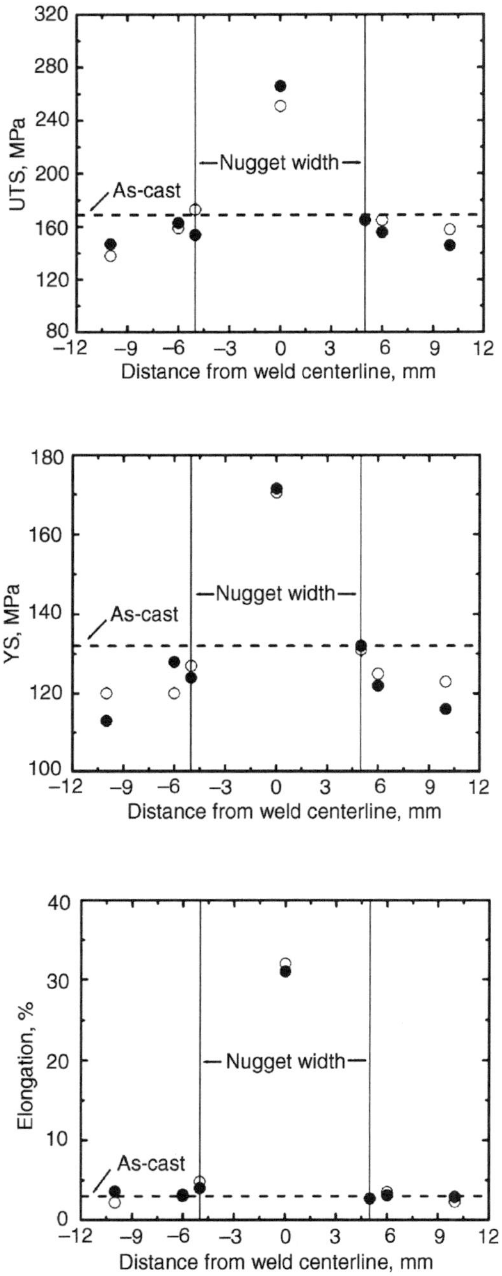


**Fig. 14.37** Variation in tensile properties with distance from the nugget in the heat-affected zone for friction stir processed A356 (solid symbol for triflute pin, 700 rpm for 203 mm/min, or 8 in./min; open symbol for standard pin, 900 rpm for 203 mm/min). UTS, ultimate tensile strength; YS, yield strength

growth rates in FSP A356, compared to cast A356, a greater than 36% increase in load is required. Also, the friction stir processed alloy follows region II in the *da/dN* versus stress-intensity range ($\Delta K$) plot at higher $\Delta K$ values. The slower crack growth rates in FSP A356 are attributed to the finer microstructure developed during FSP. Results from tests conducted at higher stress ratios indicate that crack closure is the dominant mechanism in increasing crack growth resistance in FSP samples in the threshold region (Ref 99).

The upper limit of the crack-driving force was assumed to be the "pseudo"-fracture toughness of the materials. Because the compact tension specimens in this study did not meet the plane-strain fracture toughness requirements of ASTM E 399, the measured fracture toughness values are referred to as pseudo-fracture toughness. Pseudo-fracture toughness was determined using the crack length and critical load at the onset of unstable fracture. The pseudo-fracture toughness is only slightly influenced by the T6 heat treatment for the as-cast A356, while FSP samples show higher toughness than cast A356 samples (Table 14.5). The pseudo-fracture toughness of FSP samples improved by over 30% when compared to cast samples, and in the T6 condition, FSP samples showed >50% improvement in toughness.

In summary, the FSP of aluminum castings significantly improves properties, including:

- Strength increases by more than 25% in the T6 condition.
- Ductility increases by 3 to 10 times in various thicknesses.
- Fatigue life increases by many orders of magnitude, and fatigue stress increases by approximately two times.
- Toughness increases by 50%.

The implementation of FSP technology to enhance castings can lead to weight reduction in castings, performance and/or life enhancement of castings, and substitution of forgings with FSP-modified castings.

## 14.4 Modification of Fusion Welds for Increased Fatigue Resistance

It will not be possible to friction stir weld all aluminum structures and reap the benefits of this solid-state process. For example, large structures, locations inaccessible to a friction stir system, and very thick plate would be difficult to friction stir weld. However, eventually, it may be possible to friction stir process the surface of fusion welds using a portable system. By

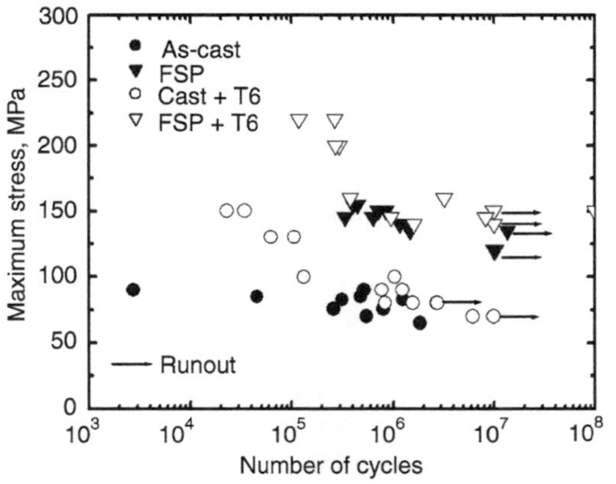


**Fig. 14.38** Plot of fatigue lifetime versus maximum stress for as-cast and friction stir processed (FSP) A356 samples. Source: Ref 87, 99

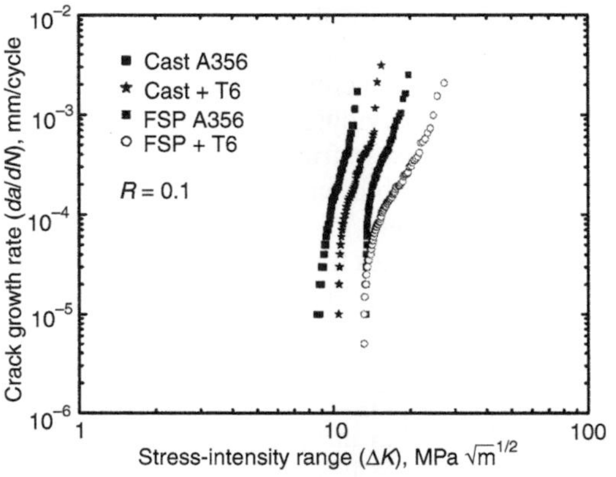


**Fig. 14.39** Crack growth rates in A356 under various processing conditions at stress ratio $R = 0.1$. Source: Ref 99

**Table 14.5 Comparison of pseudo-fracture toughness of A356 under various processing conditions**

| | Cast | Cast + T6 | FSP(a) | FSP + T6 |
|---|---|---|---|---|
| $K_Q$(MPa $\sqrt{m}$) | 14.6 ± 2 | 15.8 ± 4 | 19.5 ± 1 | 24.4 ± 1 |

(a) FSP, friction stir processing

friction stir processing the surface, a cast fusion weld microstructure will be converted to a fully recrystallized, fine grain, and weld defects near the surface will be eliminated. Potential benefits include both increased corrosion resistance and fatigue life. The following illustrates an example whereby the crown or toes of a fusion weld are friction stir processed and subsequent fatigue life increased.

Past research on structural aluminum alloys demonstrated lower fatigue resistance in gas metal arc welds (GMAW) when compared to base-metal (BM) properties (Ref 102, 103). Fatigue behavior of GMAW can be accommodated by increasing the reinforcement at the arc weld location, thereby increasing component weight. However, there is an emphasis to decrease the cost or weight of a given structure. Friction stir processing is a technique that produces local microstructural modification, and when applied to GMAW, improves the microstructure and corresponding mechanical properties at the weld toe and crown locations (Ref 104, 105). Reasons commonly cited for lower fatigue resistance of full-penetration GMAW include a weaker filler metal than the BM (an undermatched weld); defects within the weld nugget, such as solidification porosity; and stress concentrations at the weld bead (Ref 106, 107). Stress concentrations at the weld toe are the most important factor influencing the fatigue behavior of aluminum GMAW; thus, removal of the weld bead increases fatigue resistance (Ref 108).

Referring to the work of Fuller et al., gas metal arc welds were produced on 6 mm thick 5083-H321 Al plates with automated metal inert gas welds operating at 180 A, 26 V, and a head travel speed of 22.8 cm min$^{-1}$ (Ref 104, 105). Two different FSP approaches were examined, including weld toe FSP and weld crown FSP. The weld toe is defined as the interface between the arc weld nugget and BM on the top surface. Weld toe FSP was performed with a small tool containing an 11 mm (0.4 in.) diameter shoulder and a 3 mm (0.12 in.) long conical probe (6.35 mm, or 0.25 in., diameter tapering to 4.6 mm, or 0.18 in.) operating at 1600 rpm and 40.6 cm · min$^{-1}$. This tool traversed along each of the two arc weld toes for a total of two FSP passes per plate. Weld crown FSP used a probeless 28.6 mm (1.13 in.) diameter scrolled shoulder tool operating at 400 rpm and 20.3 cm · min$^{-1}$ and was traversed across the arc weld crown in a single pass. All FSP tools were operated with counterclockwise rotation, with the Z-axis in position control, and were manufactured from MP159 alloy. Figure 14.40 illustrates the four FSP approaches evaluated, that is, as-arc welded, weld toe FSP with arc weld nugget on the advancing side of the tool, weld toe FSP with arc weld nugget on the retreating side of the tool, and weld crown FSP. Metallography illustrates the fusion weld microstructures, both before and after the different FSP procedures (Fig. 14.41).

Transverse tensile property results of 5083-H321 Al/5356 Al GMAW as a function of FSP approach were established using microtensile samples and are listed in Table 14.6 (Ref 105). As-welded 5083-H321/5356 Al had the lowest strength values and elongation. Both FSP approaches were observed to provide small increases in the yield strength and tensile strength, and significant increases in elongation of GMAW 5083-H321/5356 Al. The 5083-H321 Al BM tensile properties are higher than any of the experimental strength data (Ref 109). The strength differences are due to a reduction in strain hardening as a result of thermal exposure produced from GMAW and FSP.

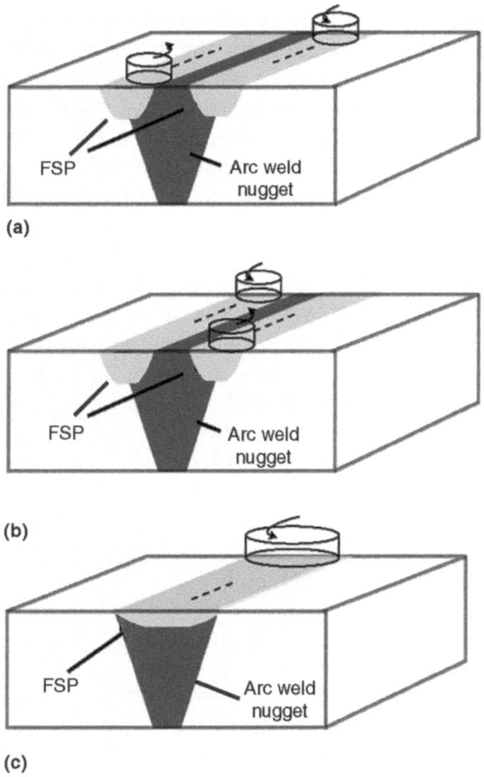


**Fig. 14.40** Schematic of friction stir processing (FSP) approaches in relation to the arc weld nugget. (a) Weld toe FSP with arc weld nugget on advancing side of tool. (b) Weld toe FSP with arc weld nugget on retreating side of tool. (c) Weld crown FSP

Figure 14.42 presents four-point bending fatigue results for the arc-welded, weld crown FSP, and weld toe FSP, with the arc weld on the retreating side, where the number of cycles to failure are plotted as a function of the maximum applied load. All specimens were orientated such that the crown surface was in tension. The as-arc-welded approach has the lowest four-point bending fatigue resistance. The addition of FSP improves the four-point bending fatigue resistance, with no significant difference in fatigue resistance as a function of FSP approach. The as-arc-welded sample loaded to 60 kg failed after $6.7 \times 10^5$ cycles, but none of the friction stir processed samples loaded to 60 kg failed, even after $1.4 \times 10^7$ cycles. This fatigue improvement represents greater than a 20 times improvement in fatigue life. A runout specimen (no failure after $10^7$ cycles) for the as-arc-welded condition was reached at 46 kg, while the friction stir processed conditions produced runouts at 60 to 61 kg, a 30% increase in applied load.

Chapter 8 in this volume presents FSP of copper alloys, including NiAl bronze, an alloy frequently used to fabricate ship propellers. Data in Chapter 8 show mechanical and fatigue properties to be improved considerably by FSP. However, as shown previously for an aluminum alloy, the need may arise to friction stir process a prior-fusion repair within an NiAl bronze propeller. Thus, studies were initiated to evaluate procedures and properties for this unique combination of prior processing. Fusion welds were made at the Naval Surface Warfare Center, Carderock Division, using standard Navy weld procedures for NiAl bronze. Figure 14.43 illustrates a multipass 12.7 mm (0.5 in.) penetration fusion weld using Ampcotrode 46 weld wire of composition 8.5–9.5Al, 3.0–5.0Fe, 0.6–3.5Mn, 4.5 Ni, bal Cu, with typical elongation of 23%. A typical fusion weld defect is shown on the left side (arrow), which appears to be associated with the interface between passes. The microstructure in the fusion zone is a fine Widmanstätten (Fig. 14.44). This fusion weld was friction stir processed without removing the weld crown. The FSP parameters included a step-spiral 12.7 mm (0.5 in.) deep Densimet tool at 1000 rpm and 102 mm/min (4 in./min), with 6.4 mm (0.25 in.) translation between passes.

Figure 14.45(a) shows a cross section of the microstructure following FSP. The microstructure following FSP is mixed, including regions of fine grain and regions where a composite of morphologies is found, including both fine grain and Widmanstätten (Fig. 14.45b). Average tensile properties (six samples) for the longitudinal orientation include a yield strength of 415 MPa (60 ksi), tensile strength of 760 MPa (110 ksi), and elongation of 28%. From this brief study of FSP over a fusion weld in NiAl bronze, the following conclusions are made:

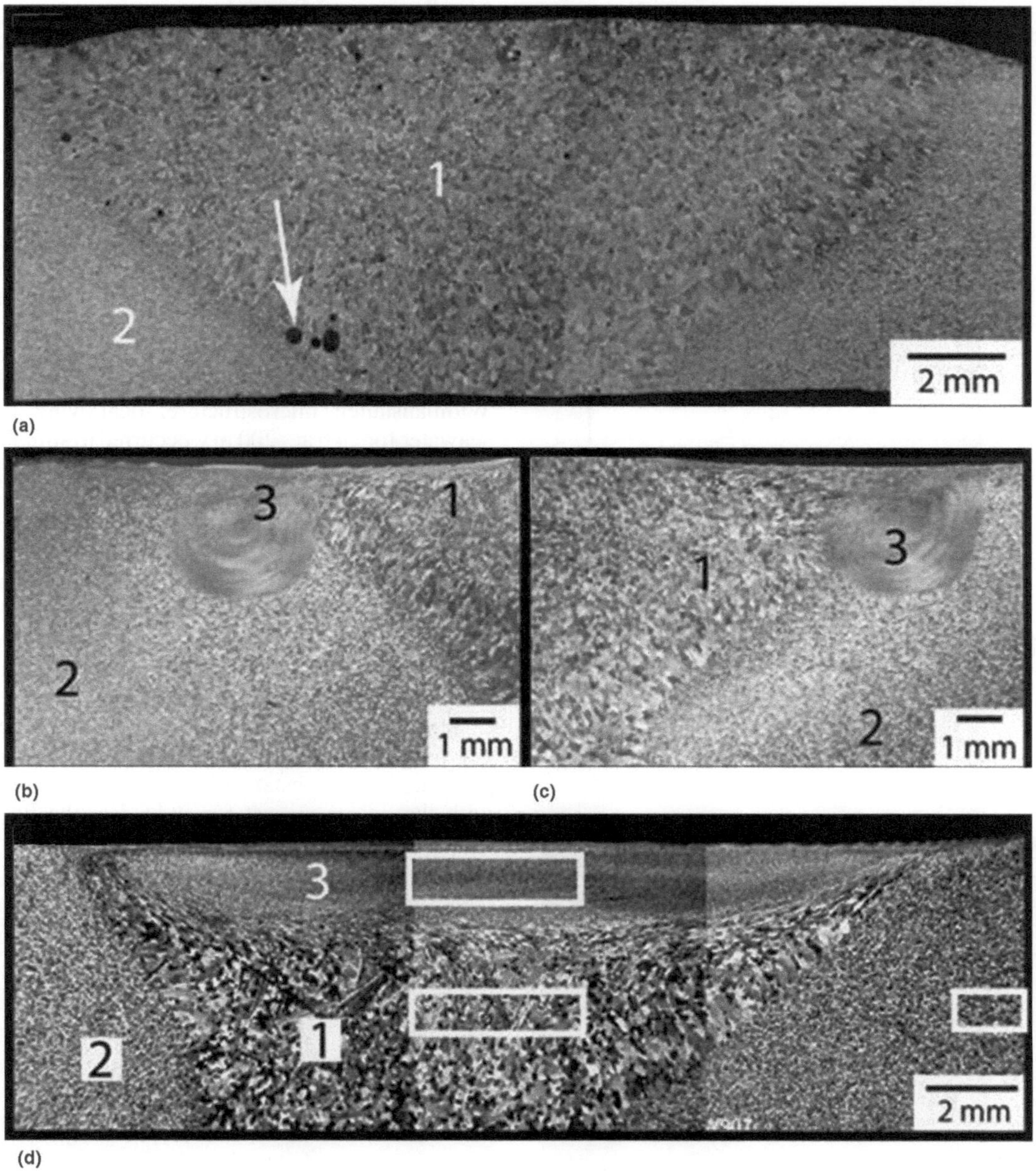


**Fig. 14.41** Light macrographs of 5083-H321 Al/5356 Al arc weld in the following conditions: (a) as-arc welded, (b) weld toe friction stir processing (FSP) with arc weld nugget on advancing side, (c) weld toe FSP with arc weld nugget on retreating side, and (d) weld crown FSP. Different microstructural regions within the micrographs are indicated by: (1) arc weld nugget (5336 Al), (2) base metal (5083-H321 Al), and (3) fine-grain FSP. The arrow in (a) indicates porosity within the arc weld nugget, and the boxes in (d) indicate the locations of microtensile specimens. For all macrographs, the right side is the advancing side of the FSP tool, and tool travel is into the page.

**Table 14.6 Tensile properties of arc-welded 5083/5356 Al as a function of friction stir processing (FSP) modification**

| FSP approach | YS(a), MPa | UTS(b), MPa | Elongation, % |
|---|---|---|---|
| As-GMAW(c)(d) | 117 ± 1 | 259 ± 8 | 10.8 ± 3.1 |
| Weld toe FSP(c) | 132 ± 8 | 275 ± 11 | 15.8 ± 1.0 |
| Weld crown FSP(c) | 125 ± 1 | 283 ± 11 | 19.5 ± 6.2 |
| 5083-H321 | 228 | 317 | 16 |

(a) YS, yield strength. (b) UTS, ultimate tensile strength. (c) Each value represents the average of three samples. (d) GMAW, gas metal arc welded

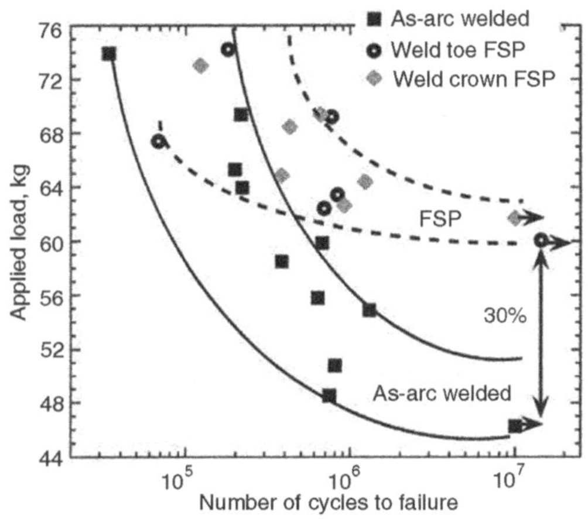


**Fig. 14.42** Four-point bending fatigue results as a function of friction stir processing (FSP) approach

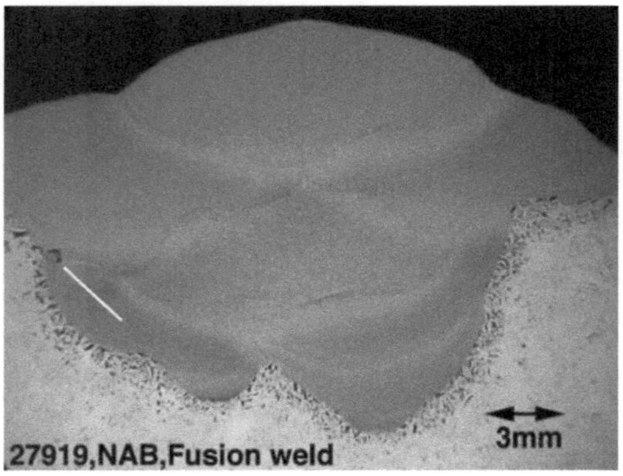


**Fig. 14.43** Micrograph illustrating a multipass 13 mm (½ in.) penetration fusion weld using Ampcotrode 46 weld wire of composition 8.5–9.5Al, 3.0–5.0Fe, 0.6–3.5Mn, 4.5Ni, bal Cu; typical elongation = 23%

- FSP eliminates fusion weld defects.
- FSP creates a mixed microstructure of fine grains and Widmanstätten.
- Mechanical properties in the FSP zone, longitudinal direction, are excellent.

## 14.5 Corrosion Resistance in Friction Stir Processed Sonoston

Friction stir processing was applied to cast Sonoston, a 52Mn-4Al-3Fe-1.5Ni-39Cu alloy used in a seawater environment when high damping is required, to improve corrosion resistance (Ref 110). The cast Sonoston microstructure is relatively coarse and suffers from selective corrosion. Friction stir processing was evaluated to determine if refining the microstructure could increase corrosion resistance. In Sonoston, a variety of microstructures are created by FSP (Fig. 14.46). For FSP material (~0.1 mm, or 0.004 in., below the FSP surface) with a refined Widmanstätten microstructure, dealloying in seawater for 24 h at –200 mV occurred to similar depths to as-cast material. However, for the FSP material, much more severe cracking (delamination parallel to the surface as well as normal to the surface) occurred, and surface layers flaked off readily (Fig. 14.47). Specimens with surfaces exhibiting a very fine-grained microstructure (~4 mm below the original FSP surface) were also dealloyed and cracked to a similar depth after exposure to seawater for 24 h at –200 mV.

After stress-relieving heat treatments, the depths of dealloying for the refined FSP microstructures were substantially reduced compared with the coarse as-cast structure. For the fine

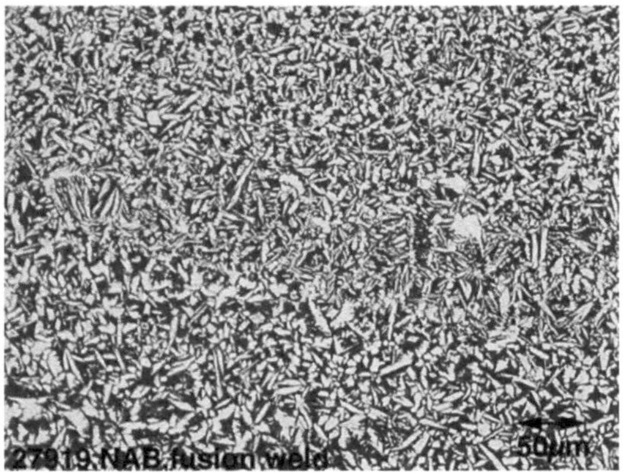

**Fig. 14.44** Fine Widmanstätten microstructure in the fusion zone of the weld shown in Fig. 14.43

Widmanstätten microstructure just below the FSP surface, stress relieving for various times and temperatures showed that 8 h at 500 °C (930 °F) or 2 h at 600 °C (1110 °F) were required for improved corrosion resistance. For specimens with the fine-grained region at the surface, 24 h at 450 °C (840 °F) was sufficient to dramatically decrease the depth of dealloying to only 5 to 10 μm (Fig. 14.48). The stress-relief heat treatments appeared to have little effect on the depth of dealloying for the coarse as-cast microstructures.

The stress-relieved and refined FSP microstructures have shallower dealloyed layers than the coarse as-cast microstructures, because dealloying is confined to manganese-rich regions

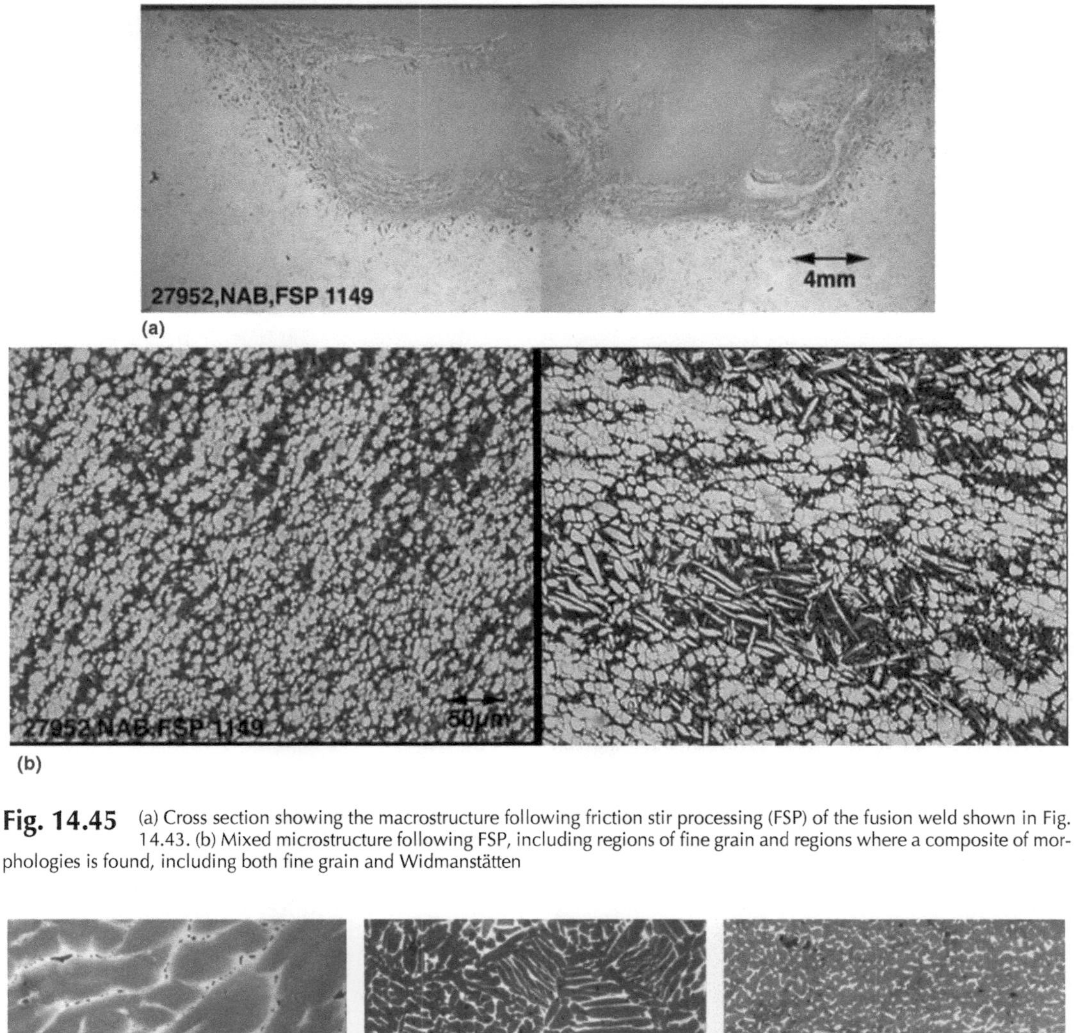


**Fig. 14.45** (a) Cross section showing the macrostructure following friction stir processing (FSP) of the fusion weld shown in Fig. 14.43. (b) Mixed microstructure following FSP, including regions of fine grain and regions where a composite of morphologies is found, including both fine grain and Widmanstätten

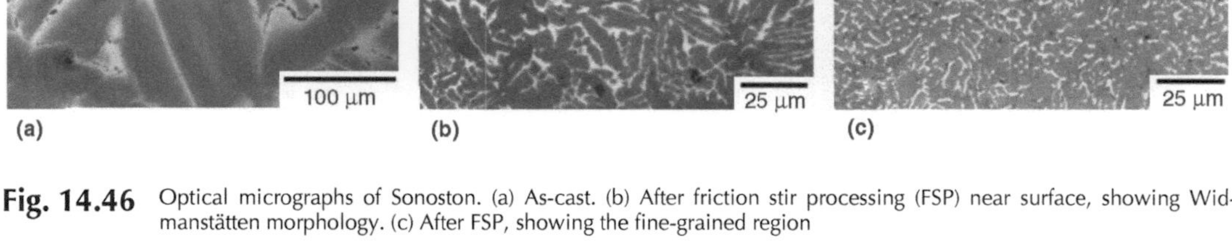


**Fig. 14.46** Optical micrographs of Sonoston. (a) As-cast. (b) After friction stir processing (FSP) near surface, showing Widmanstätten morphology. (c) After FSP, showing the fine-grained region

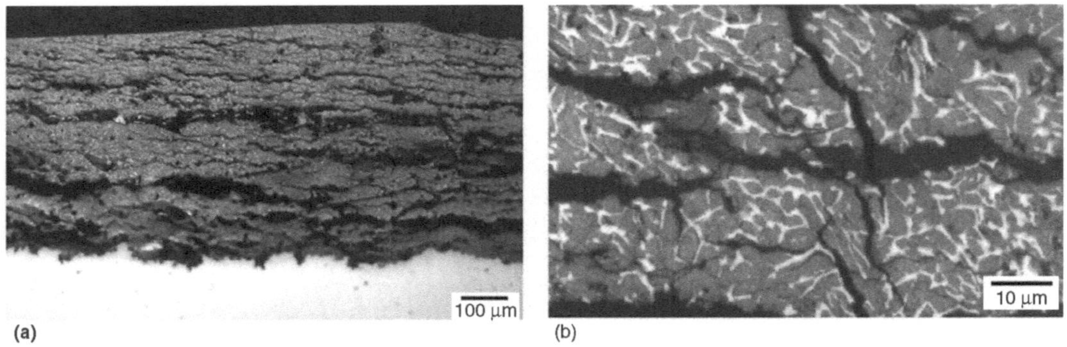


**Fig. 14.47** Optical micrographs of unetched sections normal to the surface of specimens with the cubic boron nitride friction stir processed fine Widmanstätten microstructure dealloyed for 24 h at –200 mV in seawater. (a) At low magnification, showing heavily cracked dealloyed layer. (b) At high magnification, showing dealloyed manganese-rich areas and uncorroded copper-rich areas

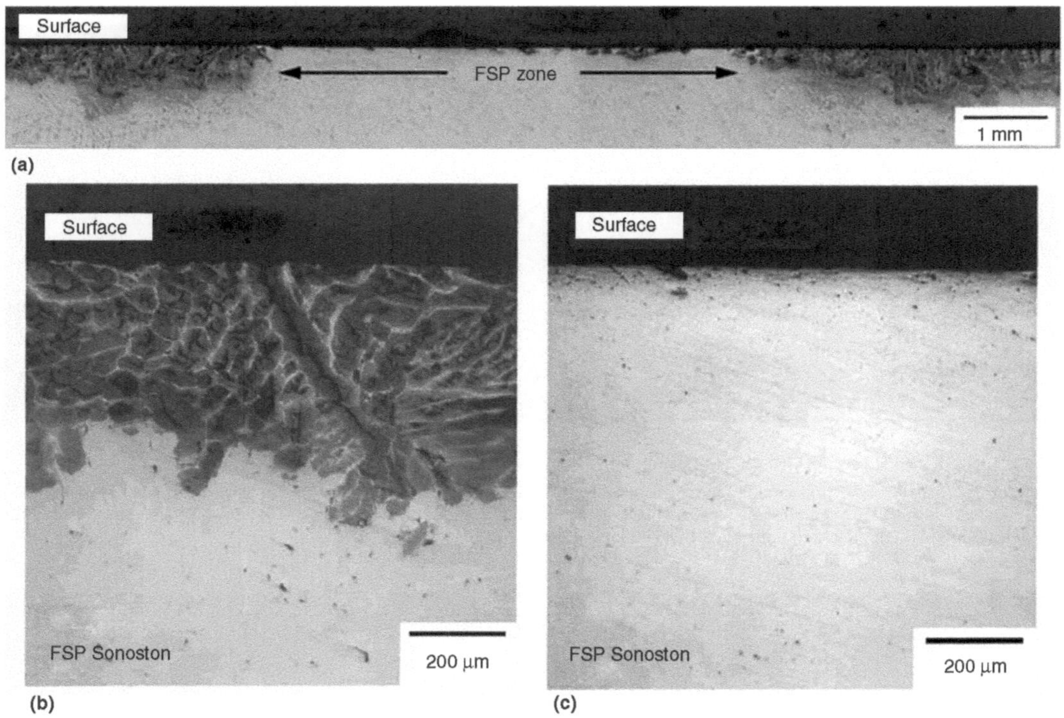


**Fig. 14.48** Optical micrographs of unetched sections normal to surfaces with cubic boron nitride friction stir processed (FSP) fine-grained globular structures, and adjacent as-cast structures dealloyed for 24 h at –200 mV (versus saturated calomel electrode) for specimens stress relieved for 24 h at 450 °C (840 °F). (a) FSP zone and adjacent as-cast zones at low magnification. (b)(c) FSP zones and adjacent zones at a higher magnification

that are connected to the surface, and such regions occur to shallower depths following FSP. However, when high residual tensile stresses are present (created by FSP), stress-corrosion cracking occurs through the copper-rich areas, thereby allowing the environment to penetrate to manganese-rich areas not otherwise connected to the surface, so that dealloying continues to occur.

## 14.6 Friction Stir Processing for Surface Composite Fabrication and Microstructural Homogenization

Compared to unreinforced metals, metal-matrix composites (MMCs) reinforced with ceramic phases exhibit high strength, high elastic modulus, and improved resistance to wear, creep, and fatigue. These properties make MMCs promising structural materials for aerospace and automobile industries. However, MMCs also suffer from a great loss in ductility and toughness due to incorporation of nondeformable ceramic reinforcements, and they are relatively costly. These restrictions limit their wider application. For many applications, the useful life of components often depends on surface properties such as wear resistance. In these situations, only the surface layer needs to be reinforced by ceramic phases, while the bulk of the component should retain the original composition and structure with higher toughness. There is also an emphasis on added functionality. For example, a structural component can be designed to serve additional nonstructural functions. This approach has the possibility of integrating subsystems.

In recent years, several surface-modification techniques, such as high-energy laser melt treatment (Ref 111–118), high-energy electron beam irradiation (Ref 119), plasma spraying (Ref 120), cast sinter (Ref 121, 122), and casting (Ref 123), have been developed to fabricate surface MMCs. Among these techniques, the laser melt treatment (also called laser processing or laser surface engineering) is widely used for surface modification. During this process, a laser beam melts the surface of the substrate along with the deposited material, usually either carbide powder (SiC, TiC, or WC) or a combination of carbide powders and a binding material (cobalt, aluminum, or nickel). The coating material is either predeposited (or preplaced) or injected through a specific nozzle, with simultaneous laser beam radiation. In the injection technique, the powder material to be deposited is carried through a nozzle by a carrier inert gas to the surface to be treated, where it is incorporated into the laser surface melted pool.

Pantelis et al. (Ref 112) and Hu et al. (Ref 113, 114) created surface Al-SiC composites by means of laser processing techniques. Furthermore, Hu et al. (Ref 115) overlapped laser tracks on the aluminum alloy, creating a continuous surface Al-SiC composite. The thickness of the surface composite layer was limited to 30 to 50 μm when SiC particles were preplaced on the substrate (Ref 113), whereas a thickness of up to ~450 μm was obtained for the particle injection technique (Ref 111). The SiC particles were uniformly distributed in the surface layer, and the surface composite exhibited high microhardness and excellent wear resistance compared to untreated material. Pantelis et al. (Ref 112) reported a partial reaction of some SiC particles with the aluminum matrix, whereas Hu et al. (Ref 113–116) revealed partial dissolution of the SiC particles in the liquid, with subsequent reprecipitation during solidification forming a new Al-SiC during laser processing.

The existing processing techniques for forming surface composites are based on liquid-phase processing at high temperatures. In this case, it is hard to avoid an interfacial reaction between the reinforcement and metal matrix and the formation of some detrimental phases. Furthermore, critical control of processing parameters is necessary to obtain the ideal solidified microstructure in the surface layer. Obviously, if processing of a surface composite is carried out at temperatures below the melting point of the substrate, the problems mentioned previously can be avoided. In the last five years, attempts have been made to use FSP to incorporate ceramic particles into the surface layer of aluminum alloys to form a surface composite (Ref 124–133), as well as to modify the powder metallurgy processed alloys and composites (Ref 59, 134–139).

**Localized surface modification** can be a very powerful tool to achieve the right combination of properties, that is, a gradient of properties within a monolithic structure. The potential exists to broaden design possibilities using

MMC surfaces. Some examples of properties that can be influenced are listed in Table 14.7. A number of these approaches require particles of a stoichiometric nature. The properties of these particles can degrade or change if they undergo chemical reaction with the matrix. The short thermal cycle and relatively low temperature during FSP can help to avoid or eliminate reaction products. Table 14.8 provides a summary of various efforts to date (Ref 124–133). The initial results are very encouraging and clearly demonstrate the viability of FSP.

Figures 14.49(a) and (b) show examples of SiC distributed using the surface-addition method (Ref 124, 125). The uniform SiC distribution is demonstrated, and a reaction and defect-free composite/matrix interface illustrated. Figure 14.49(c) shows the fracture surface of a single-wall carbon nanotube/aluminum composite tested in tension (Ref 131). The survivability following large processing strains and the thermal cycle is noteworthy. This illustrates the possibility of developing sensors and actuators by locally embedding functional particles. In another attempt to embed functional particles, Dixit et al. (Ref 133) have observed clean Al-NiTi interfaces after FSP (Fig. 14.49d).

**Processing of Powder Metallurgy Alloys.** Powder metallurgy processed aluminum alloys suffer from three major microstructural problems that limit their full potential: prior-particle boundaries with an aluminum oxide film, microstructural inhomogeneity, and remnant porosity. These microstructural features particularly hamper the ductility in very high-strength aluminum alloys. Berbon et al. (Ref 59, 134)

**Table 14.7 Some examples of properties that can be tailored by localized surface modification**

| Property | Approach |
|---|---|
| Elastic modulus | Addition of ceramic particles or intermetallic particles |
| Wear resistance | Addition of second-phase particles and microstructural refinement can enhance wear properties. |
| Fatigue | Addition of shape-memory particles can alter the residual stresses, thereby influencing the fatigue properties. |
| Magnetic | Magnetic particles can be added in local regions to obtain magnetic properties in otherwise nonmagnetic materials. |
| Electrical conductivity | Second-phase additions can be used to enhance or lower the electrical conductivity. |
| Thermal conductivity | Second-phase particles can be used to enhance or lower thermal conductivity based on the thermal conductivity of matrix and reinforcement. |
| Damping | Shape-memory particles and piezoelectric particles can be added to enhance the damping capabilities. |

**Table 14.8 Summary of surface modification and in situ composite efforts**

| Material system | Remarks |
|---|---|
| 5083 Al-SiC (Ref 124) | SiC particles were put on the surface and stirred into the matrix. |
| A356 Al-SiC (Ref 125) | SiC particles were put on the surface and stirred into the matrix. |
| 7050 Al-WC (Ref 126) | WC particles were put on a machined surface slot and stirred. |
| 1100 Al-$SiO_2$ and $TiO_2$ (Ref 127) | Introduced the concept of reaction processing during FSP. The reaction product was placed subsurface with a three-layer setup and friction stir processed. |
| 7050 Al and 6061 Al-WC, SiC, $Al_2O_3$, $MoS_2$, Fe, Zn, Cu (Ref 128) | Powders were placed in subsurface drilled holes. The hole geometry provided good control of the volume fraction. A number of ceramic and metallic phases were explored, including a combination of SiC and $MoS_2$. |
| AZ61-$SiO_2$ (Ref 129, 130) | Distributed nanoparticles by using repeated runs |
| Al-SWCNT (Ref 131) | Demonstrated the survivability of single-wall carbon nanotubes (SWCNT) during friction stir processing. The nanotubes were placed subsurface by drilling a hole from the top and using a plug. |
| AZ31-MWCNT (Ref 132) | Multiwall carbon nanotubes (MWCNT) were distributed in a magnesium alloy. |
| Al-NiTi (Ref 133) | Shape-memory alloy (NiTi) was distributed using the hole method without any interfacial reaction with aluminum. |

have shown that FSP can be used as a homogenization tool. Figure 14.50 shows the microstructural difference in an Al-Ti-Cu alloy processed by extrusion and by FSP. The FSP microstructure is remarkably different from the as-extruded microstructure. This leads to an excellent combination of strength and ductility. Spowart et al. (Ref 135). have highlighted the effect of spatial heterogeneity on mechanical properties. They used FSP to modify the homogeneity of three aluminum-matrix composites produced with controlled inhomogeneity. Figure 14.51 shows the relationship between homogeneous length scale and ductility in aluminum-matrix composites. Results clearly demonstrate that FSP can be a very useful tool to enhance the mechanical properties of high-strength alloys and composites. Combining the trends observed by various studies cited in this section, the potential of FSP as a tool to create homogeneous composites on a local scale can be visualized. Designers and fabricators can take this approach to design components and subsystems that take advantage of localized property enhancements to augment conceptual design elements.

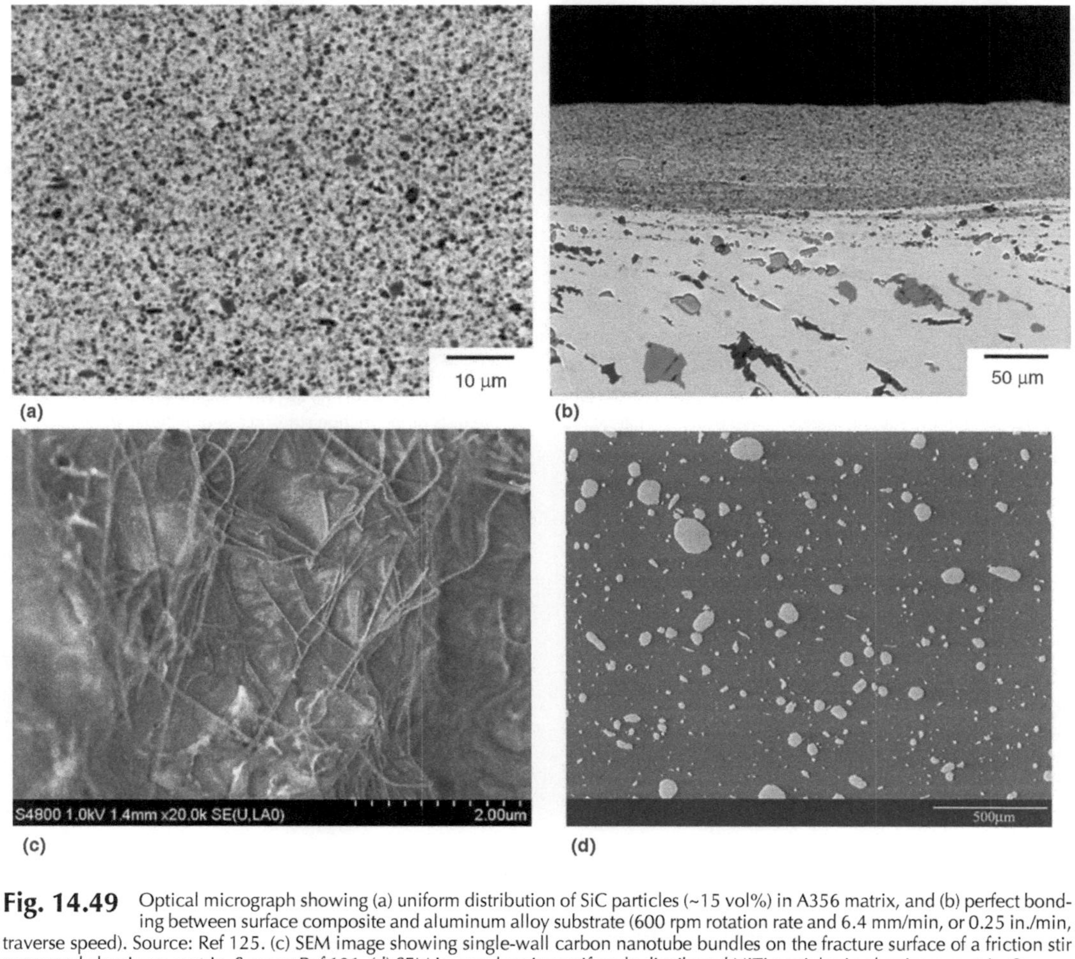


**Fig. 14.49** Optical micrograph showing (a) uniform distribution of SiC particles (~15 vol%) in A356 matrix, and (b) perfect bonding between surface composite and aluminum alloy substrate (600 rpm rotation rate and 6.4 mm/min, or 0.25 in./min, traverse speed). Source: Ref 125. (c) SEM image showing single-wall carbon nanotube bundles on the fracture surface of a friction stir processed aluminum matrix. Source: Ref 131. (d) SEM image showing uniformly distributed NiTi particles in aluminum matrix. Source: Ref 133

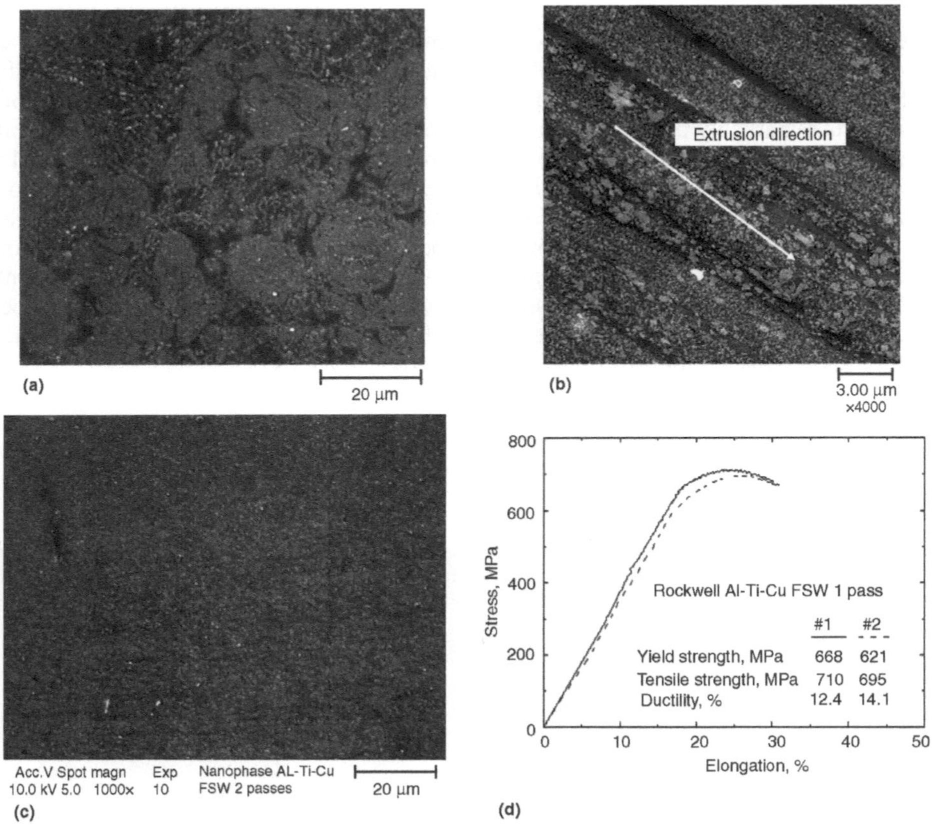


**Fig. 14.50** (a) Typical microstructure in the as-hot isostatic pressed condition. Dark regions consist of pure aluminum, gray regions consist of fine intermetallics dispersed in an aluminum matrix, and light regions consist of coarse intermetallics in an aluminum matrix. (b) Typical as-extruded microstructure shows the same three microstructural features, now elongated in the extrusion direction. (c) Typical microstructure observed in the friction stir processed nugget. The three different microstructural features seen in the starting material have been homogenized. (d) Tensile tests of the friction stir processed material show excellent strength and more than 10% ductility. Source: Ref 59

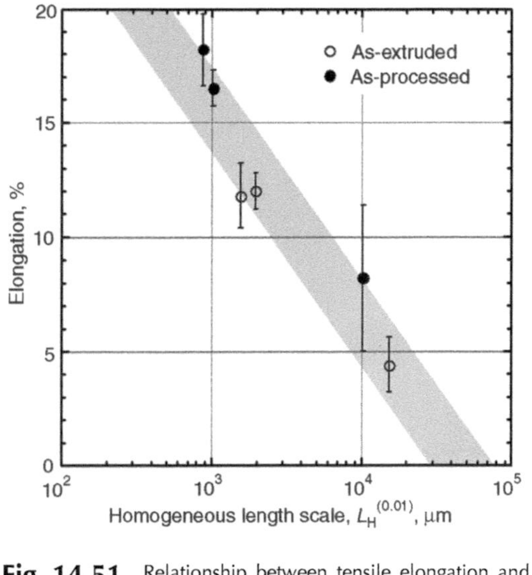


**Fig. 14.51** Relationship between tensile elongation and level of spatial heterogeneity, as characterized by the homogeneous length scale, $L_H^{(0.01)}$

## REFERENCES


1. *New Materials for Next-Generation Commercial Transports,* Committee on New Materials for Advanced Civil Aircraft, National Research Council, 1966, p 33
2. *National Critical Technologies Report,* Chapter 6, March 1995, p 5
3. "Basic Research Needs for Environmentally Responsive Technologies of the Future," Jan 4–6, 1996 (New Orleans, LA), National Science Foundation and the Department of Energy
4. G. Gutowski et al., "WTEC Panel Report on Environmentally Benign Manufacturing," April 2001, p 69; http://wtec.org/loyola/ebm/(accessed Sept 28, 2006)
5. R. Martin and D. Evans, *J. Met.,* Vol 52 (No. 3), 2000, p 24
6. R.S. Mishra, T.R. Bieler, and A.K. Mukherjee, *Acta Metall. Mater.,* Vol 43, 1995, p 877
7. R.Z. Valiev, A.V. Korznikov, and R.R. Mulyukov, *Mater. Sci. Eng. A,* Vol 168, 1993, p 141
8. R.S. Mishra, W.B. Lee, Y.-W. Kim, and A.K. Mukherjee, in *International Symposia on Gamma Titanium Aluminides,* Y.-W. Kim, R. Wagner, and M. Yamaguchi, Ed., TMS, 1995, p 571
9. T.G. Nieh, J. Wadsworth, and O.D. Sherby, *Superplasticity in Metals and Ceramics,* Cambridge University Press, Cambridge, U.K., 1997
10. R.S. Mishra and M.W. Mahoney, Friction Stir Processing: A New Grain Refinement Technique to Achieve High Strain Rate Superplasticity in Commercial Alloys, *Mater. Sci. Forum,* Vol 357–359, 2001, p 507–514
11. T.R. McNelley and M.E. McMahon, Analyzing Superplastic Microstructures Using Interactive EBSP Methods, *J. Met.,* Vol 48 (No. 2), 1996, p 58
12. S.D. Terhune, Z. Horita, M. Nemoto, Y. Li, T.G. Langdon, and T.R. McNelley, in *Recrystallization and Related Phenomena,* T. Sakai and H.G. Suzuki, Ed., Japan Inst. of Metals, Japan, 1999, p 515
13. R.S. Mishra and Z.Y. Ma, Friction Stir Welding and Processing, *Mater. Sci. Eng. R,* Vol 50, 2005, p 1–78
14. C.G. Rhodes, M.W. Mahoney, W.H. Bingel, R.A. Spurling, and C.C. Bampton, *Scr. Mater.,* Vol 36, 1997, p 69
15. G. Liu, L.E. Murr, C.S. Niou, J.C. McClure, and F.R. Vega, *Scr. Mater.,* Vol 37, 1997, p 355
16. K.V. Jata and S.L. Semiatin, *Scr. Mater.,* Vol 43, 2000, p 743
17. Z.Y. Ma, R.S. Mishra, and M.W. Mahoney, *Acta Mater.,* Vol 50, 2002, p 4419
18. Y.S. Sato, M. Urata, and H. Kokawa, *Metall. Mater. Trans. A,* Vol 33, 2002, p 625
19. B. Heinz and B. Skrotzki, *Metall. Mater. Trans. B,* Vol 33 (No. 6), 2002, p 489
20. L.E Murr, G. Liu, and J. C. McClure, *J. Mater. Lett.,* Vol 16, 1997, p 1081
21. G.S. Frankel and Z. Xia, *Corrosion,* Vol 55, 1999, p 139
22. Y.S. Sato, S.H.C. Park, and H. Kokawa, *Metall. Mater. Trans. A,* (Vol 32), 2001, p 3023
23. S.H. Kazi and L.E. Murr, *Friction Stir Welding and Processing,* K.V. Jata, M.W. Mahoney, R.S. Mishra, S.L. Semiatin, D.P. Field, Ed., TMS, 2001 p 139
24. H.G. Salem, A.P. Reynolds, and J.S. Lyons, *Scr. Mater.,* Vol 46, 2002, p 337
25. R. Braun and L. Litynska-Dobrzynska, *Mater. Sci. Forum,* Vol 396–402, 2002, p 1531
26. A.F. Norman, I. Brough, and P.B. Prangnell, *Mater. Sci. Forum,* Vol 331–337, 2000, p 1713
27. K.A.A. Hassan, A.F. Norman, and P.B. Prangnell, *Mater. Sci. Forum,* Vol 396–402, 2002, p 1549
28. J.Q. Su, T.W. Nelson, R.S. Mishra, and M.W. Mahoney, *Acta Mater.,* Vol 51, 2003, p 713
29. Z.Y. Ma, R.S. Mishra, M.W. Mahoney, and R. Grimes, *Mater. Sci. Eng. A.,* Vol 351, 2003, p 148
30. I. Charit and R.S. Mishra, *Mater. Sci. Eng. A.,* Vol 359, 2003, p 290
31. I. Charit, R.S. Mishra, and M.W. Mahoney, *Scr. Mater.,* Vol 47, 2002, p 631
32. I. Charit, Z.Y. Ma, and R.S. Mishra, *Hot Deformation of Aluminum Alloys III,* Z. Jin, A. Beaudoin, T.R. Bieler, and B. Radhakrishnan, Ed., TMS, 2003, p 331–342
33. P.S. Pao, E. Lee, C.R. Feng, H.N. Jones, and D.W. Moon, *Friction Stir Welding and Processing II,* K.V. Jata, M.W. Mahoney, R.S. Mishra, S.L. Semiatin, and T. Lienert, Ed., TMS, 2003, p 113

34. R.S. Mishra, M.W. Mahoney, S.X. McFadden, N.A. Mara, and A.K. Mukherjee, High Strain Rate Superplasticity in a Friction Stir Processed 7075 Al Alloy, *Scr. Mater.,* Vol 42 (No. 2), Dec 31, 1999, p 163–168
35. R.S. Mishra, Processing Commercial Aluminum Alloys for High Strain Rate Superplasticity, *JOM,* Vol 53 (No. 3), March 2001, p 23–26
36. M.W. Mahoney, R. Mishra, and T. Nelson, High Strain Rate Superplasticity in Thick Section 7050 Aluminum Created by Friction Stir Processing, *Proc. Third International Symposium on Friction Stir Welding,* Sept 27–28, 2001 (Kobe, Japan), TWI, 2002
37. I. Charit, R.S. Mishra, and K.V. Jata, Superplastic Behavior of Friction Stir Processed Aluminum-Lithium Alloy, *Friction Stir Welding and Processing,* Nov 4–8, 2001 (Indianapolis, IN), TMS, 2001, p 225–234
38. K.V. Jata, V. Seetharaman, and S.L. Semiatin, High Temperature Deformation of Friction-Stir Processed Aluminum-Lithium Alloy AF/C458, *Friction Stir Welding and Processing,* Nov 4–8 2001 (Indianapolis, IN), TMS, 2001, p 195–204
39. M. Mahoney, R.S. Mishra, T. Nelson, J. Flintoff, R. Islamgaliev, and Y. Hovansky, High Strain Rate, Thick Section Superplasticity Created via Friction Stir Processing, *Friction Stir Welding and Processing,* Nov 4–8 2001 (Indianapolis, IN), TMS 2001, p 183–194
40. Z.Y. Ma, R.S. Mishra, and M.W. Mahoney, Superplastic Deformation Behaviour of Friction Stir Processed 7075Al Alloy, *Acta Mater.,* Vol 50 (No. 17), Oct 9, 2002, p 4419–4430
41. Z.Y. Ma and R.S. Mishra, Cavitation in Superplastic 7075Al Alloys Prepared via Friction Stir Processing, *Acta Mater.,* Vol 51 (No. 12), July 16, 2003, p 3551–3569
42. M. Mahoney, A.J. Barrnes, W.H. Bingel, and C. Fuller, Superplastic Forming of 7475 Al Sheet after Friction Stir Processing (FSP), *Mater. Sci. Forum,* Vol 447–448, 2004, p 505–512
43. H.G. Salem, A.P. Reynolds, and J.S. Lyons, Structural Evolution and Superplastic Formability of Friction Stir Welded AA 2095 Sheets, *J. Mater. Eng. Perform.,* Vol 13 (No. 1), Feb 2004, p 24–31
44. R.S. Mishra, Friction Stir Processing for Superplasticity, *Adv. Mater. Process.,* Vol 162 (No. 2), Feb 2004, p 45–47
45. I. Charit, Z.Y. Ma, and R.S. Mishra, High Strain Rate Superplasticity in Aluminum Alloys via Friction Stir Processing, *Advances in Superplasticity and Superplastic Forming,* 2004 TMS Annual Meeting, March 14–18, 2004 (Charlotte, NC), 2004, p 201–209
46. Z.Y. Ma, R.S. Mishra, and M.W. Mahoney, Superplasticity in Cast A356 Induced via Friction Stir Processing, *Scr. Mater.,* Vol 50 (No. 7), April 2004, p 931–935
47. I. Charit and R.S. Mishra, Evaluation of Microstructure and Superplasticity in Friction Stir Processed 5083 Al Alloy, *J. Mater. Res.,* Vol 19 (No. 11), Nov 2004, p 3329–3342
48. M.M. Attallah and H.G. Salem, Influence of Process Parameters on Superplasticity of Friction Stir Processed Nugget in High Strength Al-Cu-Li Alloy, *Mater. Sci. Technol.,* Vol 20 (No. 11), Nov 2004, p 1370–1376
49. Y. Takayama, T. Nagai, T. Shibayanagi, H. Kato, and H. Watanabe, Superplasticity and Crystallographic Orientation Distribution in Friction Stir Processed Sheet of an Al-Mg-Mn Alloy, *Mater. Sci. Forum,* Vol 475–479 (Part 4), 2005, p 3029–3032
50. L.B. Johannes, I. Charit, and R.S. Mishra, High Strain Rate Superplastic Properties of 7075 Aluminum via Friction Stir Processing, *Friction Stir Welding and Processing III,* 2005 TMS Annual Meeting, Feb 13–17, 2005 (San Francisco, CA), p 123–130
51. A. Dutta, I. Charit, L.B. Johannes, and R.S. Mishra, Deep Cup Forming by Superplastic Punch Stretching of Friction Stir Processed 7075 Al Alloy, *Mater. Sci. Eng. A,* Vol 395 (No. 1–2) March 25, 2005, p 173–179
52. Z.Y. Ma, R.S. Mishra, M.W. Mahoney, and R. Grimes, Effect of Friction Stir Processing on the Kinetics of Superplastic Deformation in an Al-Mg-Zr Alloy, *Metall. Mater. Trans. A,* Vol 36 (No. 6), June 2005, p 1447–1458
53. Z.Y. Ma and R.S. Mishra, Development

of Ultrafine-Grained Microstructure and Low Temperature (0.48 $T_m$) Superplasticity in Friction Stir Processed Al-Mg-Zr, *Scr. Mater.*, Vol 53 (No. 1), July 2005, p 75–80
54. P. Cavaliere and A. Squillace, High Temperature Deformation of Friction Stir Processed 7075 Aluminum Alloy, *Mater. Charact.*, Vol 55 (No. 2), Aug 2005, p 136–142
55. I. Charit and R.S. Mishra, Low Temperature Superplasticity in a Friction-Stir-Processed Ultrafine Grained Al-Zn-Mg-Sc Alloy, *Acta Mater.*, Vol 53 (No. 15), Sept 2005, p 4211–4223
56. R.S. Mishra and M.W. Mahoney, Metal Superplasticity Enhancement and Forming Process, U.S. Patent 6,712,916, March 30, 2004
57. I. Charit and R.S. Mishra, in *Materials Processing Handbook,* J.R. Groza, J.F. Shackelford, E.J. Lavernia, and M.T. Powers, Ed., CRC Press, 2006, in press
58. Y. Wang, R.S. Mishra, and Z.Y. Ma, *Mater. Sci. Eng. A,* 2006, in press
59. P.B. Berbon, W.H. Bingel, R.S. Mishra, C.C. Bampton, and M.W. Mahoney, Friction Stir Processing: A Tool to Homogenize Nanocomposite Aluminum Alloys, *Scr. Mater.*, Vol 44, 2001, p 61
60. R. Grimes and R.G. Butler, in *International Conference on Superplasticity in Advanced Materials,* S. Hori, M. Tokizane, and N. Furushiro, Ed., (Osaka, Japan), Japan Society for Research in Superplasticity, 1991, p 771
61. M. Mahoney, C. Fuller, and W. Bingel, Rockwell Scientific, unpublished research
62. G.J. Grant, D. Herling, W. Arbegast, C. Allen, and C. Degen, paper presented at 2006 International Conference on Superplasticity in Advanced Materials, June 23, 2006 (Chengdu, China)
63. R.S. Mishra, R.K. Islamgaliev, T.W. Nelson, Y. Hovansky, and M.W. Mahoney, Abnormal Grain Growth during High Temperature Exposure in Friction Stir Processed 7050 and 2519 Aluminum Alloys, *Friction Stir Welding and Processing,* K.V. Jata, M.W. Mahoney, R.S. Mishra, S.L. Semiatin, and D.P. Field, Ed., TMS, 2001, p 205
64. F.J. Humphreys and P.S. Bate, *Mater. Sci. Forum,* Vol 357–359, 2001, p 477
65. R.D. Schelleng and G.H. Reynolds, *Metall. Trans.*, Vol 4, 1973, p 2199
66. N. Ridley, D.W. Livesey, and A.K. Mukherjee, *Metall. Trans. A,* Vol 15, 1984, p 1443
67. P. Bompard, J.Y. Lacroix, and A. Varloteaux, *Aluminum,* Vol 64, 1988, p 162
68. C.L. Chen and M.J. Tan, *Mater. Sci. Eng. A,* Vol 298 (No. 1–2), Jan 31, 2001 p 235–244
69. M. Mahoney and W. Bingel, Thick Section Bending Patent, U.S. Patent 6,866, 180, March 15, 2005
70. M. Miles, M. Mahoney, T. Nelson, and R. Mishra, Finite Element Simulation of Plane-Strain Thick Plate Bending of Friction-Stir Processed 2519 Aluminum, *TMS Proceedings Friction Stir Welding and Processing II,* March 2–6, 2003, p 253–258
71. M. Mahoney, C. Fuller, and W. Bingel, Rockwell Scientific, unpublished research
72. M. Mahoney, C. Fuller, M. Miles, and W. Bingel, Thick Plate Bending of Friction Stir Processed Aluminum Alloys, *Friction Stir Welding and Processing III,* K. Jata, M. Mahoney, R. Mishra, and T. Lienert, Ed., TMS, 2005, p 131–137
73. C. Fuller and M. Mahoney, Rockwell Scientific, unpublished research
74. T. Din and J. Campbell, High Strength Aerospace Aluminum Casting Alloys: A Comparative Study, *Mater. Sci. Technol.*, Vol 12, 1996, p 644–650
75. D.L. Zhang and L. Zheng, The Quench Sensitivity of Cast Al-7wt.%Si-0.4wt.%Mg Alloy, *Metall. Mater. Trans. A,* Vol 27, 1996, p 3983–3991
76. Y.B. Yu, P.Y. Song, S.S. Kim, and J.H. Lee, Possibility of Improving Tensile Strength of Semi-Solid Processed A356 Alloy by a Post Heat Treatment at an Extremely High Temperature, *Scr. Mater.*, Vol 41, 1999, p 767–771
77. D.L. Zalensas, *Aluminum Casting Technology,* 2nd ed., AFS Inc., 1993, p 77
78. S. Kumai, J. Hu, Y. Higo, and S. Nunomura, Effects of Dendrite Cell Size and Particle Distribution on the Near-Threshold Fatigue Crack Growth Behavior of Cast Al-SiCp Composites, *Acta Mater.*, Vol 44, 1996, p 2249–2257
79. B. Zhang, D.R. Poirier, and W. Chen, Microstructural Effects on High-Cycle

Fatigue-Crack Initiation in A356.2 Casting Alloy, *Metall. Mater. Trans. A,* Vol 30, 1999, p 2659–2666
80. M.E. Seniw, J.G. Conley, and M.E. Fine, The Effect of Microscopic Inclusion Locations and Silicon Segregation on Fatigue Lifetimes of Aluminum Alloy A356 Castings, *Mater. Sci. Eng. A,* Vol 285, 2000, p 43–48
81. G. Atxaga, A. Pelayo, and A.M. Irisarri, Effect of Microstructure on Fatigue Behavior of Cast Al-7Si-Mg Alloy, *Mater. Sci. Technol.,* Vol 17, 2001, p 446–450
82. K. Oh-Ishi, A.M. Cuevas, D.L. Swisher, and T.R. McNelley, The Influence of Friction Stir Processing on Microstructure and Properties of a Cast Nickel Aluminum Bronze Material, *Mater. Sci. Forum,* Vol 426–432 (Part 4), 2003, p 2885–2890
83. W.A. Palko, R.S. Fielder, and P.F. Young, Investigation of the Use of Friction Stir Processing to Repair and Locally Enhance the Properties of Large Ni-Al Bronze Propellers, *Mater. Sci. Forum,* Vol 426–432 Part 4, 2003, p 2909–2914
84. M.W. Mahoney, W.H. Bingel, S.R. Sharma, and R.S. Mishra, Microstructural Modification and Resultant Properties of Friction Stir Processed Cast NiAl Bronze, *Mater. Sci. Forum,* Vol 426–432 (Part 4), 2003, p 2843–2848
85. Z.Y. Ma, S.R. Sharma, R.S. Mishra, and M.W. Mahoney, Microstructural Modification of Cast Aluminum Alloys via Friction Stir Processing, *Mater. Sci. Forum,* Vol 426–432 (Part 4), 2003, p 2891–2896
86. Z.Y. Ma, R.S. Mishra and M.W. Mahoney, Friction Stir Processing for Microstructural Modification of an Aluminum Casting, *Friction Stir Welding and Processing II,* 2003 TMS Annual Meeting, March 2–6, 2003 (San Diego, CA), p 221–230
87. S.R. Sharma, Z.Y. Ma, and R.S. Mishra, Effect of Friction Stir Processing on Fatigue Behavior of A356 Alloy, *Scr. Mater.,* Vol 51 (No. 3), Aug 2004, p 237–241
88. K. Oh-Ishi and T.R. McNelley, Microstructural Modification of As-Cast Ni-Al Bronze by Friction Stir Processing, *Metall. Mater. Trans. A,* Vol 35 (No. 9), Sept 2004, p 2951–2961
89. K. Oh-Ishi, A.P. Zhilyaev, R. Williams, and T.R. McNelley, Friction Stir Processing for Microstructure Modification of Nial Bronze, *Friction Stir Welding and Processing III,* 2005 TMS Annual Meeting, Feb 13–17, 2005 (San Francisco, CA), p 107–114
90. R.S. Mishra, M.R. Sharma, and W. Mahoney, Embedding a Wrought Microstructure in a Cast Component: A Novel Application of Friction Stir Processing, *Trans. Am. Foundry Soc.,* Vol 113, 2005, p 139–143
91. K. Oh-Ishi and T.R. McNelley, The Influence of Friction Stir Processing Parameters on Microstructure of As-Cast NiAl Bronze, *Metall. Mater. Trans. A,* Vol 36 (No. 6), June 2005, p 1574–1585
92. M.L. Santella, T. Engstrom, D. Storjohann, and T.-Y. Pan, Effects of Friction Stir Processing on Mechanical Properties of the Cast Aluminum Alloys A319 and A356, *Scr. Mater.,* Vol 53 (No. 2), July 2005, p 201–206
93. K. Oh-Ishi, A.P. Zhilyaev, and T.R. McNelley, Enhancement of Ductility and Strength through Microstructural Refinement by FSP of Nickel Aluminum Bronze, *Mater. Sci. Forum,* Vol 503–504, 2006, p 161–168
94. M.L. Santella, T. Engstrom, D. Storjohann, and T. Pan, Effects of Friction Stir Processing on Mechanical Properties of the Cast Aluminum Alloy A356, *SAE Trans. J. Mater. Manuf.,* Vol 114, 2006, p 599–603
95. Z.Y. Ma, S.R. Sharma, and R.S. Mishra, Effect of Multiple-Pass Friction Stir Processing on Microstructure and Tensile Properties of a Cast Aluminum-Silicon Alloy, *Scr. Mater.,* Vol 54 (No. 9), May 2006, p 1623–1626
96. K. Oh-Ishi, A.P. Zhilyaev, and T.R. McNelley, A Microtexture Investigation of Recrystallization during Friction Stir Processing of As-Cast NiAl Bronze, *Metall. Mater. Trans. A,* Vol 37 (No. 7), July 2006, p 2239–2251
97. Z.Y. Ma, S.R. Sharma, and R.S. Mishra, *Mater. Sci. Eng. A,* 2006, in press
98. Z.Y. Ma, S.R. Sharma, and R.S. Mishra, *Metall. Mater. Trans. A,* 2006, in press
99. S.R. Sharma and R.S. Mishra, unpublished research, 2006
100. W.B. Lee, Y.M. Yeon, and S.B. Jung, *Mater. Sci. Eng. A,* Vol 355, 2003, p 154

101. Y. Murakami and M. Endo, *Fatigue,* Vol 16, 1994, p 163
102. W.W. Sanders, Jr. and F.V. Lawrence, Jr., *Fatigue Behavior of Aluminum Alloy Weldments,* STP 648, American Society for Testing and Materials, 1978, p 22
103. U. Brandt, F.V. Lawrence, and C.M. Sonsino, *Fatigue Fract. Eng. Mater. Struct.,* Vol 24, 2001, p 117
104. C. Fuller, M. Mahoney, and W. Bingel, Friction Stir Processing of Aluminum Fusion Welds, *Proceedings of the Fourth International Symposium on Friction Stir Welding,* May 14–16, 2003 (Park City, UT), TWI
105. C. Fuller and M. Mahoney, The Effect of Friction Stir Processing on 5083-H321/5356 Al Arc Welds: Microstructural and Mechanical Analysis, *Metall. Trans.,* 2006, in press
106. P.G. Forrest, *Fatigue of Metals,* Pergamon Press, Oxford, U.K., 1962, p 280
107. T.R. Gurney, *Fatigue of Welded Structures,* Cambridge University Press, Cambridge, U.K., 1971, p 215
108. R.P. Newman, *Br. Weld. J.,* Vol 6, 1959, p 324
109. *Aluminum Standards and Data,* The Aluminum Association, Washington D.C., 2000, p 2-2
110. S.P. Lynch, D. Edwards, A. Majumdar, S. Moutsos, and M. Mahoney, Friction-Stir Processing of a High-Damping Mn-Cu Alloy Used for Marine Propellers, *THERMEC 2003,* Trans Tech Publications
111. G. Ricciardi, M. Cantello, G. Mollino, W. Varani, and E. Garlet, *Proc. Second Int. Seminar on Surface Engineering with High Energy Beam, Science and Technology* (Lisbon, Portugal), CEMUL-IST, 1989, p 415–423
112. D. Pantelis, A. Tissandier, P. Manolatos, and P. Ponthiaux, *Mater. Sci. Technol.,* Vol 11, 1995, p 299
113. C. Hu and T.N. Baker, *J. Mater. Sci.,* Vol 30, 1995, p 891
114. C. Hu, H. Xin, and T.N. Baker, *J. Mater. Sci.,* Vol 30, 1995, p 5985
115. C. Hu, H. Xin, and T.N. Baker, *Mater. Sci. Technol.,* Vol 12, 1996, p 227
116. C. Hu and T.N. Baker, *J. Mater. Sci.,* Vol 32, 1997, p 5047
117. T.C. Lei, J.H. Ouyan, Y.T. Pei, and Y. Zhou, *Mater. Sci. Technol.,* Vol 11, 1995, p 520
118. L.R. Katipelli and N.B. Dahotre, *Mater. Sci. Technol.,* Vol 17, 2001, p 1061
119. S.-H. Choo, S. Lee, and S.-J. Kwon, *Metall. Mater. Trans. A,* Vol 30, 1999, p 1211, 3131
120. M.C. Gui and S.B. Kang, *Mater. Lett.,* Vol 46, 2000, p 296
121. Y. Wang, X. Zhang, G. Zeng, and F. Li, *Mater. Des.,* Vol 21, 2000, p 447
122. Y.S. Wang, X.Y. Zhang, G.T. Zeng, and F.C. Li, *Compos. Part A: Appl. Sci. Manuf.,* Vol 32, 2001, p 281
123. A.N. Attia, *Mater. Des.,* Vol 22, 2001, p 451
124. R.S. Mishra, Z.Y. Ma, and I. Charit, Friction Stir Processing: A Novel Technique for Fabrication of Surface Composite, *Mater. Sci. Eng. A,* Vol 341, 2003, p 307
125. Z.Y. Ma and R.S. Mishra, Friction Stir Surface Composite Fabrication, *Surface Engineering: In Materials Science II,* TMS, 2003, p 243
126. J.W. Newkirk, R. Mishra, J. Thomas, and J.A. Hawk, Friction Stir Processing to Create Surface Composites, *Advances in Powder Metallurgy and Particulate Materials-2003,* MPIF, p 6.60–6.70
127. S.M. Howard, B.K. Jasthi, W.J. Arbegast, G.J. Grant, and D.R. Herling, Friction Surface Reaction Processing in Aluminum Substrates, *Friction Stir Welding and Processing III,* 2005 TMS Annual Meeting, 2005, p 139–146
128. U. Ramadorai, J.W. Newkirk, R.S. Mishra, and J.A. Hawk, "Surface Modification of Aluminum Alloys to Create in situ Surface Composites," Fourth ASM International Surface Engineering Congress and 19th International Conference on Surface Modification Technologies, Aug 2005
129. C.J. Lee, J.C. Huang, and P.J. Hsieh, Mg Based Nano-Composites Fabricated by Friction Stir Processing, *Scr. Mater.,* Vol 54, 2006, p 1415–1420
130. C.J. Lee, J.C. Huang, and P.L. Hsieh, *Compos. Mater.,* Vol 313, 2006, p 69
131. L.B. Johannes, L.L. Yowell, E. Sosa, S. Arepalli, and R.S. Mishra, Survivability of Single-Walled Carbon Nanotubes during Friction Stir Processing, *Nanotechnology,* Vol 17, 2006, p 3081–3084
132. Y. Morisada, H. Fujii, T. Nagaoka, and M. Fukusumi, *Mater. Sci. Eng. A: Struc.*

*Mater. Prop. Microstruct. Process.,* Vol 419, 2006, p 344

133. M. Dixit, J.W. Newkirk, and R.S. Mishra, *Scr. Mater.,* 2006, submitted for publication
134. J. Zheng, R.S. Mishra, P.B. Berbon, and M.W. Mahoney, Microstructure and Mechanical Behavior of Friction Stir Processed Al-Ti-Cu Alloy, *Friction Stir Welding and Processing,* Nov 4–8, 2001 (Indianapolis, IN), TMS, 2001, p 235–242
135. J.E. Spowart, Z.-Y. Ma, and R.S. Mishra, The Effect of Friction Stir Processing (FSP) on the Spatial Heterogeneity of Discontinuously-Reinforced Aluminum (DRA) Microstructures, *Friction Stir Welding and Processing II,* 2003 TMS Annual Meeting, March 2–6, 2003 (San Diego, CA), 2003, p 243–252
136. C.J. Hsu, P.W. Kao, and N.J. Ho, *Scr. Mater.,* Vol 53, 2005, p 341
137. J.C.H. Chuang, J.C. Huang, and P.J. Hsieh, *Scr. Mater.,* Vol 53, 2005, p 1455
138. J.W. Newkirk, R. Mishra, and J. Ply, Local Modification of P/M Parts to Fully Densify and Enhance Properties, *Advances in Powder Metallurgy and Particulate Materials-2005,* MPIF, 2005, p 6.62–6.70
139. A. Tewari, J.E. Spowart, A.M. Gokhale, R.S. Mishra, and D.B. Miracle, *Mater. Sci. Eng. A: Struct. Mater. Prop. Microstruct. Process.,* Vol 428, 2006, p 80

# CHAPTER 15

# Future Outlook for Friction Stir Welding and Processing

Rajiv S. Mishra, Center for Friction Stir Processing
University of Missouri-Rolla
Murray W. Mahoney, Rockwell Scientific Company

FRICTION STIR WELDING (FSW) has made a significant impact on the welding community in a relatively short time, in terms of both the volume of research activities and the growing number of commercial applications. Because of the significant benefits provided by FSW, we believe growth will continue, and likely at an accelerated pace. Currently, FSW is much more developed than friction stir processing (FSP). This is evident from the emphasis on FSW in this book (Chapters 1 to 13), compared to only one chapter (Chapter 14) on FSP. However, FSP is a very new concept, and although directly linked to FSW, such a new and unique metallurgical tool would be expected to take longer to mature. Although FSP is in its infancy, we believe both research and commercial applications will continue to grow, because considerable benefits will also be realized.

In the first ten years of FSW, the technological breakthroughs and early adoption by industry are very evident. This has been accomplished without recognized industry-wide standards. Further, albeit considerable research has been completed in this short time, a scientific fundamental understanding is still lagging. For our final thoughts, we briefly outline four key areas we believe will require considerable attention for continued and efficient growth in applying FSW and FSP to commercial structures.

## 15.1 Scientific Knowledge Gaps

As highlighted in Chapter 1, two key major and fundamental aspects of FSW are still not well understood: thermal input through frictional heating and deformation, and material flow and subsequent consolidation. These fundamental features of FSW remain controversial. Chapters 3 and 10 highlight some modeling aspects, and Chapters 4 to 9 present resultant microstructures and properties. Data in these chapters provide empirical observations but not an understanding of the process itself. Even a simple question such as "What is the friction coefficient at the tool/material interface during friction stir processes?" is difficult to answer because of the number of variables and the dynamic nature of the overall process. Without a fundamental understanding, predicting the resultant microstructure and defect-free nugget for a given set of parameters will not be achieved. A concerted effort is needed to address the basic components of the process, with subsequent integration into a complete process description. Early in the incubation years of FSW, we believed this new welding procedure to be relatively simple. In fact, in application, FSW is simple, but the metallurgical fundamentals that result in these remarkable postweld properties have been found to be quite complex.

## 15.2 Lack of Process Specifications

Specifications or standards are the backbone of consistency. Currently, a few professional societies have technical committees entrusted with developing FSW process specifications. In addition, some large industries have generated their own welding specifications. However, specifications have not yet emerged for general use. Lack of process specifications is a considerable barrier for industry-wide technological implementation. A broader acceptance of FSW will be feasible only after common specifications are readily available. This will also boost confidence of potential new users who rely on manufacturing supply chains for components and subsystems. Development of specifications is also a sign of a mature technology.

## 15.3 Lack of Design Guidelines

The early adopters of a new technology are generally technology enthusiasts who convince the powers-to-be to adopt a new technology. In reality, only a very small percentage of companies have the technical staff that can provide this leadership role. Most companies run in a follower mode, and rightfully so. That is, they wait until a significant number of technology adoption cases emerge to lower the risk and uncertainty. Further, the choice to implement a new technology requires a risk-versus-reward decision. In circumstances where the new technology is a concept-enabler, the leading companies are willing to assume risk and pay the higher premium for a new technology. In the United States, this was certainly the case for the National Aeronautics and Space Administration (NASA) and Boeing for space applications. Most of the early application examples are of technology pull, where the new technology provided solutions to well-recognized problems, shortcomings, or achieved significant cost-savings. Chapter 13 highlights various applications where these benefits were recognized and FSW was adopted. However, for broader use, standard design guidelines need to be developed to enable the rapid introduction of new technologies, such as FSW and FSP. Further, most information on new technologies, such as FSW and FSP, is initially scattered throughout the welding and metallurgical literature and within international technical proceedings. Thus, a designer's access to information on new technologies is often not easy to obtain or, from a practical perspective, even impossible to obtain, especially for smaller organizations. We hope this first reference volume on FSW/FSP will help to alleviate this information-access difficulty.

## 15.4 Design and Designers: Education and Implementation

This is an overlapping theme with the previous topic. Designers design based on their knowledge and experience, using the tools in their "toolbox." The development of concepts into a formal design generally locks in the usable technology. Most often, designers specify the material and process in the embodiment of the concepts. The best chance to introduce a new technology is for the designer to specify its use. In practice, this requires visionary designers who are not restricted and are allowed to explore the limits beyond their comfort zone. In reality, the practitioners of new technologies need to educate the designers of the possible benefits, and they themselves understand the designer's information needs. This is not an easy or commonly traveled path. As the knowledge of and comfort with FSW increases and designers more frequently implement this technology, the opportunities for FSP will also increase. The research community can help the process by building demonstrative prototypes, establishing design data, and presenting this information to new audiences. In addition, engineering considerations, such as reliability and statistical variance of properties, need to be published to enhance technology-push opportunities. At this time, there is simply an insufficient quantity of hard data available in the literature for many to become comfortable with FSW and especially FSP. Considerably more data are necessary. This is a challenge for both the research and engineering communities. Finally, from a practical perspective, the cost and design of machines will dictate the affordability of this technology. For the widespread use of FSW and FSP, it is imperative to develop low-cost machines and flexible platforms.

# Index

## A

## R



## S

## T

FSC
www.fsc.org
MIX
Papier | Fördert
gute Waldnutzung
FSC® C083411

Kolibri 360 GmbH
Ferdinand-Jühlke-Straße 7
99095 Erfurt, Deutschland
produktsicherheit@kolibri360.de